# Lecture Notes on Data Engineering and Communications Technologies 180

Series Editor

Fatos Xhafa, *Technical University of Catalonia, Barcelona, Spain*

The aim of the book series is to present cutting edge engineering approaches to data technologies and communications. It will publish latest advances on the engineering task of building and deploying distributed, scalable and reliable data infrastructures and communication systems.

The series will have a prominent applied focus on data technologies and communications with aim to promote the bridging from fundamental research on data science and networking to data engineering and communications that lead to industry products, business knowledge and standardisation.

Indexed by SCOPUS, INSPEC, EI Compendex.

All books published in the series are submitted for consideration in Web of Science.

Zhengbing Hu · Qingying Zhang · Matthew He
Editors

# Advances in Artificial Systems for Logistics Engineering III

Set 2

*Editors*
Zhengbing Hu
Faculty of Applied Mathematics
National Technical University of Ukraine
"Igor Sikorsky Kyiv Polytechnic Institute"
Kyiv, Ukraine

Qingying Zhang
College of Transportation and Logistics
Engineering
Wuhan University of Technology
Wuhan, China

Matthew He
Halmos College of Arts and Sciences
Nova Southeastern University
Fort Lauderdale, FL, USA

ISSN 2367-4512   ISSN 2367-4520 (electronic)
Lecture Notes on Data Engineering and Communications Technologies
ISBN 978-3-031-36114-2   ISBN 978-3-031-36115-9 (eBook)
https://doi.org/10.1007/978-3-031-36115-9

© The Editor(s) (if applicable) and The Author(s), under exclusive license to Springer Nature Switzerland AG 2023

This work is subject to copyright. All rights are solely and exclusively licensed by the Publisher, whether the whole or part of the material is concerned, specifically the rights of translation, reprinting, reuse of illustrations, recitation, broadcasting, reproduction on microfilms or in any other physical way, and transmission or information storage and retrieval, electronic adaptation, computer software, or by similar or dissimilar methodology now known or hereafter developed.
The use of general descriptive names, registered names, trademarks, service marks, etc. in this publication does not imply, even in the absence of a specific statement, that such names are exempt from the relevant protective laws and regulations and therefore free for general use.
The publisher, the authors, and the editors are safe to assume that the advice and information in this book are believed to be true and accurate at the date of publication. Neither the publisher nor the authors or the editors give a warranty, expressed or implied, with respect to the material contained herein or for any errors or omissions that may have been made. The publisher remains neutral with regard to jurisdictional claims in published maps and institutional affiliations.

This Springer imprint is published by the registered company Springer Nature Switzerland AG
The registered company address is: Gewerbestrasse 11, 6330 Cham, Switzerland

# Preface

The development of artificial intelligence (AI) systems and their applications in various fields is one of the modern science and technology's most pressing challenges. One of these areas is AI and logistics engineering, where their application aims to increase the effectiveness of AI generation and distribution for the world's population's life support, including tasks such as developing industry, agriculture, medicine, transportation, and so on. The rapid development of AI systems necessitates an increase in the training of an increasing number of relevant specialists. AI systems have a lot of potential for use in education technology to improve the quality of training for specialists by taking into account the personal characteristics of these specialists as well as the new computing devices that are coming out.

As a result of these factors, the 3rd International Conference on Artificial Intelligence and Logistics Engineering (ICAILE2023), held in Wuhan, China, on March 11–12, 2023, was organized jointly by Wuhan University of Technology, Nanning University, the National Technical University of Ukraine "Igor Sikorsky Kyiv Polytechnic Institute", Huazhong University of Science and Technology, the Polish Operational and Systems Society, Wuhan Technology and Business University, and the International Research Association of Modern Education and Computer Science. The ICAILE2023 brings together leading scholars from all around the world to share their findings and discuss outstanding challenges in computer science, logistics engineering, and education applications.

Out of all the submissions, the best contributions to the conference were selected by the program committee for inclusion in this book.

March 2023

Zhengbing Hu
Qingying Zhang
Matthew He

# Organization

## General Chairs

Q. Y. Zhang	Wuhan University of Technology, China
Ivan Dychka	National Technical University of Ukraine "Igor Sikorsky Kyiv Polytechnic Institute", Ukraine

## Online conference Organizing Chairs

Z. B. Hu	National Technical University of Ukraine "Igor Sikorsky Kyiv Polytechnic Institute", Ukraine
C. L. Wang	Anhui University, China
Y. Wang	Wuhan University of Science and Technology, China

## Program Chairs

Q. Y. Zhang	Wuhan University of Technology, China
Matthew He	Nova Southeastern University, USA
G. E. Zhang	Nanning University, China

## Publication Chairs

Z. B. Hu	National Technical University of Ukraine "Igor Sikorsky Kyiv Polytechnic Institute", Ukraine
Q. Y. Zhang	Wuhan University of Technology, China
Matthew He	Nova Southeastern University, USA

## Publicity Chairs

Y. Z. Pang	Nanning University, China
Q. L. Zhou	Shizuoka University, Japan

O. K. Tyshchenko   University of Ostrava, Czech Republic
Vadym Mukhin      National Technical University of Ukraine "Igor Sikorsky Kyiv Polytechnic Institute", Ukraine

## Program Committee Members

| | |
|---|---|
| Margherita Mori | University of L'Aquila, Italy |
| X. H. Tao | School of Intelligent Manufacturing of Nanning University, China |
| Felix Yanovsky | Delft University of Technology, Netherlands |
| D. Y. Fang | Beijing Technology and Business University, China |
| Ivan Izonin | Lviv Polytechnic National University, Ukraine |
| L. Luo | Sichuan University, China |
| Essa Alghannam | Tishreen University, Syria |
| L. Xu | Southwest Jiaotong University, China |
| H. F. Yang | Waseda University, Japan |
| Y. C. Zhang | Hubei University, China |
| A. Sachenko | Kazimierz Pułaski University of Technology and Humanities in Radom, Poland |
| S. Gnatyuk | National Aviation University, Ukraine |
| X. J. Zhou | Wuhan Textile University, China |
| Oleksandra Yeremenko | Kharkiv National University of Radio Electronics, Ukraine |
| Rabah Shboul | Al-albayt University, Jordan |
| W. B. Hu | Wuhan University, China |
| Yurii Koroliuk | Chernivtsi Institute of Trade and Economics, Ukraine |
| O. K. Tyshchenko | University of Ostrava, Czech Republic |
| H. L. Zhong | South China University of Technology, China |
| J. L. Zhang | Huazhong University of Science and Technology, China |
| G. K. Tolokonnikova | FNAT VIM of RAS, Moscow, Russia |

# Contents

**Advances of Computer Algorithms and Methods**

AI Chatbots for Banks: Evolving Trends and Critical Issues .............. 3
   *Margherita Mori and Lijing Du*

Traffic Flow Characteristics and Vehicle Road Coordination
Improvement in Subterranean Interweaving ........................... 14
   *Enshi Wang, Bihui Huang, and Bing Liu*

Influential Factors and Implementation Path of Talent Digital Evaluation
Based on ISM Model: Taking Electric Power Enterprises as an Example ...... 25
   *Wei Luo, Jiwei Tang, Saixiao Huang, and Yuan Chen*

Knowledge Associated with a Question-Answer Pair ..................... 35
   *Igor Chimir*

Research on Low Complexity Differential Space Modulation Detection
Algorithm .......................................................... 45
   *Shuiping Xiong and Xia Wu*

Mathematical Model of the Process of Production of Mineral Fertilizers
in a Fluidized Bed Granulator ....................................... 55
   *Bogdan Korniyenko and Andrii Nesteruk*

Simulation Study on Optimization of Passenger Flow Transfer
Organization at Nanning Jinhu Square Metro Station Based on Anylogic ..... 65
   *Yan Chen, Chenyu Zhang, Xiaoling Xie, Zhicheng Huang,
   and Jinshan Dai*

Engine Speed Measurement and Control System Design Based
on LabVIEW ......................................................... 79
   *Chengwei Ju, Geng E. Zhang, and Rengshang Su*

Research and Design of Personalized Learning Resources Precise
Recommendation System Based on User Profile ......................... 90
   *Tingting Liang, Zhaomin Liang, and Suzhen Qiu*

Mixed Parametric and Auto-oscillations at Nonlinear Parametric
Excitation ......................................................... 101
   *Alishir A. Alifov*

Spectrum Analysis on Electricity Consumption Periods by Industry
in Fujian Province .................................................. 109
   *Huawei Hong, Lingling Zhu, Gang Tong, Peng Lv, Xiangpeng Zhan,
   Xiaorui Qian, and Kai Xiao*

Determination of the Form of Vibrations in Vibratory Plow's Moldboard
with Piezoceramic Actuator for Maximum Vibration Effect ............... 120
   *Sergey Filimonov, Sergei Yashchenko, Constantine Bazilo,
   and Nadiia Filimonova*

Evaluating Usability of E-Learning Applications in Bangladesh:
A Semiotic and Heuristic Evaluation Approach ......................... 129
   *Samrat Kumar Dey, Khandaker Mohammad Mohi Uddin,
   Dola Saha, Lubana Akter, and Mshura Akhter*

Perceptual Computing Based Framework for Assessing Organizational
Performance According to Industry 5.0 Paradigm ....................... 141
   *Danylo Tavrov, Volodymyr Temnikov, Olena Temnikova,
   and Andrii Temnikov*

Petroleum Drilling Monitoring and Optimization: Ranking the Rate
of Penetration Using Machine Learning Algorithms ..................... 152
   *Ijegwa David Acheme, Wilson Nwankwo, Akinola S. Olayinka,
   Ayodeji S. Makinde, and Chukwuemeka P. Nwankwo*

Application of Support Vector Machine to Lassa Fever Diagnosis ....... 165
   *Wilson Nwankwo, Wilfred Adigwe, Chinecherem Umezuruike,
   Ijegwa D. Acheme, Chukwuemeka Pascal Nwankwo,
   Emmanuel Ojei, and Duke Oghorodi*

A Novel Approach to Bat Protection IoT-Based Ultrasound System
of Smart Farming ..................................................... 178
   *Md. Hafizur Rahman, S. M. Noman, Imrus Salehin,
   and Tajim Md. Niamat Ullah Akhund*

Synthesis and Modeling of Systems with Combined Fuzzy P + I -
Regulators ........................................................... 187
   *Bohdan Durnyak, Mikola Lutskiv, Petro Shepita, Vasyl Sheketa,
   Nadiia Pasieka, and Mykola Pasieka*

Protection of a Printing Company with Elements of Artificial
Intelligence and IIoT from Cyber Threats ............................. 197
   *Bohdan Durnyak, Tetyana Neroda, Petro Shepita, Lyubov Tupychak,
   Nadiia Pasieka, and Yulia Romanyshyn*

| | |
|---|---|
| AMGSRAD Optimization Method in Multilayer Neural Networks <br> *S. Sveleba, I. Katerynchuk, I. Kuno, O. Semotiuk, Ya. Shmyhelskyy, S. Velgosh, N. Sveleba, and A. Kopych* | 206 |
| Regional Economic Development Indicators Analysis and Forecasting: Panel Data Evidence from Ukraine <br> *Larysa Zomchak, Mariana Vdovyn, and Olha Deresh* | 217 |
| An Enhanced Performance of Minimum Variance Distortionless Response Beamformer Based on Spectral Mask <br> *Quan Trong The and Sergey Perelygin* | 229 |
| Modelling Smart Grid Instability Against Cyber Attacks in SCADA System Networks <br> *John E. Efiong, Bodunde O. Akinyemi, Emmanuel A. Olajubu, Isa A. Ibrahim, Ganiyu A. Aderounmu, and Jules Degila* | 239 |
| Program Implementation of Educational Electronic Resource for Inclusive Education of People with Visual Impairment <br> *Yurii Tulashvili, Iurii Lukianchuk, Valerii Lishchyna, and Nataliia Lishchyna* | 251 |
| A Version of the Ternary Description Language with an Interpretation for Comparing the Systems Described in it with Categorical Systems <br> *G. K. Tolokonnikov* | 261 |
| A Hybrid Centralized-Peer Authentication System Inspired by Block-Chain <br> *Wasim Anabtawi, Ahmad Maqboul, and M. M. Othman Othman* | 271 |
| Heuristic Search for Nonlinear Substitutions for Cryptographic Applications <br> *Oleksandr Kuznetsov, Emanuele Frontoni, Sergey Kandiy, Oleksii Smirnov, Yuliia Ulianovska, and Olena Kobylianska* | 288 |
| Dangerous Landslide Suspectable Region Forecasting in Bangladesh – A Machine Learning Fusion Approach <br> *Khandaker Mohammad Mohi Uddin, Rownak Borhan, Elias Ur Rahman, Fateha Sharmin, and Saikat Islam Khan* | 299 |
| New Cost Function for S-boxes Generation by Simulated Annealing Algorithm <br> *Oleksandr Kuznetsov, Emanuele Frontoni, Sergey Kandiy, Tetiana Smirnova, Serhii Prokopov, and Alisa Bilanovych* | 310 |

Enriched Image Embeddings as a Combined Outputs from Different Layers of CNN for Various Image Similarity Problems More Precise Solution .................................................... 321
    *Volodymyr Kubytskyi and Taras Panchenko*

A Novel Approach to Network Intrusion Detection with LR Stacking Model ........................................................ 334
    *Mahnaz Jarin and A. S. M. Mostafizur Rahaman*

Boundary Refinement via Zoom-In Algorithm for Keyshot Video Summarization of Long Sequences .................................. 344
    *Alexander Zarichkovyi and Inna V. Stetsenko*

Solving Blockchain Scalability Problem Using ZK-SNARK ............... 360
    *Kateryna Kuznetsova, Anton Yezhov, Oleksandr Kuznetsov, and Andrii Tikhonov*

Interactive Information System for Automated Identification of Operator Personnel by Schulte Tables Based on Individual Time Series .............. 372
    *Myroslav Havryliuk, Roman Kaminskyy, Kyrylo Yemets, and Taras Lisovych*

DIY Smart Auxiliary Power Supply for Emergency Use ................ 382
    *Nina Zdolbitska, Mykhaylo Delyavskyy, Nataliia Lishchyna, Valerii Lishchyna, Svitlana Lavrenchuk, and Viktoriia Sulim*

A Computerised System for Monitoring Water Activity in Food Products Using Wireless Technologies ................................ 393
    *Oksana Honsor and Roksolana Oberyshyn*

Reengineering of the Ukrainian Energy System: Geospatial Analysis of Solar and Wind Potential .......................................... 404
    *Iryna Doronina, Maryna Nehrey, and Viktor Putrenko*

Complex Approach for License Plate Recognition Effectiveness Enhancement Based on Machine Learning Models ....................... 416
    *Yakovlev Anton and Lisovychenko Oleh*

The Analysis and Visualization of CEE Stock Markets Reaction to Russia's Invasion of Ukraine by Event Study Approach ................. 426
    *Andrii Kaminskyi and Maryna Nehrey*

The Same Size Distribution of Data Based on Unsupervised Clustering Algorithms ...................................................... 437
    *Akbar Rashidov, Akmal Akhatov, and Fayzullo Nazarov*

Investigation of Microclimate Parameters in the Industrial Environments ..... 448
*Solomiya Liaskovska, Olena Gumen, Yevgen Martyn, and Vasyl Zhelykh*

Detection of Defects in PCB Images by Separation and Intensity
Measurement of Chains on the Board ........................................... 458
*Roman Melnyk and Ruslan Tushnytskyy*

Tropical Cyclone Genesis Forecasting Using LightGBM .................. 468
*Sabbir Rahman, Nusrat Sharmin, Md. Mahbubur Rahman,
and Md. Mokhlesur Rahman*

Optimization of Identification and Recognition of Micro-objects Based
on the Use of Specific Image Characteristics ............................. 478
*Isroil I. Jumanov and Rustam A. Safarov*

Development and Comparative Analysis of Path Loss Models Using
Hybrid Wavelet-Genetic Algorithm Approach ........................... 488
*Ikechi Risi, Clement Ogbonda, and Isabona Joseph*

**Mathematical Advances and Modeling in Logistics Engineering**

Risk Assessment of Navigation Cost in Flood Season of the Upper
Reaches of the Yangtze River Based on Entropy Weight Extension
Decision Model ....................................................... 503
*Jun Yuan, Peilin Zhang, Lulu Wang, and Yao Zhang*

Influencing Factors and System Dynamics Analysis of Urban Public
Bicycle Projects in China Based on Urban Size and Demographic
Characteristics ....................................................... 514
*Xiujuan Wang, Yong Du, Xiaoyang Qi, and Chuntao Bai*

Cross-border Logistics Model Design e-Tower Based on Blockchain
Consensus Algorithm ................................................. 524
*Shujun Li, Anqi He, Bin Li, Fang Ye, and Bin Cui*

Research on the Service Quality of JD Daojia's Logistics Distribution
Based on Kano Model ................................................. 536
*Yajie Xu and Xinshun Tong*

CiteSpace Based Analysis of the Development Status, Research
Hotspots and Trends of Rural E-Commerce in China .................... 547
*Yunyue Wu*

The Optimization and Selection of Deppon Logistics Transportation
Scheme Based on AHP ................................................. 558
  Long Zhang and Zhengxie Li

Pharmacological and Non-pharmacological Intervention in Epidemic
Prevention and Control: A Medical Perspective ......................... 573
  Yanbing Xiong, Lijing Du, Jing Wang, Ying Wang, Qi Cai,
  and Kevin Xiong

Passenger Flow Forecast of the Section of Shanghai-Kunming
High-Speed Railway from Nanchang West Station to Changsha South
Station ............................................................... 583
  Cheng Zhang, Puzhe Wei, and Xin Qi

The Effect of Labor Rights on Mental Health of Front-Line Logistics
Workers: The Moderating Effect of Social Support ...................... 598
  Yi Chen and Ying Gao

An Investigation into Improving the Distribution Routes of Cold Chain
Logistics for Fresh Produce ........................................... 608
  Mei E. Xie, Hui Ye, Lichen Qiao, and Yao Zhang

Development of Vulnerability Assessment Framework of Port Logistics
System Based on DEMATEL ............................................... 618
  Yuntong Qian and Haiyan Wang

Application Prospect of LNG Storage Tanks in the Yangtze River Coast
Based on Economic Model ............................................... 628
  Jia Tian, Hongyu Wu, Xi Chen, Xunran Yu, and Li Xv

Multi-depot Open Electric Truck Routing Problem with Dynamic
Discharging .......................................................... 640
  Xue Yang and Ning Chen

Analysis on the Selection of Logistics Distribution Mode of JD Mall
in the Sinking Market ................................................. 651
  Weihui Du, Xiaoyu Zhang, Saipeng Xing, and Can Fang

Research on Cruise Emergency Organization Based on Improved
AHP-PCE Method ........................................................ 664
  Long Zhang and Zhengxie Li

Fresh Agricultural Products Supplier Evaluation and Selection
for Community Group Purchases Based on AHP and Entropy Weight
VIKOR Model .................................................... 681
   *Gong Feng, Jingjing Cao, Qian Liu, and Radouani Yassine*

Logistics of Fresh Cold Chain Analysis of Joint Distribution Paths
in Wuhan .......................................................... 696
   *Weihui Du, Donglin Rong, Saipeng Xing, and Jiawei Sun*

Optimization of Logistics Distribution Route Based on Ant Colony
Algorithm – Taking Nantian Logistics as an Example ................ 708
   *Zhong Zheng, Shan Liu, and Xiaoying Zhou*

Optimization Research of Port Yard Overturning Operation Based
on Simulation Technology .......................................... 719
   *Qian Lin, Yang Yan, Ximei Luo, Lingxue Yang, Qingfeng Chen,
   Wenhui Li, and Jiawei Sun*

A Review of Epidemic Prediction and Control from a POM Perspective ...... 734
   *Jing Wang, Yanbing Xiong, Qi Cai, Ying Wang, Lijing Du,
   and Kevin Xiong*

Application of SVM and BP Neural Network Classification in Capability
Evaluation of Cross-border Supply Chain Cooperative Suppliers ............. 745
   *Lei Zhang and Jintian Tian*

Research on Port Logistics Demand Forecast Based on GRA-WOA-BP
Neural Network .................................................... 754
   *Zhikang Pan and Ning Chen*

Evaluation and Optimization of the Port A Logistics Park Construction
Based on Fuzzy Comprehensive Method ............................... 764
   *Xin Li, Xiaofen Zhou, Meng Wang, Rongrong Pang, Hong Jiang,
   and Yan Li*

## Advances in Technological and Educational Approaches

OBE Oriented Teaching Reform and Practice of Logistics Information
System Under the Background of Emerging Engineering Education .......... 777
   *Yanhui Liu, Jinxiang Lian, Xiaoguang Zhou, and Liang Fang*

Teaching Practice of "Three Integration" Based on Chaoxing Learning
Software – Taking the Course of "Complex Variable Function
and Integral Transformation" as an Example .......................... 787
   *Huiting Lu and Xiaozhe Yang*

Transformation and Innovation of E-Commerce Talent Training in the Era of Artificial Intelligence ... 801
*Lifang Su and Ke Liu*

Talent Training Mode Based on the Combination of Industry-Learning-Research Under the Background of Credit System Reform ... 811
*Shanyong Qin and Minwei Liu*

Analysis of the Innovation Mechanism and Implementation Effect of College Students' Career Guidance Courses Based on Market Demand ... 822
*Jingjing Ge*

Comparative Study on the Development of Chinese and Foreign Textbooks in Nanomaterials and Technology ... 833
*Yao Ding, Jin Wen, Qilai Zhou, Li Liu, Guanchao Yin, and Liqiang Mai*

Practical Research on Improving Teachers' Teaching Ability by "Train, Practice and Reflect" Mode ... 844
*Jing Zuo, Yujie Huang, and Yanxin Ye*

College Foreign Language Teacher Learning in the Context of Artificial Intelligence ... 854
*Jie Ma, Pan Dong, and Haifeng Yang*

The Innovation Integration Reform of the Course "Single Chip Microcomputer Principle and Application" ... 865
*Chengquan Liang*

Discussion and Practice on the Training Mode of Innovative Talents in Economics and Management in Women's Colleges ... 876
*Zaitao Wang, Ting Zhao, Xiujuan Wang, and Chuntao Bai*

Cultivation and Implementation Path of Core Quality of Art and Design Talents Under the Background of Artificial Intelligence ... 886
*Bin Feng and Weinan Pang*

Reform and Innovation of International Logistics Curriculum from the Perspective of Integration of Industry and Education ... 899
*Xin Li, Meng Wang, Xiaofen Zhou, Jinshan Dai, Hong Jiang, Yani Li, Sida Xie, Sijie Dong, and Mengqiu Wang*

An Analysis of Talent Training in Women's Colleges Based on the Characteristics of Contemporary Female College Students ... 911
*Ting Zhao, Zaitao Wang, Xiujuan Wang, and Chuntao Bai*

| | |
|---|---|
| Solving Logistical Problems by Economics Students as an Important Component of the Educational Process | 921 |
| *Nataliya Mutovkina* | |
| Exploration and Practice of Ideological and Political Construction in the Course of "Container Multimodal Transport Theory and Practice" for Application-Oriented Undergraduate Majors—Taking Nanning University as an Example | 931 |
| *Shixiong Zhu, Liwei Li, and Zhong Zheng* | |
| A Study on Learning Intention of Digital Marketing Micro Specialty Learners Under the Background of New Liberal Arts—Based on Structural Equation Model | 944 |
| *Yixuan Huang, Mingfei Liu, Jiawei You, and Aiman Magde Abdalla Ahmed* | |
| Comparisons of Western and Chinese Textbooks for Advanced Electronic Packaging Materials | 954 |
| *Li Liu, Guanchao Yin, Jin Wen, Qilai Zhou, Yao Ding, and Liqiang Mai* | |
| Innovation and Entrepreneurship Teaching Design in Application-Oriented Undergraduate Professional Courses – Taking the Transportation Enterprise Management Course as an Example | 963 |
| *Yan Chen, Liping Chen, Hongbao Chen, Jinming Chen, and Haifeng Yang* | |
| The Construction of University Teachers' Performance Management System Under the Background of Big Data Technology | 974 |
| *Fengcai Qin and Chun Jiang* | |
| Curriculum Evaluation Based on HEW Method Under the Guidance of OBE Concept | 983 |
| *Chen Chen and Simeng Fan* | |
| The Relevance of a Systematic Approach to the Use of Information Technologies in the Educational Process | 995 |
| *Nataliya Mutovkina and Olga Smirnova* | |
| Construction and Practice of "CAD/CAM Foundation" Course Based on Learning Outcome | 1006 |
| *Ming Chang, Wei Feng, Zhenhua Yao, and Qilai Zhou* | |
| Research and Practice of Ideological and Political Education in the Context of Moral Education and Cultivating People | 1016 |
| *Geng E. Zhang and Liuqing Lu* | |

A Quantitative Study on the Categorized Management of Teachers'
Staffing in Colleges and Universities .................................. 1028
   *Zhiyu Cui*

Course Outcomes and Program Outcomes Evaluation
with the Recommendation System for the Students ..................... 1039
   *Khandaker Mohammad Mohi Uddin, Elias Ur Rahman,*
   *Prantho kumar Das, Md. Mamun Ar Rashid, and Samrat Kumar Dey*

Methodology of Teaching Educational Disciplines to Second (Master's)
Level Graduates of the "Computer Science" Educational Program ........... 1054
   *Ihor Kozubtsov, Lesia Kozubtsova, Olha Myronenko, and Olha Nezhyva*

Professional Training of Lecturers of Higher Educational Institutions
Based on the Cyberontological Approach and Gamification ............... 1068
   *Oleksii Silko, Lesia Kozubtsova, Ihor Kozubtsov,*
   *and Oleksii Beskrovnyi*

Exploring the Perceptions of Technical Teachers Towards Introducing
Blockchain Technology in Teaching and Learning ....................... 1080
   *P. Raghu Vamsi*

**Author Index** ........................................................ 1091

# Mathematical Advances and Modeling in Logistics Engineering

# Risk Assessment of Navigation Cost in Flood Season of the Upper Reaches of the Yangtze River Based on Entropy Weight Extension Decision Model

Jun Yuan[1,2(✉)], Peilin Zhang[1], Lulu Wang[2], and Yao Zhang[3]

[1] School of Transportation and Logistics Engineering, Wuhan University of Technology, Wuhan 430063, China
yuanjun623@126.com
[2] School of Logistics, Wuhan Technology and Business University, Wuhan 430065, China
[3] Technical University of Munich Asia Campus, Technical University of Munich (TUM), Singapore, Singapore

**Abstract.** Inland navigation is greatly affected by climate and hydrological changes, of which the flood and dry season have the greatest impact on inland navigation. Changes in water level and flow rate during the flood and dry season will have various impacts on such aspects as navigation speed, navigation ship loading rate, navigation time efficiency, etc., many of which will cause changes in navigation costs. Therefore, how to predict and evaluate the cost risk of navigation during the flood season has become an important topic. This paper takes the navigation cost risk in the flood period of the upper reaches of the Yangtze River as the research object, establishes the navigation cost risk assessment index system in the flood period of the upper reaches of the Yangtze River and the navigation cost risk assessment model in the flood period of the upper reaches of the Yangtze River based on the entropy weight extension decision model. On this basis, the feasibility of risk assessment and prevention of navigation cost in flood season of inland river shipping is discussed by using the case of relevant projects to test the evaluation model.

**Keywords:** Inland River shipping · Risk assessment · Entropy weight extension model

## 1 Introduction

Inland river shipping is one of the important components of the national comprehensive transportation system, which has the advantages of large transportation capacity, small land occupation, small energy consumption and light pollution. Vigorously developing inland shipping is of great significance to the reduction of comprehensive logistics costs and the improvement of logistics benefits, and thus to the improvement of the overall national economic level. As the most important part of China's inland river shipping system, Yangtze River shipping is of self-evident importance [1, 2].

Inland navigation is greatly affected by climate and hydrological changes, of which the flood and dry season have the greatest impact on inland navigation. Changes in water level and flow rate during the flood and dry season will have various impacts on such aspects as navigation speed, navigation ship loading rate, navigation time efficiency, etc., many of which will cause changes in navigation costs. Transportation cost is the most important factor in the production and operation of inland shipping enterprises. It will directly determine the benefits of shipping enterprises and then affect the benefits of the entire industry. Therefore, how to predict and evaluate the cost risk of navigation during the flood season has become an important topic [3–5].

This paper takes the navigation cost risk in the flood period of the upper reaches of the Yangtze River as the research object, establishes the navigation cost risk assessment index system in the flood period of the upper reaches of the Yangtze River and the navigation cost risk assessment model in the flood period of the upper reaches of the Yangtze River based on the entropy weight extension decision model. On this basis, the feasibility of risk assessment and prevention of navigation cost in flood season of inland river shipping is discussed by using the case of relevant projects to test the evaluation model.

## 2 Risk Assessment Index System of Navigation Cost in Flood Season in the Upper Reaches of the Yangtze River

### 2.1 Impact of Flood Period on the Navigation Cost of Ships in the Upper Reaches of the Yangtze River

The upstream channel of the Yangtze River is a typical inland channel. To evaluate the navigation cost risk of the upstream of the Yangtze River during the flood period, it is necessary to establish the corresponding evaluation index system. Before that, first analyze the impact of the flood period on the navigation cost of the upstream of the Yangtze River [6, 7].

The flood period of the inland waterway shows the characteristics of the overall rise of the water level of the waterway, the sharp daily fluctuation, and the increase of the flow velocity. According to this analysis, the main impact of the flood period on the navigation cost of the upper reaches of the Yangtze River is as follows.

#### 2.1.1 Impact on Navigation Speed of Navigable Ships

When the flood season comes, the water level in the inland waterway rises, the river flow and flow rate increase in varying degrees, the speed of downstream ships increases, and the speed of upstream ships decreases. Since the upper reaches of the Yangtze River are mainly downstream channels, the shipping speed of ships in the upper reaches of the Yangtze River increases during the flood season, which improves the efficiency of ship operation and has a certain impact on the navigation cost [8, 9].

#### 2.1.2 Impact on the Loading Rate of Navigable Ships

The ship loading rate is related to the ship type and channel water depth, of which large ships are greatly affected by the channel water depth, and small and medium-sized ships are relatively small. During flood period, the water level in the upper reaches of the Yangtze River changes sharply and rises as a whole. The changes in ship loading rate caused by these water level changes will have an indirect impact on ship navigation costs [10, 11].

### 2.1.3 Impact on Navigation Obstruction Time

Due to the changes in the navigation environment of the channel caused by the changes in the flood season, a large number of ships may be blocked and the waiting time may occur. The time cost caused by the blocking and waiting time will affect the overall cost of ship navigation [12, 13].

### 2.1.4 Other Impacts

Due to the special navigation environment during the flood season, it may cause certain additional losses to navigation ships. In order to adapt to the navigation environment in flood season, ships need to additionally improve the navigation adaptability of ships in flood season, which will cause changes in navigation costs [14, 15].

## 2.2 The Risk Assessment System of Ship Navigation Cost in Flood Season in the Upper Reaches of the Yangtze River

According to the above analysis, the factors that affect the risk of ship navigation costs in the flood season in the upper reaches of the Yangtze River can be divided into five categories: Factories of Ship, Factories of Channel, Factories of Crew, Factories of Natural Environment, Factories of Relevant Department management [16].

Factors of Ship includes: Index1 Tonnage of ship; Index2 Age of ship; Index3 Navigation adaptability of ships in flood season.

Factors of Channel includes: Index4 Affluence depth of channel; Index5 Width of channel; Index6 Navigation density of channel; Index7 Perfection of channel supporting facilities.

Factors of Crew includes: Index8 Basic professional level of crew; Index9 Safety navigation training level of crew; Index10 Crew's awareness of safe navigation; Index11 Average age of crew.

Factors of Natural environmental includes: Index12 Flow velocity of navigable channel; Index13 Wind speed of navigable channel; Index14 Visibility of navigation channel.

Factors of Department management includes: Index15 Management level of maritime departments; Index16 Management level of channel departments; Index17 Management level of shipping company [17].

The Risk assessment index system of navigation cost in flood season in the upper reaches of the Yangtze Rivers is showed as Table 1.

**Table 1.** The Risk assessment index system of navigation cost in flood season in the upper reaches of the Yangtze River

| Factor type | Index list |
| --- | --- |
| Factors of Ship | Index1 Tonnage of ship <br> Index2 Age of ship <br> Index3 Navigation adaptability of ships in flood season |
| Factors of Channel | Index4 Affluence depth of channel <br> Index5 Width of channel <br> Index6 Navigation density of channel <br> Index7 Perfection of channel supporting facilities |
| Factors of Crew | Index8 Basic professional level of crew <br> Index9 Safety navigation training level of crew <br> Index10 Crew's awareness of safe navigation <br> Index11 Average age of crew |
| Factors of Natural environmental | Index12 Flow velocity of navigable channel <br> Index13 Wind speed of navigable channel <br> Index14 Visibility of navigation channel |
| Factors of Management | Index15 Management level of maritime departments <br> Index16 Management level of channel departments <br> Index17 Management level of shipping company |

## 3 Extension Evaluation Model of Risk Entropy Weight for Navigation Cost in Flood Season in the Upper Reaches of the Yangtze River

Based on the construction of the evaluation index system of navigation cost risk in flood period in the upper reaches of the Yangtze River, the entropy weight extension evaluation model of navigation cost risk in flood period in the upper reaches of the Yangtze River is constructed as follows [18–21]:

### 3.1 Evaluation Object, Classical Domain and Section Domain of the Model

According to the basic theory of the entropy weight extension decision model, the matrix of the composite matter element $R_{xi}$ in the classical domain of the navigation cost evaluation model in the flood season of the upper reaches of the Yangtze River is:

$$R_{xj} = (N_{xj}, c_i, v_{xji}) = \begin{bmatrix} N_{xj}, c_1, v_{xj1} \\ \vdots \\ c_n, v_{xjn} \end{bmatrix} \quad (1)$$

Assume that there are a total of $i(i = 1, 2, \cdots, n)$ indicators in the navigation cost risk assessment system in the upper reaches of the Yangtze River during the flood period, and there are a total of $j(j = 1, 2, \cdots, m)$ categories of risk ratings of the assessment objects.

The navigation cost risk is a matter element $R_{xj}$ representing the navigation cost risk in the flood period of the upper reaches of the Yangtze River; $N_{xj}$ represents the $j$ risk rating in the navigation cost risk assessment system of the upper reaches of the Yangtze River during flood season; $C_i$ represents the evaluation index $i$, and its classical domain is $v_{xyi} = (a_{xjn}, b_{xjn})$.

Construct the corresponding node $R_p(R_p \supset R_j)$ for the above classical domain, Assume that the value range of the $i$th characteristic $c_i$ of matter-element $R_p$ is $(a_{pi}, b_{pi})$. Then $R_p$ is recorded as:

$$R_p = (p_0, c_i, v_i) = \begin{bmatrix} P, c_1, v_{p1} \\ \vdots \vdots \\ c_n \ v_{pn} \end{bmatrix} = \begin{bmatrix} P, c_1, (a_{p1}, b_{p1}) \\ \vdots \vdots \\ c_n \ (a_{pn}, b_{pn}) \end{bmatrix} \quad (2)$$

The matter element of the risk assessment object of navigation cost in flood period of the upper reaches of the Yangtze River to be assessed is recorded as $R_x$, Let the actual value of the $i$th index $c_i$ of matter element $R_x$ be $v_i$.

Then $R_p$ is recorded as:

$$R_x = (p_x, c_i, v) = \begin{bmatrix} p_x, c_1, v_1 \\ \vdots \vdots \\ c_n \ v_n \end{bmatrix} \quad (3)$$

## 3.2 Correlation Function and Correlation Degree of the Model

According to the basic theory of the entropy weight extension decision model, the correlation between the $i^{\text{th}}$ evaluation index of the evaluation object matter-element R and the $j^{\text{th}}$ rating grade of the entropy weight extension evaluation model of the navigation cost risk in the flood period in the upper reaches of the Yangtze River is:

$$K_{xj}(v_i) = \begin{cases} \frac{\gamma(v_i, v_{xji})}{\gamma(v_i, v_{pi}) - \gamma(v_i, v_{xji})}, & v_i \notin v_{xji} \\ -\frac{\gamma(v_i, v_{xji})}{|a_{xji} - b_{xji}|}, & v_i \in v_{xji} \end{cases} \quad (4)$$

In this formula:

$$\gamma(v_i, v_{xji}) = \left| v_i - \frac{1}{2}(a_{xji} + b_{xji}) \right| - \frac{1}{2}(b_{xji} - a_{xji}) \quad (5)$$

$$\gamma(v_i, v_{pi}) = \left| v_i - \frac{1}{2}(a_{pi} + b_{pi}) \right| - \frac{1}{2}(b_{pi} - a_{pi}) \quad (6)$$

## 3.3 Comprehensive Correlation Degree and Risk Level of the Model

Assume that the correlation degree between the evaluation index $v_i$ of the subject matter element $R_x$ and the risk rating grade $j$ is $K_{xj}(v_i)$, then take the weighted value of the correlation degree between each index and each evaluation grade as the comprehensive correlation degree $K_{xj}(R_x)$, and record it as:

$$K_{xj}(R_x) = \sum_{i=1}^{n} \omega_i K_{xj}(v_i) \tag{7}$$

Calculate the maximum value of the comprehensive correlation degree $K_{xj}(R_x)$ between all the material elements $R_x$ and the evaluation level to determine the final cost risk value level of the current case. The calculation formula is expressed as follows:

$$K'_{xj}(R_x) = \max_{j=1,2,\cdots,m} K_{xj}(R_x) \tag{8}$$

# 4 Test of Navigation Cost Risk Assessment Model in Flood Period of Upper Yangtze River Based on Entropy Weight Extension Decision Model

## 4.1 Data Source of Test Case

Based on relevant projects, this paper studies the risk of ship navigation cost of a ship company A passing through a section B in the upper reaches of the Yangtze River during flood season. The project investigated the natural environment of the relevant channel, the basic condition of the channel, the basic condition of the navigable ships and the crew. The investigation method is field investigation, and the relevant information is collected by inviting the management personnel of relevant channel management departments, shipping companies and port companies to have a discussion. The information involves all indicators in the navigation cost risk assessment system of the upper reaches of the Yangtze River in flood season in this paper.

## 4.2 Optimization of Evaluation Index Date

The data collected in this paper relate to the parameters of the ship, the age and training status of the crew, the weather data of the route section, the basic condition of the channel, etc., while 8 of which are qualitative indicators and 9 are quantitative indicators, and the actual value is directly taken. Invite relevant water transport cost risk assessment experts and internal management personnel of the shipping company to form a 10-person assessment team to score each risk assessment index in turn. The score adopts a five-point system. The higher the score, the less likely the index will cause cost risk.

According to the characteristics of inland river shipping, this paper divides the risk of navigation costs in the flood season of the upper reaches of the Yangtze River into five levels, which are respectively recorded as high risk (Level A), high risk (Level B), general risk (Level C), low risk (Level D), and very low risk (Level E), corresponding to expert scores of 1, 2, 3, 4, and 5.

## 4.3 Determination of Case Evaluation Index Weight

In risk assessment research, there are many methods to determine the weight of evaluation indicators. In order to conform to the characteristics of entropy weight extension evaluation research, this paper adopts the information entropy method to determine the weight of various indicators. The specific calculation method is as follows:

If the evaluation index system has i index and j evaluation levels, the evaluation matrix is:

$$X = \begin{bmatrix} X_{11} & \cdots & X_{1y} \\ \vdots & \vdots & \vdots \\ X_i & \cdots & X_{ij} \end{bmatrix} \quad (9)$$

The calculation formula of entropy $H(X_n)$ of the index $\eta$ $(n = 1, 2, \cdots, i)$ is

$$H(X_n) = -\sum_{n=1}^{i} p_{iy} \times \ln p_{iy} \quad (10)$$

In this formula:

$$p_{iy} = \frac{X_{ij}}{\sum_{x=1}^{y} X_{iy}}, i = 1, 2, \cdots, n \quad (11)$$

The calculation formula of the difference coefficient h of the index is

$$h_n = 1 - H(X_n) \quad (12)$$

Then the calculation formula of the corresponding index weight $\omega_n$ is

$$\omega_n = \frac{h_n}{\sum_{n=1}^{i} h_n} \quad (13)$$

According to the above formula and the collected data, it can be calculated that the weight of the cost risk assessment index of the navigation section B of the ship company A during the flood period is showed as Table 2.

**Table 2.** The Weight of case navigation cost risk assessment index

| Index  | 1     | 2     | 3     | 4     | 5     | 6     | 7     | 8     | 9     |
|--------|-------|-------|-------|-------|-------|-------|-------|-------|-------|
| Weight | 0.071 | 0.067 | 0.061 | 0.052 | 0.044 | 0.048 | 0.086 | 0.072 | 0.054 |
| Index  | 10    | 11    | 12    | 13    | 14    | 15    | 16    | 17    |       |
| Weight | 0.058 | 0.038 | 0.066 | 0.062 | 0.064 | 0.044 | 0.051 | 0.062 |       |

## 4.4 Determine of the Classical Domain and Matter Element to be Evaluated

According to the constructed entropy weight extension evaluation model of navigation cost risk in the flood period of the upper reaches of the Yangtze River, the classical domain included in the evaluation system in this case is:

$$R_{xd} = \begin{vmatrix} A & \text{Tonnage of ship} & (0,1) \\ & \vdots & \vdots \\ & \text{Management level of shipping company} & (0,1) \end{vmatrix} \quad (14)$$

$$R_{xB} = \begin{vmatrix} B & \text{Tonnage of ship} & (1,2) \\ & \vdots & \vdots \\ & \text{Management level of shipping company} & (1,2) \end{vmatrix} \quad (15)$$

$$R_{xC} = \begin{vmatrix} C & \text{Tonnage of ship} & (2,3) \\ & \vdots & \vdots \\ & \text{Management level of shipping company} & (2,3) \end{vmatrix} \quad (16)$$

$$R_{xD} = \begin{vmatrix} D & \text{Tonnage of ship} & (3,4) \\ & \vdots & \vdots \\ & \text{Management level of shipping company} & (3,4) \end{vmatrix} \quad (17)$$

$$R_{xE} = \begin{vmatrix} E & \text{Tonnage of ship} & (4,5) \\ & \vdots & \vdots \\ & \text{Management level of shipping company} & (4,5) \end{vmatrix} \quad (18)$$

The object element to be evaluated in this case is

$$R_x = (p_x, c_n, v_n) = \begin{bmatrix} p_x, & \text{Tonnage of ship}, & v_1 \\ & \vdots & \vdots \\ & \text{Management level of shipping company} & v_n \end{bmatrix} \quad (19)$$

According to the data brought in by formulas (4), (5), (6) and (7), it can be calculated that the comprehensive correlation between all indicators of the case and all evaluation levels is showed as Table 3.

Table 3. Comprehensive correlation between each evaluation index and each risk level of the case

| Risk value | 1 | 2 | 3 | 4 | 5 |
|---|---|---|---|---|---|
| Index 1 | −0.0395 | −0.5245 | −0.0563 | 0.0556 | −0.3151 |
| Index 2 | −0.2523 | −0.0656 | −0.0875 | −0.2351 | −0.8622 |
| Index 3 | −0.0026 | −0.0876 | 0.0862 | −0.1253 | 0.0314 |

(*continued*)

**Table 3.** (continued)

| Risk value | 1 | 2 | 3 | 4 | 5 |
|---|---|---|---|---|---|
| Index 4 | −0.0532 | −0.3251 | −0.0723 | −0.0321 | −0.0882 |
| Index 5 | −0.0499 | 0.0209 | −0.0351 | −0.0831 | −0.0542 |
| Index 6 | −0.0323 | 0.0624 | −0.0251 | −0.0885 | −0.0231 |
| Index 7 | −0.0872 | −0.0731 | −0.0031 | −0.0724 | −0.0312 |
| Index 8 | −0.0621 | 0.0213 | −0.0241 | −0.0631 | −0.0821 |
| Index 9 | −0.0012 | −0.0063 | −0.0321 | −0.0516 | −0.0851 |
| Index 10 | 0.0231 | −0.0826 | −0.0351 | −0.0420 | −0.0671 |
| Index 11 | −0.0301 | −0.0821 | −0.0842 | −0.0316 | −0.0696 |
| Index 12 | −0.0216 | 0.0251 | −0.0061 | −0.0698 | −0.0286 |
| Index 13 | −0.0761 | −0.0071 | −0.0761 | −0.0781 | −0.0862 |
| Index 14 | −0.0871 | −0.0316 | −0.0224 | −0.0361 | −0.0372 |
| Index 15 | −0.0051 | −0.0871 | −0.0774 | −0.0765 | −0.0351 |
| Index 16 | 0.0324 | −0.0368 | −0.0875 | −0.0614 | −0.0881 |
| Index 17 | −0.0031 | −0.3159 | −0.0616 | −0.0375 | −0.0135 |

According to the above table, the maximum correlation degree max $K_{xj}(c_1)$ = 0.0556 between index1 (Tonnage of ship) and each risk level can be calculated, Calculate the risk grade correlation degree and final evaluation grade of all indicators is showed as Table 4.

**Table 4.** Maximum correlation degree and final risk level between case evaluation index and each risk level

| Index | 1 | 2 | 3 | 4 | 5 | 6 | 7 | 8 | 9 |
|---|---|---|---|---|---|---|---|---|---|
| max $K_{xj}(c_n)$ | −0.0395 | −0.0655 | 0.0862 | −0.0321 | 0.0209 | 0.0624 | −0.0031 | 0.0213 | −0.0012 |
| Risk level | 4 | 2 | 3 | 4 | 2 | 2 | 3 | 2 | 1 |
| Index | 10 | 11 | 12 | 13 | 14 | 15 | 16 | 17 | |
| max $K_{xj}(c_n)$ | 0.0231 | −0.0301 | 0.0251 | −0.0071 | −0.0224 | −0.0051 | 0.0324 | −0.0031 | |
| Risk level | 1 | 1 | 2 | 2 | 3 | 1 | 1 | 1 | |

In this case, evaluation indicators 1 and 4 are at high risk; Evaluation indicators 3, 7 and 14 are at general risk; Evaluation indicators 2, 5, 6, 8, 12 and 13 are at low risk; The evaluation indicators 9, 10, 11, 15, 16 and 17 are at very low risk.

Analyzing the distribution of main risk levels in this case, we find that the assessment index 1, ship tonnage, and the assessment index 4, channel affluence risk is at a higher risk level. The indicators 1 and 2 are mainly related to the uncertainty risk caused by the increase of water flow speed due to the rise of water level during the flood period, which needs to be improved. The evaluation index 3 Navigation adaptability of ships in flood season, index 7 Perfection of channel supporting facilities, and index 14 Visibility of navigation channel are at general risk. Although there is no major risk for the time being, it still needs the attention of relevant shipping companies and management departments. For index 3, shipping companies can improve the ship's navigation adaptability according to the navigation characteristics in flood season; For index 7, the channel management department should strengthen the construction of channel supporting facilities to reduce the probability of navigation risk accidents in flood season; For index 14, it is necessary for the shipping company and the channel management department to take measures to deal with the low visibility of the navigable river section in advance. Other indicators are at low risk or very low risk, and there is no need for obvious improvement for the time being.

## 5 Conclusion

Based on the analysis of the main factors affecting the navigation cost in the flood period of the upper reaches of the Yangtze River, this paper constructs the navigation risk evaluation index system in the flood period of the upper reaches of the Yangtze River, and establishes the entropy weight extension evaluation model of the navigation cost in the flood period of the upper reaches of the Yangtze River. Using this model, the cost risk of a ship company's navigation in a certain section of the upper reaches of the Yangtze River during the flood period is evaluated, and the main indicator factors that may cause cost risk in this case are evaluated, and the corresponding improvement suggestions are given.

The navigation risk assessment model of the upper reaches of the Yangtze River constructed in this paper can also be applied to the assessment of navigation risk in dry season or the navigation risk in the upper, middle and lower reaches of other inland waterways after modifying the scoring criteria of some evaluation indicators. For example, the impact of ship tonnage on the cost in dry season is just opposite to that in flood season, and the relevant research is not repeated due to space limitation.

The evaluation model built in this paper has certain pertinence. If it needs to be widely used, it needs to further optimize the evaluation indicators and models, and relevant research needs to be further carried out in the future.

## References

1. Huang, Q.: Yangtze river shipping development report 2020. Changjiang Navigation Administration, pp. 6364 (2021, in Chinese)
2. Liu, N.: Theory, method and application of transportation project operation cost evaluation. Shanxi People's Publishing House, pp. 1821 (2003, in Chinese)
3. Lei, X., Wu, Y., Ye, S.: Regional food security early warning based on entropy weight extension decision model. J. Agri. Eng. **28**(5), 233-239 (2012, in Chinese)
4. Shi, K., Sun, X.: Evaluation of Water Resources Carrying Capacity in Chongqing Three Gorges Reservoir Area Based on Entropy Weight Extension Decision Model **33**(2), 609-616 (2013, in Chinese)
5. Hu, E.: Practical techniques and methods for environmental risk assessment. Beijing: China Environmental Science Press, pp. 3334 (2000, in Chinese)

6. Peng, Z., Perch, S.: Transportation and Climate Change. McGraw-Hill, New York, U.S. A, pp. 5455 (2011)
7. Zhang, H., Zhao, X.: Analysis of organizational behavior of container shipping in the upper and middle reaches of the Yangtze River based on hub-and-spoke network. J. Coastal Res. **73**, 119–125 (2015)
8. Bernstein, L., Roy, J., Delhotel, K.C.: Contribution of Working Group Three To Fourth Assessment Report of the Intergovernmental Panel on Climate Change, pp. 1–22. Cambridge University Press, UK (2007)
9. Greater London Authority. Climate Change and London's Transport Systems. Greater London Authority, London, UK, pp. 4446 (2005)
10. Cai, W., Yang, C., He, B.: Preliminary Extension Logic. Science and Technology Literature Press, Beijing, pp. 20–21 (2015, in Chinese)
11. Luo, Z.: Research on logistics security system. Logistics Technol. **12**(10), 8-10 (2005, in Chinese)
12. Liu, T.: Study on the Influence Mechanism and Model of Water Level Change on Shipping Logistics Cost in the Middle Reaches of the Yangtze River. Wuhan University of Technology, Wuhan, pp. 5458 (2015, in Chinese)
13. Zhang, H., He, Q.-M., Zhao, X.: Balancing herding and congestion in service systems: A queueing perspective. Inf. Syst. Operations Res. **58**(3), 511–536 (2020)
14. Vanem, E., Skjong, R.: Designing for safety in passenger ships utilizing advanced evacuation analyses: a risk based approach. Saf. Sci. **44**(2), 111–135 (2006)
15. Wu, Z.: Marine Traffic Engineering. Dalian Maritime University Press, Dalian, pp. 66-68 (2014)
16. Pillay, A., Wang, J.: Technology for Safety of Marine Systems. Elservier Ocean Engineering Book Series, pp. 8185 (2013)
17. Cai, W.: Matter-Element Model and its Application. Science Press, Beijing, pp. 89 (2014, in Chinese)
18. Abdullah, L., Otheman, A.: A new entropy weight for sub-criteria in interval type-2 fuzzy topsis and its application. Int. J. Intelligent Systems and Appl. (IJISA) **5**(2), 2533 (2013)
19. He, Q.-M., Zhang, H., Ye, Q.: An M/PH/K queue with constant impatient time. Mathematical Methods of Operations Res. **18**(11), 139-168 (2018)
20. Hadiwijaya, N.A., Hamdani, H., Syafrianto, A.: The decision model for selection of tourism site using analytic network process method. Int. J. Intell. Syst. Appl. (IJISA) **10**(9), 2331 (2018)
21. Thayananthan, V., Shaikh, R.A.: Contextual risk-based decision modeling for vehicular networks. Int. J. Comput. Network Inf. Security (IJCNIS) **8**(9), 19 (2016)

# Influencing Factors and System Dynamics Analysis of Urban Public Bicycle Projects in China Based on Urban Size and Demographic Characteristics

Xiujuan Wang[1], Yong Du[1], Xiaoyang Qi[1], and Chuntao Bai[2(✉)]

[1] Fundamental Science Section in Department of Basic Courses, Chinese People's Armed Police Force Logistics College, Tianjin 300309, China

[2] Party and Government Office, Tianjin Beichen Science and Technology Park Management Co., Ltd,, Tianjin 300405, China

wxj1701@126.com

**Abstract.** The public bicycle system can be seen as a product of the development and progress of a low-carbon society. With the continuous development of urban scale, the problem of traffic congestion is becoming increasingly prominent. In order to reduce carbon emissions, "bicycle priority" has gradually become an important concept of urban transportation and travel. In some major cities in China, such as Hangzhou, the growth rate of public bicycles is extremely fast. However, at the same time, the research on China's urban public bicycle system is relatively insufficient. In response to this issue, this paper analyzes the main characteristics of Chinese cities, and uses system dynamics methods to analyze the factors that affect the advancement of the public bicycle system, such as the size of Chinese cities and the main characteristics of the urbanization process, the division of transportation modes, population characteristics, bicycle infrastructure and right-of-way, and institutional structure. The results of these analyses is capable to help planners design feasible public bicycle development strategies. The paper summarizes the existing problems and solutions of the public bicycle rental system, providing a certain reference for further improving the public bicycle rental system.

**Keywords:** Public bike · Characteristics of Chinese cities · Transportation system · Bicycle · System dynamics · Engineering of communication

## 1 Introduction

Public bike system is a short term leasing system, in which user can pick the bike up and return it in any bike station, and do not need to bear the cost of buying one nor worry about its safety or parking [1].

In 1965, the first generation of public bike system began to operate in Holland, Amsterdam, which has experienced three generations after over 50 years' development. At present, the vast majority of public bicycle projects in the world belong to the third generation public bicycle system. By 16 September 2018, 1780 cities worldwide have hosted the advanced bike-sharing programs (Meddin and DeMaio, 2018) [2]. Once known as "the kingdom of bicycles", China has a long development history of bicycle, which has served as the one of the most important means of transportation for urban and rural residents and is widely used [3]. Even today, in most of the cities, bicycle is still one of the most important means of transportation. Even so the bicycle traffic in China has been neglected for a long time, so has the relevant research, not to mention related research on public bicycle Public bicycle has a long development time in foreign countries, and the related research is also relatively rich. According to the diffusion theory, its experience and policy are transferable [4]. In China, the history of public bicycle project is only 13 years. The actual projects basically learn from and even copy the experience of foreign countries'. There is no denying that the developed countries in the process of development of public bicycle project accumulated a lot of successful experience, but Chinese cities have their own characteristics and need to combine the experiences with the characteristics of local development to develop public bicycle projects.

## 2 Relevant Research on Public Bike

Along with the extensive development of public bicycle at home and abroad, the research of public bicycle begins to emerge. Take published public bicycle related articles in Transportation Research Record (TRR), the vane in traffic field, for example, before 2009 there is no article in this field; after 2009, it increases year by year, 2 in 2010, 4 in 2011, 5 in 2012, and 5 in 2013. As of May, 2018, a total of 345 articles were presented by Web of Science, which indicates that public bicycle has attracted many researchers' attention and is becoming a hot research topic. In general, the study of public bicycle shows a similar pattern, which has been in the stage of rapid development in the last more than ten years [5, 6]. From 2002, public bicycle research literature began to appear abroad, which can be divided into the following aspects: a. the introduction of public bicycle development [7]; b. feasibility analysis on public bicycle project [8]; c. public bicycle helmet use and safety of users [9]; d. research on public bicycle station layout optimization and public bicycle vehicle balancing and scheduling [10]; e. public bicycle project replacing other modes of transportation [11]; f. evaluation after the public bicycle project has been carried out [10].

Compared with foreign countries, the time of China's public bicycle project is relatively late. In May 2008, Hangzhou launched China's first public bicycle project - Hangzhou public bicycle system. Since then the public bike development in China's cities went onto the fast lane. By April 24, 2015, a total of 237 public bicycle projects had been launched in 29 provinces and municipalities in mainland China [2]. Public bicycle has been paid more attention to in China, which has experienced a faster development process. However, China's existing public bicycle study focused on introduction of public bicycle concept, technology, experience, etc., while analysis on problems and countermeasures of in the process of the development of public bicycle, public bicycle project investigation, public bicycle project operation evaluation, and systematic study based on the characteristics of China's urban public bicycle have not yet been carried out. Compared with cities in developed countries, Chinese cities have many characteristics, which will affect and even determine whether public bicycle project can be successfully implemented. Therefore, this paper will analyze the impact on the development of public bicycle key factors basing on the characteristics of Chinese cities.

## 3 Analysis on Characteristics of Chinese Cities

Factors affecting the development of urban public bicycle project can be divided into internal ones and external ones. Internal factors refer to the factors that can be adjusted according to the external environment in the short term, which can be divided into institutional design and physical design factors. The external factors are the factors that are related to certain cities and not easy to adjust in the short term. Specific division and contains of internal and external factors are in Table 1.

**Table 1.** Internal and factors influencing the development of urban public bicycle project

| Internal factors | Physical design | Hardware and technology |
|---|---|---|
|  |  | Service design |
|  | institutional design | Operator category |
|  |  | Contract and ownership |
|  |  | Sources of funds |
|  |  | Job opportunities |
| External factors |  | City scale |
|  |  | climate |
|  |  | Traffic mode division |
|  |  | Population density |
|  |  | Demographic characteristics |
|  |  | Economic development level |
|  |  | Geography and terrain |
|  |  | Existing infrastructure conditions |
|  |  | Financial situation |
|  |  | Political conditions |

**Resource:** According to Büttner J, Petersen T. Optimising Bike Sharing in European Cities-A Handbook.

Considering China's vast territory and different cities have their own characteristics, this paper only selected general characteristics which can represent Chinese city but significantly different with foreign ones when discussing the external factors affecting he public bicycle. Specifically, it includes city size, process of urbanization, traffic patterns, demographic characteristics, existing infrastructure conditions, political conditions, etc.

## 3.1 City Size and Process of Urbanization

Usually, there are two kinds of indicators of measuring the size of the city--the demographic indicator and the geographical indicator. From the aspect of regional scale, the size of China's cities continues to expand. From 1990 to 2000, the built-up area of China's urban area increased from 12.2 thousand square kilometers to 21.8 thousand square kilometers, an increase of 78.3%. By 2010, this number increased to 40.5 square kilometers, and increased by another 85.5%. The number in 2010 is more than twice of that in 1990 [13]. With the constant enlargement of the city scale, people's travel distance will be increased accordingly.

The urban population spatial distribution is further concentrated, and the regional imbalance is further intensified. According to the Chinese city status report (2012–2013), taking the three major cities--Beijing, Shanghai and Guangzhou for example, in 2010, the total population of three cities in the proportion of 268 prefecture level cities was 4.5%. The population density of the Yangtze River Delta, Pearl River Delta and Beijing-Tianjin-Hebei region were 739 people per square kilometer, 608 people per square kilometer and 481 people per square kilometer, respectively, far higher than the average population density of China (140 people/km$^2$) and the western region (53 people/km$^2$).

Public bicycle project has scale benefit and network benefit, and with the increasing of the scale and scope of the network, the operation cost is lower and the service level is higher, namely, to provide services through the construction of the scale and network services; therefore, to carry out the project requires certain local urban traffic demand. If the local population and travel demand are small, in order to ensure the convenience of the public bicycle project, it is necessary to maintain a certain scale of facilities, it will not be economic and the benefit cannot be guaranteed. if the operation depends only on the government's subsidy, it is difficult to ensure that economic projects can be sustained. Compared with foreign cities, Chinese cities have large scale, high population density; big cities, especially mega cities have objective needs to carry out the public bicycle project, and even larger public bicycle plan.

## 3.2 Traffic Mode Division

As a new mode of traffic travel in the city, public bicycle is bound to have an impact on the structure of urban transportation system. At the same time, the structure of the urban transportation system will affect the development and operation of the project, so it is necessary to understand the structure of urban traffic system, and determine the proportion of different traffic modes.

So far, bicycle traffic is still a very important travel method in many Chinese cities, its proportion is much higher than that in Germany and Holland where public bicycle project is carried out smoothly. At present, both production and export amount of Chinese bicycles ranked the or above 60% all over the world, and the consumption ranked the first. Taking Beijing for example, the proportion of traveling by bicycle has been dropping since1986, but by 2010 it is still the important method of traveling in the city traffic system, which is shown in Fig. 1.

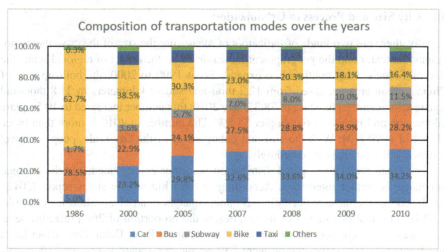

**Data recourse**: Beijing traffic development research center.
Note: Walking not included

**Fig. 1.** Traffic mode proportion of Beijing

## 3.3 Demographic Characteristics

### 3.3.1 Age

Considering China's rapid urbanization, there will be more and more senior citizens. Since public bicycle needs human power to drive and there is a certain requirement of physical strength and techniques. The quality of the individual and riding speed limit their travel distances. The elder's physical is relatively poor, so travel ways consume too much physical strength are not applicable. Therefore, in order to achieve the sustainable development of public bicycle which is a green traffic mode in urban China, the aging problem must be taken into consideration.

Compared with western countries, China is facing a more serious problem of population aging. In 2000, China's over 60 years old population is 1.3 billion, accounting for 10.71% of the total population at that time. According to international aging society standard (population over 60 years old accounted for the proportion of the total population of 10%, or over the age of 65 population accounted for total population proportion reached 7%), China has become an aging country since 2000.

### 3.3.2 Incomes

Income of urban residents continued to increase and the growth speed is relatively fast, but the low-income groups still account for a higher proportion. The per capital disposable income of urban residents increased from 1510 yuan in 2011 to 21810 yuan in 1990, an increase of 14.5 times. The proportion of lower income households, whose average income is less than 6566 yuan, is 40%. The middle-income households, whose average income is 13178 yuan, accounted for 20%. The proportion of low-income groups is larger. In China's urban transport system, the bicycle traffic is a major way to travel.

### 3.4 Bicycle Infrastructure and Right of Way Characteristics

Bicycle infrastructure and security are very closely linked. A United States survey shows that bicycle traffic safety and travel rate increase with the improvement of bicycle infrastructure. But for the country where bicycle infrastructure is relatively common, the new facilities have less obvious influence on improving the performance of the level of bicycle travel. The existence of bicycle infrastructure cannot directly determine the proportion of bicycle travel, while a higher proportion of bicycle travel can stimulate the construction of bicycle infrastructure.

Bicycle traffic has always been an important part of the urban transportation system in China, and perfect bicycle infrastructure has been built up in most cities. In recent years, for the purpose of developing sustainable transportation, many cities in China are planning and constructing bicycle traffic network system in all directions in the urban areas to meet the development of bicycle traffic.

In recent years, the standard of many city's bike lanes and road mileage has been improved, but there is still uneven allocation of urban road space, bicycle driving space constantly being squeezed and unguaranteed traffic safety, lacking effective laws and regulations to protect the rider's right of way.

### 3.5 Regime Structure

Public bicycle is a typical quasi-public goods, which has the characteristics of public welfare and economy, which needs the participation of the government to ensure its public welfare. In order to strengthen the efficiency of management, government is divided into several departments, each of whose responsibilities and rights are only limited to the scope of the higher authorities. There is no special and independent organization to manage the public bicycle, which is a new thing, so its operators need to deal with all departments. Taking the site selection of the public bike station as an example, it involves the different departments concerning traffic management, urban management, urban planning, urban construction, greening, electricity and others. For example, in Shenzhen, when developing public bicycle project, setting rental set near the bus station needs the approval of the city transportation department; setting rental set near the subway station requires the approval of the Shenzhen Metro Group; setting rental set on the pavement needs the approval of Urban Management Department.

## 4 System Dynamics Analysis on the Relationship Between Bicycle Travel and Public Bicycle

In the short term, public bicycle development can ease the traffic problems caused by motor vehicles, but it also occupies public resources for the development of bicycle traffic, such as funds, land and so on. At present, the funding of the development of public bicycle project is not sustainable, which needs long-term capital from government and other investment funds to operate. When transport related investment is not sufficient, public bicycle will occupy funds used for other bicycle related projects, such as funds for bicycle related infrastructure construction, and cycling promoting related policies. In most Chinese cities, the bicycle is an important means of transportation, which attracts a large number of users; but resources input for improving bicycle safety and convenience is far from enough. If the resources are used to carry out public bicycle project, attracting the original bicycle users to the use of public bike, it will not be very useful to solve the urban traffic problems. Therefore, the effect of settling the urban transport problems by whether using capital, land and other public resources to carry out public bicycle project, or directly to improve the use of bicycles need to be examined in a long-term aspect. In this paper, this issue will be examined with the system dynamics method in urban traffic system.

### 4.1 Reason for Using the System Dynamics Method

System dynamics method was first proposed by Forrester, which is based on feedback control theory and takes computer simulation technology as the main method. It is mainly used in the study of complex social economic system. It was first used in the urban areas, and was then quickly spread to economics, ecology and other fields. In the early stages of the traffic field was used to analyze the relationship between traffic and land use.

Urban transportation system includes a large number of elements, and the feedback mechanism between elements is complex. There is a dynamic and complex nonlinear relationship and feedback between the subsystems of different levels and the variables. It is difficult to comprehensively grasp the characteristics of urban traffic system by using the common method of quantitative research, while the system dynamics method can be used to conduct systematic comprehensive quantitative analysis. The public bicycle is a systematic problem, whose related variables have a dynamic and complex nonlinear relationship and feedback effect, and it is suitable to be analyzed with systematic dynamic method.

### 4.2 System Dynamics Analysis of the Relationship Between Bicycle and Public Bicycle

China's rapid urbanization process has increased the size of the city, which in turn has expanded the average travel distance of people. Considering suitable trip distance of different urban travel mode, cycling ratio decreases with people' s average travel distance increasing, and the proportion of motor vehicle travel increases with people' s average travel distance increasing. The convenience of vehicle travel provides the conditions to further expanding the scale of cities. On the other hand, this issue can be considered from the perspective of characteristics of Chinese urban land use. Before the Reform and Opening-up, more than 90% of urban population belongs to a neighboring work unit, with suitable housing, education, health care and other institutions around, and community function is perfect. While after the Reform and Opening up, China has implemented the land administrative allocation system for a long time and due to the scarcity of urban land, Chinese cities improve the utilization efficiency of urban land by increasing intensity of land, and integrating a variety of different land types. High mixture of the multi-function area is still the main characteristic of Chinese city. Therefore, the function structure and land use type of Chinese cities are different from western industrialized countries, which have the characteristics of compact structure and land use. Chinese cities' this characteristic shorten people's travel distance, making bicycles and other modes suitable for short distance travel play a greater role in the transport, reducing the demand for motorized travel mode. Mixed land use, the process of urbanization and motor vehicle travel form a dynamic feedback loop.

Motorized travel growth rate exceeds the service capacity of the urban road, and traffic congestion reduces the vehicle speed. The rush hour traffic speed of certain road sections is even lower than the bicycle speed. Taking the cost into account, people will turn to the use of bicycles, thereby reducing the amount of motor vehicles. The aging of urban population, the increase of people's income level and the bicycle travel forms a dynamic feedback loop.

For travelers, the biggest advantage of public bike is exempting bike purchase and maintenance cost, but bicycles cost is not the main factor deciding bicycle use level. In urban mixed traffic environment, the most major factor affecting people using bicycle is safety. When the rider's safety is guaranteed, more travelers will use bicycle. The more riders, the more attention motor vehicle drivers will pay, which can further improve the rider's safety. Public resources devoted to cycling safety and the use of bicycle form a dynamic feedback loop [13].

Decision makers understanding the importance of bicycle infrastructure will benefit the investment of related public resources, but even if public resources are devoted into the bicycle infrastructure construction, the construction need some time. So there will not immediately be positive effect. In most cities in China, with the rapid increase of motor vehicles, bicycles and motor vehicles compete fiercely on the limited road, the right of road, capital and other public resources. That results in a lack of cycling facilities investment, and further affects the safety of bicycle use. Once security becomes a significant problem, there will be widespread social concern and decision makers' awareness of the importance of bicycle infrastructure will be improved. A dynamic feedback loop is formed between the structure of Chinese cities and the safety of bicycle.

Experience of carrying out public bike project at home and abroad shows that public bicycle increases the number of riders in the short term, which in turn requires more bicycle related facilities. So it improves the decision makers' awareness of the importance of bicycle facilities and increases investment of cycling facilities. Due to that in a certain period a city's resources of developing bicycle traffic is limited, the bicycle infrastructure construction and public bicycle project competition in the aspect of investment of public resources. Increasing bicycle facilities investment means reducing investment for public bicycle project resources, which is not conducive to the improvement of public bicycle project expansion and the service level. The public bicycle project and bicycle infrastructure development form a dynamic feedback loop [13–15].

According to the shortage of investment and growth based model, there is time delay between public resources input and bicycle related facilities level improvement. When the demand for bicycle related facilities is low, the government should increase the public resources investment in case of use decline due to the lack of related infrastructure. In general, there is bicycle related facilities supply shortage in most Chinese cities, this not an appropriate time to promote the growth ring, that is not to take measures to continue to increase bicycle travel. The correct management policy should aim to improve the level of bicycle infrastructure.

## 5 Conclusion

Public bicycle can ease the traffic problems caused by motor vehicles in the short term, but it also occupies the public resources for the development of bicycle traffic. The development of public bicycle is generally in the city center, occupying the valuable land resources, caused the public bicycle project is non-sustainable. In most cities of China, bicycle is an important means of transportation with huge number of users, but resources for improving bicycle safety and convenient is far from enough. Therefore, whether public resources such as funds and land should be used to develop public bicycle projects or directly used to increase the utilization rate of bicycles needs to be examined from the perspective of long-term solution to urban traffic problems.

Public bike is a systematic problem, with complex dynamic nonlinear relationship and feedback effect between its related variables. The characteristics of urban traffic system cannot be comprehensively grasped by the usual quantitative research methods, while systematic dynamics method can carry out systematic comprehensive quantitative analysis of it. Therefore, this paper, taking the characteristics of Chinese cities into consideration, uses the systematic dynamics method to analyze the relationship between bicycle and public bicycle. The results showed that characteristics of Chinese cities determine that when most Chinese cities solving urban traffic problems through alternative means of transportation, especially the bicycle travel, the most pressing matter of the moment is not immediately launching large-scale public bicycle project, but taking measures to create a good external environment for bicycle travel, introducing relevant laws and regulations to protect the bicycle rider's right of way, ensure the safety and convenience of bicycle travel.

**Acknowledgment.** This project is supported by Basic Research Project of Logistics College of Armed Police Force References (WHJ202101).

## References

1. Shaheen, S.A., Guzman, S., Zhang, H.: Bikesharing in Europe, the Americas, and Asia past, present, and future. Transp. Res. Rec. **2143**, 159–167 (2010)
2. Meddin, R., DeMaio, P.: The Bike-Sharing World Map [EB/OL]. http://www.bikesharingworld.com. Accessed 26 Sept 2018
3. Lin, X., Wells, P., Sovacool, B.K.: The death of a transport regime? The future of electric bicycles and transportation pathways for sustainable mobility in China. Technol. Forecast. Soc. Chang. **132**, 255–267 (2018)
4. Hua, Z.: Research on Consumer Adoption Behavior of Low Carbon Transportation Mode Innovation. Lanzhou University, Lanzhou (2011). (in Chinese)
5. Zhengke, Y., Kailing, D., Xuemei, Z.: Review of research on shared bicycles at home and abroad. J. Chengdu Univ. Soc. Sci. Ed. **2**, 7 (2018). (in Chinese)
6. Zhang, S., Chen, L., Li, Y.: Shared bicycle distribution connected to subway line considering citizens' morning peak social characteristics for urban low-carbon development. Sustainability **13**, 9263 (2021). https://doi.org/10.3390/su13169263
7. Ji, Y., Fan, Y., Ermagun, A., et al.: Public bicycle as a feeder mode to rail transit in China: the role of gender, age, income, trip purpose, and bicycle theft experience. Int. J. Sustain. Transp. **11**(4), 308–317 (2017)

8. El-Assi, W., Mahmoud, M.S., Habib, K.N.: Effects of built environment and weather on bike sharing demand: a station level analysis of commercial bike sharing in Toronto. Transportation **44**(3), 589–613 (2017)
9. Mooney, S.J., Lee, B., O'Connor, A.W.: Free-floating bikeshare and helmet use in Seattle, WA. J. Community Health **24**, 1–3 (2018)
10. Haider, Z., Nikolaev, A., Kang, J.E., et al.: Inventory rebalancing through pricing in public bike sharing systems. Eur. J. Oper. Res. **270**(1), S0377221718302030 (2018)
11. Campbell, K.B., Brakewood, C.: Sharing riders: how bikesharing impacts bus ridership in New York City. Transp. Res. Part A Policy Pract. **100**, 264–282 (2017)
12. Yin, J., Qian, L., Singhapakdi, A.: Sharing sustainability: how values and ethics matter in consumers' adoption of public bicycle-sharing scheme. J. Bus. Ethics **149**(2), 313–332 (2018)
13. Porag, F., Hossain, S.: Intelligent tour planning system using crowd sourced data. Int. J. Educ. Manage. Eng. (IJEME) **01**(08), 22–29 (2018)
14. Fadel, M.A., Elrefaei, L.A.: Adjustments of methodology planning and assessment activities of senior projects in the computer science program. Int. J. Mod. Educ. Comput. Sci. (IJMECS) **10**(2), 16–25 (2018)
15. Biswas, D., Samsuddoha, M.: Determining proficient time quantum to improve the performance of round robin scheduling algorithm. Int. J. Mod. Educ. Comput. Sci. **11**(10), 33–40 (2019). https://doi.org/10.5815/ijmecs.2019.10.04

# Cross-border Logistics Model Design e-Tower Based on Blockchain Consensus Algorithm

Shujun Li[1(✉)], Anqi He[1], Bin Li[1], Fang Ye[1], and Bin Cui[2]

[1] School of Economics and Management, Wuhan Railway Vocational College of Technology, Wuhan 430205, China
462400086@qq.com

[2] Faculty of Art, Dongguk University, Seoul 100715, South Korea

**Abstract.** Logistics plays an important role in the development of cross-border e-commerce. Cross border e-commerce enterprises have to choose appropriate logistics channels. This paper discusses the typical application models of blockchain technology in cross-border logistics, and proposes specific implementation paths to improve the performance of cross-border logistics management. This paper first analyzes the structure framework and core technology (common consensus algorithms) of the blockchain. Taking Shenzhen ZK cross-border logistics company as an example, a cross-border logistics mode ZK e-Tower based on cross-border blockchain has been proposed. The key issues in the system design, such as access control and identity management, consensus algorithm, blockchain based transaction process, its application in cross-border customs clearance, security and privacy protection have been analyzed in detail, and a prototype system was built for simulation experiments thus solving the problems of traditional information systems, such as excessive power of central organizations and information tracking difficulties.

**Keywords:** Blockchain Technology · Cross-border Logistics · Application Mode

## 1 Introduction

According to the report issued by the National Bureau of Statistics, China's total import and export of goods in 2021 will be 391009.4 billion yuan, an increase of 21.4% year on year. Looking at China's foreign economic data in the past five years, the total import and export of goods has shown a sustained growth trend, with the total import and export exceeding 35000 billion yuan in 2021. However, there is a contradiction between the slow development of cross-border logistics and the rapid development of cross-border e-commerce. The low transparency of information, long logistics time and space cycle, high cost, numerous and miscellaneous logistics participants, low logistics standardization level and low logistics efficiency have seriously affected consumers' shopping experience. The Business transformation and improvements in operation cost have contributed to the development of the Supply chain analytics [1]. Combining the characteristics of blockchain technology, such as decentralization and traceability, to establish a new import and export cross-border logistics model -- building a blockchain logistics application platform will help solve the problems of opaque cross-border logistics information, difficult returns and exchanges, and low service level in transaction, and help improve the efficiency of import and export logistics [2].

Blockchain technology has been extended from the original Bitcoin to all walks of life, especially in logistics transportation, logistics supply chain and other service transactions [3]. It also becomes an important task for state controllers to discover, prevent and pre-control Emergency Logistics Risks (ELR) [4]. In essence, blockchain technology is based on peer-to-peer untrusted networks. Network nodes can communicate directly without a central node. It integrates new applications of computer technologies such as data encryption, storage, communication, consensus mechanism, and network structure [5]. Its characteristics of decentralization and trustworthiness overcome the insecurity of centralized networks. The traditional transaction model of logistics independent certification center has information security problems [6]. First, it is vulnerable to network attacks, which can lead to the disclosure of confidential information; Second, it is impossible to verify the authenticity of the user's identity and review the traceability, and the unique identity signs of both parties to the transaction cannot be guaranteed [7, 8]; Third, credit rating is carried out through the past experience and qualification review of both parties to the transaction. Credit rating may require enterprises to spend corresponding costs, which allows the decision-maker to have knowledge of economic, ecological and social cost before making a decision [9]. And references [9, 10] have also discussed relevant blockchain issues on other areas. It is very difficult to collect comprehensive, complete and systematic logistics transaction information. Therefore, the information security of logistics trading activities and the consensus trust of upstream and downstream customers need to be solved urgently. Reference [11] have also pointed relevant security issues, and the algorithm used in the modes can be verified. Therefore, it is of great scientific significance and economic value to integrate cloud computing and blockchain technology and use the blockchain consensus algorithm to solve the problem of decentralization of logistics transactions and trust between users.

## 2 Common Consensus Algorithms in Blockchain Technology

Now, there are many consensus algorithms used in the blockchain, such as: Proof of Work (PoW) algorithm, Proof of Stake (PoS) algorithm, Delegated Proof of Stake (DPoS) algorithm, and practical Byzantine Fault Tolerance (PBFT) algorithm [12]. Common consensus algorithms in blockchain technology are as follows.

### 2.1 PoW Algorithm

PoW algorithm is a mechanism to prevent distributed service resources from being abused and denial of service attacks, which requires nodes to perform complex operations that consume time and resources in an appropriate amount, and its results can be quickly verified by other nodes to ensure that services and resources are used by real demand.

Hash algorithm is the basic technical principle of PoW algorithm. Suppose the hash value Hash (r) is calculated. If the original data is $r$, the operation result is $R$ (Result).

1) $R = Hash(r)$

For any input value $r$, the result $R$ is obtained, and it cannot be inferred back from $R$. When the input original data $r$ changes by 1 bit, the resulting $R$ value changes completely. In the PoW algorithm of Bitcoin, the algorithm difficulty $d$ and the random value $n$ are introduced, and the following formula is obtained:

2) $Rd = Hash(r + n)$

This formula requires that when the random value $n$ is filled in, the first $d$ bytes of the calculation result Rd must be $0$. Due to the unknown result of the hash function, each node needs to do a lot of operations to get the correct result. After the calculated result is broadcast to the whole network, other nodes only need to perform a hash operation to verify. PoW algorithm uses this method to make computing consume resources, and the verification only needs one time.

## 2.2 PoS Algorithm

The PoS algorithm requires that the node verifier must pledge a certain amount of funds to qualify for mining and packaging, and the regional chain system uses a random method when selecting packaging nodes. When the node pledges more funds, the greater the probability that it will be selected to package blocks.

For example, if a node owns 5% of the entire blockchain system, it will have a 5% probability of packaging blocks in the next block out cycle.

The process of node block out through PoS algorithm is as follows: To become a block out node, an ordinary node first needs to pledge its assets. When it is the node's turn to block out, the block is packaged, and then broadcast to the whole network. Other verification nodes will verify the validity of the block.

## 2.3 DPoS Algorithm

DPoS algorithm is similar to PoS algorithm, which also uses shares and equity pledge. The difference is that the DPoS algorithm uses the entrusted pledge method, which is similar to the method of electing representatives by the whole people to select large numbers of nodes for blocking.

Voters cast their votes to a node. If a node is elected as an accounting node, the accounting node can return their votes to other voters in any way after obtaining a block of rewards. Those nodes will take turns to generate blocks, and the nodes will supervise each other. If they commit crimes, the pledge deposit will be deducted.

## 2.4 PBFT Algorithm

We believe that PBFT algorithm can solve more problems with higher efficiency [13]. Figure 1 shows the basic consensus process of PBFT algorithm.

In 1982, Leslie Lamport proved in the paper The Byzantine Generals Problem that when the total number of generals is more than $3f$ and the number of traitors is $f$ or less, loyal generals can reach consensus on orders, that is, $3f + 1 \leq n$, and the algorithm complexity is $O\ (nf+1)$.

In 1999, Miguel Castro and Barbara Liskov published a paper, Practical Byzantine Fault Tolerance, proposing the PBFT algorithm [14]. The number of Byzantine fault tolerance of the algorithm also meets $3f + 1 \leq n$, and the algorithm complexity is $O\ (n2)$. This algorithm can provide high-performance computing, so that the system can process thousands of requests per second, which is faster than the old system.

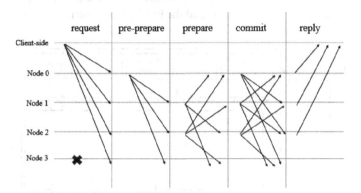

**Fig. 1.** Consensus process of PBFT algorithm

The consensus process of the PBFT algorithm is as follows: the client initiates a message request, broadcasts and forwards it to each replica node (Replica), and one of the master nodes (Leader) initiates a pre-prepare proposal message and broadcasts it. Other nodes obtain the original message and send the prepare message after verification. Each node receives $2f + 1$ prepare message, which means that the preparation is completed, and sends a commit message [15]. When the node receives $2f + 1$ commit message, it is considered that the message has been confirmed to be completed (reply).

## 3 e-Tower Design for Shenzhen ZK Company Cross-border Logistics

According to the actual situation of ZK company, this paper designs the cross-border logistics information platform ZK e-Tower based on blockchain technology, which mainly involves the following key issues: access control and identity management; security and privacy protection of users and transaction data in, as well as the consensus algorithm used in the blockchain network. The system adopts alliance chain architecture and introduces PKI system in the outer layer, so as to achieve the purpose of strengthening system identity management and access control. At the same time, BFT-SMART protocol based on fault-tolerant is introduced as the consensus algorithm [16, 17]. On the premise of ensuring that the system performance meets the actual needs, the consensus algorithm of conflict fault-tolerant commonly used in the existing chain system is raised to the level that can tolerate 31% Byzantine errors. Although the system discards the completely decentralized characteristics of the traditional blockchain, it can meet the needs of information platform. On the contrary, the multi centralized architecture based on blockchain is conducive to giving full play to the advantages of blockchain technology in data tamper proof, traceability, multi-party verification and maintenance and other features.

### 3.1 Overall Architecture Design

The e-Tower system architecture proposed in this paper is shown in Fig. 2, which is divided into two layers: the interaction and the blockchain consensus layer.

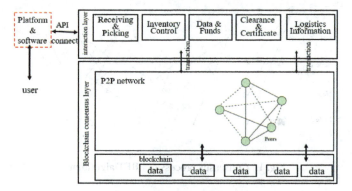

**Fig. 2.** Overall architecture design of e-Tower

The application layer is responsible for encapsulating user data and operations into standard digital assets, and then sending various requests to blockchain nodes in a predefined standard transaction form to complete user operations, and finally storing the data on the blockchain [17].

The blockchain consensus layer includes the P2P node network that maintains the blockchain and the only blockchain jointly maintained by the nodes in the system. The main participants are responsible for running the physical server as the node, verifying the transactions from the upper layer and composing the transactions into blocks, running the consensus algorithm and finally updating the data in the blockchain copy of each node.

### 3.2 Composition of Underlying P2P Network

The e-Tower system adopts modular design and divides the nodes in the P2P network into three roles:

1) Endorsement Node: the endorsement node is responsible for verifying the legitimacy of the transaction and signing it. After the validation, it is responsible for executing the transaction.

2) Organization Node: the organization node receives the transaction signed by the endorsement node, runs the consensus algorithm, and organizes the transaction into blocks to deliver to the submission node after ensuring consistency.
3) Submission node: submit the node to verify the block completed by the organization, and then update the blockchain with the block.

### 3.3 Application of PBFT Algorithm

Integrating the cross-border logistics transaction process, mechanism and related theories of blockchain and cloud computing, and using the consensus mechanism of the practical fault tolerance algorithm (PBFT) of blockchain, the definition of logistics blockchain and cloud logistics blockchain is proposed, see Fig. 3.

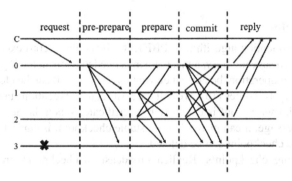

**Fig. 3.** PBFT algorithm of e-Tower

The main steps of the algorithm's application are as follows:
1) The client sends a request to the master node.
2) The master node broadcasts the request to other nodes, and the node executes the three-stage consensus process of the PBFT algorithm.
3) After the node processes the three-stage process, it returns a message to the client.
4) After the client receives the same message from the $f + 1$ node, the consensus has been completed correctly [17].

### 3.3.1 Definitions

For e-Tower, six definitions are proposed:

Definition 1: There are at least $3f + 1$ nodes in the logistics blockchain, which are adjacent to each other and form a basic Byzantine unit (UB), where $f$ represents a normal logistics node and UB represents a consensus network unit.

Definition 2: In the logistics blockchain, the PBFT mechanism is implemented according to the View, which is numbered and recorded as $n$. In a View with $3f + 1$ nodes in total, there is only one master node, recorded as $m$, and the rest of the nodes are backup nodes called replicas. Each node is represented by an integer and is $\{0, 1, \ldots 3f + 1\}$ in order, satisfying the following: $m = n \bmod 3f + 1$, where $n$ represents consensus network initialization, $m$ represents the origination block.

Definition 3: All nodes(Nodei) in the logistics blockchain use double chains to store blockchain information of other nodes. Nodei = {a, b}, a and b represent blockchain header index and blockchain content respectively.

Definition 4: Blockchain information B-Info stores blockchain data information $D$, and records the node name N, data key K, and update time T. B-Info structure: B-Info = (BD, BN, BK, BT).

Definition 5: On the basis of logistics blockchain, e-Tower virtual logistics network node mapping: Node: UB → $UB^C$. C represents a collection of virtual nodes.

Definition 6: Logistics blockchain achieves virtual node $U$ through certificate transfer $U_B^C$ Consensus process [18]. The certificate message format is: Message = (REPLY, b, v, i), where REPLY is the message type. Depending on the certificate type, $b$ may be a block or a block hash, $v$ represents the view number, and $i$ represents the node number.

### 3.3.2 Recycle Bin

With the state replication algorithm of PBFT, Replicas will record executed messages in the local log. To save memory, a mechanism is needed to clean the log. It is unwise to execute after each operation, because it consumes resources. It can be cleaned regularly, such as once every 100 times [19]. We call the post request execution status Checkpoint; whose checkpoint with proof is called stable certificate. When the node receives $2f + 1$ checkpoint message, it can prove that the stable checkpoint is correct. Log messages before the stable checkpoint can be deleted.

When clearing checkpoints, Replica broadcasts a checkpoint protocol to other replicas:

$$< checkpoint, n, d, i >_{sig_i} \qquad (1)$$

$n$ is the serial number of the last correct execution request, and $d$ is the summary of its current status. If each replica receives $2f + 1$ checkpoint message with the same sequence number $n$ and summary $d$, then each replica can clear the sequence number $< n, d, v >$ whose log information less than or equal to $n$. Checkpoint protocols are also used to update logistics routes. The normal line $h$ is equal to the serial number of the nearest stable logistics line, the abnormal line $H = h + l$, and $l$ is the log size.

### 3.3.3 Information View Change

When the master node of logistics is invalid, or some nodes receive $2f + 1$ commits in the commit phase, and some nodes do not receive $2f + 1$ commits, resulting in inconsistent states, the information view needs to be changed to provide system activity and security.

When the request times out, the backup node enters view $v + 1$ and broadcasts the view change message:

$$< new - change, v + 1, n, C, P, i > \qquad (2)$$

$n$ checks serial number, $C$ checks certificate, $P$ is a set that contains the relevant message set $P_m$ for request $m$ (the request serial number is greater than $n$). $P_m$ contains $2f + 1$ identical preparation message.

When the master node of view $v + 1$ receives $2f$ change messages of the same view, it broadcasts new view messages to other replicas:

$< new - view, v + 1, V, O >$, $V$ is $2f + 1$ view change message. The calculation rules of $O$ are as follows:

1) Determine the serial numbers $min_s$ and $max_s$, where $min_s$ is equal to the stable checkpoint serial number in $V$, and $max_s$ is equal to the maximum prepare message serial number in $V$.
2) The master node allocates pre-prepare messages for each sequence number $n$ between $min_s$ and $max_s$. If the $V$ contains the $P$ combination corresponding to $n$, the corresponding pre-prepared message:

$< pre - prepare, v + 1, n, d >$, that is, there are $2f + 1$ prepare messages for the request corresponding to serial number $n$, and the request is still submitted in the new view). If $V$ does not contain the $P$ combination corresponding to $n$, the submitted null message:

$< pre - prepare, v + 1, n, d < null >>$, that is, no processing is done. After receiving the new view message, the replica broadcasts a prepare message, enters $v + 1$, and the view change is completed [19, 20].

### 3.4 Transaction Process

In the e-Tower system, a standard transaction process consists of three stages: transaction endorsement, generation block and confirmation submission. The specific transaction info-flow process is shown in Fig. 4.

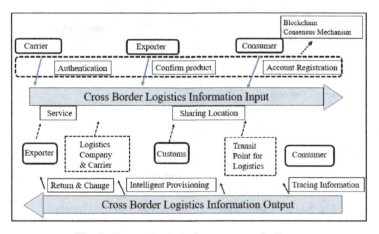

**Fig. 4.** Transaction info-flow process of e-Tower

The client sends a request to any legal endorsement node to start a transaction. The endorsement node simulates the smart contract that executes the transaction, reads the status of the current blockchain, signs it, and returns it to the client program. After receiving the data, the client program will package it with the original transaction data and broadcast it to all organization nodes [20]. The organization node cluster implements the improved consensus algorithm based on BFT-SMART. After receiving a certain number of transactions or reaching a certain time, the transactions that have passed the consensus algorithm are organized into a block. When a block is completed, the organization node broadcasts the block to all submission nodes. The submission node verifies the timeliness and legitimacy of the block, and updates the local blockchain copy with the transaction data in the block after passing. Finally, all submission nodes separately notify the client whether the transaction is successfully submitted [21]. A complete transaction process is completed.

### 3.5 Application in Cross-Border Customs Clearance

Customs clearance is one of the core links of cross-border logistics. It is not only necessary to strictly manage customs, but also to improve the speed of customs clearance and promote trade facilitation. These two aspects have been difficult problems for customs agencies [21]. The key and difficult point is to ensure the reliability of the information data of entry-exit goods. Generally, enterprises need to provide complete documents and materials to verify the accuracy of information, including on-site inspection of goods, which makes it difficult to improve the speed and experience of customs clearance [22]. As shown in Fig. 5, the combination of blockchain technology and cross-border customs clearance has achieved results.

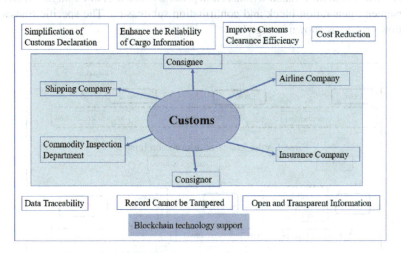

**Fig. 5.** Blockchain support in cross-border customs clearance

## 4 Privacy and Security Verification of e-Tower

The e-Tower system guarantees the security of transaction data and user information in two aspects of data storage and access control. The classic blockchain structure proposed by Satoshi Nakamoto is adopted for data storage. All data generated are encapsulated in various standardized transactions. After being regularly packaged into blocks by organization nodes, they are finally put into the only blockchain owned by the trading network and become an immutable part of the blockchain.

At the system access control level, the system introduces the X.509 certificate specification and Merkle tree to manage the identity and authority in the system. Any participant who joins the network must first obtain a digital certificate from the authority as the only legal identity for activities in the network. As shown in Fig. 6, Merkle tree is introduced to ensure the privacy and security of the system based on cryptography principles, which is used to design and realize the confidentiality, integrity, authentication and non-repudiation of the blockchain.

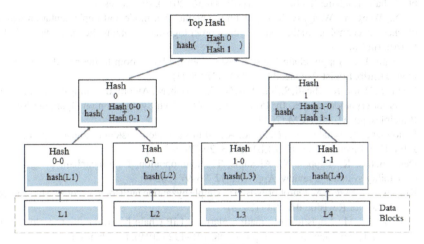

**Fig. 6.** Merkle tree

## 5 Conclusion

Based on the consensus algorithm of blockchain technology, a cross-border logistics blockchain model e-Tower has been designed for Shenzhen ZK cross-border logistics company. Especially the PBFT algorithm based on the practical Byzantine consensus algorithm realizes the decentralization and non-tampering of the system. At the same time, the distributed storage characteristics of cloud computing are applied to solve the computing power problem of large-scale consensus computing, providing a blockchain solution to a series of problems existing in the current cross-border logistics industry, such as opaque transactions, high security risks, and poor interactivity. Experiments show that the model has good performance in security, stability and throughput. However, the openness and transparency of the blockchain also means that cross-border logistics information is exposed in the logistics network. While users trace the source of information flow, some information that is not suitable for disclosure will also be exposed. Therefore, we need to further apply decentralized construction, optimization, and higher-level privacy security technology to better ensure the system operation focusing on users' information security.

# References

1. Anitha, P., Pati, M.M.: A review on data analytics for supply chain management: a case study. Int. J. Inf. Eng. Electron. Bus. (IJIEEB) **11**(05), 30–39 (2018)
2. Zhang, Y.: Building China-EU cross-border e-commerce ecosystem with blockchain technology. China's Circ. Econ. **32**(2), 66–72 (2018). (in Chinese)
3. Wang, C., Yidi, W., Qin, Q.: Supply chain logistics information ecosystem model based on blockchain. Intell. Theor. Pract. **40**(7), 115–120 (2017). (in Chinese)
4. Cheng, Q., Lei, Y.: Operational mechanism and evaluation system for emergency logistics risks. Int. J. Intell. Syst. Appl. (IJISA) **02**, 25–32 (2010)
5. Hu, J., Ge, C., Sun, Y.: Research on logistics information management framework based on blockchain. Logistics Technology **41**(10), 34–36 (2018). (in Chinese)
6. Li, X., Wang, Y., Wang, F.: Research on the application mode and implementation path of blockchain technology in the field of cross-border logistics. Contemp. Econ. Manage. **07**, 1–2 (2020). (in Chinese)
7. Cavinato, J.L.: Supply chain logistics risks from the back room to the board room. Int. J. Phys. Distrib. Logist. Manag. **34**(5), 383–387 (2004)
8. Gilad, Y., Hemo, R., Micali, S., Vlachos, G., Zeldovich, N.: Algorand: scaling byzantine agreements for cryptocurrencies. In: Proceedings of the 26th Symposium on Operating Systems Principles, pp. 51–68. ACM (2017)
9. Benotmane, Z., Belalem, G., Neki, A.: A cost measurement system of logistics process. Int. J. Inf. Eng. Electron. Bus. (IJIEEB) **05**, 23–29 (2018)
10. Demestichas, K., Peppes, N., Alexakis, T., Adamopoulou, E.: Blockchain in agriculture traceability systems: a review. Appl. Sci. **12**, 96–97 (2020)
11. Kiayias, A., Russell, A.: Ouroboros-BFT: a simple byzantine fault tolerant consensus protocol. Preprint Minor Revision **01**, 1049 (2018)
12. Ning, Z., Li, M.: Logistics information platform LIP Chain based on alliance blockchain. Computer Technology and Development **08**, 191–192 (2019). (in Chinese)
13. David, B., Gaži, P., Kiayias, A., Russell, A.: Ouroboros praos: an adaptively-secure, semi-synchronous proof-of-stake blockchain. In: Nielsen, J.B., Rijmen, V. (eds.) Advances in Cryptology – EUROCRYPT 2018: 37th Annual International Conference on the Theory and Applications of Cryptographic Techniques, Tel Aviv, Israel, April 29 - May 3, 2018 Proceedings, Part II, pp. 66–98. Springer International Publishing, Cham (2018). https://doi.org/10.1007/978-3-319-78375-8_3
14. Zhou, J., Li, W.: Research on consensus algorithm of logistics blockchain based on cloud computing. Comput. Eng. Appl. **54**(19), 239 (2018)
15. Ma, J.: Research on the new mode of blockchain and import logistics. Technol. Market **25**, 23–24 (2018). (in Chinese)
16. Yao, L., Hao, D., Jianhua, M.: Optimization of logistics supply chain based on blockchain technology and genetic algorithm. Process Automation Instrumentation (07), 97 (2022). (in Chinese)
17. Zhao, G., Dong, J., Liu, M., Zhai, K.: Evolution game of banks and enterprises in digital supply chain finance driven by blockchain. In: The 34th Chinese Control and Decision Conference (CCDC), pp. 147–151. Hefei Press, Hefei (2022). (in Chinese)
18. Liang, W.: Research on service quality improvement of third-party logistics based on blockchain technology. Logistics Technol. Appl. **02**, 101–102 (2020)
19. Saberi, S., Kouhizadeh, M.: Blockchain technology and its relationships to sustainable supply chain management. Int. J. Prod. Res. **57**, 7–8 (2019)
20. Issaoui, Y., Khiat, A., Bahnasse, A., Ouajji, H.: Smart logistics: study of the application of blockchain technology. Procedia Comput. Sci. **160**, 266–271 (2019)

21. Wang, Y.: Strategies and suggestions for coordinated development of cross-border e-commerce and cross-border logistics in China. Economics Management **05**, 14–15 (2019)
22. Zhu, G., Zhao, Y.: Application and development of blockchain technology in logistics. J. Suzhou Univ. Sci. Technol. (Eng. Technol.) **12**, 34–35 (2020). (in Chinese)

# Research on the Service Quality of JD Daojia's Logistics Distribution Based on Kano Model

Yajie Xu[1,3(✉)] and Xinshun Tong[1,2]

[1] Business School, Zhongyuan Institute of Science and Technology, Zhengzhou 45000, China
`verxyj@163.com`
[2] College of Economics and Management, Zhengzhou University of Light Industry, Zhengzhou 45000, China
[3] Henan Port Hub and Port Economic Research Center, Zhengzhou 45000, China

**Abstract.** As an important support for the development of new retail, logistics distribution is the closest logistics service format to consumers. Improving the quality of logistics distribution service, improving logistics efficiency and consumer satisfaction are the necessary conditions for the development of new retail. In order to effectively improve the service quality and customer satisfaction problems in the logistics and distribution of JD Daojia, based on Kano model, establish JD Daojia logistics service quality improvement model, determine the priority of the improvement of JD Daojia's service quality elements, and guide enterprises to achieve maximum customer satisfaction with minimum investment. Through empirical research, the customer's perceived importance and satisfaction with JD Daojia's service quality elements were analyzed, and the service quality factors with the highest priority for improvement were obtained, which verified the feasibility and effectiveness of the model, It can provide some practical guidance for improving the service quality of JD Daojia.

**Keywords:** Kano model · Service quality · JD Daojia

## 1 Introduction

According to the Chinese New Year Consumption Report released by JD Daojia, quality replaces price to become shopping consumption first choice factor, "Self-pleasing consumption" is leading the new trend of mass online shopping. According to the consumption data released by JD Daojia, In the past few years, "One-Hour Shopping" is officially moving towards all categories, all scenes, and all customers, and in addition to the first-tier cities also obtained rapid development, The GMV growth rate of second- and third-tier cities has exceeded that of first-tier cities. At the same time of its rapid development, Consumers' satisfaction with their service quality is becoming more and more important. Delivery service quality is an important indicator reflecting customer's perceived satisfaction. This paper is based on the Kano model, established JD Daojia logistics service quality improvement model. And through empirical research, analyze customers' perceived importance and satisfaction with the elements of JD Daojia's logistics service quality, obtain the most prioritized stream service quality elements for improvement, it provides certain decision basis for enterprise logistics management.

## 2 The Connotation of JD Daojia and Kano Model

### 2.1 JD Daojia

JD Daojia is a brand-new business model developed by JD Group based on the traditional B2C business model to the higher frequency and sub-commodity service field. It is the O2O life service platform that JD.com focused on in 2015 and is an important improvement based on the traditional B2C model to the high frequency field [1]. It is based on the advantages of JD logistics system and logistics management, and at the same time, relying on the "Internet+" technology to vigorously develop "crowd sourcing logistics" under the promotion of the popular sharing economy, integrate various O2O life categories, cooperate with Dada, and provide consumers The distribution of fresh food and supermarket products is quickly delivered within 2 h based on LBS positioning, creating an integrated application platform for life services.

Based on the change in consumption trends, in order to better meet consumer demand, JD Daojia proposed to promote the accelerated development of O2O instant consumption channels with "deep empowerment, deepening of channels, and deepening of products" as its core strategy. JD Daojia provides several types of home services, including supermarket home, takeout home, quality life, door-to-door service, and healthy home, etc. JD Daojia forms the prototype of "social e-commerce" in China through leveraging social resources [2]. In terms of commodity and service sources, JD Daojai provides logistics delivery through cooperation with large shopping malls. In terms of life services, JD Daojia can provide thousands of products and services such as supermarket products, fresh food, and takeout; In terms of the logistics distribution system of JD Daojia, there are not only JD's self-operated logistics distribution, but also social logistics such as "JD Crowdsourcing" [3]. The following figure shows the information index of JD Daojia in the past three months, as shown in Fig. 1.

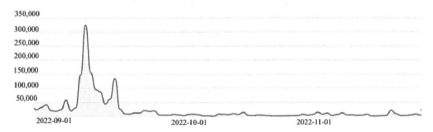

**Fig. 1.** JD Daojia Information Index

### 2.2 Kano Model

In 1974, Herzberg put forward the two-factor theory when researching employee satisfaction. He criticized the traditional concept that "satisfaction and dissatisfaction are completely opposite". He believed that even if the dissatisfaction factor is removed, employees will Not necessarily satisfied, and vice versa. Inspired by the two-factor theory, Noriaki Kano proposed the Kano demand model. As shown in Fig. 2, the abscissa means the quality characteristics of the product relative to the degree of satisfaction of consumers [4]. The more to the right, the better the quality characteristics of the product can meet the expectations of consumers; The ordinate represents the level of consumer satisfaction, and the higher it is, the higher the consumer's satisfaction with the product [5].

Charm attribute belongs to the attribute of "icing on the cake", which means that if the product provides this quality feature, the user's feeling can get "surprise" and "pleasant" experience, and satisfaction will increase; However, if this quality feature is not provided, customers will not be disappointed and sad, they will feel "taken for granted" and will not reduce their satisfaction [6]. One-dimensional attributes are also called expected attributes, which means that if the product provides the quality characteristics, the user's satisfaction will be very high, otherwise it will be very low. The necessary attributes are "must have" attributes. When the quality characteristics of the product are insufficient, it will greatly reduce the user's experience and satisfaction; But if you find ways to optimize and improve this quality feature, user satisfaction will not increase significantly; The indifference attribute belongs to the attribute of "optional", which means that no matter whether the product has the quality characteristic, the user's feeling is not too strong, and the user satisfaction will not be affected by it [7]. The reverse attribute is a "backfire" attribute, which means that users have no demand for this quality characteristic. If the product provides this quality characteristic, it will make customers "very dislike" and "boring" and reduce user satisfaction [8].

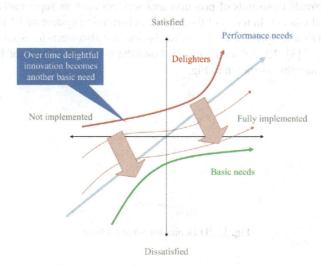

**Fig. 2.** Kano two-dimensional attribute model

The Kano model is a useful tool to prioritize the classification of user needs. It is based on analyzing the impact of user needs on user satisfaction, and reflects the nonlinear relationship between product performance and user satisfaction [9]. In the Kano model, "M" means necessary demand, "A" means attractive demand, "O" means one-dimensional (expected) demand, "R" means reverse demand, "I" means indifferent demand, and "Q" Means that the surveyor's answers are contradictory and are marked as "questionable answers". See Table 1.

**Table 1.** Classification of Kano model evaluation results

|  |  | Not providing service quality elements (negative problem) |  |  |  |  |
|---|---|---|---|---|---|---|
|  |  | I like this | It must be like this | I do not mind | I can stand it | I hate that |
| Provide service quality elements (positive issues) | I like this | Q | A | A | A | O |
|  | It must be like this | R | I | I | I | M |
|  | I do not mind | R | I | I | I | M |
|  | I can stand it | R | I | I | I | M |
|  | I hate that | R | R | R | R | Q |

## 3 JD Daojia Service Quality Improvement Model Based on Kano Model

### 3.1 Questionnaire Design and Data Collection

Firstly, design the questionnaire of the Kano model. For each service, the service quality elements of JD Daojia are shown in Table 2.

**Table 2.** JD Daojia's logistics service quality indicators

| Number | Content of service quality index |
|---|---|
| 1 | Delivered on time |
| 2 | Goods in good condition |
| 3 | Convenient query |
| 4 | Wide distribution range |
| 5 | Night delivery |
| 6 | Personalized service |
| 7 | Promotional offers |
| 8 | Pick up offline |
| 9 | Packaging and Message |
| 10 | Information push |

The forward question and the reverse question are set separately from the two aspects of whether they are satisfied. The structure refers to the Likert five-dimensional scale, from "I like this", "It must be like this", "I do not mind", "I can stand it" to "I hate that". Applying it to the "JD Daojia Customer Satisfaction Survey", the Kano questionnaire design sample form shown in Fig. 3 can be obtained. Then, issue a questionnaire and conduct a survey.

1. What would you say if JD Daojia can deliver the goods within the promised time?
   ○ I like this  ○ It must be like this  ○ I do not mind  ○ I can stand it  ○ I hate that

2. What would you say if JD Daojia can't deliver the goods within the promised time?
   ○ I like this  ○ It must be like this  ○ I do not mind  ○ I can stand it  ○ I hate that

**Fig. 3.** Sample questionnaire

This questionnaire survey adopts the method of online survey, sending electronic questionnaires online to collect the survey results. A total of 108 questionnaires were distributed and 90 were recovered. The invalid data were sorted through the questionnaires. There were 66 valid questionnaires, and the effective recovery rate was 73%. The relevant data collected by the questionnaire is mainly analyzed through Excel and SPSS 25.

Collect data through fuzzy Kano questionnaires, and perform calculations and statistics, cross-analyze positive and negative questions, and get the result, for example, as shown in Table 3, thereby deriving the category of each service quality indicator.

**Table 3.** Example diagram of service indicator type judgment

| Delivered on time | | | Negative | | | | | | |
|---|---|---|---|---|---|---|---|---|---|
| | | | I like this | It must be like this | I do not mind | I can stand it | I hate that | SUM | 66 |
| Positive | I like this | Count | 0 | 0 | 4 | 19 | 19 | | |
| | | Percentage (%) | 0.00 | 0.00 | 9.52 | 45.24 | 45.24 | A | 34.85% |
| | It must be like this | Count | 0 | 0 | 5 | 2 | 11 | O | 28.79% |
| | | Percentage (%) | 0.00 | 0.00 | 27.78 | 11.11 | 61.11 | M | 16.67% |

(*continued*)

## Table 3. (continued)

| Delivered on time | | | Negative | | | | | | |
|---|---|---|---|---|---|---|---|---|---|
| | | | I like this | It must be like this | I do not mind | I can stand it | I hate that | SUM | 66 |
| | I do not mind | Count | 0 | 0 | 3 | 3 | 0 | I | 19.70% |
| | | Percentage (%) | 0.00 | 0.00 | 50.00 | 50.00 | 0.00 | R | 0.00% |
| | I can stand it | Count | 0 | 0 | 0 | 0 | 0 | Q | 0.00% |
| | | Percentage (%) | 0.00 | 0.00 | 0.00 | 0.00 | 0.00 | | |
| | I hate that | Count | 0 | 0 | 0 | 0 | 0 | Better/SI | 0.63 |
| | | Percentage (%) | 0.00 | 0.00 | 0.00 | 0.00 | 0.00 | Worse/DSI | −0.45 |

### 3.2 Data processing of Better-Worse coefficient

According to the questionnaire survey made by the Kano model, the data is statistically sorted and divided to complete the demand classification for quality characteristics, and the next step is the Kano model analysis. The Kano model analysis is to determine the sensitivity of consumers to changes in the level of these service quality factors through the analysis of the satisfaction and dissatisfaction influence of various service quality factors, then determine the key factors for improving those service quality characteristics that are highly sensitive and more conducive to improving consumer satisfaction. According to the quality improvement coefficient theory proposed by Matzler (1998), the increase in satisfaction coefficient SI (Better coefficient) and the decrease in dissatisfaction coefficient DSI (Worse coefficient)—Better-Worse coefficient are calculated to show that the attribute of achieving this factor increases Satisfaction or elimination of the impact of dissatisfaction, Better's value is usually positive, indicating that if the product provides a certain function or service, user satisfaction will increase [10].

The formula is as follows:
Increase the coefficient of satisfaction SI

$$SI(better\ coefficient) = (A + 0)/(A + 0 + M + I),$$

Reduce the coefficient of dissatisfaction DSI

$$DSI(worse\ coefficient) = -1 * (0 + M)/(A + 0 + M + I)$$

The Kano model method is used to summarize and summarize the questionnaire. According to the relationship between different types of quality characteristics and customer satisfaction, the quality characteristics of products and services are divided into five categories: basic demand, expected demand, exciting demand, and no Differential demand, reverse demand. The letters appearing in the Kano model satisfaction table represent the satisfaction between JD Daojia logistics services and customers, and the specific types are judged based on the data obtained, as shown in Table 4.

Table 4. Kano model attribute type judgment

| number | Service quality factors | A(%) | O(%) | M(%) | I(%) | R(%) | Q(%) | Type |
|---|---|---|---|---|---|---|---|---|
| 1 | Delivered on time | 34.85 | 28.79 | 16.67 | 19.70 | 0.00 | 0.00 | A |
| 2 | Goods in good condition | 13.64 | 39.39 | 24.24 | 22.73 | 0.00 | 0.00 | O |
| 3 | Convenient query | 25.76 | 34.85 | 15.15 | 24.24 | 0.00 | 0.00 | O |
| 4 | Wide distribution range | 43.94 | 30.30 | 1.52 | 19.70 | 4.55 | 0.00 | A |
| 5 | Night delivery | 45.45 | 3.03 | 1.52 | 42.42 | 6.06 | 1.52 | A |
| 6 | Personalized service | 57.58 | 7.58 | 4.55 | 27.27 | 3.03 | 0.00 | A |
| 7 | Promotional offers | 53.03 | 18.18 | 4.55 | 21.21 | 1.52 | 1.52 | A |
| 8 | Pick up offline | 31.82 | 10.61 | 7.58 | 43.94 | 6.06 | 0.00 | I |
| 9 | Packaging and Message | 51.52 | 6.06 | 3.03 | 39.39 | 0.00 | 0.00 | A |
| 10 | Information push | 21.21 | 3.03 | 1.52 | 59.09 | 15.15 | 0.00 | I |

Calculate the Better-Worse coefficient to show the degree of influence of achieving this factor attribute on increasing satisfaction or eliminating dissatisfaction. The value of Better is usually positive, which means that if JD Daojia provides a certain service, consumer satisfaction will increase [11]. The greater the positive value, the stronger the effect of improving consumer satisfaction and the faster the increase in satisfaction; The value of worse is usually negative, which means that if JD Daojia does not provide a certain service, consumer satisfaction will decrease. The larger the negative value, the stronger the effect of reducing consumer satisfaction, and the faster the decrease in satisfaction. According to the classification results of service factor attributes in Table 3, based on the Kano model, JD Daojia's logistics distribution service types mainly include attractive attributes (A), expected attributes (O), and indifference attributes (I). Therefore, according to the Better-Worse coefficient, projects with higher absolute scores should be implemented first. As shown in Table 5.

### 3.3 Data Analysis and Conclusion

In Kano model, the Better-Worse matrix mainly refers to the four-quadrant matrix graph based on the Better coefficient and the Worse coefficient. Through Kano research and analysis, according to the Better-Worse coefficient, a Better-Worse matrix is constructed as shown in Fig. 4. The Better-Worse matrix diagram more intuitively shows the classification and importance of JD Daojia's logistics and distribution services. It can be seen that this data post-processing method is different from the two-dimensional attribute scale.

Research on the Service Quality of JD Daojia's Logistics Distribution 543

Table 5. Better-Worse coefficient analysis

| Number | Service quality factors | Better(SI) coefficient(%) | Worse(DSI) coefficient(%) |
|---|---|---|---|
| 1 | Delivered on time | 63.64 | 45.45 |
| 2 | Goods in good condition | 53.03 | 63.64 |
| 3 | Convenient query | 60.61 | 50.00 |
| 4 | Wide distribution range | 77.78 | 33.33 |
| 5 | Night delivery | 52.46 | 4.92 |
| 6 | Personalized service | 67.19 | 12.50 |
| 7 | Promotional offers | 73.44 | 23.44 |
| 8 | Pick up offline | 45.16 | 19.35 |
| 9 | Packaging and Message | 57.58 | 9.09 |
| 10 | Information push | 28.57 | 5.36 |

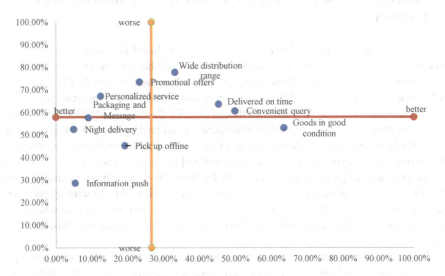

Fig. 4. Better-Worse matrix

The first quadrant represents the better coefficient value is high, and the absolute value of the worst coefficient is also high. The factors that fall into this quadrant are called expectation factors (one-dimensional factors). The wide range of delivery, on-time delivery, and convenient inquiries fall into this quadrant, which means that JD Daojia provides this service and consumer satisfaction will increase. When this function is not provided, customer satisfaction will decrease;

The second quadrant represents the situation where the better coefficient is high and the absolute value of the worst coefficient is low. The factors that fall into this quadrant are called glamour factors. Packaging and message services, personalized services and promotional services fall into this quadrant, which means that this service is not provided, and consumer satisfaction will not decrease. When JD Daojia When this service is provided, consumer satisfaction will be greatly improved;

The third quadrant represents the better coefficient value is low, and the absolute value of the worst coefficient is also low. The factors that fall into this quadrant are called indifferent factors. Night distribution, offline pick-up and information push services fall into this quadrant, that is, whether these services are provided or not, consumer satisfaction will not change, these services are services that consumers don't care about;

The fourth quadrant represents the situation where the better coefficient is low and the absolute value of the worst coefficient is high. The factors that the good product service elements fall into this quadrant are called essential factors, which means that when JD Daojia provides this service, user satisfaction will not increase, and when this function is not provided, user satisfaction will be greatly reduced; Services that fall into this quadrant are the most basic delivery services.

## 4 Simulation Case JD Daojia's Service Quality Improvement Strategy

According to the data analysis structure, it can be concluded that in the actual service process, JD Daojia should first try its best to meet the most basic needs of consumers, that is, the intact factors of goods. These needs are what consumers believe JD Daojia has the obligation to do. This type of delivery service quality factor is also the main factor that causes customers to be dissatisfied with the service. JD Daojia should rationally allocate resources, make more perfect product classification and packaging, and use green environmentally friendly materials to package products to increase consumer awareness of environmental protection and achieve regeneration [12]. Utilize resources and promote the development of green logistics. Meet the basic needs of consumers, reduce the rate of customer complaints, and improve customer satisfaction. After realizing the most basic needs, JD Daojia should try its best to meet the expectations of customers, that is, the expectation factor expressed in the first quadrant, which is a competitive factor of service quality.

Provide consumers with additional service functions that make their services superior to and different from competitors, and guide consumers to strengthen their good impression of JD Daojia. Finally, we strive to realize the glamorous needs of consumers, increase customer stickiness, which is the glamour factor represented by the second quadrant, and enhance consumer loyalty [13]. The consumption level is upgraded, the pace of life is accelerated, and convenient services can improve customer satisfaction.

Therefore, the Better-Worse coefficient value calculated according to the Kano model shows that JD Daojia needs to optimize the wide range of delivery, on-time delivery service, and convenient query services, and then satisfy consumers for packaging and message services, personalized services and promotional offers Service demand, providing such new retail real-time logistics services is the key to improving customer satisfaction. JD Daojia can improve the corresponding distribution area division according to the actual situation in order to tap potential customers and expand the market. Carry out merchandise promotion and full discount activities, rationally use the membership system and coupons to attract users, and improve the market competitiveness of JD Daojia [14]. With the help of big data, artificial intelligence, 5G and other information technologies, we will improve the digital level of distribution services, rationalize the delivery time, improve the convenience of order inquiry, improve the online service platform, expand personalized business, and create high-quality logistics and distribution services.

However, for consumers, night service, offline self-service and information push service, there is no difference in demand. Offline self-collection has less competitive advantages when facing retail stores in residential areas, and information push services are dispensable for online shopping, which is a highly purposeful shopping method. Therefore, when resources are limited under circumstances, there is no need to make great efforts to achieve it. Through the establishment of logistics distribution service quality improvement model, JD Daojia can set different weights for different service attributes in the process of improving logistics service quality, concentrate resources to develop customer-focused service attributes, and maximize customer satisfaction.

## 5 Conclusion

Further improve the quality of JD Daojia home logistics and distribution services, and give priority to improving the important quality factors that affect consumer satisfaction. This paper uses the Kano model method and based on the data obtained from the questionnaire survey to establish a JD Daojia logistics distribution service quality improvement model to obtain the priority of the improvement of the delivery service quality factors, and obtains the distribution service quality elements of JD Daojia, which provides certain practical guidance for improving the service quality of JD Daojia logistics.

Nowadays, in real life, there are more complicated and uncertain factors that affect the quality of JD Daojia's logistics and distribution services. This article is just a few attempts to improve the quality of JD Daojia's logistics and distribution services, and some preliminary results have been obtained. Therefore, Further research on the influence of JD Daojia's delivery service quality is of practical significance to the study of factors that new retail enterprises improve the quality of logistics services. Under the conditions of continuous advancement of policies, continuous advancement of technology, and continuous growth of consumer demand, the integration of online and offline, new retail online platforms such as JD Daojia will usher in a golden period of development, and will better satisfy people's consumption diversified demands on the Internet and bring better user experience.

## References

1. Chen, K., Jin, J., Luo, J.: Big consumer opinion data understanding for Kano categorization in new product development. J. Ambient. Intell. Humaniz. Comput. **13**(4), 2269–2288 (2021). https://doi.org/10.1007/s12652-021-02985-5
2. 2019 China Real-time Logistics Industry Research Report [DB/OL]. http://report.iresearch.cn/report_pdf.aspx?id=3415
3. kano model[DB/OL]. https://en.wikipedia.org/wiki/Kano_model
4. Cho, J.H., Kim, B.S.: Determining via the Kano Model the Importance of Quality Characteristics in Construction Management. KSCE J. Civ. Eng. **2022**(26), 2555–2566 (2022)

5. Chen, W.-K., Chang, J.-R., Chen, L.-S., Hsu, R.-Y.: Using refined kano model and decision trees to discover learners' needs for teaching videos. Multimedia Tools and Applications 81(6), 8317–8347 (2022). https://doi.org/10.1007/s11042-021-11744-9
6. Ponnam, A., Sahoo, D., Balaji, M.: Satisfaction-based segmentation: Application of Kano model in Indian fast food industry. J. Target Meas Anal. Mark 2011(19), 195–205 (2011)
7. Florez-Lopez, R., Ramon-Jeronimo, J.M.: Managing logistics customer service under uncertainty: An integrative fuzzy Kano framework. Inf. Sci. 2012(202), 41–57 (2012)
8. Shi, Y., Peng, Q.: Enhanced customer requirement classification for product design using big data and improved Kano model. Adv. Eng. Inf. 2021(49), 101340 (2021)
9. Choudhury, D.K., Gulati, U.: Product attributes based on customer's perception and their effect on customer satisfaction: the Kano analysis of mobile brands. Decision 47(1), 49–60 (2020). https://doi.org/10.1007/s40622-020-00233-x
10. Chen, M.-C., Hsu, C.-L., Lee, L.-H.: Investigating pharmaceutical logistics service quality with refined Kano's model. J. Retailing Consumer Serv. 2020(57), 102231 (2020)
11. Suh, Y., Woo, C., Koh, J., Jeon, J.: Analysing the satisfaction of university–industry cooperation efforts based on the Kano model: a Korean case. Technol. Forecasting Soc. Change 2019(148), 119740 (2019)
12. Gulc, A.: Models and Methods of Measuring the Quality of Logistic Service. Procedia Eng. 2017(182), 255–264 (2017)
13. Zheng, B., Wang, H., Golmohammadi, A.-M., Goli, A.: Impacts of logistics service quality and energy service of Business to Consumer (B2C) online retailing on customer loyalty in a circular economy. Sustainable Energy Technol. Assessments 2022(52), 102333 (2022)
14. Do, Q.H., Kim, T.Y., Wang, X.: Effects of logistics service quality and price fairness on customer repurchase intention: the moderating role of cross-border e-commerce experiences. J. Retailing Consumer Serv. 2023(70), 103165 (2023)

# CiteSpace Based Analysis of the Development Status, Research Hotspots and Trends of Rural E-Commerce in China

Yunyue Wu[✉]

School of Transportation, Nanning University, Nanning 530200, China
W17736621947@163.com

**Abstract.** Rural e-commerce is an important way to promote the circulation of rural economy and increase the income of rural population, and it is also a new important economic growth point in China. This paper uses Citespace 6.1 software and CNKI database as the literature source to retrieve 471 core journals and CSSCI journal papers related to rural e-commerce research in China. It analyzes the time distribution, authors, spatial distribution, keywords, etc. and forms a knowledge map, so as to obtain the research status and research trends in rural e-commerce in China. The following results are obtained through analysis: China's rural e-commerce research is on the rise, and the current research focuses are rural revitalization, e-commerce, rural economy, targeted poverty alleviation and e-commerce poverty alleviation. In the future, the research focus will tend to farmers' income, coordinated development, rural development, trade circulation, informatization and some other aspects.

**Keywords:** Rural e-commerce · Rural vitalization · CiteSpace · Rural commercial circulation

## 1 Introduction

The vast countryside provides a large market for commercial circulation, and the development of e-commerce is no exception [1, 2]. Rural modernization cannot exist without digital construction, and the development of the digital economy requires digital rural development as well. Rural e-commerce development can drive agricultural modernization through informatization, transform the mode of rural economic growth, raise rural residents' income levels, and encourage migrant workers to return home for employment and entrepreneurship [3]. It is critical to improve the quality and efficiency of agricultural economic operations and to better understand the market's decisive role in resource allocation [4, 5]. Previous studies have fully recognized the significance of rural e-commerce development for China's economic and social development. However, the foundation for China's rural e-commerce development is very uneven, with the "digital divide" between urban and rural areas, the "last mile" of logistics, rural e-commerce talents, and other issues remaining unresolved [6, 7]. As a result, using the Citespace visualization tool and content analysis research methods, this paper combs the research literature of rural e-commerce in China in recent years and creates a visual chart, with the goal of analyzing the current research status and research hotspots of rural e-commerce and providing a reference for rural e-commerce research.

## 2 Research Design

### 2.1 Research Methods

CiteSpace software is used to conduct quantitative analysis on the reference data of rural e-commerce research in China. Starting from the number of annual papers published, the number of source journals and colleges, the number of documents cited, and the cooperative research of multiple authors, the research status of a selected topic is explored. Then, through keyword co-occurrence, clustering, highlighting and other analysis functions, the hot spots and development trends of rural e-commerce research in China are found.

### 2.2 Data Source

Aiming at the theme of rural e-commerce development in China, this research uses the CNKI database as a literature source and conducts quantitative analysis on the literature using literature retrieval. The following are the specific operations: In the CNKI database, choose the "Advanced Search" type and the "Subject" search. The search term is "Rural E-commerce," and the source of literature is academic journals. There is no time constraint. A total of 5239 journal papers were searched. Based on the quality of the literature, this study only selected the core journals of Peking University and CSSCI literature, removing irrelevant literature such as book reviews, news, enrollment brochures, and so on, to obtain 471 research papers.

## 3 Analysis of Research Status

### 3.1 Time Distribution Analysis

The search revealed that Song Jian, the first core journal paper published on rural e-commerce development, was Research on the Impact of Broadband. Potential Development of China's Rural E-commerce Logistics Market. According to statistics (Fig. 1), there were only one paper issued in 2014, and 62 in 2017. According to the trend, the number of documents issued in recent years has increased, which is consistent with the current domestic and international situation. The development of rural e-commerce has sparked heated debate in the academic community, particularly with the national development of the digital economy and the implementation of digital rural policies.

CiteSpace Based Analysis of the Development Status 549

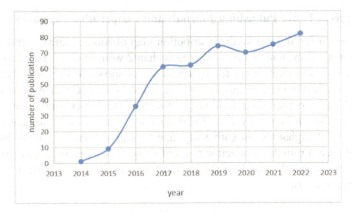

**Fig. 1.** Time distribution analysis of rural e-commerce in China

## 3.2 Author Statistics and Cooperative Network Analysis

CiteSpace is used to generate the author statistics and cooperation network diagram, select the authors whose threshold is greater than 2, and finally get the information of 29 authors, as shown in Fig. 2. It can be seen from the figure that Cao Lingling, Wang Yihuan and Nie Zhaoying have the largest number of articles, all of which are 4. At the same time, from the connection between the authors, Wang Yihuan and Nie Zhaoying were the two scholars who cooperated most closely.

**Fig. 2.** Author statistics and cooperative network analysis of rural e-commerce in China

## 3.3 Analysis of Paper Publishing Journals and Cooperation Networks

CiteSpace was used to generate the distribution map of source journals, and the paper publishing institutions with a threshold greater than 2 were selected, and the published papers of 27 organizations were finally obtained, as shown in Fig. 3. It can be seen from the figure that the largest number of papers published is China Agricultural University, with 9 papers published, followed by South China Agricultural University, with 5 papers published, followed by Yiwu Industrial and Commercial Vocational and Technical College, Shanghai Ocean University, and Shanghai Jianqiao University, with 3 papers published. From the connection point of view, the cooperation between publishing institutions is mostly based on regions, and universities in the same province and city cooperate closely.

**Fig. 3.** Cooperative network of publishing institutions of rural e-commerce papers in China

## 4 Research Hotspot and Trend Analysis

### 4.1 Keyword Co-occurrence Analysis

The keyword co-occurrence analysis method is a direct statistics of the currently published literature. It seeks the topics focused on by the current papers and reflects the focus and hot spots after the trend is formed [8]. Through the co-occurrence analysis of keywords in rural e-commerce research literature, Fig. 4 is obtained. "Rural e-commerce" has the highest frequency and the largest nodes, covering the whole process of literature research. However, the source of this paper is "rural e-commerce" as the search subject, so it will not be discussed in this paper. After excluding "rural e-commerce", the top five keywords are selected for ranking analysis according to the two indicators of keyword frequency and centrality, as shown in Table 1 and Table 2.

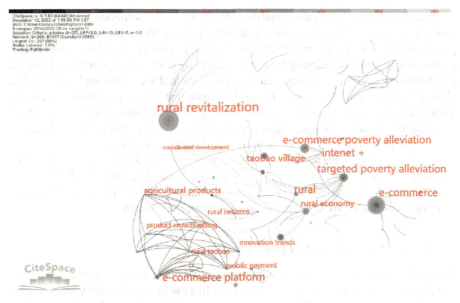

**Fig. 4.** Key words of rural e-commerce research in China

**Table 1.** High frequency keywords of rural e-commerce research in China

| Serial number | Hot words | Frequency of occurrence | Middle coefficient |
|---|---|---|---|
| 1 | rural revitalization | 57 | 0.26 |
| 2 | e-commerce | 37 | 0.19 |
| 3 | rural economy | 21 | 0.16 |
| 4 | targeted poverty alleviation | 17 | 0.41 |
| 5 | e-commerce poverty alleviation | 16 | 0.06 |

**Table 2.** Key words in the research of rural e-commerce in China

| Serial number | Hot words | Frequency of occurrence | Middle coefficient |
|---|---|---|---|
| 1 | intenet+ | 8 | 0.88 |
| 2 | rural development | 2 | 0.65 |
| 3 | taobao village | 14 | 0.64 |
| 4 | pattern | 3 | 0.62 |
| 5 | development | 4 | 0.42 |

It can be achieved from Table 1 that the high-frequency keywords are "rural revitalization", "e-commerce", "rural economy", "targeted poverty alleviation" and "e-commerce poverty alleviation". However, not all high-frequency keywords have high center coefficients, so we cannot judge research hotspots simply by keyword frequency [9]. From the perspective of center coefficient, "Internet plus", "Rural Development" and "Taobao Village" are the network centers of rural e-commerce. Rural e-commerce development needs to rely on the development and upgrading of Internet technology. Rural e-commerce development will drive rural development. One of the important forms of rural e-commerce is the construction of Taobao Village. The central coefficient of "Internet plus" is 0.88, with the highest degree of support for the network, followed by "rural development" with the central coefficient of 0.65, and the third "Taobao Village" with the central coefficient of 0.64. Therefore, "Internet plus", "Rural Development" and "Taobao Village" are hot spots in rural e-commerce research.

### 4.2 Keyword Emergence Analysis

Emergent words refer to keywords that increase suddenly or appear more frequently in a certain period. The research frontier of the discipline can be determined by analyzing the emergence of keywords [8]. Through keyword emergence analysis of rural e-commerce research field (Figs. 5 and 6), 26 emerging words were obtained: e-commerce platform, mode, rural shop, rural Internet users, rural Taobao, development, development status, development mode, targeted poverty alleviation, Taobao village, industrial evolution, development path, dynamic mechanism, joint distribution, e-commerce poverty alleviation, shared logistics, rural poverty alleviation, innovative development, influencing factors, three rural issues, countermeasures Farmers' income, coordinated development, rural development, trade circulation, and informatization. The research on rural e-commerce in China can be divided into three stages according to the emergence of key words.

(1) Stage 1

The first stage is from 2014 to 2016. The research on rural e-commerce in China is still at the initial stage, with a weak foundation. In the process of development, it faces many difficulties, such as farmers' low digital technology, inconsistent ideas with the development of rural e-commerce, and inadequate circulation channels, which restrict the development of rural e-commerce [10]. However, a lot of explorations have been carried out in various places. Most of these explorations are based on e-commerce platforms. For example, Alibaba, JD, Suning and other e-commerce platforms have their own rural e-commerce plans [11]. Rural Taobao has gradually developed into the most important form of rural e-commerce [12]. Taobao Village has become a typical case of rural e-commerce development mode, and the industry of "Taobao Village", It takes only a few years to complete the industrial process that traditionally takes hundreds of years [13].

(2) Stage 2

The second stage is from 2017 to 2019. With the deepening of targeted poverty alleviation policies, scholars have put forward a lot of suggestions on how rural e-commerce can promote rural revitalization and achieve targeted poverty alleviation, such as the government should specifically support featured e-commerce projects, rural e-commerce talent introduction strategies, improve the talent security mechanism, and create new e-commerce brands [14]. However, to achieve sustainable development, rural e-commerce needs to be driven by market demand, national policy guarantee, farmers' willingness, resource supply regeneration and other forces [15]. However, rural e-commerce also faces the problem of distribution, especially in the context of sharing economy, how to optimize distribution will be the focus of future research [16].

(3) Stage 3

The third stage is from 2020 till now. In 2020, China will build a moderately prosperous society in an all-round way, and rural development will focus on rural revitalization. At this stage, rural e-commerce focuses on issues such as farmers' income, coordinated development, rural development, trade circulation, and informatization. Rural e-commerce has had a positive impact on improving farmers' income, driving rural employment, attracting e-commerce entrepreneurship, reducing local unsalable agricultural products, targeted poverty alleviation and other aspects [17], maintaining the stable growth of farmers' income, which is conducive to promoting the development of rural e-commerce [18]. The development of rural e-commerce can not be separated from the development of rural industries, and more importantly, it needs to be coordinated with the development of rural commercial infrastructure and rural consumption structure [1+]. In the context of rapid development of rural e-commerce, the development of rural e-commerce directly promotes the development of trade circulation industry. However, rural e-commerce can form a virtuous circle only when rural e-commerce enterprises,

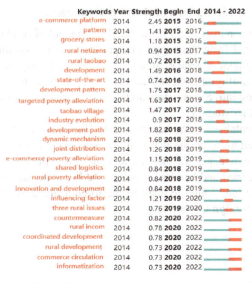

**Fig. 5.** Key words of rural e-commerce research in China

trade circulation industry and farmers are in a "win-win" situation [20, 21]. Especially in the context of the digital revolution, how to achieve industrial upgrading of rural e-commerce will be the focus of future research.

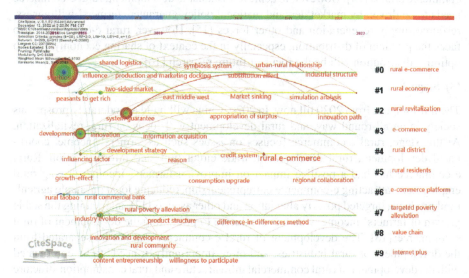

**Fig. 6.** Time zone chart of research keywords of rural e-commerce in China

### 4.3 Analysis Results and Development Strategies of Rural E-Commerce

#### 4.3.1 Analysis Results and Limitations of Rural E-Commerce

From the above analysis, it can also be seen that the development of rural e-commerce in China cannot be separated from the support of policies and the guarantee of systems. However, the development of rural e-commerce can not only rely on policies and systems, but the key is the ability of rural self-development. So, how to improve farmers' awareness of integrating into e-commerce and improve farmers' ability to develop e-commerce should be further studied.

This paper uses CNKI database as the core of Peking University and CSSCI journal paper data source. The limitations of the research are: first, the literature source does not include other databases and general papers, and the domestic literature has not been comprehensively screened and combed. Second, the literature only analyzed the domestic research situation, and failed to analyze the foreign literature, and failed to compare the domestic and foreign literature.

### 4.3.2 Important Problems to be Solved in Rural E-Commerce Development

There are two important problems that need to be solved in the process of rural e-commerce development.

(1) How to integrate small farmers into e-commerce

The integration of small farmers into e-commerce can not only timely obtain market demand, consumer feedback and other information, but also play a positive role in improving the economic income of small farmers and promoting the promotion of agricultural products. However, small farmers are limited by their skills and knowledge, so it is worth further studying how to integrate into e-commerce.

(2) Public service system of rural e-commerce

The development of rural e-commerce requires the construction of infrastructure and the provision of corresponding public services by the government, such as personnel training, comprehensive agricultural services, and rural e-commerce planning. In the process of vigorously developing rural e-commerce, the country should improve the public service system of rural e-commerce and create a good entrepreneurial environment.

## 5 Research Conclusions and Prospects

### 5.1 Research Conclusion

In this paper, Cite Space software is used to analyze and visualize the knowledge atlas and related data of literature generation on rural e-commerce of core journals and CSSCI journals in the CNKI database at different levels. The research draws the following conclusions:

(1) From the perspective of time distribution, the research on rural e-commerce in China started in 2014 and started in 2017, showing a continuous upward trend in recent years.
(2) From the perspective of spatial distribution, most of the researchers focus on rural e-commerce research in colleges and universities, but there is still a lack of research institutions and leaders with sufficient influence. Cooperation among research institutions is not close enough, especially cross regional cooperation is rare.
(3) From the perspective of keyword co-occurrence, "Internet plus", "rural development" and "Taobao Village" are the current research hotspots.
(4) From the perspective of keyword emergence, rural e-commerce research has shifted from e-commerce platforms, models, rural stores, rural Internet users, and rural Taobao to farmers' income, coordinated development, rural development, trade circulation, and informatization, which is consistent with the current social development trend.

### 5.2 Prospect of Future Research on Rural E-Commerce

With the rapid development of We Media, farmers can also become the protagonists of rural e-commerce. The development of rural e-commerce requires more talents who understand technology and business. Future research can be carried out in the following directions. First, the cultivation of rural e-commerce talents, including farmers' digital skills, rural e-commerce service talents, etc. Only by improving people's ability can rural e-commerce be better developed. Second, the development path of rural e-commerce, especially in the era of popular short videos, how to give full play to the main role of rural Internet users and farmer anchors will be the focus of relevant research.

**Acknowledgment.** This project is supported by 2022 Guangxi Vocational Education Teaching Reform Research Project "Effective Connection of Transportation and Logistics Major Group Undergraduate Program and Research and Practice of Curriculum System Construction" (GXGZZJG2022B181).

# References

1. Koniew, M.: Classification of the user's intent detection in e-commerce systems–survey and recommendations. Int. J. Inf. Eng. Electron. Bus. (IJIEEB) **12**(6), 1–12 (2020)
2. Haji, K.: E-commerce development in rural and remote areas of BRICS countries. J. Integrative Agric. **20**(4), 979–997 (2021)
3. Nurhuda, A., Khoirunnita, A., Rusli, A., et al.: Development E-commerce information system of agriculture in Samarinda. Int. J. Inf. Eng. Electron. Bus. (IJIEEB) **14**(6), 46–54 (2022)
4. Kshetri, N.: Rural e-commerce in developing countries. IT Professional **20**(2), 91–95 (2018)
5. Qin, N.: Quantitative research on China's rural e-commerce policy text – content analysis based on policy tools and business ecosystem. Econ. Syst. Reform **04**, 25–31 (2016). (in Chinese)
6. Lei, Z., Lei, H.: Does the development of e-commerce economy expand the income gap between urban and rural residents? Econ. Manage. Res. **38**(05), 3–13 (2017). (in Chinese)
7. Baako, I., Umar, S.: An integrated vulnerability assessment of electronic commerce websites. Int. J. Inf. Eng. Electron. Bus. (IJIEEB) **12**(5), 24–32 (2020)
8. Azam, A., Ahmed, A., Wang, H., et al.: Knowledge structure and research progress in wind power generation (WPG) from 2005 to 2020 using CiteSpace based scientometric analysis. J. Clean. Prod. **295**, 126496 (2021)
9. Yue, C., Liu, Z.: The rising map of scientific knowledge. Sci. Sci. Res. **02**, 149–154 (2005). (in Chinese)
10. Zhu, S.: The impact of rural E-commerce development on the logistics industry and the construction of rural logistics system. Price Monthly **03**, 75–78 (2016). (in Chinese)
11. Dong, K., Hou, W., Ding, H., Wang, P.: Research on innovation oriented rural E-commerce cluster development - analysis based on Suichang Model And Shaji Model. Agricultural Econ. Issues **37**(10), 60–69+111 (2016). (in Chinese)
12. Guo, C.: Analysis of rural E-commerce mode - based on the research of taobao village. Econ. Syst. Reform. **05**, 110–115 (2015). (in Chinese)
13. Liu, Y., Chu, X.: Research on the industrial evolution of China's "Taobao Village." China Soft Sci. **02**, 29–36 (2017). (in Chinese)
14. Lu, B.: Research on the dynamic mechanism and development path of rural e-commerce promoting rural revitalization. Agric. Econ. **2019**(12), 129–130 (2019). (in Chinese)
15. Li, X.: Discussion on the dynamic mechanism of the sustainable development of rural E-Commerce in the view of industrial Chain. Bus. Econ. Res. **2019**(02), 73–75 (2019). (in Chinese)

16. Song, L.: Research on rural E-Commerce joint distribution operation mode from the perspective of shared logistics. Bus. Econ. Res. **08**, 132–135 (2019). (in Chinese)
17. Yi, F., Sun, Y., Cai, Y.: Evaluation of the policy effect of the government on promoting rural e-commerce development - empirical research from the "comprehensive demonstration of E-commerce in rural areas." Nankai Econ. Res. **03**, 177–192 (2021). (in Chinese)
18. Tang, H., Li, S.: E-commerce, poverty alleviation and rural revitalization: role and path. J. Guangdong Univ. Finance Econ. **35**(06), 65–77 (2020). (in Chinese)
19. Liu, Y.: Research on the difference of the impact of rural E-commerce development on rural residents' consumption structure from the perspective of double cycle. Bus. Econ. Res. **09**, 64–68 (2021). (in Chinese)
20. Liu, Y.: Research on the price threshold effect of rural E-commerce development on the commercial circulation industry. Commercial Econ. Res. **01**, 25–29 (2022). (in Chinese)
21. Aparco, R.H., Del Carmen Delgado Lim, M., Tadeo1, F.T., et al.: Sustainability of rural agribusiness through e-commerce information systems. IOP Conf. Ser. Earth Environ. Sci. (968), 012002 (2022)

# The Optimization and Selection of Deppon Logistics Transportation Scheme Based on AHP

Long Zhang[1] and Zhengxie Li[2(✉)]

[1] School Logistics, Wuhan Technology and Business University, Wuhan 430065, China
[2] Philippine Christian University Center for International Education, 1004 Manila, Philippines
435840200@qq.com

**Abstract.** The selection and optimization of logistics transportation scheme is crucial to the operation of enterprises. Deppon Logistics Company is a leading enterprise in the logistics industry and occupies a pivotal position in the logistics industry. In recent years, Deppon Logistics performance is outstanding, but there are still many logistics transportation problems in the end of the distribution link, such as Poor timeliness of transportation, the express damaged, and cargo lost, restricting development of the company. In order to maintain the long-term stable development of enterprises, enterprises should start from the details, combing the existing transportation routes, planning optimize the logistics and transportation scheme, reduce logistics costs, improve logistics and transportation benefits. In this paper, the analytic hierarchy process is used to analyze the logistics transportation scheme of Deppon Logistics. Starting from the influential factors such as timeliness, economy and reliability of logistics transportation, the logistics transportation model is built to help Deppon Logistics find the most appropriate transportation scheme for the enterprise.

**Keywords:** Logistics transportation · AHP · Transportation scheme

## 1 Introduction

In the face of human society in the 21st century, e-commerce is developing rapidly and vigorously with the support of the Internet. The logistics industry closely related to e-commerce has deeply integrated into people's daily life and become an important pillar industry of the national economy. The definition of the term "logistics" in China is "The physical flow of goods from the place of supply to the place of receipt. According to the actual needs, the basic functions of transportation, storage, handling, packaging, circulation processing, distribution and information processing are organically combined [1]. "What people refer to as "logistics" in daily life is just the flow of goods, while the real "logistics" is a huge industry, including complicated links and specific processes, which profoundly affects and changes the production and life of modern people and occupies an important share in the gross product of the national economy.

## 1.1 Status Analysis of Deppon Logistics Transportation Scheme

Types of Deppon logistics transportation schemes, Deppon has logistics transportation schemes such as land transportation and air transportation, and its main services include the following main forms: precise air transportation, Precision Kahang, precise motor transport. Problems of Deppon logistics transportation, Unreasonable logistics and transportation handover links harm the interests of customers, In order to protect their own interests and avoid the responsibility for goods damage, some employees of Deppon repackaged the damaged goods and asked the consignee to sign for acceptance [2]. If the goods are found damaged after the customer signs, Deppon Logistics staff will shift the responsibility to the sender, which will disrupt the delivery link and damage the interests of the customer [3].

# 2 Deppon Logistics Transportation Plan Optimization and Selection Analysis

## 2.1 Introduction to the Analytic Hierarchy Process

T.L.saaty first proposed analytic hierarchy process (AHP) in the 1970s, which is a qualitative and quantitative analysis method. It divides key factors in a problem into hierarchical structures such as target layer, criterion layer and scheme layer for qualitative and quantitative comprehensive analysis [4]. Use the analytic hierarchy process to conduct in-depth analysis and research on things, draw the hierarchy chart of things, use the judgment matrix containing the key factors to calculate the hierarchy weight, repeatedly carry out hierarchical iteration, according to the evaluation results, and finally get the best decision plan or the highest target ranking, shown in Fig. 1.

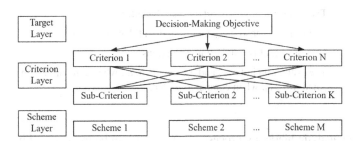

**Fig. 1.** Hierarchical structure model of analytic hierarchy process

The advantages of AHP are as follows:

(1) Systematic. When the analytic hierarchy process is used to solve the problem, the problem is often regarded as a whole and analyzed from the point of view of the system.
(2) Less information. According to the operation steps of analytic hierarchy process, the evaluation of a problem is not based on specific numbers, but on the importance standard identified by people, and converted into the calculation results of weight,

through this process to solve some problems that are difficult to express data and statistics.
(3) Simple and practical. Analytic hierarchy process has great advantages in processing uncertain and subjective information [5].

At the same time, AHP also has many disadvantages:

1) Unable to provide new solutions. Analytic hierarchy process can only find the optimal solution in the existing scheme.
2) Qualitative data contains subjective factors. Analytic hierarchy process (AHP) combines qualitative and quantitative methods to analyze data, which inevitably brings in subjective factors.
3) Over-emphasis on consistency testing. In the process of using the analytic hierarchy process, it is necessary to check the consistency of the matrix [6].

## 2.2 The Mathematical Model of Analytic Hierarchy Process

In the analytic hierarchy process, Table 1 is usually used to represent the weight comparison of two factors in the judgment matrix scaling table [7]. The nine-level scale table is shown in Table 1.

**Table 1.** 9-level scale table of judgment matrix.

| Relative ratios | Weight ratio meaning |
|---|---|
| 1 | Two evaluation indicators are equally important |
| 3 | The former is slightly more important than the latter |
| 5 | The former is significantly more important than the latter |
| 7 | The former is more important than the latter |
| 9 | The former is absolutely more important than the latter |
| 2, 4, 6, 8 | The median of two adjacent evaluations |
| Reciprocal | The evaluation index $a_i$ and $a_j$ is equal to $a_{ij}$ when compared with $a_j$, then $a_j$ is equal to $1 / a_{ij}$ when compared with $a_i$ |

Through analytic hierarchy process, the problem can be transformed into a matrix containing the listed factors:

$$A = \begin{bmatrix} a_{11} & \cdots & a_{1n} \\ \vdots & \ddots & \vdots \\ a_{n1} & \cdots & a_{nn} \end{bmatrix} \quad (1)$$

where n is the number of factors, and $a_{ij}$ is the weight ratio of $a_i$ and $a_j$ corresponding to this position. Then the sum product method is used to normalize the matrix A by column to get the matrix Z. According to the matrix Z, the relative weight ω of each factor can be obtained. Then, the consistency test of the matrix is carried out, and the formula is used to calculate the maximum eigenvalue λ:

$$\lambda max = \sum_{i=1}^{n} \frac{[A_\omega]i}{n\omega_i} \tag{2}$$

Consistency index calculation formula:

$$Ci = \frac{\lambda - n}{n - 1} \tag{3}$$

Consistency index calculation formula:

$$CR = \frac{C_i}{R_i} \tag{4}$$

where n is the order of matrix Z, $R_i$ is the random consistency index, which can be obtained by referring to the table, as shown in Table 2.

**Table 2.** Reference table of average random consistency index Ri

| n | 1 | 2 | 3 | 4 | 5 | 6 | 7 | 8 | 9 |
|---|---|---|---|---|---|---|---|---|---|
| $R_i$ | 0 | 0 | 0.52 | 0.9 | 1.12 | 1.26 | 1.36 | 1.41 | 1.4 |

When the calculated $C_R$ is less than *0.1*, it is considered that the judgment matrix has consistency, and the normalized feature vector can be used as the weight vector; otherwise, the matrix needs to be rebuilt. A matrix is constructed for the weight of each factor compared with each scheme to form the matrix $A_1$, $A_2$, $A_3$... The An. Using the above method to calculate the weight of each matrix $\omega_i$, check the consistency of the matrix, this process is called hierarchical single sorting. Finally, the consistency test of total hierarchy ranking is carried out to complete the analysis of the problem by AHP [8].

### 2.3 Analysis of Influencing Factors of Deppon Logistics Transportation Scheme

#### 2.3.1 Transport Route Selection Principle

The selection of transport path is to plan out one or more reasonable vehicle routes for all the places through [9]. Logistics transportation route selection is directly related to logistics cost and transportation convenience, which affects the transportation efficiency of logistics enterprises and customer service experience. According to the service characteristics of Deppon, it is concluded that the selection of logistics transportation routes should follow the four principles of timeliness, economy, stability and convenience [10].

(1) Timeliness. The timeliness of goods transportation is crucial for logistics enterprises. People have higher and higher requirements for service quality. The goods can not only be delivered, but also need to be delivered as soon as possible. Timeliness has become the focus of competition of logistics enterprises, speed is the advantage [11].

(2) Economy. Transportation cost has always been an important expenditure of logistics enterprises, accounting for a large proportion of the total cost of logistics enterprises. In the guarantee of service quality at the same time, as far as possible to reduce logistics transportation costs, to achieve the purpose of economic is always the goal of logistics enterprises [12].
(3) Stability. Stability refers to the probability of accidents in the process of logistics transportation. For example, on the conventional highway transport route, whether there is a high incidence of accidents on this route, which will cause traffic jams and affect the passing of vehicles; Whether the operation is closed during special periods or festivals, and the traffic is prohibited; Whether it will be renovated and expanded in the short term [13].
(4) Convenience. Convenience refers to whether the route is convenient for logistics vehicles to transport goods [14]. Logistics freight vehicles are mostly medium and large heavy trucks, with long body, high roof, heavy cargo, wide road, height and weight limits, which lead to difficult vehicles, and will not be able to meet the principle of convenient transportation. Convenient transportation also requires the choice of a scientific and reasonable route, does not deviate from any point of the driving route, avoid secondary loading and unloading transportation, control time and economic costs [15].

### 2.3.2 Transport Route Selection Principle

According to the transport route selection principle, the influencing factors required by the analytic hierarchy process (AHP) are selected, namely the criterion layer of the analytic hierarchy process (AHP) [16]. Combined with practical investigation and theoretical analysis, the initial influencing factors selected are shown in Table 3.

**Table 3.** Initial influencing factors of Deppon transport scheme selection

| Serial number | Influencing factors |
|---|---|
| 1 | The distance of the starting point and destination from the transport route |
| 2 | Vehicle operation and maintenance costs |
| 3 | Road smoothness |
| 4 | Distance of transport line |
| 5 | Cargo quantity |
| 6 | Type of goods and requirements |
| 7 | Consistency of transport |

### 2.3.3 Analysis of Influencing Factors

The selection of too many factors is not conducive to the analytic hierarchy process to solve the problem. Under normal circumstances, it is common to select 3–5 factors for evaluation. Therefore, the above initial influencing factors are summarized or screened, and the following four criteria are obtained [17].

(1) Transportation cost: refers to the total cost of logistics vehicles during the operation and maintenance period.
(2) Time spent: it refers to the total time spent by logistics vehicles from the transfer field to the destination for unloading the goods.
(3) Anti-risk ability: it refers to the ability of logistics enterprises to cope with emergencies and transport stability.
(4) Cargo loading rate: refers to the cargo loading rate of the vehicle, that is, the utilization rate of the carriage space [18].

## 2.4 Application of Analytic Hierarchy Process in Deppon Logistics Transportation Scheme Optimization

In the analytic hierarchy process, we usually use Table 1 to represent the weight comparison of two factors in the judgment matrix scaling table. The nine-level scale table is shown in Table 1.

(1) Use analytic hierarchy process (AHP) to establish hierarchy structure

Logistics enterprises should first consider the problem of profit, the second is to serve the public. Therefore, logistics enterprises need to control logistics costs, but also to grasp the transportation time. In order to obtain the optimal transport scheme, logistics enterprises should consider transport path, transport time, transport stability and other factors to comprehensively screen out and optimize the logistics transport scheme [19]. This paper has selected four evaluation indexes of transportation cost, time spent, risk resistance ability and cargo loading rate as evaluation indexes of Deppon logistics transportation scheme optimization. Combining with the structural model and evaluation index system of analytic hierarchy process, the hierarchical structure model of Deppon logistics is obtained as shown in Fig. 1.

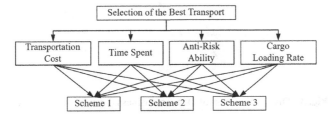

**Fig. 2.** Transport scheme selection hierarchy model

(2) Expert consultation method to determine the index weight

The weight reflects the importance of each factor in the system. This paper uses the expert consultation method to determine the weight. Generally let the industry experts score the indicators, according to the results decide the weight [20]. In Deppon logistics optimization problem, after consulting 13 authoritative managers in the logistics industry, according to their opinions, the 9-level scale table is listed, and the final integration is as follows: Table 4.

**Table 4.** Weight table of AHP criterion layer

| Index weight | Transportation cost | Time spent | Anti-risk ability | Cargo loading rate |
|---|---|---|---|---|
| Transportation cost | 1 | 1/3 | 2/3 | 4 |
| Time spent | 3 | 1 | 2 | 5 |
| Anti-risk ability | 2 | 1/2 | 1 | 3 |
| Cargo loading rate | 1/4 | 1/5 | 1/3 | 1 |

## 3 Solution and Application of Numerical Examples

### 3.1 General Situation of Logistics Transportation in a Certain Area of Wenzhou

Through the understanding of analytic hierarchy process, we will select an example to analyze its application. For example, there are three Deppon business offices (hereinafter collectively referred to as Part A,B and C) in L Town,C County, Wenzhou City. Their location distribution is shown in Fig. 3. The logistics shuttles to the three business departments are independent of each other, that is, they all have their own logistics shuttles.

**Fig. 3.** Distribution map of Deppon business department in 1L town

Due to the small scale and small cargo volume of Part A, in order to avoid the excessive empty load rate, it has been sharing a vehicle with another small department in the neighboring town, tentatively called Part D, and its geographical location is shown in Fig. 4.

**Fig. 4.** Location map of part D

**Fig. 5.** Bus route map

The starting point of the logistics shuttle bus is the transfer center of L Island, and its route is shown in Fig. 5.

From the figure, it can be clearly seen that the shuttle bus starts from L Island, drives into the highway, and then you can choose to go down the highway Speed, drive west to Part A first, then east to Part D after unloading the goods of Part A, or get off the highway, directly to Part D, and then to Part A. In this process, the company does not have hard and fast rules about which department to go to first, the department manager can communicate with the driver which department to go to first, but it is also restricted by the loading order of the transfer field, because the goods of the two departments will be loaded first and then, distributed in the front and back sections of the bus carriage, there will be no mixed loading, and the first arrival is more conducive to unloading.

### 3.2 Examples of Three Logistics Transportation Schemes

#### 3.2.1 Single Transport Scheme

The scheme has a single shuttle bus for the transport of Part A. After the change of scale, the cargo volume of Part A is maintained at about 1 to 1.5 carloads. Without transporting the goods of Part D, one carload still cannot be fully loaded, but the remaining quantity is not much. The plan is to ship only one truckload per day, and the rest of the goods will be shipped the next day, but the goods are overstocked. One temporary vehicle will be added every 3–5 days, and two truckloads will be shipped to Part A on another day. The feasibility of this scheme lies in the priority level of the goods.

### 3.2.2 Multi-vehicle Combined Transport Scheme

This plan is a temporary plan adopted by Deppon after the change of the department. It remains unchanged with the same car of Department D. If the car of the main department cannot be loaded, the shuttle bus of the other department will be loaded with additional part. As a result, there is an extra train and all the goods can be sent through the transfer yard to the terminal delivery department on the same day.

### 3.2.3 Multi-vehicle Single Transport Scheme

In this scheme, two buses are directly assigned to Part A, so that the goods cannot be loaded, and there is no need to share the bus with other departments to cause some potential risks, such as wrong unloading and leakage of the goods. In the time arrangement more freedom, do not have to consider the shuttle bus arrival time of other departments, the disadvantage is high transportation costs.

## 3.3 Comparative Analysis of Three Logistics Transportation Schemes

### 3.3.1 Transportation Cost Analysis

After the per-selection scheme is defined, the analytic hierarchy process is used to analyze the different schemes one by one. The first is the single transport scheme (hereinafter referred to as Scheme 1). From the perspective of transport cost, Scheme 1 uses A shuttle bus specifically for the transport of Part A, and then sends an extra car to clean up the surplus every once in a while. If the cost of a car is expressed as c, the transport cost of Scheme 1 is $(1 + 0.25)$ c.

Compared with scheme 1, multi-vehicle combined transport scheme (hereinafter referred to as Scheme 2) adopts the combined transport with Part D. The actual transport cost borne by Part A is 0.1c, excluding part of the cost of the shuttle bus going to Part D normally, and the transport cost is $(1 + 0.1c)$ with the addition of one additional bus in Part A.

Multi-vehicle single transport scheme (hereinafter referred to as Scheme 3) directly add two shuttle buses, the cost of which is 2c.

### 3.3.2 Take the Time to Analyze

In scenario 1, if only one bus arrives, the usual shuttle bus will leave at 9 am, take 1.5 h and arrive before 11 am, plus it takes an average of 1 h for the Courier to unload a load of goods, so the total time is about 2.5 h.

Scheme two due to the intervention of the car, it will inevitably produce time conflict, such as the first bus unloaded, the second bus has not arrived, the Courier must wait in the department, cannot start delivery in time, otherwise the shuttle bus arrived no delivery. Or two buses arrive at the same time, unloading busy. These situations do occur in real work. Therefore, in terms of time cost, even in the best case, the second bus will arrive immediately after the first bus is unloaded. In this way, apart from the 2.5 h required in scheme 1, it still needs 30 min for the second bus to unload (because the second bus is a combined bus, with only half or even less cargo), which takes about 3 h.

Similarly, Plan 3 and Plan 2 take similar time, but because the bus is not combined, the shuttle bus can stay in the department until the evening before departure, so the time spent is not as urgent as Plan 2, relatively better than plan 2, but the absolute time spent is still higher than plan 1.

### 3.3.3 Risk Resistance Analysis

Obviously, the weakest risk prevention is the first plan, which is prone to customer dissatisfaction due to not receiving goods on time. In the special period, when the goods increase abruptly, a large amount of goods will be piled up and cannot be transferred in time, resulting in loss of customer experience. Scheme 2 and Scheme 3 have strong anti-risk ability and can alleviate these problems to varying degrees. Scheme 3 has the strongest anti-risk ability because both vehicles belong to Part A with large freedom.

### 3.3.4 Cargo Loading Rate Analysis

Scheme one has only one car, which can be filled with the whole car every time, and there is no insufficient utilization of space. In scheme 2, the shuttle bus shared with Part D can also guarantee the full load, but another shuttle bus separately transporting the goods of Part A cannot be fully loaded, so there is a waste of space. Scheme three wastes more space.

## 4 Introduction of Calculation

### 4.1 Model

After sorting out the above analysis, the weight scoring table of each factor is obtained through pairwise comparison, as shown in Table 5, 6, 7 and 8.

**Table 5.** Comparison of the relative importance of transport costs

| A1 | Scheme 1 | Scheme 2 | Scheme 3 |
|---|---|---|---|
| Scheme 1 | 1 | 1/2 | 3 |
| Scheme 2 | 2 | 1 | 5 |
| Scheme 3 | 1/3 | 1/5 | 1 |

**Table 6.** Comparison of the relative importance of time spent

| A2 | Scheme 1 | Scheme 2 | Scheme 3 |
|---|---|---|---|
| Scheme 1 | 1 | 4 | 3 |
| Scheme 2 | 1/4 | 1 | 1/2 |
| Scheme 3 | 1/3 | 2 | 1 |

**Table 7.** Comparison of the relative importance of risk resistance

| A3 | Scheme 1 | Scheme 2 | Scheme 3 |
|---|---|---|---|
| Scheme 1 | 1 | 1/3 | 1/7 |
| Scheme 2 | 3 | 1 | 1/3 |
| Scheme 3 | 7 | 3 | 1 |

**Table 8.** Comparison of the relative importance of cargo loading rate

| A4 | Scheme 1 | Scheme 2 | Scheme 3 |
|---|---|---|---|
| Scheme 1 | 1 | 1/2 | 5 |
| Scheme 2 | 2 | 1 | 7 |
| Scheme 3 | 1/5 | 1/7 | 1 |

Firstly, the information obtained by expert scoring method is used to establish a judgment matrix for the pairwise comparison of transportation cost, time spent, risk resistance ability and cargo loading rate.

$$A = \begin{bmatrix} 1 & 1/3 & 2/3 & 4 \\ 3 & 1 & 2 & 5 \\ 2 & 1/2 & 1 & 3 \\ 1/4 & 1/5 & 1/3 & 1 \end{bmatrix} \quad (5)$$

And the weight judgment matrix between each factor:

$$A_1 = \begin{bmatrix} 1 & 1/2 & 3 \\ 2 & 1 & 5 \\ 1/3 & 1/5 & 1 \end{bmatrix} A_2 = \begin{bmatrix} 1 & 4 & 3 \\ 1/4 & 1 & 1/2 \\ 1/3 & 2 & 1 \end{bmatrix}$$

$$A_3 = \begin{bmatrix} 1 & 1/3 & 1/7 \\ 3 & 1 & 1/3 \\ 7 & 3 & 1 \end{bmatrix} A_4 = \begin{bmatrix} 1 & 1/2 & 5 \\ 2 & 1 & 7 \\ 1/5 & 1/7 & 1 \end{bmatrix} \quad (6)$$

## 4.2 Weight Calculation

First, calculate the maximum average value of the factor matrix of the criterion layer and verify the consistency of the matrix.

Step 1: Calculate the weight order corresponding to the pairwise comparison

$$A = \begin{bmatrix} 1 & 1/3 & 2/3 & 4 \\ 3 & 1 & 2 & 5 \\ 2 & 1/2 & 1 & 3 \\ 1/4 & 1/5 & 1/3 & 1 \end{bmatrix} \quad \omega = \begin{bmatrix} 0.1996 \\ 0.4641 \\ 0.2617 \\ 0.0747 \end{bmatrix} \rightarrow A_\omega = \begin{bmatrix} 0.8273 \\ 1.9594 \\ 1.1168 \\ 0.3046 \end{bmatrix} \quad (7)$$

Step 2: Calculate the maximum eigenvalue

$$\lambda max = \sum_{i=1}^{n} \frac{[A_\omega]i}{n\omega_i} = 4.1789 \quad (8)$$

Step 3: Calculate the consistency index $C_i$ and the consistency ratio $C_R$

$$Ci = \frac{\lambda - n}{n - 1} = (4.1789 - 4)/(4 - 1) = 0.0596 \quad (9)$$

$C_R = 0{,}0596/0.9 = 0.0662 < 0.1$, meet consistency check.

**Table 9.** Transport cost index comparison matrix, weight and consistency test table

| Transportation Cost | Scheme 1 | Scheme 2 | Scheme 3 | Weight |
|---|---|---|---|---|
| Scheme 1 | 1 | 1/2 | 3 | 0.3092 |
| Scheme 2 | 2 | 1 | 5 | 0.5813 |
| Scheme 3 | 1/3 | 1/5 | 1 | 0.1096 |
| Consistency Check | $\lambda = 3.0036$ $C_i = 0.0018$ $R_i = 0.52$ $C_R = 0.0035$ | | | Test Passed |

**Table 10.** Comparison matrix, weight and consistency test table of time spent index

| Time Spent | Scheme 1 | Scheme 2 | Scheme 3 | Weight |
|---|---|---|---|---|
| Scheme 1 | 1 | 4 | 3 | 0.6232 |
| Scheme 2 | 1/4 | 1 | 1/2 | 0.1373 |
| Scheme 3 | 1/3 | 2 | 1 | 0.2395 |
| Consistency Check | $\lambda = 3.0183$ $C_i = 0.0091$ $R_i = 0.52$ $C_R = 0.0176$ | | | Test Passed |

In the same way, the other items are checked for consistency, as shown in Table 9, 10, 11 and 12.

**Table 11.** Comparison matrix, weight and consistency test table of risk resistance indicators

| Anti-risk Ability | Scheme 1 | Scheme 2 | Scheme 3 | Weight |
|---|---|---|---|---|
| Scheme 1 | 1 | 1/3 | 1/7 | 0.0882 |
| Scheme 2 | 3 | 1 | 1/3 | 0.2431 |
| Scheme 3 | 7 | 3 | 1 | 0.6687 |
| Consistency Check | $\lambda = 3.0070$ $C_i = 0.0035$ $R_i = 0.52$ $C_R = 0.0067$ | | | Test Passed |

**Table 12.** Comparison matrix, weight and consistency check table of cargo loading rate indicators

| Cargo Loading Rate | Scheme 1 | Scheme 2 | Scheme 3 | Weight |
|---|---|---|---|---|
| Scheme 1 | 1 | 1/2 | 5 | 0.3338 |
| Scheme 2 | 2 | 1 | 7 | 0.5907 |
| Scheme 3 | 1/5 | 1/7 | 1 | 0.0755 |
| Consistency Check | $\lambda = 3.0141$ $C_i = 0.0070$ $R_i = 0.52$ $C_R = 0.0136$ | | | Test Passed |

**Table 13.** Calculation table of total hierarchical sorting

| Z | $\omega$ | Scheme 1 | Scheme 2 | Scheme 3 |
|---|---|---|---|---|
| Transportation Cost | 0.1996 | 0.3092 | 0.5813 | 0.1096 |
| Time Spent | 0.4641 | 0.6232 | 0.1373 | 0.2395 |
| Anti-risk Ability | 0.2617 | 0.0882 | 0.2431 | 0.6687 |
| Cargo Loading Rate | 0.0747 | 0.3338 | 0.5907 | 0.0755 |
| Final Weight | | 0.3989405 | 0.287432 | 0.31362756 |

**Table 14.** Consistency check of the total hierarchical order

| | $C_i$ | $\omega$ |
|---|---|---|
| A1 | 0.0018 | 0.1996 |
| A2 | 0.0092 | 0.4641 |
| A3 | 0.0035 | 0.2617 |
| A4 | 0.0071 | 0.0747 |
| Total Hierarchical Sort | $C_i = 0.0061$ $R_i = 0.52$ $C_R = 0.0116 < 0.1$ | |

## 4.3 Hierarchical Sorting and Order

The overall hierarchical ranking of the comparison of all influencing factors can be obtained as follows, shown in Table 13 and Table 14.

As can be seen from Table 13 above, Scheme 1 has the greatest weight, and through calculation in Table 14, its consistency test passes and meets the requirements.

In summary, this paper presents a solution to the selection of Deppon logistics terminal transportation scheme in Wenzhou by using analytic hierarchy process (AHP), and makes a specific analysis according to the special situation of the region. Finally, it is calculated that scheme 1 has the highest weight and is the preferred scheme among the current three schemes.

## 5 Conclusion

Logistics transportation scheme is of great importance to the development of logistics enterprises. Choosing the appropriate logistics transportation scheme is conducive to the logistics enterprises to reduce fixed asset investment, avoid operating risks, improve profits, reduce operating costs, and improve the competitiveness of enterprises.

This paper takes the AHP as the main tool, Deppon terminal transportation scheme as the target body, uses the AHP to determine the index weight, Combination of qualitative and quantitative analysis, builds a set of evaluation index system, combines the case analysis, provides a scientific basis for logistics enterprises to make decisions. However, due to the expert scoring method used in the selection of index weights, it is inevitable that there are subjective factors, which lead to inaccurate data analysis and affect the analysis results. The author will focus on how to improve and avoid the influence of human factors on the analysis results, so as to make the analysis results more accurate and convincing, provide more scientific basis for logistics enterprises to make decisions, and really help enterprises to reduce costs and increase efficiency.

## References

1. Li, W.: Logistics and Transportation Management, p. 218. Science Press, Beijing (2021)
2. Ke, W.: Research on the location of Ya Long Logistics Park based on AHP. Kunming University of Science and Technology, Kunming (2017)
3. Cheung, Y., et al.: Adaptive decision evaluation method based on analytic hierarchy process. J. Beijing Univ. Aeronautics Astronautics **9**, 79–80 (2016)
4. Bi Xin Hua et al.: Modern Logistics Management, p. 148. Science Press, Beijing (2021)
5. Lu Hua Wei: Study on transportation route optimization of international engineering bulk logistics. Dalian Maritime University, Dalian (2017)
6. Jin, Y.: Study on the production and countermeasures of logistics cost "Benefit reversal." Hebei Enterprise **4**, 14–15 (2016)
7. Gu, Q.: Application of Analytic Hierarchy Process in Supplier Selection of a Company. Shenzhen University, Shenzhen (2017)
8. Li, B.: Construction and implementation of logistics Business performance management system. Discuss Study **4**, 138–139 (2021)
9. Yaling, X.: Ahp-based Evaluation of Zhoushan River-sea Combined Transport Public Information Platform. Zhejiang Ocean University, Zhoushan (2017)

10. Le, W.: Research on Subject Service Evaluation of University Library Based on Analytic Hierarchy Process. Anhui University, Hefei (2017)
11. Aguezzoul, A.: Third-party logistics selection problem: a literature review on criteria and methods. Omega **16**(5), 24–34 (2014)
12. Huang, M., Cui, Y., Yang, S., et al.: Fourth party logistics routing problem with fuzzy duration time. Int. J. Prod. Econ. **145**(1), 2–3 (2013)
13. Liu, Y., Zhou, P., Li, L., Feng, Z.: Interactive decision-making method for third-party logistics provider selection under hybrid multi-criteria. Symmetry **11**(5), 11–13 (2020)
14. Yong, W., Jie, Z., Guan, X., et al.: Collaborative multiple centers fresh logistics distribution network optimization with resource sharing and temperature control constraints. Expert Syst. Appl. **17**(3), 165–166 (2021)
15. Fan, Z., Wang, L.: Evaluation of university scientific research ability based on the output of sci-tech papers: A D-AHP approach. PLoS ONE **17**(2), 3–4 (2017)
16. Fan, G., Zhong, D., Yan, F., Yue, P.: A hybrid fuzzy evaluation method for curtain grouting efficiency assessment based on an AHP method extended by D numbers. Expert Syst. Appl. **11**(3), 14–15 (2016)
17. Deng, X., Yong, H., Deng, Y., Mahadevan, S.: Supplier selection using AHP methodology extended by D numbers. Expert Syst. Appl. **17**(1), 8–9 (2014)
18. Rabinovich, E., Bailey, J.P.: Physical distribution service quality in Internet retailing: service pricing, transaction attributes, and firm attributes. J. Oper. Manage. **11**(6), 13–14 (2003)
19. Saaty, T.L.: Decision-making with the AHP: why is the principal eigenvector necessary. Eur. J. Oper. Res. **5**(1), 6–7 (2002)
20. Zhong, X., Wen, Z., Wei, L., Wenjie, X.: Analysis of influencing factors of cold chain logistics cost of dairy products. Ind. Eng. Innov. Manage. **2**(5), 3–4 (2022)

# Pharmacological and Non-pharmacological Intervention in Epidemic Prevention and Control: A Medical Perspective

Yanbing Xiong[1], Lijing Du[1,2,3]((✉)), Jing Wang[1], Ying Wang[4], Qi Cai[1], and Kevin Xiong[5]

[1] School of Safety Science and Emergency Management, Wuhan University of Technology, Wuhan 430070, China
dulijing@whut.edu.cn
[2] School of Management, Wuhan University of Technology, Wuhan 430072, China
[3] Research Institute of Digital Governance and Management Decision Innovation, Wuhan University of Technology, Wuhan 430072, China
[4] School of Business, Wuchang University of Technology, Wuhan 430223, China
[5] Information Technology Consulting Services, Ontario Limited, Markham, ON 1750351, Canada

**Abstract.** Since COVID-19 broke out in Wuhan, Hubei Province in 2020, COVID-19 has spread rapidly worldwide. Until today, the epidemic situation in some countries has not been effectively controlled. Therefore, the research on epidemic prevention and control has become a key research direction of global concern. At present, according to the classification and summary of this article, medical research on epidemic prevention and control measures can be divided into pharmacological intervention research and non-pharmacological intervention research. Researchers mainly study epidemic prevention and control from these two research directions. This paper briefly introduces the relevant research on epidemic prevention and control measures in medicine in the past five years, summarizes the current research status of epidemic prevention and control in medicine, classifies all papers, points out the shortcomings of existing research in the field of medicine, and provides future research ideas for researchers in the field of medicine. At the same time, draw some useful experience in operation management to help the government and enterprises improve production and supply chain management, so as to better promote economic recovery under the condition of ensuring epidemic prevention and control.

**Keywords:** COVID-19 · Epidemic prevention and control · Vaccine Intervention · Social restrictions · Isolation

## 1 Introduction

Since the outbreak of novel coronavirus in 2020, the death toll has been increasing rapidly. This public health emergency has caused varying degrees of damage to the economy, society and politics of various regions. Therefore, in order to reduce the negative impact of the epidemic on social and economic losses, it is necessary to intervene in the development and spread of the epidemic and prevent the further spread of the epidemic. A large number of researchers began to study the epidemic prevention and control measures in the medical field, from the aspects of pharmacological intervention and non-pharmacological intervention, to study the role of various vaccines and drugs, as well as other non-pharmacological measures for epidemic prevention and control. This paper collects and collates the research on epidemic prevention and control in the medical field in the past five years, summarizes the current research status of epidemic prevention and control, and provides reference for future research directions of epidemic prevention and control.

## 1.1 Selection of Journals

The time span of this review is nearly 5 years. The journals included in this part of the study are listed below in alphabetical order: Cell, Nature, Science, MECS, the Journal of American Medical Association, The Lancet, The New England journal of medicine.

In this study, for each journal and paper, there is at least one of the following keywords in the research text, which is considered as the potential research object of this review: "pandemic" "infection" "epidemic" "influenza" "flu" "epidemiological" "infectious" "contagious" "SARS" "H1N1" "MERS" "Ebola" "COVID-19" "cholera" "plague" "intervention" "control" "prevention" "mitigate" "mitigation" "mitigating" "evolve" "evolution". In the process of preliminary search and screening, 551 research papers with the above keywords were published in six journals. Each paper was reviewed by three authors, and after repeated screening, 94 related papers were finally obtained. Through specific reading of the papers, 72 research papers highly related to epidemic prevention and control were obtained, including 54 in the past five years.

## 1.2 Chronology of Growth in Epidemic Prevention and Control Research

Figure 1 shows the number of papers by journal and year. From this table, we can see that the journal with the largest number of papers is The Lancet (17), followed by The Journal of American Medical Association (10), Science (9), The New England journal of medicine (7), Nature (7) MECS (3) and Cell (1). The numbers in brackets indicate the number of papers.

In order to study the growth rate of research literature on epidemic prevention and control in the medical field, we also used a three-year moving average. Table 1 and Fig. 2 describe the three-year moving average of research literature on epidemic prevention and control. From the figure, we can see that the three-year moving average in 2021 is 8.00 (higher than any previous year), which is due to the publication of 21 papers in 2020. By 2022, the three-year moving average will continue to rise to 14.00, and this also takes into account that we only collected papers in the first half of 2022. These figures clearly show that researchers in the medical field are increasingly interested in epidemic prevention and control.

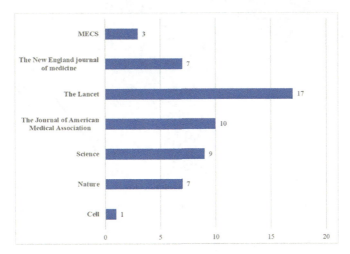

**Fig. 1.** Count of journals and papers

**Table 1.** Year and count of papers

| Year | Count | Three-year moving average |
|---|---|---|
| 2018 | 3 | 0.33 |
| 2019 | 0 | 1.00 |
| 2020 | 21 | 1.00 |
| 2021 | 23 | 8.00 |
| 2022 | 7 | 14.67 |
| Total | 54 | |

## 2 Pharmacological Intervention

Table 2 lists all pharmacological intervention papers by year of publication, disease type and control measures. From the perspective of disease types, since the outbreak of COVID-19 in 2020, almost all the papers related to epidemic prevention and control are about COVID-19. Only before 2020 have scholars studied other types of diseases, but few [1–3]. Specifically, pharmacological intervention can be divided into vaccines and other drugs.

The research of vaccine intervention mainly focuses on the effectiveness and safety of vaccine treatment for various diseases, which is verified by randomized controlled trials [4, 5], or based on data analysis [6–15]. Other pharmacological intervention studies are aimed at other drugs other than vaccines, and other drug studies are basically verified by means of controlled trials [16–30]. In addition, there are studies on the use of ECMO in the treatment of COVID-19 [31]. Through these studies, the effectiveness and safety of various vaccines and drugs against different diseases have been verified. Among them, the RECOVERY Collaborative Group represented by Peter W Horby published a

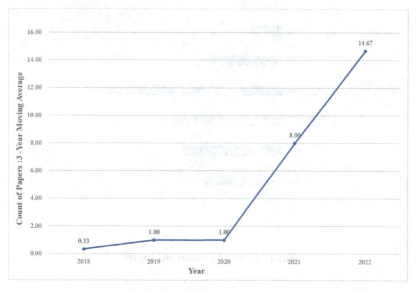

**Fig. 2.** Three-Year moving average

**Table 2.** Research related to pharmacological intervention

| Year | Disease Type | Control Measures | Authors |
|---|---|---|---|
| 2022 | COVID-19 | vaccination | Analía Rearte et al. (2022)<br>Makrufa Sh. Hajirahimova et al. (2022) |
|  |  | medicine | Peter W Horby et al. (2022)<br>Mark N Polizzotto et al. (2022)<br>Peter W Horby et al. (2022)<br>Eduardo Ramacciotti et al. (2022)<br>M.J. Levin et al. (2022) |
| 2021 | COVID-19 | vaccination | Caroline E. Wagner et al. (2021)<br>Raphael Sonabend et al. (2021)<br>Nathan D. Grubaugh et al. (2021)<br>Alasdair P S Munro et al. (2021)<br>Alice Cho et al. (2021)<br>M.G. Thompson et al. (2021)<br>Ioannis Katsoularis et al. (2021)<br>Ewen Callaway(2021) |
| 2021 | COVID-19 | medicine | Peter W Horby et al. (2021)<br>Ly-Mee Yu et al. (2021)<br>Christopher C Butler et al. (2021)<br>Peter W Horby et al. (2021)<br>Peter W Horby et al. (2021)<br>Daniel R. Kuritzkes (2021)<br>O. Mitjàet al. (2021)<br>Courtney Temple et al. (2021) |
|  |  | ECMO | Ryan P Barbaro et al. (2021) |

(*continued*)

**Table 2.** (*continued*)

| Year | Disease Type | Control Measures | Authors |
|---|---|---|---|
| 2020 | COVID-19 | vaccination | Saad B. Omer et al. (2020) |
|  |  |  | Eric J. Rubin et al. (2020) |
|  |  | medicine | Peter W Horby et al. (2020) |
|  |  |  | Mary Marovich et al. (2020) |
|  | Tuberculosis Infection |  | D. Ganmaa et al. (2020) |
| 2018 | Mosquito-borne diseases | vaccination | Neil M. Ferguson (2018) |
|  | M. tuberculosis Infection |  | E. Nemes et al. (2018) |

total of six papers related to it from 2020 to 2022, respectively verifying the effectiveness of aspirin, Tocilizumab, Azithromycin, Lopinavir ritonavir, convalescent plasma of COVID-19 patients, and Casivivimab and imdevimab in the treatment of COVID-19 [15, 17, 20, 23, 24, 28]. Moreover, the trials in Casivivimab and imdevimab proved for the first time that antiviral treatment can reduce the mortality of patients hospitalized for COVID-19.

Therefore, the current research of pharmacological intervention is mainly to evaluate the effectiveness of vaccines and drugs against the epidemic situation through control experiments or actual data, which is also the focus of this research. Through this research, we can find out the targeted vaccines and drugs as soon as possible to control the epidemic.

## 3 Non-pharmacological Intervention

Table 3 lists all non-pharmacological intervention papers by year of publication, disease type, control type and control measures. Like pharmacological intervention research, non-pharmacological intervention related research is also mainly focused on COVID-19. It is mainly concentrated in 2020, the year when COVID-19 just broke out. Before 2020, the research on this aspect was seriously insufficient. After 2020, due to the effective control of the epidemic, it gradually lost interest.

Of the 22 non-pharmacological intervention related studies we collected, 20 were about COVID-19, and only 2 were about other infectious diseases [32, 33]. The research on COVID-19 basically adopts the method of establishing mathematical models or analyzing data to evaluate the effect of various control measures [8, 34–52]. Only two studies have different directions, and study the detection and tracking of epidemic situation [53, 54]. In addition, according to the different functions of the control measures, we divide the control measures into the following four categories: isolation (case isolation and contact isolation), social restrictions (closing schools, entertainment places, prohibiting large gatherings, etc.), traffic restrictions (travel restrictions and traffic restrictions, etc.) and personal protection (wearing masks, washing hands, etc.).

The current research of non-pharmacological intervention is mainly to evaluate the impact of different control measures on the evolution of the epidemic by analyzing the measures taken in the epidemic area and the epidemic related data. This research needs to analyze the data generated by different control measures, but due to the particularity of the epidemic, researchers are difficult to control the epidemic area and implement a single measure, so it is impossible to evaluate the effect of a specific measure. Through this research, we can evaluate the effect of different prevention and control measures, find out the effective epidemic prevention and control measures, and inhibit the evolution of the epidemic.

**Table 3.** Research related to non-pharmacological intervention.

| Year | 2021 | | | | | 2020 | | | |
|---|---|---|---|---|---|---|---|---|---|
| Disease Type | COVID-19 | | | | | COVID-19 | | | |
| Control Type | Personal protection Social restrictions | Social restrictions | Social restrictions | Personal protection | Social restrictions Traffic restrictions | Quarantine Personal protection Social restrictions | Quarantine | Social restrictions | Traffic restrictions |
| Control Measures | staying at home, using masks, washing hands | maintaining the social distancing | Limit the number of parties, close high-risk enterprises, close schools | wearing masks | reducing community transmission, travel restrictions | Community isolation, travel restrictions, wearing masks | shelter in place | case isolation, contact tracing and quarantine, physical distancing, decontamination, hygiene measures. | Improve detection strength, pertinence and efficiency, make detection easier to obtain | Cancel large public activities Close schools, childcare facilities and unnecessary stores, Strict access ban | enhanced social distancing | travel restrictions |
| Authors | Akalu Abriham et al(2021) | Anurag Tatkare et al(2021) | Jan M. Brauner et al(2021) | Mark W. Tenforde et al (2021) John T. Brooks et al (2021) | Nathan D. Grubaugh et al (2021) | Matthew A. Crane et al (2021) | Rochelle P. Walensky et al (2020) | Luca Ferretti et al (2020) | Giulia Pullano et al (2020) | Jonas Dehning (2020) | Patrick G. T. Walker et al (2020) | Moritz U. G. Kraemer et al (2020) |

| Year | 2021 | | 2020 | | | | | | | 2018 |
|---|---|---|---|---|---|---|---|---|---|---|
| Disease Type | COVID-19 | | COVID-19 | | | | | | infection | Cholera |
| Control Type | Personal protection | Traffic restrictions Social restrictions | Social restrictions Personal protection | Quarantine Social restrictions Traffic restrictions | Quarantine Social restrictions Traffic restrictions | | | Personal protection Social restrictions | Personal protection |
| Control Measures | wearing masks | Close schools, supermarkets, travel restrictions | Isolate cases, suspend traffic, close schools and, entertainment places, check the floating population. | Personal protective, Case isolation | Community isolation, wearing masks, shutting down unnecessary community services. | Isolation of cases and contacts, travel restrictions, closure of schools and workplaces, cancellation of meetings, hand washing | intensive intracity and intercity traffic restriction, social distancing measures, home isolation and centralized quarantine, improvement of medical resources | Social alienation, border closure, school closure, isolation of cases and contacts, population blockade, travel restrictions. | isolation of confirmed and suspected cases, suspension of intra- and intercity travel, keep social distance | hand hygiene and staying away from others | safe water, appropriate sanitation, hygiene (WaSH) oral cholera vaccines (OCVs) |
| Authors | Andrea M. Lerner et al (2020) | Darlan S. Candido et al (2020) | Huaiyu Tian et al (2020) | Stephen M. Parodi et al (2020) | Kathy Leung et al (2020) | Shengjie Lai et al (2020) | An Pan et al (2020) | Seth Flaxman et al (2020) | Kristina Campbell (2020) | Justin Lessler et al (2018) |

## 4 Limitations and Directions for Future Research

The above research also has some limitations. First of all, because the data used by researchers are mainly provided by the government or relevant departments, some data may inevitably be lost or distorted due to the severity of the epidemic or the effectiveness of statistical capacity, leading to deviation of research conclusions. Secondly, the research of prevention and control measures needs to be based on the measures implemented by the government. Most governments take multiple measures to jointly implement the epidemic situation. The data obtained by researchers are generated by the joint implementation of multiple measures. It is difficult to obtain the data of the implementation of a single measure.

At present, the research on epidemic prevention and control mainly focuses on the evolution of the epidemic. In the future, we can try to study how the government should support and encourage enterprises to produce drugs and vaccines, and how to reasonably allocate limited medical resources in the face of the shortage of medical supplies at the beginning of the epidemic. In the case of insufficient medical supplies, how to promote enterprises to expand production and how to deal with the oversupply market situation after the epidemic. At the same time, how to reduce the impact of the epidemic on the economy and minimize the overall risk of the evolution of the epidemic and economic fluctuations of the impact of non-pharmacological interventions (such as home orders, traffic restrictions, etc.) on the social economy.

## 5 Conclusions

In this study, we reviewed the epidemic prevention and control research published in six major medical journals in the past five years. The purpose of this study is to evaluate and present the current research situation in this field from a macro perspective, rather than simply analyze a single paper in depth. We hope to find out the limitations of current research and future research directions from these studies, and get some reference experience in operation and management, which can be used to improve production and supply chain management, so as to promote epidemic prevention and control and economic recovery.

First of all, the research on epidemic prevention and control measures in the medical field analyzed in this paper can be divided into two categories: pharmacological intervention measures and non-pharmacological intervention measures. The research of pharmacological intervention measures can be divided into vaccines and other drugs. The research on vaccines is mainly to evaluate the effectiveness of different vaccines for COVID-19 treatment by analyzing the test data generated by the government's vaccination activities. Different from the research of vaccines, the research of other drugs is mainly to judge the effects of different drugs by means of control experiments. The research of non-pharmacological intervention measures is mainly to evaluate the impact of different control measures on the evolution of the epidemic situation by analyzing the measures taken in the epidemic area and the epidemic related data, so as to provide reference for the epidemic prevention and control in all regions.

Secondly, because various prevention and control measures (such as traffic restrictions, social restrictions, etc.) during the epidemic have had a huge impact on the global supply chain, leading to enterprises facing supply chain interruption, labor shortage and other situations, enterprises have to close down or even close down (especially manufacturing industry), resulting in market economic depression. Therefore, we give some suggestions for future research directions:

(1) Combine epidemic prevention and control with operation management, and study how to improve production and supply chain management during epidemic prevention and control to reduce the impact of prevention and control measures on enterprises;

(2) With the end of the epidemic, various prevention and control measures have begun to be lifted. How can the government effectively help enterprises speed up the pace of resumption of work and production, thus promoting economic recovery.
(3) In the early stage of the epidemic, how to promote enterprises to expand the production of medical materials in the case of insufficient medical materials, and how to deal with the market situation of oversupply of medical materials in the case of overcapacity.

**Acknowledgment.** This research was supported by the National Natural Science Foundation of China (No. 72104190) and the Humanities and Social Sciences Program of the Ministry of Education (No. 20YJC630018).

# References

1. Davaasambuu, G., Buyanjargal, U., Xin, Z., et al.: Vitamin D supplements for prevention of tuberculosis infection and disease. N. Engl. J. Med. **383**(4), 359–368 (2020)
2. Ferguson, N.M.: Challenges and opportunities in controlling mosquito-borne infections. Nature **559**(7715), 490–497 (2018)
3. Nemes, E., Geldenhuys, H., Rozot, V., et al.: Prevention of M. Tuberculosis infection with H4:IC31 vaccine or BCG revaccination (Article). New England J. Med. **379**(2), 138–149 (2018)
4. Rearte, A., et al.: Effectiveness of rAd26-rAd5, ChAdOx1 nCoV-19, and BBIBP-CorV vaccines for risk of infection with SARS-CoV-2 and death due to COVID-19 in people older than 60 years in Argentina: a test-negative, case-control, and retrospective longitudinal study. Lancet **399**(10331), 1254–1264 (2022)
5. Munro Alasdair, P.S., Janani, L., et al.: Safety and immunogenicity of seven COVID-19 vaccines as a third dose (booster) following two doses of ChAdOx1 nCov-19 or BNT162b2 in the UK (COV-BOOST): a blinded, multicentre, randomised, controlled, phase 2 trial. The Lancet **398**(10318), 2258–2276 (2021)
6. Wagner, C.E., Saad-Roy, C.M., Morris, S.E., et al.: Vaccine nationalism and the dynamics and control of SARS-CoV-2. Science **373**(6562), 1488 (2021)
7. Sonabend, R., Whittles, L.K., et al.: Non-pharmaceutical interventions, vaccination, and the SARS-CoV-2 delta variant in England: a mathematical modelling study. Lancet **398**(10313), 1825–1835 (2021)
8. Grubaugh, N.D., Hodcroft, E.B., et al.: Public health actions to control new SARS-CoV-2 variants. Cell **184**(5), 1127–1132 (2021)
9. Cho, A., Muecksch, F., et al.: Anti-SARS-CoV-2 receptor-binding domain antibody evolution after mRNA vaccination. Nature **600**(7889), 517–522 (2021)
10. Thompson Mark, G., et al.: Prevention and attenuation of COVID-19 with the BNT162b2 and mRNA-1273 vaccines. New England J. Med. **385**(4), 320–329 (2021)
11. Katsoularis, I., Fonseca-Rodríguez, O., et al.: Risk of acute myocardial infarction and ischaemic stroke following COVID-19 in Sweden: a self-controlled case series and matched cohort study. Lancet **398**(10300), 599–607 (2021)
12. Callaway, E.: Beyond Omicron: what's next for COVID's viral evolution. Nature **600**(7888), 204–207 (2021)
13. Omer, S.B., Yildirim, I., Forman, H.P.: Herd immunity and implications for SARS-CoV-2 control. JAMA **324**(20), 2095–2096 (2020)

14. Rubin Eric, J., Longo Dan, L.: SARS-CoV-2 vaccination - an ounce (actually, much less) of prevention. New England J. Med. **383**(27), 2677–2678 (2020)
15. Hajirahimova, M.Sh., Aliyeva, A.S.: Analyzing the impact of vaccination on COVID-19 confirmed cases and deaths in azerbaijan using machine learning algorithm. Int. J. Educ. Manage. Eng. (IJEME) **12**(1), 1–10 (2022)
16. RECOVERY Collaborative Group: Aspirin in patients admitted to hospital with COVID-19 (RECOVERY): a randomised, controlled, open-label, platform trial. Lancet **399**(10320), 143–151 (2022)
17. ITAC (INSIGHT 013) Study Group. Hyperimmune immunoglobulin for hospitalised patients with COVID-19 (ITAC): a double-blind, placebo-controlled, phase 3, randomised trial. Lancet (London, England) **399**(10324), 530–540 (2022)
18. RECOVERY Collaborative Group: Casirivimab and imdevimab in patients admitted to hospital with COVID-19 (RECOVERY): a randomised, controlled, open-label, platform trial. Lancet **399**(10325), 665–676 (2022)
19. Eduardo, R., et al.: Rivaroxaban versus no anticoagulation for post-discharge thromboprophylaxis after hospitalisation for COVID-19 (MICHELLE): an open-label, multicentre, randomised, controlled trial. Lancet **399**(10319), 50–59 (2022)
20. Levin, M.J., Ustianowski, A., De Wit, S., et al.: Intramuscular AZD7442 (tixagevimab-cilgavimab) for prevention of COVID-19. N. Engl. J. Med. **386**(23), 2188–2200 (2022)
21. RECOVERY Collaborative Group: Tocilizumab in patients admitted to hospital with COVID-19 (RECOVERY): a randomised, controlled, open-label, platform trial. Lancet **397**(10285), 1637–1645 (2021)
22. Ly-Mee, Yu., Bafadhel, M., et al.: Inhaled budesonide for COVID-19 in people at high risk of complications in the community in the UK (PRINCIPLE): a randomised, controlled, open-label, adaptive platform trial. Lancet **398**(10303), 843–855 (2021)
23. Butler, C.C., Dorward, J., Ly-Mee, Yu., et al.: Azithromycin for community treatment of suspected COVID-19 in people at increased risk of an adverse clinical course in the UK (PRINCIPLE): a randomised, controlled, open-label, adaptive platform trial. Lancet **397**(10279), 1063–1074 (2021)
24. RECOVERY Collaborative Group: Azithromycin in patients admitted to hospital with COVID-19 (RECOVERY): a randomised, controlled, open-label, platform trial. Lancet **397**(10274), 605–612 (2021)
25. RECOVERY Collaborative Group: Convalescent plasma in patients admitted to hospital with COVID-19 (RECOVERY): a randomised controlled, open-label, platform trial. Lancet **397**(10289), 2049–2059 (2021)
26. Kuritzkes, D.R.: Bamlanivimab for prevention of COVID-19. JAMA **326**(1), 31–32 (2021)
27. Oriol, M., Marc, C.-M., et al.: A cluster-randomized trial of hydroxychloroquine for prevention of COVID-19. N. Engl. J. Med. **384**(5), 417–427 (2021)
28. Temple, C., Hoang, R., Hendrickson, R.G.: Toxic effects from ivermectin use associated with prevention and treatment of COVID-19. N. Engl. J. Med. **385**(23), 2197–2198 (2021)
29. RECOVERY Collaborative Group: Lopinavir-ritonavir in patients admitted to hospital with COVID-19 (RECOVERY): a randomised, controlled, open-label, platform trial. Lancet **396**(10259), 1345–1352 (2020)
30. Marovich, M., Mascola, J.R., Cohen, M.S.: Monoclonal antibodies for prevention and treatment of COVID-19. JAMA **324**(2), 131–132 (2020)
31. Barbaro, R.P., MacLaren, G., Boonstra, P.S., et al.: Extracorporeal membrane oxygenation for COVID-19: evolving outcomes from the international Extracorporeal Life Support Organization Registry. Lancet **398**(10307), 1230–1238 (2021)
32. Campbell, K.: The art of infection prevention. Nature **586**(7830), S53–S54 (2020)

33. Lessler, J., Moore, S.M., Luquero, F.J., et al.: Mapping the burden of cholera in sub-Saharan Africa and implications for control: an analysis of data across geographical scales. Lancet **391**(10133), 1908–1915 (2018)
34. Brauner, J.M., Mindermann, S., Sharma, M., et al.: Inferring the effectiveness of government interventions against COVID-19. Science **371**(6531), 802 (2021)
35. Tenforde, M.W., Fisher, K.A., Patel, M.M.: Identifying COVID-19 risk through observational studies to inform control measures. JAMA **325**(14), 1464–1465 (2021)
36. Brooks, J.T., Butler, J,C.: Effectiveness of mask wearing to control community spread of SARS-CoV-2. JAMA **325**(10), 998–999 (2021)
37. Crane, M.A., Shermock, K.M., Omer, S.B., Romley, J.A.: Change in reported adherence to nonpharmaceutical interventions during the COVID-19 pandemic, April-November 2020. JAMA **325**(9), 883–885 (2021)
38. Walensky Rochelle, P., del Rio, C.: From mitigation to containment of the COVID-19 pandemic: putting the SARS-CoV-2 genie back in the bottle. JAMA **323**(19), 1889–1890 (2020)
39. Dehning, J., Zierenberg, J., Spitzner, F P., et al.: Inferring change points in the spread of COVID-19 reveals the effectiveness of interventions. Science **369**(6500), 160 (2020)
40. Walker Patrick, G.T., et al.: The impact of COVID-19 and strategies for mitigation and suppression in low- and middle-income countries. Science **369**(6502), 413–422 (2020)
41. Kraemer Moritz, U.G., Yang, C.-H., Bernardo, G., et al.: The effect of human mobility and control measures on the COVID-19 epidemic in China. Science **368**(6490), 493–497 (2020)
42. Lerner, A.M., Folkers, G.K., Fauci, A.S.: Preventing the spread of SARS-CoV-2 with masks and other "low-tech" interventions. JAMA **324**(19), 1935–1936 (2020)
43. Candido Darlan, S., et al.: Evolution and epidemic spread of SARS-CoV-2 in Brazil. Science **369**(6508), 1255–1260 (2020)
44. Huaiyu, T., Yonghong, L., Yidan, L., et al.: An investigation of transmission control measures during the first 50 days of the COVID-19 epidemic in China. Science **368**(6491), 638–642 (2020)
45. Parodi, S.M., Liu, V.X.: From containment to mitigation of COVID-19 in the US. J. Am. Med. Assoc. **323**(15), 1441–1442 (2020)
46. Leung, K., Wu, J.T., Liu, D., Leung, G.M.: First-wave COVID-19 transmissibility and severity in China outside Hubei after control measures, and second-wave scenario planning: a modelling impact assessment. Lancet **395**(10233), 1382–1393 (2020)
47. Lai, S., Ruktanonchai, N.W., Zhou, L., et al.: Effect of non-pharmaceutical interventions for containing the COVID-19 outbreak in China. Nature **585**, 410–413 (2020)
48. An, P., Li, L., Chaolong, W., et al.: Association of public health interventions with the epidemiology of the COVID-19 outbreak in Wuhan, China. JAMA **323**(19), 1915–1923 (2020)
49. Flaxman, S., Gandy, A., Zhu, H., et al.: Estimating the effects of non-pharmaceutical interventions on COVID-19 in Europe. Nature **584**(7820), 257–261 (2020)
50. Ali, S.T., Wang, L., Lau, E.H.Y., et al.: Serial interval of SARS-CoV-2 was shortened over time by nonpharmaceutical interventions. Science **369**(6507), 1106–1109 (2020)
51. Abriham, A., Dejene, D., Abera, T., et al.: Mathematical modeling for COVID-19 transmission dynamics and the impact of prevention strategies: a case of ethiopia. Int. J. Math. Sci. Comput. (IJMSC) **7**(4), 43–59 (2021)
52. Tatkare, A., Patil, H., Salunke, T., et al.: COVID-19 patient health monitoring system. Int. J. Eng. Manuf. (IJEM) **11**(5), 48–55 (2021)
53. Ferretti, L., Wymant, C., Kendall, M., et al.: Quantifying SARS-CoV-2 transmission suggests epidemic control with digital contact tracing. Science **368**(6491), 619 (2020)
54. Pullano, G., Di Domenico, L., Sabbatini, C.E., et al.: Underdetection of cases of COVID-19 in France threatens epidemic control. Nature **590**(7844), 134–139 (2021)

# Passenger Flow Forecast of the Section of Shanghai-Kunming High-Speed Railway from Nanchang West Station to Changsha South Station

Cheng Zhang, Puzhe Wei[✉], and Xin Qi

School of Transportation Engineering, East China Jiaotong University, Nanchang 330000, China
969258986@qq.com

**Abstract.** Passenger flow is an important basis for the formulation of transportation organization schemes of a high-speed railway. A reasonable method of passenger flow forecasting can not only accurately predict the passenger flow of railway in the coming years, but also greatly reduce the difficulty and workload of forecasting. Thanks to its adaptive ability and self-learning ability, artificial neural network can be well applied to the passenger flow forecasting of high-speed railways. This study constructed an artificial neural network forecasting model and collected historical data of population, GDP, per capita disposable income, tourism and passenger flow of cities along the section of Shanghai-Kunming High-speed Railway from Nanchang West Station to Changsha South Station for the training of the forecasting model. After comparative analysis with the exponential smoothing method and the time series method, this study used the regression analysis method to forecast the population, GDP, per capita disposable income, tourism and other data of cities along this route, and then forecast the middle and long-term passenger flow between Nanchang West Station and Changsha South Station, hoping to provide reference for the transportation organization of this section.

**Keywords:** Passenger flow forecasting · Artificial neural network · Exponential smoothing · Shanghai-Kunming high-speed railway

## 1 Introduction

In 2008, China opened its era of high-speed railways with the opening of the Beijing-Tianjin Intercity High-speed Railway. In the second decade of this century, China has built the Beijing-Shanghai, Shanghai-Kunming, Xuzhou - Lanzhou, Harbin - Dalian, Beijing - Zhangjiakou high-speed railways in succession. Passenger flow forecasting is the main basis for railway operation departments to coordinate their work, and is the basis for supporting the fine development of product design of railway passenger transportation. As more and more high-speed railways are opened, feasible methods for passenger flow forecasting and accurate passenger flow forecasting results are paid more and more attention by the organizers of high-speed railway operation [1]. By analyzing the passenger flow of Shanghai-Kunming High-speed Railway in previous years and other related indicators, this paper intends to forecast the overall trend of passenger flow in the medium and long term, so as to enable transportation organization personnel to formulate targeted transportation organization schemes for the high-speed railway, and tap the potential of the line and meet the travel needs of residents along the line [2].

## 2 Influencing Factors and Forecasting Methods of Passenger Flow

### 2.1 Analysis of Influencing Factors of Passenger Flow Forecast

Passenger flow forecast must be related to the overall development of the cities. This paper will be based on the local population, regional gross national product, per capita disposable income of urban residents, tourism and other factors for an overall analysis [3].

1) Local population
2) Level of economic development
3) Tourism

### 2.2 Commonly Used Methods of Passenger Flow Forecasting and Their Applicability Analysis

Passenger flow refers to the number of passengers passing through a section of a line. It plays an important role in the planning of line networks, the setting of passenger station scale, the selection of passenger transport means and the formulation of operation schemes [4]. It depends on such factors as the industrial and agricultural conditions, the distribution of tourism resources, the living and income levels of urban and rural residents, as well as the developed degree of transportation network in the regions along the route of various means of transportation [5].

(1) The time series method

The time series method refers to the application of the data analysis methods to forecast future trends through data series of a specific period. This method is particularly suitable for forecasting events that are in a continuous state. It requires a large amount of historical data, which is sequenced in chronological order. In other words, it uses the past to forecast the future [6]. Because many of the data are subject to certain uncertainties, the trends are further muddled and the accuracy of the forecasts is compromised.

(2) The regression analysis method

After simple processing of a large number of data, a regression equation of dependent variable and independent variable is obtained by the regression analysis method, and then known relationships are used to forecast the result [7]. The advantage of regression analysis is that it can show the significant relationships between independent variables and dependent variables. The relationship between independent and dependent variables can be many-to-one. However, for passenger flow forecast, the overall process of this algorithm is relatively simple, and the forecasting process is difficult to control. Most of the time, it is used to forecast with simply processed data, so as to approximately reflect the relationships between variables.

(3) The exponential smoothing method

The exponential smoothing method was proposed by Brown. He believed that the tendency of time series has stability or regularity, so time series can be reasonably extended along the tendency. He believed that the tendency in the recent past would to some extent affect the future [8]. The exponential smoothing method is a sort of moving average method. Its feature is to give different weights to the past observations. The weights of the recent observations are higher than those of the observations obtained earlier. According to the different smoothing times, exponential smoothing method can be divided into first exponential smoothing method, second exponential smoothing method and third exponential smoothing method. But they share the same idea: the forecast value is the weighted sum of previously observed values, and different weights are given to different data, with more weight given to new data and less weight given to old data [9].

(4) Artificial neural network.

Artificial neural network is a mathematical model of distributed and parallel information processing based on the behavior characteristics of biological neural network. It is used to estimate or approximate a function [10]. Depending on the complexity of the system, the artificial neural network algorithm achieves the purpose of information processing by adjusting the interconnections between a large number of internal nodes. It is especially suitable for complex nonlinear structures. It is characterized by distributed storage and parallel cooperative management of information. While individual neurons are extremely simple in structure and limited in function, a large number of neurons can achieve a wide variety of behaviors. The unique ability of nonlinear adaptive processing of artificial neural network overcomes the defects of traditional artificial intelligence methods in intuitions, such as patterns, speech recognition, and unstructured information. As a result, it has been successfully applied in such fields as neural expert systems, pattern recognition, intelligent control, combination optimization, and forecasting [11]. Its adaptability depends on the results of a lot of training, and according to different external characteristics, the internal structure of the neural network will also change accordingly, which is a self-adapting ability [12].

The generation of railway passenger flow is mostly based on the data set of stations. Factors affecting passenger flow are changing with the development of society, and passengers' travel behavior will also change accordingly. In addition, there are many internal and external factors affecting railway passenger flow, and the relationship between these factors is difficult to grasp accurately [13]. The entire transportation system is in a changeable and complicated state. In order to accurately combine these factors which are not closely related with the passenger flow forecasting, this paper selects the neural network method to forecast the passenger flow, and uses the mapping relationship between input and output data to forecast the future passenger flow.

## 3 Passenger Flow Forecasting for the Section of Shanghai-Kunming High-Speed Railway from Nanchang West Station to Changsha South Station

### 3.1 An Overview of the Shanghai-Kunming High-speed Railway

Shanghai-Kunming High-speed Railway is referred to as Hu-Kun High-speed Railway for short. Making it the longest east-west high-speed railway in China and the one that passes through the most provinces among east-west high-speed railways. The Shanghai-Kunming High-speed Railway runs through three economic circles on the upper, middle and lower reaches of the Yangtze River. The whole line consists of the Shanghai-Hangzhou section, the Hangzhou-Changsha Section and the Changsha-Kunming section. This paper has selected the section from Nanchang West Station to Changsha South Station for a case analysis (Table 1).

**Table 1.** Information of stations in the section from Nanchang West Station to Changsha South Station

| Station | Grade | Scale of the Station | Population | GDP | Per Capita Disposable Income |
|---|---|---|---|---|---|
| Nanchang West | Special grade | 12 platforms and 26 lines | 644 | 6651 | 5.04 |
| Gaoan | Third grade | 2 platforms and 4 lines | 74.5 | 530 | 4.04 |
| Xinyu North | Second grade | 2 platforms and 5 lines | 120 | 1155 | 4.57 |
| Yichun | Second grade | 4 platforms and 10 lines | 497 | 3191 | 3.99 |
| Pingxiang North | Second grade | 3 platforms and 7 lines | 181 | 1108 | 4.34 |
| Liling East | Third grade | 2 platforms and 4 lines | 88.6 | 825 | 4.84 |
| Changsha South | Special grade | 13 platforms and 28 lines | 1024 | 13300 | 6.21 |

In this section, there are 7 stations in total, with 21 OD. According to the comprehensive information in the table and the administrative level of the cities, the seven stations in the table have been divided into three types: large, medium and small, and six sections have been selected as the training sample: Nanchang West-Changsha South (large-large), Nanchang West-Yichun (large-medium), Nanchang West-Gaoan (large-small), Xinyu North-Pingxiang North (medium-medium), Pingxiang North-Liling East (medium-small), Gaoan-Liling East (small-small).

## 3.2 A Passenger Flow Forecasting Method Based on the Artificial Neural Network

Artificial neural network is a nonlinear function reflecting the mapping relationship between input and output. This relationship goes through a method similar to black-box processing, and after weight adjustment and the determination of the optimal hidden number, the accuracy of the mapping relationship is guaranteed [14]. At the same time, the training errors of the whole neural model will also be reflected through the mathematical model, so as to facilitate the subsequent adjustment and processing. Assuming that a total of K samples are input into the network, and the output of each sample is N, the overall error of the network can be constructed according to formula (1).

$$\delta ANN = \frac{\sum_{K=1}^{K} \sqrt{\sum_{n=1}^{N} (\delta_{kn})^2}}{2} \tag{1}$$

The number of hidden layer nodes can be determined by formula (2).

$$n_1 = \sqrt{nm} + \frac{N}{2} \tag{2}$$

Generally speaking, a neural network with 3 or more layers of neurons includes input layer(s), hidden layer(s) and output layer(s). The upper and lower layers are fully connected, the generated signals are transmitted forward, and the feedback error signals are propagated backward. The hidden layer(s) can be either single or multiple. For each training sample, the signal is propagated in turn, from input to output to error generation.

### 3.2.1 Preparation of the Input Data for Passenger Flow Forecast

Several factors affecting the passenger flow have been mentioned above: population, economy, tourism and distance. In terms of the population, this study selects the total number of permanent residents along the routes; the most direct reflection of economic level is the regional GDP; in terms of the overall development of tourism, this paper quantifies it with the total tourism revenue; in passenger flow forecasting, distance is different from the other three input indicators, for it is a fixed value. After selecting the high-speed railway section for forecasting, this paper makes overall statistics on the distance between different places. The above four aspects serve as input indexes of the artificial neural model. The transfer process is as shown in Fig. 1.

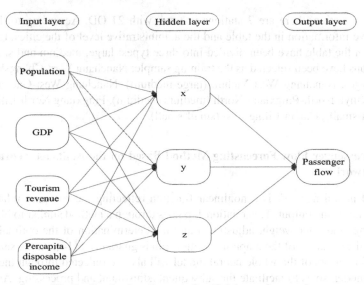

**Fig. 1.** Conduction diagram of artificial neural network

There are a total of 7 stations between Nanchang West Station and Changsha South Station. Limited by the length of this paper, between 2015 and 2019, only the input data of year 2019 is taken as the example, and the population, Gross Domestic Product (GDP), per capita disposable income, tourism revenue, and distances between stations are as shown in the table Table 2.

**Table 2.** Input data of cities in the section from Nanchang West Station to Changsha South Station in 2019

| Station | Population (ten thousand) | GDP (ten thousand yuan) | Per capita disposable income (yuan) | Tourism revenue (100 million yuan) |
| --- | --- | --- | --- | --- |
| Nanchang West | 560 | 5596 | 44136 | 1869 |
| Gaoan | 88 | 449 | 26790 | 90 |
| Xinyu North | 119.6 | 972 | 30046 | 521 |
| Yichun | 505.4 | 2688 | 25597 | 1016 |
| Pingxiang North | 181.1 | 930 | 29019 | 520 |
| Liling East | 105 | 716 | 35679 | 83 |
| Changsha South | 839 | 11574 | 55211 | 2029 |

Note: The data in the table come from the statistical communiques of national economic and social development of different regions

### 3.2.2 The Training of the Artificial Neural Network Forecasting Model

(1) The determination of hidden layer nodes

The hidden layer nodes cover weight parameters. The neural network plays a crucial role in data processing and storage. It is worth exploring how to give the appropriate layer value for the forecast data and forecast model. According to the summary of previous experts and scholars, researchers derive the number of hidden layer nodes mainly through formula method and with their experience. Generally, the number of hidden layers should not be too small, but too many hidden layers will also lead to a longer training time of the model and slower convergence.

(2) The choice of the transfer function

The transfer function is a mathematical model, which is mainly an algorithm to express the differential equation of the output variables and the input variables in the neural network. This study selects the transig function from the neural network toolbox. This function can improve the nonlinear mapping ability between input and output and enhance the accuracy of the algorithm.

(3) The training of artificial neural forecasting models

In the second stage, the researcher used the MATLAB artificial neural toolbox to input the original data from 2015 to 2018 to train the artificial neural network model. Population, GDP, per capita disposable income, tourism income and distance between a pair of cities are taken as nine input variables. The number of hidden layers is determined as 3 by the formula mentioned above, and the output variable is the passenger flow. The schematic diagram of the artificial neural network is as shown in Fig. 2.

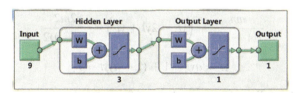

**Fig. 2.** Schematic diagram of artificial neural network

Due to the constraint of the length, this paper only lists the passenger flow in 2019 forecast by the neural network model, and compares it with other methods in Table 3.

**Table 3.** The forecast passenger flow between Nanchang West Station-Changsha South Station in 2019

|  | Nanchang West | Gaoan | Xinyu North | Yichun | Pingxiang North | Liling East | Changsha South |
|---|---|---|---|---|---|---|---|
| Nanchang West |  | 446520 | 649014 | 811889 | 723946 | 30297 | 1131922 |
| Gaoan | 353332 |  | 44141 | 151390 | 25309 | 3305 | 75573 |
| Xinyu North | 612719 | 55202 |  | 50673 | 49017 | 4049 | 237401 |
| Yichun | 756952 | 149252 | 85626 |  | 126311 | 12134 | 426250 |
| Pingxiang North | 669652 | 28464 | 60708 | 102982 |  | 23571 | 782000 |
| Liling East | 30720 | 3041 | 4500 | 8237 | 15733 |  | 496998 |
| Changsha South | 1142427 | 81586 | 269834 | 434763 | 887162 | 535655 |  |

### 3.3 Passenger Flow Forecast Based on the Exponential Smoothing Method

The exponential smoothing algorithm is exactly referred to as the "three exponential smoothing". Through a time window function, it gradually fits from history to the present to realize the forecast of the future.

$$yt + m = \left(3yt' - 3yt + yt\right) + \left[(6 - 5a)yt' - (10 - 8a)yt + (4 - 3a)yt\right].$$

$$\frac{am}{2(1-a)2} + \left(yt' - 2yt + yt'\right) \cdot \frac{a2m2}{2(1-a)2} \tag{3}$$

$$yt = ayt - 1 + (1-a)yt - 1 \tag{4}$$

Upper confidence limit: The upper limit of future trends does not exceed this line
Lower confidence limit: The lower limit of future trends does not exceed this line
Confidence interval: The interval in which future trends fluctuate
Trend line: Future trends are most likely to follow the trend of this line

At the same time, the three exponential smoothing algorithm can also analyze and follow the volatility and periodicity presented by historical data. It should be noted that seasonality is not necessarily cyclical in human life and work.

This study plans to use the passenger flow data of 2015–2018 to forecast the passenger flow data of 2019 and it plans to compare and verify the forecasting effect. In addition, six representative intervals out of twenty OD intervals between Nanchang West and Changsha South will be selected for a case analysis (Tables 4, 5, 6, 7, 8 and 9).

**Table 4.** Passenger flow sample between Nanchang West-Gaoan (large-small)

| Year | Value | Trend Forecast | Lower Confidence Limit | Upper Confidence Limit |
|---|---|---|---|---|
| 2015 | 180952 | | | |
| 2016 | 307089 | | | |
| 2017 | 342143 | | | |
| 2018 | 418250 | 418250 | 418250 | 418250 |
| 2019 | 444791 | 499041 | 448145 | 549937 |

**Table 5.** Passenger flow sample between Nanchang West-Yichun (large-medium)

| Year | Value | Trend Forecast | Lower Confidence Limit | Upper Confidence Limit |
|---|---|---|---|---|
| 2015 | 398515 | | | |
| 2016 | 532044 | | | |
| 2017 | 604617 | | | |
| 2018 | 711889 | 711889 | 711889 | 711889 |
| 2019 | 808374 | 815322 | 783469 | 847174 |

**Table 6.** Passenger flow sample between Nanchang West-Changsha South (large-large)

| Year | Value | Trend Forecast | Lower Confidence Limit | Upper Confidence Limit |
|---|---|---|---|---|
| 2015 | 750107 | | | |
| 2016 | 861616 | | | |
| 2017 | 937563 | | | |
| 2018 | 1031922 | 1031922 | 1031922 | 1031922 |
| 2019 | 1144047 | 1125809 | 1106710 | 1144908 |

**Table 7.** Passenger flow sample between Xinyu North-Pingxiang North (medium-medium)

| Year | Value | Trend Forecast | Lower Confidence Limit | Upper Confidence Limit |
|---|---|---|---|---|
| 2015 | 23111 | | | |
| 2016 | 31644 | | | |
| 2017 | 36556 | | | |
| 2018 | 43017 | 43017 | 43017 | 43017 |
| 2019 | 48826 | 49744 | 47692 | 51797 |

### 3.4 Passenger Flow Forecast Based on the Time Series Method

The passenger flow fitting figure calculated according to the time series is shown in Fig. 3.

**Table 8.** Passenger flow sample between Pingxiang North-Liling East (medium-small)

| Year | Value | Trend Forecast | Lower Confidence Limit | Upper Confidence Limit |
|------|-------|----------------|------------------------|------------------------|
| 2015 | 5694  |                |                        |                        |
| 2016 | 10959 |                |                        |                        |
| 2017 | 14442 |                |                        |                        |
| 2018 | 20571 | 20571          | 20571                  | 20571                  |
| 2019 | 24111 | 25064          | 23999                  | 26129                  |

**Table 9.** Passenger flow sample between Gaoan-Liling East (small-small)

| Year | Value | Trend Forecast | Lower Confidence Limit | Upper Confidence Limit |
|------|-------|----------------|------------------------|------------------------|
| 2015 | 3107  |                |                        |                        |
| 2016 | 3454  |                |                        |                        |
| 2017 | 3250  |                |                        |                        |
| 2018 | 3305  | 3305           | 3305                   | 3305                   |
| 2019 | 3225  | 3378           | 3074                   | 3682                   |

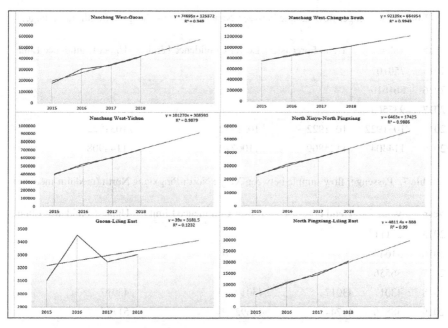

**Fig. 3.** Passenger flow fitting diagram of time series

According to the fitting analysis of time series, the passenger flow forecasting formula of each representative interval can be obtained as follows:

$$\text{Nanchang West - Gaoan}: y = 74695x + 125372 \ (R^2 = 0.949) \tag{5}$$

$$\text{Nanchang West - Yichun}: y = 101270x + 308593 \ (R^2 = 0.9879) \quad (6)$$

$$\text{Nanchang West - Changsha South}: y = 92139x + 664954 \ (R^2 = 0.9949) \quad (7)$$

$$\text{Xinyu North - Pingxiang North}: y = 6463x + 17425 \ (R^2 = 0.9886) \quad (8)$$

$$\text{Pingxiang North - Liling East}: y = 4811.4x + 888 \ (R^2 = 0.99) \quad (9)$$

$$\text{Gaoan - Liling East}: y = 39x + 3181.5 \ (R^2 = 0.1232) \quad (10)$$

### 3.5 Comparative Verification of the Forecasting Effect

Taking the section of Shanghai-Kunming High-speed Railway between Nanchang West and Changsha South in 2019 as an example, the specific error calculation results are as shown in Table 10.

**Table 10.** Comparison of passenger flow forecasting errors of the exponential smoothing method, the time series method and the artificial neural network method

| Interval | Actual value | Forecast value of exponential smoothing method | Relative error | The forecast value of the time series method | Relative error | The forecast value of the artificial neural network method | Relative error |
|---|---|---|---|---|---|---|---|
| Nanchang West-Gaoan | 444791 | 448145 | 0.75% | 468847 | 5.41% | 446520 | 0.39% |
| Nanchang West-Yichun | 808374 | 815321 | 0.86% | 814943 | 0.81% | 811889 | 0.43% |
| Nanchang West-Changsha South | 1144047 | 1125809 | 1.59% | 1125649 | 1.61% | 1131922 | 1.06% |
| Xinyu North-Pingxiang North | 48826 | 49744 | 1.88% | 49740 | 1.87% | 49017 | 0.39% |
| Pingxiang North-Liling East | 24111 | 25064 | 3.95% | 24945 | 3.46% | 23571 | 2.24% |
| Gaoan-Liling East | 3225 | 3378 | 4.74% | 3377 | 4.71% | 3305 | 3.1% |

Note: The actual passenger flow in the table comes from the railway operation departments

By comparing the passenger flow forecast data in 2019, it can be seen that the artificial neural network forecasting model is significantly superior to the other two methods. Its error rate is basically controlled within 3%, in line with the standard error rate of engineering forecasting ($\leq 10\%$). According to Zhang Bomin's research on the passenger flow forecast of Shanghai-Nanjing Intercity Railway, it can be seen that the forecast results of this paper are not much different from those in Zhang's research. The error accuracy can also meet the requirements of operation agencies in passenger transportation organization and marketing [15].

## 4 Forecast of Passenger Flow in Coming Years Based on the Artificial Neural Network

### 4.1 Input Data Forecast in the Forecast of Passenger Flow in Coming Years

(1) The regression analysis method

The regression analysis method is a method used to determine the relationship between independent variables and dependent variables. It determines a suitable mathematical model to approximately express the mean change relationship between variables. It performs simple processing on large amounts of data and establishes correlation equations. If dependent variables are represented in the expression as a first-order function of the independent variables, the equation is called a linear regression equation, otherwise it is called a nonlinear regression equation. According to the number of independent variables and dependent variables, regression can be divided into unary regression or multiple regression. The regression analysis method aims to find a relationship between variables and express this relationship through mathematical expressions. The general model of regression is:

$$y = h(x) = \omega x + b + \varepsilon \tag{11}$$

where, $\omega$ is the correlation coefficient, b is the offset, and $\varepsilon$ is the error term. When unknown parameters $\omega$ and b are obtained, it means that the model is established. That is, given x, the corresponding y (input) value can be forecast.

Using the method of regression analysis, this study will forecast the sample values of four input data for each region in the medium (2025) and long (2035) terms.

(2) Input data forecast in the forecast of passenger flow in coming years.

After randomly selecting four input indicators from 2015 to 2019, this study establishes four corresponding regression models: population, GDP, per capita disposable income, and tourism income. Their analysis of the forecast results are as shown in Table 11.

In Table 11, R Error represents Residual Error, S R Error represents Standard Residual Error. Regression statistical results of the model: "With an observed value of 20, the linear regression coefficient was 0.999953, the fitting coefficient was 0.999905", showing a high degree of fitting. In view of this, this model was used to forecast the input data of the forecast of passenger flow in 2025 and 2035. The forecast results are as shown in Table 12.

**Table 11.** The analysis of prediction results of input index regression model from 2015 to 2019

| Observed Value | | Forecast Y | R Error | S R Error | Observed Value | | Forecast Y | R Error | S R Error |
|---|---|---|---|---|---|---|---|---|---|
| Population | 2015 | 914.3 | 4.65 | 1.22 | Per capita disposable income | 2015 | 27907.4 | −114.43 | −0.53 |
| | 2016 | 109.5 | −1.48 | −0.39 | | 2016 | 23702.5 | −90.51 | −0.42 |
| | 2017 | 127.7 | 0.28 | 0.07 | | 2017 | 27017.3 | −129.32 | −0.6 |
| | 2018 | 484.5 | −3.54 | −0.93 | | 2018 | 33326.3 | 97.67 | 0.45 |
| | 2019 | 39.5 | 0.47 | 0.12 | | 2019 | 51024.2 | −232.15 | −1.07 |
| GDP | 2015 | 237 | 7.01 | 0.09 | Tourism revenue | 2015 | 1548.3 | −5.29 | −0.1 |
| | 2016 | 1405.1 | −25.1 | −0.33 | | 2016 | 98.6 | −5.56 | −0.1 |
| | 2017 | 312.6 | −10.62 | −0.14 | | 2017 | 609.7 | −13.69 | −0.26 |
| | 2018 | 167 | 4.97 | 0.07 | | 2018 | 136.4 | −3.38 | −0.06 |
| | 2019 | 2043.9 | 11.12 | 0.15 | | 2019 | 208.6 | 21.42 | 0.4 |

**Table 12.** Predicted value of input samples of Hangzhou-Changsha high speed railway in 2025 and 2035

| Station | Population (ten thousand) | | GDP (ten thousand yuan) | | Per capita disposable income (yuan) | | Tourism revenue (hundred million yuan) | |
|---|---|---|---|---|---|---|---|---|
| | 2025 | 2035 | 2025 | 2035 | 2025 | 2035 | 2025 | 2035 |
| Nanchang West | 607 | 682 | 8135 | 12247 | 62297 | 92910 | 3881 | 7246 |
| Gaoan | 88 | 89 | 750 | 1345 | 37735 | 56146 | 195 | 378 |
| Xinyu North | 123 | 129 | 1431 | 2207 | 49710 | 80924 | 1386 | 2630 |
| Yichun | 569 | 587 | 3966 | 6106 | 36119 | 53875 | 2059 | 3821 |
| Pingxiang North | 200 | 210 | 1015 | 1052 | 40641 | 60213 | 1029 | 1846 |
| Liling East | 104 | 102 | 967 | 1404 | 50471 | 75065 | 172 | 320 |
| Changsha South | 984 | 1226 | 16435 | 24242 | 77640 | 115638 | 2989 | 4616 |

## 4.2 Results of Passenger Flow Forecast in the Coming Years

In the third stage, this study forecast the passenger flow of the section of Shanghai-Kunming High-speed Railway from Nanchang West to Changsha South in the middle and long term. The four input indicators have been identified above, and the forecast passenger flows in 2025 and 2035 are as shown in the table below 13 (Tables 13 and 14).

**Table 13.** Passenger flow forecast OD of the section from Nanchang West to Changsha South in 2025

| Station | Nanchang West | Gaoan | Xinyu North | Yichun | Pingxiang North | Liling East | Changsha South |
|---|---|---|---|---|---|---|---|
| Nanchang West | | 896947 | 981161 | 1176463 | 1004914 | 89642 | 1673852 |
| Gaoan | 738006 | | 153862 | 305199 | 55217 | 5085 | 155066 |
| Xinyu North | 740370 | 78543 | | 72603 | 75069 | 48294 | 437030 |
| Yichun | 1104007 | 128467 | 93117 | | 167196 | 165350 | 777483 |
| Pingxiang North | 1068171 | 158856 | 185081 | 218917 | | 169011 | 1460813 |
| Liling East | 72599 | 2972 | 16430 | 8858 | 24513 | | 810114 |
| Changsha South | 2002808 | 150992 | 470274 | 926393 | 1473751 | 1370406 | |

**Table 14.** Passenger flow forecast OD of the section from Nanchang West to Changsha South in 2035

| Station | Nanchang West | Gaoan | Xinyu North | Yichun | Pingxiang North | Liling East | Changsha South |
|---|---|---|---|---|---|---|---|
| Nanchang West | | 1491838 | 2156428 | 2291762 | 1460454 | 152597 | 2370887 |
| Gaoan | 953341 | | 256156 | 591543 | 108256 | 10638 | 281150 |
| Xinyu North | 802187 | 84795 | | 81650 | 84158 | 76781 | 671027 |
| Yichun | 1272821 | 205987 | 120876 | | 228968 | 225614 | 978126 |
| Pingxiang North | 1559070 | 293177 | 505512 | 381671 | | 159362 | 2084259 |
| Liling East | 210563 | 10229 | 56010 | 20782 | 44662 | | 1578272 |
| Changsha South | 5229507 | 436082 | 1198976 | 1683285 | 3255387 | 2628266 | |

## 5 Conclusion

Based on the historical passenger flow data of the section of Shanghai-Kunming High-Speed Railway from Nanchang West to Changsha South from 2015 to 2018, three exponential smoothing method, time series method and artificial nerve method are used for a comparative study on the methods of passenger flow forecasting. Through error analysis, it is found that the artificial neural network model fits the original passenger flow data from 2015 to 2018 well, and the error between the forecast data in 2019 and the real value is small. This proves the validity of the artificial neural network model in forecasting the passenger flow of the section of Shanghai-Kunming High-Speed Railway from Nanchang West to Changsha South.

On this basis, this study uses the artificial neural network model trained by historical passenger flow data from 2015 to 2019 to forecast the passenger flow in the above section in 2025 and 2035, which can serve as a reference for the later expansion of the high-speed railway stations in the above-mentioned cities and the formulation of the high-speed train operation schemes. Due to the numerous influencing factors of passenger flow generation and the complex influencing relationship among each other, the accuracy of high-speed railway passenger flow prediction can be further improved by further subdividing the influencing factors of passenger flow generation in future studies.

## References

1. Teng, J., Li, J.Y.: A forecasting method of short-term intercity railway passenger flow considering date attribute and weather factor. China Railway Sci. **41**(05), 136–144 (2020). (in Chinese)
2. Abisoye Blessing, O., Umar, A., Abisoye Opeyemi, A.: Challenges of airline reservation system and possible solutions (a case study of overland airways). Int. J. Inf. Technol. Comput. Sci. (IJITCS) **9**(1), 34–45 (2017)
3. Bahmani, Z., Ghasemi, M.R., Mousaviamjad, S.S., Gharehbaghi, S.: Prediction of performance point of semi-rigid steel frames using artificial neural networks. Int. J. Intell. Syst. Appl. (IJISA) **11**(10), 42–53 (2019)
4. Awad, M., Zaid-Alkelani, M.: Prediction of water demand using artificial neural networks models and statistical model. Int. J. Intell. Syst. Appl. (IJISA) **11**(9), 40–55 (2019)
5. Su, C.C.: Passenger Flow OD Prediction of Urban Rail Transit Based on Passenger Flow Trend Characteristics. Beijing Jiaotong University, Beijing (2020). (in Chinese)
6. Peng, J.Y.: Research on time sequence of rail passenger flow based on self-organizing mapping of neural network. Smart City **7**(03), 123–124 (2021). (in Chinese)
7. Hassan, M.M., Mirza, T.: Using time series forecasting for analysis of GDP Growth in India. Int. J. Educ. Manage. Eng. (IJEME) **11**(3), 40–49 (2021)
8. Vuuren, D.V.: Optimal pricing in railway passenger transport: theory and practice in The Netherlands. Transp. Policy **9**(2), 95–106 (2002)
9. Friesz, T.L.: Transportation network equilibrium, design and aggregation: key developments and research opportunities. Transp. Res. Part B **19**(5–6), 413–427 (1985)
10. Zhan, S., Wong, S.C., Lo, S.M.: Social equity-based timetabling and ticket pricing for high-speed railways. Transp. Res. Part A: Policy Practice **137**, 165–186 (2020)
11. Patriksson, M.: The traffic assignment problem: models and methods. VSP BV, Utrecht, The Netherlands (1994). 1
12. Al-Maqaleh, B.M., Al-Mansoub, A.A., Al-Badani, F.N.: Forecasting using artificial neural network and statistics models. Int. J. Educ. Manage. Eng. (IJEME) **6**(3), 20–32 (2016)
13. Mgandu, F.A., Mkandawile, M., Rashid, M.: Trend analysis and forecasting of water level in mtera dam using exponential smoothing. Int. J. Math. Sci. Comput. (IJMSC) **6**(4), 26–34 (2020)
14. Ojo, J.S., Ijomah, C.K., Akinpelu, S.B.: Artificial neural networks for earth-space link applications: a prediction approach and inter-comparison of rain-influenced attenuation models. Int. J. Intell. Syst. Appl. (IJISA) **14**(5), 47–58 (2022)
15. Zhang, B.M.: On short-term prediction of passenger flow of Shanghai-Nanjing intercity railway. China Railway (09), 29–33+42 (2014), (in Chinese)

# The Effect of Labor Rights on Mental Health of Front-Line Logistics Workers: The Moderating Effect of Social Support

Yi Chen and Ying Gao(✉)

School of Business, Sichuan University, Chengdu 610065, China
alice.gy@foxmail.com

**Abstract.** Logistics industry is a typical labor-intensive industry. It is of great significance to pay attention to and improve the mental health level of front-line logistics practitioners for the sustainable and stable development of large-scale labor resources in logistics industry. This paper adopts the random sampling method, takes the front-line logistics industry employees in Chengdu, Sichuan Province as the research sample, and collects relevant data through questionnaire survey. Through statistical analysis, it can be concluded that: Firstly, labor rights protection has a positive impact on the mental health of front-line logistics workers. Secondly, labor rights infringement has a negative impact on the mental health of front-line logistics workers. Thirdly, social support has a significant positive moderating effect on the influence of labor rights of front-line logistics workers on their mental health. Based on the research conclusions, it is proposed that front-line logistics practitioners should improve their own cultural quality, enhance their understanding of their own rights, and strengthen their awareness of rights protection. Logistics enterprises should improve their own law-abiding consciousness, optimize the front-line work environment and enhance the management suggestions of enterprise humanism.

**Keywords:** Labor rights · Mental health · Social support

## 1 Introduction

Logistics industry is an important part of our national economy. It goes deep into our daily life. Since the beginning of the 21st century, China's logistics industry has maintained rapid development. In the first three quarters of 2022, the total amount of social logistics in China reached 247 trillion yuan, an increase of 3.5 percent year-on-year. With the development of the times and the promotion of economic globalization, the service level of Chinese logistics industry has been continuously improved, the technical level has been continuously improved, the development environment has been continuously improved, and the future development of logistics industry is expected. With the rapid development of the logistics industry, it has absorbed a large number of social labor employment, showing as a labor-intensive industry. By the end of 2019, the number of employment reached 774.71 million, among which the number of logistics employees was up to 51.91 million (an increase of 3.6% over 2016), accounting for 6.7%.

In the huge employment group of employees in the logistics industry, for those who work in the office every day, they have a better working environment, regular working hours, wages paid on time, etc., and the protection of labor rights is better and less infringed. However, front-line logistics practitioners are different. Their working environment is usually outdoors in cold and hot summer. Due to the particularity of the logistics industry, front-line logistics practitioners still have high work intensity, long working hours and low salary, which infringes their labor rights, thus affecting their living conditions. Therefore, there may be a direct relationship between the labor rights and mental health of front-line logistics workers.

At present, there are less researches on labor rights, social support and mental health. In addition, studies on labor rights and social support have focused less on front-line logistics workers, whose labor rights are often not well protected or infringed due to the particularity of work content and nature.

Based on this, this paper takes front-line logistics practitioners as the research object and includes labor rights, mental health and social support into the research model to explore the impact of labor rights on their mental health and whether social support plays a regulating role in this process, and then puts forward relevant management suggestions to help protect the labor rights of front-line logistics practitioners and improve their mental health level. To achieve the sustained, steady and healthy development of large-scale labor resources in the logistics industry.

## 2 Literature Review and Hypothetical Inference

### 2.1 Labor Rights and Mental Health

Zheng Guanghuai built a sociological framework to understand employees' mental health based on the two dimensions of labor rights-management system and structure-action, and pointed out that the reality of widespread violation of labor rights would undoubtedly affect employees' mental health [1]. Liu Linping et al. pointed out that labor rights, such as overtime hours, forced labor and working environment, have a significant impact on the mental health of migrant workers [2]. Yuan Huina pointed out that increasing participation in basic medical insurance can effectively guarantee migrant workers' access to medical resources, protect their labor rights, and thus improve their health status [3]. Zhu Ling pointed out that workers with low hourly wage rate, no labor contract signed, and lack of negotiation rights in terms of wages and labor protection are more likely to work overtime, and overtime work and poor working environment will have a significant adverse impact on the normal health of these workers [4].

Marx's theory of labor alienation pointed out that in order to obtain more abundant profits, capitalists constantly exploited laborers, and laborers could not feel the happiness of realizing their own value in the process of labor, but were trapped in deep misfortune, which not only damaged their bodies, but also destroyed their spirits [5]. In the process of pursuing the maximization of benefits, capitalists will reduce costs in all aspects, which leads to the overuse of limited labor force, and the labor rights of laborers are difficult to be adequately protected, which ultimately leads to the physical and mental damage of laborers. Based on Marx's explanation of labor alienation and the research findings of various scholars, we propose the following research hypotheses:

*H1*: The labor rights protection of front-line logistics workers is positively correlated with mental health, that is, the higher the degree of labor rights protection, the better the mental state.

*H2*: The violation of labor rights of front-line logistics workers is negatively correlated with mental health, that is, the higher the violation of labor rights, the worse the mental state.

## 2.2 The Moderating Effect of Social Support

Some scholars conducted research based on the overall nature of social support. The study of He Xuesong et al. on migrant children found that the mental health of migrant children was significantly related to social support, and the improvement of social support could not only improve their mental health, but also reduce their depression [6]. He Xuesong et al. also pointed out that social support (such as support from family, friends and fellow villagers) is an important protective factor for mental health and has a positive and significant effect on mental health [7].

Another part of scholars focused on a certain aspect of social support. Liu Linping et al. pointed out that social support, especially social communication, has a significant impact on the mental health of migrant workers [2]. Ding Tenghui, based on the study of residents in five provinces, and pointed out that each dimension of interpersonal relationship has a significant impact on their mental health [8]. Guo Xinghua and Cai Fengwei conducted an empirical study on the new generation of farmers in Beijing and the Pearl River Delta region and pointed out that social interaction factors, especially group interaction, would affect the mental health of the group [9].

Frances Cullen, the author of social support theory, points out that the more developed and powerful a person's social support network is, the better able he or she is to cope with challenges in the environment. Social support can positively regulate the mental health of individuals to a certain extent. Based on the social support theory and the research findings of various scholars, we propose the following hypothesis:

*H3*: Social support has a positive moderating effect on the influence of labor rights of front-line logistics workers on their mental health.

## 3 Data Collection

### 3.1 Questionnaire Design and Chosen Scale

On the basis of referring to the existing research and literature, the two variables of "mental health" and "social support" are measured, which are authoritative and proved to be suitable for our national conditions. For the variable of "labor rights", it is divided into two variables, "labor rights protection" and "labor rights infringement", neither of which has a systematic and mature scale at present. Therefore, this study compiled the measurement scale of this variable according to relevant research and combined with the relevant regulations of China's Labor Law. The details are as follows:

1. Labor rights protection scale: The labor rights protection scale consists of two items, namely, the situation of signing labor contract and the situation of social insurance. The scale adopted Likert four-point scoring method, "1 = very bad situation", "2 = poor situation", "3 = good situation", "4 = very good situation".

2. Labor rights infringement scale: The labor rights infringement scale consists of 4 items, including salary, unpaid wages, overtime work and safety and health. The scale scoring rule is the same as 1.
3. Mental health scale: The mental health scale adopted the GHQ-12 scale, which was compiled by Goldberg D. et al. [10] and verified by Yang Tingzhong et al. and applicable to Chinese people, with a total of 12 items [11].
4. Social support scale: The SSRS scale was developed by Professor Xiao Shuiyuan based on China's basic national conditions [12], and verified by Liu Jiwen et al., which is suitable for Chinese research [13]. The scale consists of 10 items in three dimensions: objective support, subjective support and utilization of support.

In addition, the questionnaire also involved the respondent's gender, age, marital status, education level, annual income, job type and other personal characteristic variables.

### 3.2 Respondents

This study randomly selected several logistics companies in Chengdu, Sichuan Province, and distributed questionnaires to front-line logistics workers in the companies. In order to improve the completeness of the questionnaire, five questionnaires including "Basic Information Questionnaire for front-line logistics Practitioners", "Labor Rights protection Questionnaire", "Labor rights infringement questionnaire", "GHQ-12" and "SSRS" were combined into one questionnaire in order. Secondly, in order to improve the efficiency of questionnaire distribution, this study adopts the form of online questionnaire distribution, through the company's project department manager to issue questionnaires to the work group for data collection.

A total of 118 questionnaires were collected in this survey, and 107 valid questionnaires were obtained by deleting the non-research subjects and those with outliers, with an effective recovery rate of 90.68%.

## 4 Data Analysis and Research Hypothesis Testing

### 4.1 Data Analysis

#### 4.1.1 Reliability and Validity Analysis of Questionnaire

Reliability is used to test whether the measured results of the scale have good consistency and reliability. According to the test results in Table 1, the Cronbach's alpha of the scale is 0.759, 0.882 and 0.842, with values greater than 0.7, so the scale has high reliability.

Validity is used to test the extent to which the scale can accurately measure the real situation. KMO and Bartlett spherical tests are carried out on the scale, and the test results are shown in Table 1. The KMO of the scale were all greater than 0.7. Meanwhile, the significance of the Bartlett spherical test was 0.000. Therefore, the scale used in this study was suitable for factor analysis, and the effect was good.

Principal component analysis method was used to extract the scale, and Caesar's normalized maximum variance method was used to rotate them. The analysis results showed that all item loads in the variable scale were higher than 0.5. Therefore, all scales have good validity.

**Table 1.** Reliability and validity analysis of questionnaire

| Questionnaire | Cronbach's alpha | KMO | Significance |
|---|---|---|---|
| Labor rights | 0.759 | 0.770 | 0.000 |
| Mental health | 0.882 | 0.846 | 0.000 |
| Social support | 0.842 | 0.789 | 0.000 |

### 4.1.2 Analysis of Correlation

In order to further explore the interaction between variables, Pearson correlation analysis was carried out on variables, and the analysis results are shown in Table 2. According to the data in the table, in terms of the score, the relationship between labor rights protection and mental health ($r = -0.372$, $P < 0.01$), labor rights infringement and mental health ($r = -0.582$, $P < 0.01$), social support and mental health ($r = -0.585$, $P < 0.01$) had a significant negative correlation.

**Table 2.** The mean value, standard deviation and correlation of each variable

| Variable | Mean value | Standard deviation | Labor rights protection | Labor rights infringement | Mental health | Social support |
|---|---|---|---|---|---|---|
| Labor rights protection | 6.86 | 1.47 | **0.600** | | | |
| Labor rights infringement | 12.96 | 2.03 | - | **0.770** | | |
| Mental health | 17.30 | 4.80 | $-0.372^{**}$ | $-0.582^{**}$ | **0.882** | |
| Social support | 41.63 | 9.13 | $0.300^{**}$ | $0.438^{**}$ | $-0.585^{**}$ | **0.842** |

** stands for $P < 0.01$ (two-tailed test), the bolded data on the diagonal are the Cronbach's alpha values of the scale.

## 4.2 Research Hypothesis Testing

### 4.2.1 The Relationship Between Labor Rights Protection and Mental Health

Model 1 and Model 2 were constructed to conduct linear regression analysis, with gender, age, marital status, annual income, education background and job category as control variables, labor rights protection as independent variable and mental health as dependent variable. The results are shown in Table 3.

Model 1 is the relationship between control variables and mental health, and all control variables have no significant impact on mental health. In Model 2, the protection of labor rights was added on the basis of Model 1. The result ($R^2$) shows that the protection of labor rights could explain 13.8% of the fluctuation of mental health. In addition, the labor rights protection of front-line logistics practitioners can significantly affect their mental health. In terms of the scores of the two variables, the influence is negative ($\beta = -0.469, P = 0.000$). H1 is assumed to be supported.

Table 3. Analysis of regression

| Variable | Dependent variable: mental health | | |
|---|---|---|---|
|  | Model 1 | Model 2 | Model 3 |
| Control variables |  |  |  |
| Male (Female = 0) | −0.065 | −0.091 | −0.011 |
| 18–35 years old (51–60 = 0) | −0.199 | −0.164 | −0.119 |
| 36–50 years old | 0.014 | 0.007 | −0.065 |
| Married (Divorced = 0) | 0.291 | 0.078 | 0.165 |
| Unmarried | 0.275 | 0.066 | 0.185 |
| 40,000 yuan and below (over 100,000 yuan = 0) | 0.023 | −0.005 | 0.034 |
| 40,000–60,000 yuan | 0.093 | 0.041 | 0.166 |
| 60,000–80,000 yuan | 0.055 | 0.033 | 0.037 |
| 80,000–100,000 yuan | 0.009 | −0.043 | 0.013 |
| Junior College (Bachelor's degree or above = 0) | −0.057 | −0.419 | −0.120 |
| High school | 0.006 | −0.234 | −0.017 |
| Junior High School and below | −0.213 | −0.140 | −0.089 |
| General worker (Mechanic = 0) | 0.106 | 0.062 | 0.093 |
| Independent variables |  |  |  |
| Labor rights protection |  | −0.469** |  |
| Labor rights infringement |  |  | −0.582** |
| $R^2$ | 0.091 | 0.229 | 0.389 |
| $\Delta R^2$ |  | 0.138 | 0.298 |
| F | 0.716 | 1.952 | 4.188 |

** stands for P < 0.01

### 4.2.2 The Relationship Between Labor Rights Infringement and Mental Health

Similarly, Model 1 is the relationship between control variables and mental health, and Model 3 is based on model 1 by adding labor rights infringement. The results in Table 3 show that $R^2$ changes from 0.091 to 0.389, indicating that labor rights infringement can explain 29.8% of mental health fluctuations. In addition, the violation of labor rights and interests of front-line logistics practitioners can significantly affect their mental health. In terms of the scores of the two variables, the influence is negative ($\beta = -0.582, P =$

$0.000$), that is, the higher the violation of labor rights of front-line logistics practitioners, the worse their mental state. H2 is assumed to be supported.

### 4.2.3 The Moderating Effect of Social Support

According to the analysis results shown in Table 4, as far as the scores of each variable are concerned, labor equity has a significant negative impact on mental health ($\beta = -0.615$, $P = 0.000$), and social support also has a significant negative impact on mental health ($\beta = -0.652$, $P = 0.000$). The interaction variable (the product of labor equity and social support) has a positive moderating effect on the relationship between labor equity and mental health, and the effect is significant ($\beta = 0.166$, $P = 0.047$), so the hypothesis H3 is also supported.

Table 4. Analysis of regulating effect

| Variable | Control variables: gender, age, marital status, salary, education background, job type | | |
|---|---|---|---|
| | $\beta$ | T | P |
| Labor rights | −0.615 | −6.803 | 0.000 |
| Social support | −0.652 | −7.722 | 0.000 |
| Labor rights*Social support | 0.166 | 2.015 | 0.047 |

## 5 Conclusion and Management Suggestions

### 5.1 Conclusion and Discussion

According to the research results and the above discussion, this paper draws the following conclusions: Conclusion 1, labor rights protection of front-line logistics workers has a significant positive impact on mental health. Conclusion 2, Labor rights infringement of front-line logistics workers has a significant negative impact on mental health. Conclusion 3, social support can significantly affect the mental health of front-line logistics workers. Conclusion 4, social support increases the positive impact of labor rights of front-line logistics workers on mental health.

#### 5.1.1 Labor Rights and Mental Health

According to the research results, labor rights protection of front-line logistics workers has a significant positive impact on their mental health, while labor rights infringement has a significant negative impact on their mental health, which is consistent with the research results of scholars such as Liu Linping et al. [2], Sun Zhongwei et al. [14] and Zheng Guanghuai [1].

First of all, the unreasonable salary and salary arrears of front-line logistics workers, on the one hand, will increase their economic burden, so that they have to find ways to maintain a normal life, whether it is to reduce the quality of life or borrow money from friends and relatives; On the other hand, their delayed payment of wages will make them produce a sense of insecurity and anxiety, and the resulting life and psychological pressure will significantly affect their mental health.

Secondly, the daily workload of most front-line logistics practitioners is not small, and overtime work will make their body in a long-term fatigue or overload state, their body and mind can not relax, mental condition is difficult to develop well, and even may lead to industrial injuries.

In addition, fewer or no labor contracts and social insurance will reduce the sense of belonging and security of front-line logistics workers, which will cause psychological stress and lead to some mental health problems. Therefore, the establishment of a sound social security system to protect the vital interests of front-line logistics practitioners is conducive to improving their mental health.

Therefore, based on the discussion of the above results, it is of great significance to put forward relevant management suggestions to improve the labor rights and interests of front-line logistics practitioners.

### 5.1.2 The Moderating Effect of Social Support

The above results confirmed the positive moderating effect of social support, that is, social support increased the positive impact of labor rights on mental health. This is consistent with the research results of scholars such as He Xuesong et al. [7], Zhang Lei and Chang Yuanyuan [15].

According to the buffer model of social support, social support can buffer the intermediate link between stressful events and disease acquisition. Based on the analysis of the research results of this paper, front-line logistics workers will have a certain degree of psychological pressure when their labor rights are violated, which will lead to some mental problems. If they can get some social support at this time, they will underestimate the harm caused by the violation of labor rights, so as to reduce the pressure. Maintain good mental health.

Whether at work or at home, giving them support from others, such as help when they are in trouble and an enlightened ear when they are bored, helps them to solve problems and cope with difficulties and challenges. At the same time, encouraging them to interact with colleagues and neighbors may also help them maintain good mental health to some extent.

### 5.2 Management Suggestions

From the perspective of front-line logistics practitioners and logistics enterprises, this paper puts forward the following suggestions:

For front-line logistics practitioners, first of all, improve the cultural quality of front-line logistics practitioners. Secondly, improve the group's understanding of their own rights, and clarify the legitimate rights they should enjoy as workers. In addition, mass media should be used to strengthen the awareness of rights protection of front-line logistics practitioners and provide them with ways and means of rights protection.

Finally, front-line logistics practitioners should strengthen social communication and communicate with people more.

For logistics enterprises, firstly, raise the awareness of enterprises to abide by the law. At the same time, enterprises should also recognize the importance of front-line logistics workers to the enterprise. They are also a member of the company and the creator of enterprise wealth, and their legitimate labor rights as workers should be protected. Secondly, optimize the enterprise's front-line working environment. This not only benefits employees' physical and mental health, but also promotes sustainable development. Finally, promote the thought of enterprise humanism. On the one hand, this is beneficial to the work and mental health of front-line logistics practitioners; on the other hand, it also has positive significance for the long-term healthy development of enterprises.

# References

1. Zheng, G.: Towards the sociological understanding of employees' mental health. J. Soc. Sci. Res. **25**(6), 201–222+245–246 (2010). (in Chinese)
2. Liu, L., Zheng, G., Sun, Z.: Labor rights and mental health: based on the questionnaire survey of migrant workers in Yangtze River Delta and Pearl River Delta. Sociological Res. **26**(04), 164–184+245–246 (2011). (in Chinese)
3. Yuan, H.: Health and income of migrant workers: Evidence from the survey of migrant workers in Beijing. Manage. World **05**, 56–66 (2009). (in Chinese)
4. Zhu, L.: Labor hours and occupational health of rural migrant workers. Soc. Sci. China **2009**(01), 133–149+207 (2009). (in Chinese)
5. Karl, M.: Economic and philosophical manuscripts of 1844. People's Publishing House, 50 (2014)
6. He, X., Wu, Q., Huang, F., Xiao, L.: School environment, social support and migrant children's mental health. Contemp. Youth Res. **09**, 1–5 (2008). (in Chinese)
7. He, X., Huang, F., He, S.: Urban and rural migration and mental health: An empirical study based on Shanghai. Sociological Res. **25**(01), 111–129+244–245 (2010). (in Chinese)
8. Ding, T., Xiao, H.: The impact of interpersonal relationship on mental health: based on the empirical survey of residents in Hunan, Guangxi, Shanghai, Shandong and Gansu Provinces. Sci. Theory **31**, 119–120 (2010). (in Chinese)
9. Guo, X., Cai, F.: Social interaction and mental health of the new generation of migrant workers: an empirical analysis based on the survey data in Beijing and the Pearl River Delta region. Gansu Soc. Sci. **04**, 30–34 (2012). (in Chinese)
10. Goldberg, D.: The General Health Questionnaire, pp. 225–237. Oxford University Press, New York (1996)
11. Yang, T., Li, H., Wu, Z.: Study on the suitability of Chinese health questionnaire for screening psychological disorders in Chinese mainland population. Chinese J. Epidemiol. **09**, 20–24 (2003). (in Chinese)
12. Xiao, S.: The theoretical basis and research application of the social support rating scale. J. Clin. Psychiatry **02**, 98–100 (1994)
13. Liu, J., Li, F., Lian, Y.: Research on reliability and validity of social support rating scale. J. Xinjiang Med. Univ. **01**, 1–3 (2008). (in Chinese)
14. Sun, Z., Li, Z., Zhang, X.: Working environment pollution, overtime work and mental health of migrant workers: from the perspective of "second strike" theory. Population Dev. **24**(05), 14–23 (2018)

15. Lei, Z., Chang, Y.: Social support and mental health: an empirical survey of the new generation of migrant workers in six cities of Guangdong Province. Northwest Population **35**(05), 102–106 (2014). (in Chinese)
16. Kayla, H., Zauszniewski Jaclene, A.: Stress experiences and mental health of pregnant women: The mediating role of social support. Issues in Mental Health Nursing, 2019, 40(7)
17. Nakao, M., Yano, E.: A comparative study of behavioral, physical and mental health status between term-limited and tenure-tracking employees in a population of Japanese male researchers. Public Health **2**, 373–379 (2006)
18. Shafiee, N.S.M., Mutalib, S.: Prediction of mental health problems among higher education student using machine learning. Int. J. Educ. Manage. Eng. (IJEME) **10**(06), 1–9 (2020)

# An Investigation into Improving the Distribution Routes of Cold Chain Logistics for Fresh Produce

Mei E. Xie[1], Hui Ye[1], Lichen Qiao[2(✉)], and Yao Zhang[3]

[1] School of Business Administration, Wuhan Business University, Wuhan, China
[2] College of Life Sciences, Northeastern University, Shenyang, China
253637228@qq.com
[3] Technical University of Munich Asia Campus, Technical University of Munich (TUM), Singapore, Singapore

**Abstract.** In today's era, with the improvement of people's living standards and national economy, people have higher requirements for the quality of fresh agricultural products. Unlike other general logistics products, fresh agricultural products (abbreviated as FAP) have unique characteristics that are prone to deterioration and damage, resulting in significant waste. From this point of view, improving the transportation, storage, and distribution of FAP has become an urgent issue for the cold chain logistics industry. Based on the current situation analysis, this article analyzes the characteristics of FAP cold chain logistics. Aiming at specific problems, this paper proposes an optimization scheme and improvement suggestions for the distribution path of FAP cold chain logistics based on mileage saving method. The purpose is to optimize the distribution path of FAP logistics, provide theoretical basis and empirical reference for the improvement of FAP cold chain logistics in China, to ensure the quality of cold chain logistics and improve the competitive characteristics of cold chain logistics.

**Keywords:** Fresh agricultural products (FAP) · Cold chain logistics · Mileage Saving Method · Distribution path

## 1 Introduction

With the growth of the economy and the development of living level, FAP has become an essential part of people's table. With the increasing material demand of people and the development of the Internet, the traditional channels for purchasing FAP can no longer meet people's needs. Therefore, the distribution service of FAP is an inevitable trend. First, Alibaba, JD, Suning and other companies entered the fresh market first, and then the Daily Fresh, Prosperous and Preferred, Dingdong to buy vegetables and so on. The distribution market of FAP accounts for a large part of the logistics industry. According to relevant data, the distribution scale of domestic agricultural products will exceed 170 billion yuan in 2022 (see Fig. 1), and the industry has broad prospects in the future [1–3].

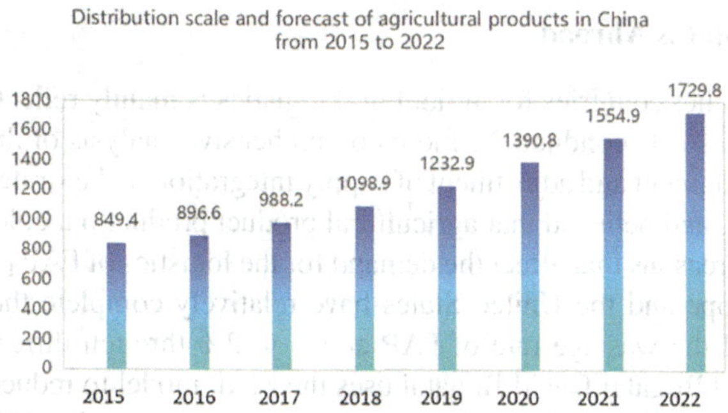

Distribution scale and forecast of agricultural products in China from 2015 to 2022

| Year | Value |
|---|---|
| 2015 | 849.4 |
| 2016 | 896.6 |
| 2017 | 988.2 |
| 2018 | 1098.9 |
| 2019 | 1232.9 |
| 2020 | 1390.8 |
| 2021 | 1554.9 |
| 2022 | 1729.8 |

**Fig. 1.** Distribution scale and forecast of agricultural products in China

In 2000s, the improvement of cold chain logistics has shown a significant upward trend. As the latest *China Cold Chain Logistics Development Report* (2022) shown by the China Federation of Logistics and Purchasing, the desire for cold chain logistics goes strong. As the report shown, the cold chain logistics market size in 2021 will reach 458.6 billion yuan, with a year-on-year growth of 19.65%. This article uses research methods such as literature research, case studies, field investigations, and mileage savings to mainly describe the characteristics of cold chain logistics for FAP, analyze the current situation of the logistics for FAP, and model and optimize the logistics distribution path for FAP. However, China has a large population, a small average cultivated area, and agricultural production and consumption are lower than the world average, making the demand for FAP even more urgent. In this context, the development of the logistics is particularly important [4–6].

## 2 Research Status at Home and Abroad

### 2.1 Domestic Research Status

For the reason to reduce the losses in the logistics and distribution of FAP, reduce the cold chain logistics costs of FAP, and improve customer satisfaction, many experts have conducted research on the logistics of FAP from various aspects.

Ma Shaohua, Zhang Feng, and Du Chengning pointed out that after agricultural products mature, they have not been properly handled in the transportation and distribution process, resulting in a lack of timeliness for agricultural products [7]. Wang Chenglin et al. (2020) designed a seamless connection of cold chain logistics networks from origin, trunk lines, warehousing to distribution through improved genetic algorithms in *Research on Location Optimization of Fresh Logistics Distribution Network* [8].

In the other aspect, the research on the development of FAP cold chain logistics and distribution network, scholars also studied the optimization of the distribution path of agricultural cold chain logistics. Tu Wenjing (2022), in *the Optimization of Agricultural Product Logistics Distribution Path Based on the Mileage Saving Method - Taking S Company as an Example*, used the mileage saving method to reduce the logistics distribution cost of FAP, improve the distribution efficiency of FAP, maximize the freshness and edible quality of enterprise FAP, and improve customer satisfaction [9].

## 2.2 Research Status Abroad

The research in other countries for agricultural logistics is mainly reflected in the influencing factors. Eksoz C conducted a more comprehensive analysis of the needs for the logistics in the judgment and adjustment of supply integration in the strategic partnership of the food chain, and believed that agricultural product production, cold chain circulation rate, etc. are reasons that affect the demand for the logistics of FAP [10]. Developed countries in Europe and the United States have relatively complete the logistics systems that control the wastage rate of FAP at 1% to 2% through differential pressure precooling. David Bogataj David Bogatai uses the MRP model to reduce losses during transportation in a network physical system based on the extended MRP model, and monitors the dynamic changes of agricultural products within their shelf life in real time to make sure the freshness of FAP in the distribution network transportation process and improve customer satisfaction [11].

In the process of reading a large number of documents, it is found that domestic and foreign scholars have conducted certain research on the logistics, but foreign scholars have conducted early research on agricultural the logistics, and their research on FAP cold chain logistics distribution strategies and optimization is comprehensive and in-depth. In recent years, as the country attaches great importance to the development of cold chain logistics, China's research on cold chain logistics distribution of agricultural products is gradually deepening.

## 3 Characteristics of Cold Chain Logistics of Fresh Agricultural Products

There are some differences between FAP and ordinary agricultural products, mainly including fruits, vegetables, meat, aquatic products, dairy products, egg products and other primary unprocessed agricultural products that are easily affected by external factors such as temperature and time. The cold chain logistics system of FAP is shown in the Fig. 2.

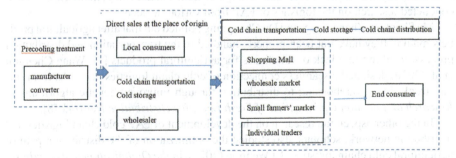

**Fig. 2.** Cold chain logistics system for fresh agricultural products

At present, the development of cold chain logistics of FAP in China is characterized by diversification, which is embodied in the following aspects.

## 3.1 High Requirements for Logistics Efficiency

The difference between FAP and ordinary commodities is that the freshness period of FAP is very short. Due to the particularity of FAP, the transportation time of the entire cold chain logistics must be minimized and the distribution must be timely, which puts forward higher requirements for distribution speed. Currently, China's fresh e-commerce market is basically operated in small batches, so the requirements for time are more stringent. The fresh agricultural product industry also has increasingly high requirements for the timeliness of cold chain logistics [12].

## 3.2 Strict Temperature Control Requirements

The logistics of FAP is also very vulnerable to temperature during transportation. On the one hand, if the temperature is too high, the product will be heated to breed bacteria, causing corruption and deterioration; On the other hand, the temperature is too low, and some foods are easy to be frozen and cannot be eaten. Therefore, the logistics of FAP has a high temperature control, which requires that the whole logistics activities, from the collection, processing, packaging, loading and unloading of raw materials to the transportation and distribution of products, should be kept within a constant temperature control range [13].

## 3.3 Wide Distribution of Supply Chain

The logistics of FAP in China is large in scale, various in types, and relatively dispersed in production and sales regions. Fresh agricultural product production workers are mainly concentrated in rural areas or urban suburbs, and the distribution of personnel is relatively scattered. However, a large amount of consumption demand for FAP is mainly concentrated in cities, and the distance between production and marketing may even exceed 1000 km. There are many links and long distances in the whole supply chain, which greatly increases the difficulty of cold chain logistics transportation and management of cold raw and FAP, and is easy to cause large product loss in each circulation link [14].

## 3.4 High Cost

FAP are divided into: first, frozen and refrigerated products. FAP need to be stored at low temperature during transportation to ensure freshness, so the requirements for storage and transportation conditions are also higher. The second is perishable goods. This kind of goods usually have a short shelf life, so attention should be paid to preservation and preservation. The third is volatile products. The value of such agricultural products is relatively low, and they are vulnerable to losses due to the impact of the natural environment. Refrigeration equipment is required in all links of the cold chain logistics of FAP, so the construction and operation costs of the logistics system are far higher than those of ordinary logistics [15].

## 4 Current Situation Analysis of Cold Chain Logistics of Fresh Agricultural Products

Now that China has become the world's second largest economy, the food issue is the top priority, so people are beginning to pay more attention to the quality of FAP.

### 4.1 Lack of Perfect Cold Chain Logistics System

The entire process of FAP from origin to circulation, processing, packaging, storage, transportation, and retail needs to be in a suitable low-temperature control environment. Currently, only a few enterprises in China can independently carry out cold chain integrated logistics services such as warehousing, transportation, and distribution. Although some small and medium-sized enterprises have a certain number of refrigeration configurations and refrigeration transportation fleets, they have not yet formed an overall logistics service network, unable to meet the needs of the Chinese market.

### 4.2 Lack of Specialized Talents in Cold Chain Logistics

At present, most internal employees in the cold chain industry are lack of professional knowledge. People engaged in the logistics industry need not only knowledge of logistics management, but also knowledge of agricultural product refrigeration technology and food characteristics. Nowadays, colleges and universities pay more attention to theoretical teaching, which leads to a certain gap between the professional knowledge and practical skills mastered by many students and the actual needs of enterprises, and cannot meet the needs of cold chain logistics management. The development of cold chain logistics needs the support of various new professional talents.

### 4.3 Backward Cold Chain Infrastructure

The FAP cold chain logistics is a new way of logistics development. Fresh cold chain logistics started late, developed for a short time, facilities and equipment are relatively backward, and the technical level is low, which restricts the development of China's market economy. However, cold chain logistics enterprises need to strictly control the storage environment of agricultural products and keep the temperature stable, otherwise it is easy to cause losses and economic losses. At present, the infrastructure and technology level of cold chain logistics in China is lower than that of developed countries significantly.

## 5 Optimization of Cold Chain Logistics Distribution Path of Fresh Agricultural Products

### 5.1 Problem Description

Generally, FAP enterprises need to have cold chain distribution centers for product distribution. The distribution centers dispatch different types of cold chain logistics vehicles to deliver the products required by each distribution point. Assumption: The distribution center can meet the different needs of each distribution point for cold goods and frozen goods; The geographical location of the distribution point remains unchanged, and the demand has been confirmed without change; On the basis of meeting the requirements of load capacity, demand and time, a logistics distribution path optimization model based on mileage saving method is established, and the feasibility of the algorithm is verified by an example.

## 5.2 Using Mileage Saving Method to Optimize Distribution Path

Specific example: A fresh agricultural product distribution center needs to distribute products to six sales regions, and the P0 logistics center is the starting point for distribution operations to the six distribution points of ABCDEF.

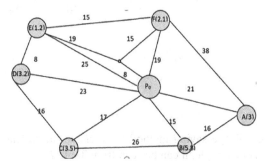

**Fig. 3.** Distribution Center and Distribution Points

Assumption: the load capacity of the vehicle is 10 tons, and the distribution distance limit is 80 km. The following figure shows the route distance (data on the straight line) between P0 and six distribution points, and the demand of each distribution point (data in parentheses), presented in Fig. 3.

Initial plan: Logistics distribution center P0 dispatches vehicles to set out and return to P0 after arriving at 6 distribution points respectively. The initial scheme is shown in Fig. 4. There are 6 distribution lines in total, the total distribution mileage is 120 km, and the number of vehicles used is 6 (10 ton vehicles).

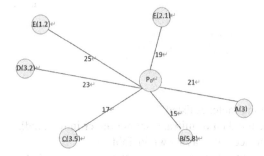

**Fig. 4.** Initial circuit diagram

Establish the model, optimize the distribution path through mileage saving method, and formulate the optimal distribution scheme.

Step 1: calculate the shortest distance between P0 and six distribution points of the FAP logistics distribution center and between each distribution point, and list the shortest distance table as shown in Table 1.

**Table 1.** Transportation Odometer

|   | P | A | B | C | D | E | F |
|---|---|---|---|---|---|---|---|
| P |   | 21 | 15 | 17 | 23 | 25 | 19 |
| A |   |   | 16 | 38 | 44 | 46 | 38 |
| B |   |   |   | 26 | 38 | 40 | 34 |
| C |   |   |   |   | 16 | 24 | 36 |
| D |   |   |   |   |   | 8 | 23 |
| E |   |   |   |   |   |   | 15 |
| F |   |   |   |   |   |   |   |

Step 2: Use the mileage saving method to calculate the mileage saving between distribution points. For example, it can be seen from Table 2.

$$LAB = PA + PB - AB = 21 + 15 - 16 = 20$$

**Table 2.** Saving Odometer

|   | A | B | C | D | E | F |
|---|---|---|---|---|---|---|
| A |   | 20 | 0 | 0 | 0 | 2 |
| B |   |   | 6 | 0 | 0 | 0 |
| C |   |   |   | 24 | 18 | 0 |
| D |   |   |   |   | 40 | 19 |
| E |   |   |   |   |   | 29 |
| F |   |   |   |   |   |   |

The mileage saving meter is listed.

Step 3: Arrange the data of mileage saving meter in descending order, and list the mileage saving sequence table, as shown in Table 3.

Step 4: Use mileage saving method to optimize the distribution route and loading capacity, and draw the final route map, shown in Fig. 5.

**Table 3.** Mileage saved is arranged in descending order

| No | Mileage | Saving mileage | No | Mileage | Saving mileage |
|---|---|---|---|---|---|
| 1 | D–E | 40 | 5 | D–F | 19 |
| 2 | E–F | 29 | 6 | C–E | 18 |
| 3 | C–D | 24 | 7 | B–C | 6 |
| 4 | A–B | 20 | 8 | A–F | 2 |

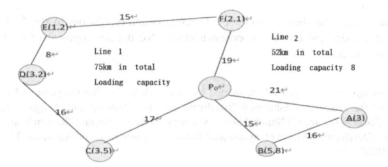

**Fig. 5.** Mileage saving and route optimization

The results are as follows:
Line 1: traffic volume $= 2.1 + 1.2 + 3.2 + 3.5 = 10T$;
Driving mileage $= 19 + 15 + 8 + 16 + 17 = 75$ km,
Then, 10t vehicle is used for loading and distribution, and the saved distance is:
$40 + 29 + 24 = 93$ km.
Line 2: traffic volume $= 5.8 + 3 = 8.8T$;
Driving mileage $= 15 + 16 + 21 = 52$ km,
Then, 10t truck will be used for loading and distribution, and the saved distance is 20 km.

## 6 Conclusion

The cold chain logistics system in our country is still at a low level and lacks market competitiveness. To achieve a higher position in the competition, we need a complete, efficient and modern cold chain logistics system. Although the total amount of agricultural production and consumption in China has increased rapidly in recent years, the seasonal and regional characteristics of agricultural products have led to large fluctuations in the price of agricultural products, and the characteristics of agricultural products are not easy to preserve, making it put forward higher standards for the timeliness of transportation. On the other hand, in order to achieve quality and safety of the product, customers also expect the quality and freshness of fresh products, which greatly increases the circulation cost. From the perspective of the development of FAP cold chain logistics in China, several suggestions are put forward:

(1) Strengthen the construction infrastructure, establish a sound refrigerated transportation system, and achieve "seamless" distribution.
(2) Accelerate the technological innovation of cold chain logistics, develop new refrigeration equipment, and adopt advanced refrigeration systems to ensure the freshness of food.
(3) Increase the support for third-party logistics enterprises, encourage the development and growth of large third-party logistics companies, and provide certain policy support.
(4) The government should introduce relevant laws and regulations to standardize the market order and create a good environment for the development of fresh food e-commerce.

**Acknowledgment.** This paper is supported by two projects: (1) The Cooperative Education Project of the Ministry of Education "Intelligent Logistics Planning and Designer Training" (220600924215841), and (2) National Social Science Foundation Program "Research on the Collaborative Development Model of Agricultural Industrial System Based on Agricultural Logistics Park" (18BJY138).

# References

1. Cui, Z.: Combining point and area with precise strategy to move towards the new journey of the "Fourteenth Five Year Plan" - Interpretation of the "Fourteenth Five Year Plan" cold chain logistics development plan. China Logist. Procurement **01**, 24–26 (2022)
2. Talib, F., Josaiman, S.K., Faisal, M.N.: An integrated AHP and ISO14000, ISO26000 based approach for improving sustainability in supply chains. Int. J. Qual. Reliab. Manag. **38**(6), 99–108 (2021)
3. Huang, Y., Cao, H.: Improved artificial bee colony algorithm to optimize support vector machine and its application. Comput. Appl. Softw. **38**(02), 258–267 (2021)
4. Gan, W., Zhang, T., Zhu, Y.: On RFID application in the information system of rail logistics center. Int. J. Educ. Manag. Eng. (IJEME), **02**(28), 52–58 (2013)
5. Luo, Q., Zhang, L.: Theoretical interpretation and realization path of high-quality development of cold chain logistics of agricultural products. China's Circulat. Econ. **35**(11), 3–11 (2021). (in Chinese)
6. Anitha, P., Malini, M.P.: A review on data analytics for supply chain management: a case study. Int. J. Inf. Eng. Electron. Bus. (IJIEEB) **09**(08), 30–39 (2018)
7. Ma, S., Zhang, F., Du, C.: Research on the current situation of cold chain logistics development of fresh agricultural products in China. China Storage Transp. **09**, 199–200 (2022)
8. Wang, C., Zheng, Y., Huangfu, Y., Hao, H.: Research on location routing optimization of fresh food logistics distribution network. Pract. Underst. Math. **50**(10), 33–43 (2020)
9. Tu, W.: Optimization of agricultural product logistics distribution path based on mileage saving method-take S company as an example. China Storage Transp. **06**, 89–90 (2022). (in Chinese)
10. Eksoz, C., Mansouri, S.A., Bourlakis, M., et al.: Judgmental adjustments through supply integration for strategic partnerships in food chains. Omega **87**(9), 20–33 (2019)

11. David, B., Marija, B., Domen, H.: Reprint of mitigating risks of perishable products in the cyber-physical systems based on the extended MRP model. Int. J. Prod. Econ. **10**(194), 113–125 (2017)
12. Zine, B., Ghalem, B., Abdelkader, N.: A cost measurement system of logistics process. Int. J. Inf. Eng. Electron. Bus. (IJIEEB) **09**(08), 23–29 (2018)
13. Zou, X., Pang, T., Zhou, H.: Research on optimization of fresh agricultural product supply chain under two-way revenue sharing cost sharing contract. Southwest Univ. J. Sci. (Nat. Sci. Ed.) **43**(11), 122–130 (2021). (in Chinese)
14. Zheng, Q., Fan, T., Zhang, L.: Income sharing contract of fresh agricultural products under the mode of "agricultural supermarket docking." J. Syst. Manag. **28**(04), 742–751 (2019)
15. Ting, P.-H.: An efficient and guaranteed cold-chain logistics for temperature-sensitive foods: applications of RFID and sensor networks. Int. J. Inf. Eng. Electron. Bus. (IJIEEB) **12**(18), 1–5 (2013)

# Development of Vulnerability Assessment Framework of Port Logistics System Based on DEMATEL

Yuntong Qian(✉) and Haiyan Wang

School of Transportation and Logistics Engineering, Wuhan University of Technology, Wuhan 430070, China
1692845903@qq.com

**Abstract.** Port is an important part of the international maritime trade and supply chain. However, in recent years, various emergencies have exposed to the port logistics system, triggered the vulnerability in it. Most of the existing studies involve the vulnerability of ports to specific threats. In order to comprehensively identify and evaluate the bottlenecks exposed by the port logistics system while facing various emergencies, this paper uses the DEMATEL (Decision Making Trial and Evaluation Laboratory) to develop a comprehensive port logistics system vulnerability assessment framework, identify the main vulnerability factors from the qualitative and quantitative perspective. A real case of port in China is investigated. The research results show that the port infrastructure system and port operation management system are the main in ternal factors to trigger the vulnerability of the port logistics system in sequence. Such findings provide useful insight to strengthen the resilience of port logistics system.

**Keywords:** Port logistics · Vulnerability assessment · DEMATEL · Resilience

## 1 Introduction

As an important node of international maritime trade, ports play an important role in ensuring the normal operation of the world economy [1]. Once affected by emergencies, seaports often fail to operate and fall into the dilemma of delay, deviation, interruption and even shutdown [2]. At the same time, due to its special terrain and key socioeconomic functions, seaports are vulnerable to natural and man-made disasters from the sea side and the land side [3]. Any interruption of the seaport will have a direct impact on the supply chain to which the seaport belongs, and will be transmitted to the supply chain network to have an indirect impact on the whole industry [4]. Therefore, with the rapid development of trade globalization, as the "engine" of economic development, port logistics has attracted more and more attention. The research on the vulnerability of port logistics system will also help us find the bottleneck that restricts the efficiency of port logistics, thus promoting the development of the global logistics industry chain.

With the occurrence of a series of port accidents caused by natural or human factors, and the impact of various events on the port supply chain, the research on port security risk and logistics vulnerability has received more and more attention. Mature research methods on vulnerability from ecology and other fields have also been applied in the port logistics field. Qualitative and quantitative research methods based on vulnerability identification have also been proposed by many scholars and have been well applied in the study of specific threats, and have achieved certain research results.

However, most of the existing researches on the vulnerability of port logistics are based on specific threats, such as earthquake, fire, tsunami, explosion, epidemic situation, terrorist attacks, etc. the established indicator system is also based on specific threats, lacking a universal and comprehensive indicator system to measure the vulnerability of port logistics [5]. Therefore, the future research direction will focus on the research on the universality and vulnerability of port logistics, which can include the consequences of various threats faced by port logistics and give a comprehensive and objective evaluation.

Hence, this paper aims to comprehensively evaluate the vulnerability of port logistics by proposing a universal vulnerability assessment framework. The framework can comprehensively assess the impact of each sub part of the port logistics system on the vulnerability of the entire logistics system, and find out the bottleneck that restricts the efficiency improvement of the port logistics system, highlight the vulnerability of the port logistics system from the side.

The vulnerability indicator system that affects port logistics is first constructed, followed by determination of index weight and vulnerability assessment. The other part of this paper is organized as following. Section 2 contains a literature review which focuses on vulnerability and port logistics impact factors, port vulnerability concept, influence of various factors on port logistics vulnerability, and vulnerability identification and research methods. Section 3 introduces the proposed port logistics vulnerability assessment framework. Section 4 presents a case study, results and discussions are also shown in this section. Finally, Sect. 5 concludes the paper with main contributions.

## 2 Literature Review

### 2.1 Vulnerability

The concept of vulnerability was first proposed in the research on ecology in the 1970s, and then applied to other fields [6]. In recent years, the research on "vulnerability" has become more prominent in the field of information and engineering [7–9]. However, the definition of "vulnerability" has not yet been clearly identified. It is still confused with reliability, risk, accessibility, and so on, even if some researchers have tried to untangle those relationships [10]. In a complex system, vulnerability refers to the collapse of a part (system) caused by internal and external interference factors of the system, while other parts (system) or the whole complex system are directly or indirectly affected, leading to the collapse. When a vulnerability factor or vulnerability source in the system is activated, it will rely on network mobility and transmissibility, leading to other vulnerability sources in a disordered state, and the vulnerability of the system is activated [11]. Therefore, the vulnerability of a complex system is reflected in each part of the system. The vulnerability of the system should be comprehensively evaluated and analyzed, and the dynamic characteristics and driving factors of the vulnerability should be explored.

Although vulnerability is proposed on the basis of risk analysis, they have different emphasis. Risk analysis often focuses on human, environmental and property losses or impacts caused by events [12]. Risk analysis often needs to find out the causes, possibilities and consequences of risk events [13]. When analyzing the vulnerability of the system, we should focus on "the extended set of threats and results", "reducing the risk of the system and restoring the system to a new stable state", "the chain breaking time before the system establishes a new stability". It should be noted that the vulnerability of the system includes the extent to which a system is exposed to external environmental disturbances, the sensitivity of the system, and the self-adaptability to external disturbances. The self-adaptability changes with the changes of external disturbances.

## 2.2 Port Logistics System

Port logistics system refers to the development of a comprehensive port service system with the characteristics of covering all links of the logistics industry chain by a central port city by making use of its own port advantages and relying on advanced software and hardware environment, strengthening its radiation capacity to logistics activities around the port, highlighting the expertise of the port in cargo collection, inventory, and distribution, taking the port-based industry as the basis, information technology as the support, and optimizing the integration of port resources as the goal [14]. Its basic elements include warehousing facilities, terminal equipment and facilities, collection and distribution equipment, handling equipment, transport vehicle equipment, communication networks and facilities, port supporting service facilities and port logistics park, port operation and management personnel, logistics distribution facilities, collection and distribution facilities and information systems. The port realizes the function of modern logistics center with compound advantage, the port multiple identities have strategic status in the international logistics, and the port provides value-added services through the logistics system.

To sum up, the port logistics system is a highly integrated and complex system. The port undertakes the conversion task of multiple transportation modes, and the internal function modules of the port are also various. Therefore, the factors that affect the development of port logistics are also diverse.

## 2.3 Vulnerability of Port Logistics System

Based on the application of vulnerability theory in other disciplines, many experts put forward their own views on the vulnerability of ports. Port logistics is not only a relatively complex system, but also closely connected with the external environment, which makes the study of port logistics vulnerability more difficult. Assessing the vulnerability of a port is very challenging, as shown in the following aspects: First, there are multidimensional definitions of port vulnerability [15], and different experts focus on different latitudes; Second, there is no statistics on the critical threshold for the occurrence of largescale disasters in ports [16]; Third, how to construct vulnerability indicators [17].

Therefore, this paper can give the concept of port logistics vulnerability: that is, under the disturbance and interference of internal and external factors, the port logistics system loses all or part of its operational capacity due to the instability and sensitivity of its own system, resulting in the decline or stagnation of the efficiency of the logistics system.

## 2.4 Vulnerability Assessment Methods

Vulnerability factors usually have strong concealment and fuzziness, and there are relatively large difficulties and challenges in collecting vulnerability factors. From domestic and foreign research literature, vulnerability identification methods in the transportation field usually refer to risk identification methods. Traditional identification methods include Delphi method, questionnaire survey method, scenario analysis method and literature sorting method. At the beginning, qualitative analysis methods were mainly used for vulnerability identification, but these traditional methods usually have complex processes and heavy tasks, The results often lack systematic ness and comprehensiveness. Later, scholars began to slowly explore the vulnerability of the system measured by quantitative methods. Quantitative vulnerability methods were first developed in the field of ecology. Me. Bar et al. [18] believed that vulnerability is the level of the critical value of disasters. They proposed an index weighted vulnerability evaluation method, gave the absolute vulnerability of a standard reference event, calculated the absolute vulnerability of the research event, and then calculated the relative vulnerability of the research event. This evaluation method is relatively reasonable. Liu et al. [19] further improved the FMEA (Failure Mode and Effects Analysis) method by combining fuzzy logic and Bayesian Network [20]. In order to model the complex environment in supply chain security management and deal with uncertain information, Yang et al. [21] further extended the classic fuzzy rule-based system by incorporating the concept of confidence into the subsequent part of the traditional IF-THEN rule. Shieh et al. [22] integrated three system analysis methods of DEMATEL, ISM (Interpretative Structural Modeling Method) and ANP (Analytic Network Process) to identify the vulnerability factors affecting the transportation system, effectively integrating the characteristics of vulnerability factors and the interaction between vulnerability factors. It can be seen that the combination of quantitative and qualitative methods can effectively overcome the shortcomings between the two, which is also the mainstream method of vulnerability research.

Although there have been many research achievements in the vulnerability of ports, most of the existing studies are aimed at the identification and evaluation of port logistics vulnerability under specific threats, lacking a research framework to explore the vulnerability of port logistics system from the overall perspective and the coupling and cooperation between subsystems. In order to fill the research gap, this paper develops a port logistics system vulnerability assessment framework to identify the potential vulnerability sources of the port logistics system, find out the important bottlenecks that affect the improvement of the port logistics efficiency, find out the vulnerable links in the port logistics system and put forward suggestions, so as to reduce the vulnerability of the system and enhance the system resilience.

## 3 Port Logistics Vulnerability Assessment Model

### 3.1 Assessment Framework

Based on the DEMETAL method, this paper developed a framework for port logistics vulnerability assessment.

Firstly, considering that the port logistics system is a complex system, and it is difficult to collect data and distort data by adopting quantitative index method, this paper establishes a vulnerability evaluation system of port logistics system based on interviews with many experts in the port field and their relevant managers (Table 1).

Table 1. Vulnerability evaluation index system of port logistics system.

| Subsystem $B_i$ | Vulnerability influencing factors $C_i$ |
|---|---|
| Natural and geographical conditions $B_1$ | The depth of water and channel conditions $C_1$ |
| | Port construction conditions $C_2$ |
| | Frequency of natural disasters $C_3$ |
| Infrastructure conditions $B_2$ | Status of port handling facilities $C_4$ |
| | Port storage conditions $C_5$ |
| | Port berth status $C_6$ |
| | Status of traffic facilities in the port $C_7$ |
| | Quality supervision and management level $C_8$ |
| | Average time of ship in port $C_9$ |
| | Personnel management ability and stuff quality $C_{10}$ |
| Logistics information system $B_3$ | Cargo information management level $C_{11}$ |
| | Management level of ship mobilization $C_{12}$ |
| | Customer relationship management level $C_{13}$ |
| Port logistics support system $B_4$ | Supporting facilities around the port $C_{14}$ |
| | Port collection and distribution capacity $C_{15}$ |
| | Development status of port industry $C_{16}$ |
| Policy and coordination factors $B_5$ | Port administration level $C_{17}$ |
| | Government supervision and coordination $C_{18}$ |

Secondly, the DEMATEL method is used to construct the overall influence matrix of the vulnerability factors of the port logistics system, which is used to characterize the comprehensive influence relationship among various factors. By using the DEMATEL method, identify the key vulnerability factors in the port logistics system, and identify the vulnerability degree of the port subsystem.

Finally, reasonable suggestions are given to reduce the vulnerability of port logistics system.

## 3.2 DEMATEL Method

DEMATEL is a method of systematic factor analysis using graph theory and matrix tools. The specific steps are as follows:

First, determine each factor with Delphi method, brainstorming method or expert interview method.

Second, determine the degree of direct influence between elements. First, use the expert scoring method to determine the direct influence matrix of vulnerability influencing factors. The relationship between the factors is divided into five grades. 0 indicates no influence relationship, 1 indicates weak influence, and 2 indicates relatively weak influence k. A score of 3 indicates a strong influence, while a score of 4 indicates a strong influence. A direct influence matrix A is established. The variable $a_{ij}$ indicates the elements in the row $i$ and column $j$ of the matrix, and the $\sum_{j=1}^{n} a_{ij}$ expression indicates the sum of rows in the matrix.

Third, normalization directly affects the matrix. Using Eq. (1) calculate the normalized direct impact matrix G.

$$G = \frac{A}{\max_{1<i<n} \sum_{j=1}^{n} a_{ij}} \quad (1)$$

Then, determine the comprehensive impact matrix. The normalized direct influence matrix is calculated by Eq. (2).

$$T = G(E - G)^{-1} \quad (2)$$

where $E$ is the identity matrix and $(E - G)^{-1}$ is the inverse matrix of $(E - G)$.

Next, obtain influence degree and affected degree. To calculate the mutual influence degree and affected degree between risk factors, according to Eq. (3), $f_i$ is the sum of row elements in the comprehensive influence matrix T, representing the direct or indirect influence degree of risk factor $i$ on risk factor $j$; $e_i$ is the sum of the elements in the comprehensive influence matrix T, and the influence value of the corresponding vulnerability factors in each row affected by other factors is called the affected degree.

$$f_i = \sum_{j=1}^{n} t_{ij}, \, e_i = \sum_{i=1}^{n} t_{ji} \quad (3)$$

Finally, determine the centrality and causality of factors.

$$R_i = f_i + e_i \quad (4)$$

$$C_i = f_i + e_i \quad (5)$$

where $R_i$ represents the centrality of vulnerability factors, and $C_i$ represents the cause of vulnerability factors.

## 4 Case Study

Several experts from a port in China were invited to give a score, and the results obtained by integrating their opinions are as follows.

**Table 2.** $f_i, e_i, R_i$ and $C_i$ of each vulnerability factor

| factors | $f_i$ | $e_i$ | $R_i$ | $C_i$ |
| --- | --- | --- | --- | --- |
| $C_1$ | 1.3450 | 0.0998 | 1.4448 | 1.2453 |
| $C_2$ | 1.7533 | 0.5965 | 2.3498 | 1.1569 |
| $C_3$ | 2.1230 | 0.0344 | 2.1574 | 2.0887 |
| $C_4$ | 1.1632 | 1.8649 | 3.0282 | 0.7017 |
| $C_5$ | 1.4826 | 1.8376 | 3.3202 | 0.3550 |
| $C_6$ | 1.1932 | 2.5972 | 3.7905 | 1.4040 |
| $C_7$ | 1.0100 | 1.4487 | 2.4588 | 0.4387 |
| $C_8$ | 1.3375 | 0.7544 | 2.0920 | 0.5831 |
| $C_9$ | 0.9347 | 3.2946 | 4.2293 | 2.3599 |
| $C_{10}$ | 2.0871 | 0.4583 | 2.5454 | 1.6288 |
| $C_{11}$ | 1.2741 | 1.3400 | 2.6141 | 0.0659 |
| $C_{12}$ | 1.1238 | 2.1953 | 3.3191 | 1.0714 |
| $C_{13}$ | 0.3883 | 0.4944 | 0.8827 | 0.1060 |
| $C_{14}$ | 1.1673 | 0.7058 | 1.8731 | 0.4615 |
| $C_{15}$ | 0.9796 | 3.2722 | 4.2517 | 2.2926 |
| $C_{16}$ | 1.0609 | 1.5954 | 2.6563 | 0.5345 |
| $C_{17}$ | 1.1938 | 0.3631 | 1.5569 | 0.8307 |
| $C_{18}$ | 1.4970 | 0.1622 | 1.6592 | 1.3348 |

According to Eqs. (1)–(5), this paper got the $f_i, e_i, R_i$ and $C_i$ of each vulnerability factor (Table 2).

It can be seen from Table 2 that the top three influencing factors are frequency of natural disasters ($C_3$), personnel management ability and staff quality ($C_{10}$), port construction conditions ($C_2$). The influencing index reflects the degree of influence of each factor on other factors. The top ranking indicates that these three factors are easy to affect other factors.

The high ranking of the affected indicators indicates that these factors are vulnerable to other factors. The factors with the highest ranking of the indicator values can be used to judge the effect of other factors. The main factors with high scores are port collection and distribution capacity ($C_{15}$), port berth status ($C_6$), management level of ship mobilization ($C_{12}$).

According to Table 2, the causality and centrality diagram of the influencing factors of port logistics vulnerability is drawn, as shown in Fig. 1.

The centrality degree indicates the importance of the influencing factors. The higher the index value, the more important the influencing factors are. The top three factors include port collection and distribution capacity ($C_{15}$), average time of ship in port ($C_9$)

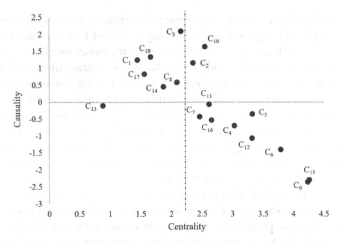

Fig. 1. Centrality-Causality diagram of the influencing factors

and port berth status ($C_6$). Improving these factors is crucial to reduce port logistics system vulnerability.

If the causality degree indicator value is less than 0, it is called a result factor; if it is greater than 0, it is a cause factor. In all result factors, the highest rank result factor is average time of ship in port ($C_9$), which indicates that it is vulnerable to other factors. When controlling the system vulnerability, it is necessary to prevent it from being interfered by other factors, leading to the emergence of system vulnerability. In all cause factors, the higher rank factor is frequency of natural disasters ($C_3$), indicating that the occurrence of natural disasters is most likely to affect other factors. When controlling the system vulnerability, we need to focus on cutting off the transmission process between it and other factors.

However, from the perspective of subsystems, the centrality score of infrastructure conditions is high, which also indicates that port infrastructure is of high importance in the whole port logistics system, and its overall causality score is negative, which indicates that the overall vulnerability of port infrastructure is high, and the system is most likely to collapse when disturbed by internal and external factors. Meanwhile, the causality score of the port operation subsystem is high while the centrality score is low, which indicates that the system is easy to directly or indirectly have an important impact on other systems, and it is also one of the more vulnerable links.

## 5 Conclusion

This paper quantifies the vulnerability risk of each part and subsystem of port logistics system by using the constructed port logistics vulnerability assessment model. Taking a port in China as an example, the most vulnerable link and the main vulnerability factors in the port logistics system are determined, and the suggestions to reduce the logistics vulnerability and improve the port operation efficiency are put forward. The main conclusions are drawn as follows: average time of ship in port, personnel management ability and staff quality and frequency of natural disasters are the key factors affecting the vulnerability of port logistics system. Managers of port logistics enterprises should pay more attention to these aspects. Therefore, port enterprises should strive to improve the efficiency of port loading and unloading, shorten the average time of ship in port as much as possible, improve the level of port management and staff quality, and do a good job in early warning and prevention and control of port natural disasters.

From the perspective of system, port infrastructure system and operation management system are the two main aspects that affect port logistics vulnerability. However, this paper also has the following shortcomings, first of all, the construction of the evaluation index system can be further improved to make the evaluation more comprehensive. Second, the coordinated development between systems has not been presented. In the future, we can try to establish a coupling coordination model to reveal the mechanism of interaction between systems and explore the source of vulnerability between systems. Third, different types of ports have different components, so the construction of the subsystem needs to be further improved. Therefore, future research can be improved and perfected in the above aspects to further verify the feasibility of the evaluation framework. In addition, the method to obtain the port logistic vulnerability inputs of ports can be expanded.

# References

1. Lam, J.S.L., Su, S.: Disruption risks and mitigation strategies: an analysis of Asian ports. Marit. Policy Manag. **42**(5), 415–435 (2015)
2. Fuchs, S., Birkmann, J., et al.: Vulnerability as assessment in natural hazard and risk analysis: current approaches and future challenges. Nat. Hazards **64**(3), 1969–1975 (2012)
3. Cao, X., Lam, J.S.L.: Catastrophe risk assessment framework of ports and industrial clusters: a case study of the Guangdong province. Int. J. Shipp. Transp. Logist. **11**(1), 1–24 (2019)
4. Cao, X., Lam, J.S.L.: Simulation-based catastrophe-induced port loss estimation. Reliab. Eng. Syst. Saf. **175**, 1–12 (2018)
5. Qi, C., Lei, Y.: Operational mechanism and evaluation system for emergency logistics risks. Int. J. Intell. Syst. Appl. **2**(2), 25–32 (2010)
6. Gaillard, J.C.: Vulnerability, capacity and resilience: perspectives for climate and development policy. J. Int. Dev. **22**(2), 218–232 (2010)
7. Mohammed, Z., Mustapha, K.A.: Evaluating and comparing size, complexity and coupling metrics as web applications vulnerabilities predictors. Int. J. Inf. Technol. Comput. Sci. **11**(7), 35–42 (2019)
8. Mumtaz, A.H., Nalin, A.G.A.: On the impact of perceived vulnerability in the adoption of information systems security innovations. Int. J. Comput. Netw. Inf. Secur. **11**(4), 9–18 (2019)
9. Rishabh, D.: Assessing vulnerability of mobile messaging apps to man-in-the-middle (mitm) attack. Int. J. Comput. Netw. Inf. Secur. **10**(7), 23–35 (2018)
10. Du, Q.Q., Kishi, K., Aiura, N., Nakatsuji, T.: Transportation network vulnerability: vulnerability scanning methodology applied to multiple logistics transport networks. Transp. Res. Rec. **2410**, 96–104 (2014)
11. Chernov, I.V.: Scenario analysis of vulnerability in control of complex systems. Autom. Remote. Control. **83**(5), 780–791 (2022)
12. Kurniawan, R., Zailani, S.H., Iranmanesh, M., Rajagopal, P.: The effects of vulnerability mitigation strategies on supply chain effectiveness: risk culture as moderator. Supply Chain Manag. Int. J. **22**(1), 1–15 (2017)

13. Aqlan, F., Lam, S.S.: A fuzzy-based integrated framework for supply chain risk assessment. Int. J. Prod. Econ. **161**, 54–63 (2015)
14. Raimbault, N.: From regional planning to port regionalization and urban logistics. The inland port and the governance of logistics development in the Paris region. J. Transp. Geogr. **78**, 205–213 (2019)
15. Brooks, N.: Vulnerability, risk and adaptation: a conceptual framework. In: Tyndall Centre Climate Change Research Working Paper, vol. 38, pp. 1–16 (2003)
16. Yang, Z.L., Wang, J., Bonsall, S., Fang, Q.G.: Use of fuzzy evidential reasoning in maritime security assessment. Risk Anal. **29**(1), 95–120 (2009)
17. Liu, J., Yang, J.B., Wang, J., SII, H.S., Wang, Y.M.: Fuzzy rule-based evidential reasoning approach for safety analysis. Int. J. General Syst. **33**(2–3), 183–204 (2004)
18. Me-Bar, Y., Valdez, F.: On the vulnerability of the ancient Maya society to natural threats. J. Archaeo Logical Sci. **32**(6), 813–825 (2005)
19. Liu, H.C., Liu, L., Lin, Q.L.: Fuzzy failure mode and effects analysis using fuzzy evidential reasoning and belief rule-based methodology. IEEE Trans. Reliab. **62**(1), 23–36 (2013)
20. Yang, Z.L., Bonsall, S., Wang, J.: Fuzzy rule-based Bayesian reasoning approach for prioritization of failures in FMEA. IEEE Trans. Reliab. **57**(3), 517–528 (2008)
21. Yang, J.B., Liu, J., Wang, J., Sii, H.S., Wang, H.W.: Belief rule-base inference methodology using the evidential reasoning approach – RIMER. IEEE Trans. Syst. Man Cybern. Part A – Syst. Hum. **36**(2), 266–285 (2006)
22. Shieh, J.-I., Wu, H.-H., Huang, K.-K.: A DEMATEL method in identifying key success factors of hospital service quality. Knowl. Based Syst. **23**(3), 277–282 (2010)

# Application Prospect of LNG Storage Tanks in the Yangtze River Coast Based on Economic Model

Jia Tian[1], Hongyu Wu[1(✉)], Xi Chen[2], Xunran Yu[1], and Li Xv[1]

[1] Water Transport Office, Transport Planning and Research Institute Ministry of Transport, Beijing 100028, China
wuhy@tpri.org.cn

[2] Hubei Institute of Logistics Technology, Xiangyang 441000, Hubei, China

**Abstract.** In order to clarify the competitiveness of LNG tank container serving the regions along the Yangtze River, and guide the layout scale of LNG terminals, based on the definition of LNG tank container, analyzed the current situation and the role of LNG tank storage in China. This paper demonstrated the reasonable ship types of LNG bulk carriers and LNG tank carriers on the Yangtze River trunk line, constructed the demonstration model of LNG multi-mode transportation organization on the Yangtze River trunk line, calculated the economy of LNG tank storage and transportation, and forecasted the LNG market scale and LNG tank market scale along the Yangtze River. The research results show that, the economy of the land and water multimodal transport of LNG tank container is equivalent to that of the traditional land and water transport mode of LNG carriers, and the maximum transfer volume of LNG tanks can reach 3.68 million tons in 2035, equivalent to about 216000 FEUs.

**Keywords:** LNG tank container · Application prospect · Ship type selection · Economic model · The Yangtze River

## 1 Introduction

Under the background of global energy green transformation, China's natural gas consumption has continued to grow rapidly [1, 2]. Restricted by the resource endowment characteristics of "rich in coal, poor in oil and less in gas" in China, China imports a large amount of natural gas every year, and has formed four major natural gas import channels of "northwest, northeast, southwest and offshore" [3]. In 2021, the imported natural gas and LNG reaches 168.7 billion m$^3$ and 78.93 million tons respectively. The increasing LNG import demand has promoted the rapid development of domestic LNG vaporization pipeline transportation, LNG tanker transportation, LNG bulk water transportation and LNG tank container transportation. Studying the application prospect of LNG tank container transportation is of great significance, for guiding the layout of LNG receiving stations, filling stations and the development of relevant ship types in the areas along the Yangtze River.

At present, the research of global LNG tank container transportation mainly focuses on two aspects. On the one hand, it is the safety of LNG tank container transportation. Meng F systematically evaluated the safety of LNG tank container transportation [4]. Peng Y A studied the dynamic gas evaporation of LNG tank container through prototype test [5]. Xiong Liansen focused on inland river transportation of LNG tanks and conducted quantitative risk assessment research on inland river transportation of LNG tank container [6, 7]. On the other hand, it is the economy of LNG tank container transportation. Economic activities promote the growth of energy consumption. [8]. Natural gas becomes one of the main fuels burned, due to easy control of flame structure and temperature distribution [9]. The global demand for LNG is strong [10, 11]. Tian YZ and Shan T analyzed the advantages and the bottlenecks restricting of LNG tank container transportation [12, 13]. Gao Zhen compared the economy of door to door supply of LNG tank container with the traditional "bulk carrier and receiving station" supply mode, and excavated the value chain of LNG tank container [14, 15]. In addition, the model and economic method of ship economic and technical demonstration and transportation organization demonstration are relatively mature, and the focus of relevant research is the application of demonstration results [16, 17].

## 2 The Role of LNG Tank Container in the Storage and Transportation Market

### 2.1 Basic Concept of LNG Tank Container

LNG tank container is a kind of special refrigerated liquid pressure vessel equipment with thermal insulation layer, which is mainly composed of tank body and frame. It can be used as both carrier of LNG transportation and storage facility. Due to its characteristics of "suitable for storage and transportation, convenient and flexible, large or small scale, mature logistics network, and multimodal transport of highway, railway and water", it has developed rapidly in recent years. At present, the production technology of LNG tank container has been mature. The natural evaporation rate of the tank container produced by domestic enterprises can reach 0.05% at the lowest, and the non-destructive cold insulation storage time is about 100–180 days.

### 2.2 Current Situation of LNG Tank Container Storage and Transportation in China

LNG tank container business has been developing for a long time abroad. As early as the end of last century, Japan began to use low-temperature ISO tanks for combined transportation [18]. The LNG tank business in China started later than that in foreign countries, but it developed rapidly. It took only a few years from the earliest import of LNG tank container to independent research, development and production of LNG tank container. Against the background of significant growth in domestic natural gas consumption scale and LNG import scale, LNG tank container quickly became a hot spot in the industry in 2018 relying on their own advantages. At the same time, LNG tank container received high attention from the National Energy Administration, the Ministry of Transport and relevant provinces and cities [19].

## 2.3 Function Analysis of LNG Tank Container

Generally, LNG tank container has many advantages. The first is the flexibility of LNG tank container storage and transportation, which can not only be directly delivered to the end user through multimodal transport, but also serve as a temporary storage facility. Second, the LNG tank container has a wide radiation range, and no land and sea transportation restrictions. Third, the utilization efficiency is high, which can be greatly improved through the drop and hook transportation mode. To sum up, in terms of "upstream supply guarantee", LNG tank container can expand more gas supply channels. In terms of "midstream transportation", it covers a wider area in the form of "virtual pipe network" through a mature container logistics network. In terms of "downstream promotion", depending on its advantages of "suitable for storage and transportation, flexible and convenient", it can develop more user resources, solve storage and transportation problems for users, and greatly increase the proportion of LNG in energy consumption.

## 3 Economic Analysis of LNG Tank Container Storage and Transportation

### 3.1 LNG Transport Ship Type Selection

Based on the maintenance water depth of each section of the Yangtze River trunk channel and the relevant bridge height, taking full account of the length to width ratio (L/B), shape to width draft ratio (B/T), shape to depth draft ratio (D/T), length to shape depth ratio (L/D), length to length draft ratio (L/T) and other ship principal dimensions, combined with the ship type dimensions recommended by the design unit, the following ship types representing bulk LNG ship types and LNG tank container transport ships in each section of the Yangtze River trunk line are analyzed:

Considering that it is difficult for the captain of inland ships to exceed 200 m, the representative ship type in Zhenjiang section is 70000m$^3$, the LNG tank container ship should choose 700FEU. Considering that the clearance height of Nanjing Yangtze River Bridge is limited to 24 m, the representative ship type of Wuhu section is 30000 m$^3$, and 400FEU is selected for LNG tank container carrier. Considering the double restrictions of headroom (24 m) and draft (6 m), the LNG carrier in Wuhan section is 25000 m$^3$, 300FEU is selected for LNG tank container carrier. Considering the double restrictions of clearance height (18 m) and maintenance water depth (4.8 m), the LNG carrier in Yueyang section is 10000 m$^3$, 200FEU is selected for LNG tank container carrier. The navigation conditions from Yueyang to Chongqing are basically the same as those from Wuhan to Yueyang, and the representative ship types are the same as those from Yueyang due to the limitation of ship lock dimensions. See Table 1 for the target ship types of each section of the Yangtze River trunk channel.

### 3.2 Construction of Economic Measurement Model

#### 3.2.1 Analytical Methods

In this paper, when evaluating the economy of different transportation modes of LNG, considering the different life cycles of transportation vehicles and the consistent output

Application Prospect of LNG Storage Tanks in the Yangtze River Coast     631

**Table 1.** Target ship type

| Destination | Reference leg | LNG bulk carrier | LNG tank container carrier |
|---|---|---|---|
| Zhenjiang | Section below Nanjing | 70000 m³ | 700FEU |
| Wuhu | Wuhu Nanjing Section | 30000 m³ | 400FEU |
| Wuhan | Wuhan Anqing Section | 25000 m³ | 300FEU |
| Yueyang | Wuhan Yueyang Section | 10000 m³ | 200FEU |

Note: FEU represents a 40-foot ISO standard tank

effect of each scheme, it is proposed to use the annual fee method. The model is as follows:

$$AC = PC(A/P, i_0, n) \quad (1)$$

$$AC = \left[\sum_{i=0}^{n} CO_i \cdot (P/F, i_0, t)\right](A/P, i_0, n) \quad (2)$$

$$(P/F, i_0, t) = 1/(1+i_0)^n \quad (3)$$

$$(A/P, i_0, n) = i_0(1+i_0)^n / ((1+i_0)^n - 1) \quad (4)$$

where: AC is the annual cost value, PC is the present value of the cost, CO is the cash flow of year t, P is the present value of the fund, F is the final value of the fund, i0 is the benchmark discount rate, n is the life cycle, and A is the serial value of equal payment.

### 3.2.2 Cost Composition of LNG Bulk Waterway Transportation

The cost of LNG carrier consists of ship investment cost and ship operation cost (including port use fee). It is assumed that the ship will carry out uninterrupted, round trip and periodic transportation within the annual operation time, and the transportation will be carried out according to 100% of the working load, as shown in Table 2.

### 3.2.3 Cost Composition of LNG Tank Waterway Transportation

The cost composition of LNG tank carrier is similar to that of LNG carrier, which is composed of ship investment cost and ship operation cost (including port use fee). At the same time, the conversion form of LNG tank cost should be considered. In this paper, LNG tank leasing will be used for analysis and calculation, as shown in Table 3.

### 3.2.4 Cost Composition of LNG Tanker Road Transportation

The road transportation cost of LNG tanker distribution includes equipment investment cost and operation cost. It is assumed that the vehicles will be transported uninterruptedly and periodically during the annual operation time, and the transportation will be carried out according to 100% of the working load, as shown in Table 4.

**Table 2.** Cost composition of LNG bulk waterway transportation

| Cost | Constituent elements | core parameters |
| --- | --- | --- |
| Investment cost | Shipbuilding costs | Ship discount rate, route distance, ship speed, residence time of ship in port, annual operating day of ship, ship operating life |
| Operation cost | Fuel cost | Fuel consumption, fuel price |
|  | Operating cost outside fuel | Insurance, maintenance, crew wages and spare parts |
|  | Port disbursement | Entering and leaving port once each time |

**Table 3.** Cost composition of LNG tank waterway transportation

| Cost | Constituent elements | core parameters |
| --- | --- | --- |
| Investment cost | Shipbuilding costs | Ship discount rate, route distance, ship speed, residence time of ship in port, annual operating day of ship, ship operating life |
| Operation cost | Fuel cost | Fuel consumption, fuel price |
|  | Operating cost outside fuel | Insurance, maintenance, crew wages and spare parts |
|  | Port disbursement | Entering and leaving port once each time |
|  | LNG tank container rental fee | Annual rental fee of 100,000 yuan |

**Table 4.** Cost composition of LNG tanker road transportation

| Cost | Constituent elements | core parameters |
| --- | --- | --- |
| Investment cost | Tractor purchase fee | Discount rate 10% |
|  | Trailer purchase fee |  |
| Investment cost | LNG tank purchase fee | Discount rate 10% |
| Operation cost | Fixed cost | Labor remuneration, parts consumption, vehicle maintenance, management costs, inspection costs, insurance costs, etc. |
| Cost | Constituent elements | core parameters |
| Operation cost | Variable cost | Fuel costs, road and bridge costs, maintenance costs, etc. |

### 3.2.5 Cost Composition of LNG Tank Container Road Transportation

The cost of road transportation of LNG tank container is basically the same as that of LNG tankers, as shown in Table 5.

Application Prospect of LNG Storage Tanks in the Yangtze River Coast 633

Table 5. Cost composition of LNG tank container road transportation

| Cost | Constituent elements | core parameters |
|---|---|---|
| Investment cost | Tractor purchase fee | Discount rate 10% |
|  | Trailer purchase fee |  |
| Operation cost | Fixed cost | Labor remuneration, parts consumption, vehicle maintenance, management costs, inspection costs, insurance costs, LNG tank container rental fee, etc. |
|  | Variable cost | Fuel costs, road and bridge costs, maintenance costs, etc. |

### 3.2.6 Cost Composition of Railway Transportation of LNG Tanks

The railway transportation cost of LNG mainly includes the basic railway freight, railway miscellaneous charges, etc., which shall be determined by the competent department of railway freight rates. The occupation cost of LNG tanks in the transportation process shall also be considered, as shown in Table 6.

Table 6. Cost composition of railway transportation of LNG tanks

| Cost | Constituent elements | core parameters |
|---|---|---|
| the basic railway freight | Railway freight tickets (base price 1, base price 2) |  |
|  | Railway construction fund |  |
| Railway miscellaneous charges | Transport fee | Other expenses include staff service fees, storage costs, etc., railway miscellaneous charges 1000 yuan/FEU |
|  | Locomotive operating costs |  |
|  | Handling fee, and Other costs |  |
| LNG tank container occupancy cost | LNG tank container rental fee |  |

### 3.3 Calculation of Storage and Transportation Economy

#### 3.3.1 Transportation Route Selection

It is assumed that LNG is imported from Bintulu Port in Malaysia to various provinces along the Yangtze River in China, such as Jiangsu, Anhui, Hubei and Hunan provinces. Zhenjiang, Wuhu, Wuhan and Yueyang are selected as waterway transit stations respectively, and the four ports are taken as the center for distribution to end users within 700 km of the province through road transportation. The relevant route selection is shown in the following Table 7.

**Table 7.** Transportation route and distance

| Place of origin | Destination | Vessel distance (nautical miles) | Vessel distance (kilometre) |
|---|---|---|---|
| Malaysia – Bintulu | Zhenjiang | 2067 | 3828 |
| | Wuhu | 2153 | 3987 |
| | Wuhan | 2482 | 4597 |
| | Yueyang | 2627 | 4865 |

### 3.3.2 Calculation of Storage and Transportation Costs

According to the actual market application and possible future application, the forms of highway, water and railway multimodal transport are combined respectively. The railway has not been put into commercial operation, and the LNG terminal and storage yard have not yet been equipped with railway facilities; In addition, railway distribution also requires highway transit cooperation, which will involve more links and costs, so railway transportation will not be considered in land and water combined transport temporarily. The calculation results are shown in Table 8.

**Table 8.** Comprehensive Cost of LNG Water land Intermodal Transportation

| Distribution area | Transport mode | LNG carrier | LNG tank container carrier | LNG receiving stations | LNG storage yards | ING tanker | LNG tank container road | Comprehensive Cost |
|---|---|---|---|---|---|---|---|---|
| Within and around Jiangsu Province | bulk + tanker | 385 | 0 | 346 | 0 | 353 | 0 | 1084 |
| | Bulk and tank container | 385 | 0 | 346 | 0 | 0 | 308 | 1039 |
| | Packaging and road | 0 | 724 | 0 | 50 | 0 | 308 | 1082 |
| Within and around Anhui Province | Bulk and tanker | 635 | 0 | 346 | 0 | 353 | 0 | 1335 |
| | Bulk and tank container | 635 | 0 | 346 | 0 | 0 | 308 | 1290 |
| | Packaging and road | 0 | 948 | 0 | 50 | 0 | 308 | 1315 |

(*continued*)

**Table 8.** (*continued*)

| Distribution area | Transport mode | LNG carrier | LNG tank container carrier | LNG receiving stations | LNG storage yards | ING tanker | LNG tank container road | Comprehensive Cost |
|---|---|---|---|---|---|---|---|---|
| Within and around Hubei Province | Bulk and tanker | 678 | 0 | 346 | 0 | 353 | 0 | 1378 |
| | Bulk and tank container | 678 | 0 | 346 | 0 | 0 | 308 | 1333 |
| | Packaging and road | 0 | 1110 | 0 | 50 | 0 | 308 | 1468 |
| Within and around Hunan Province | Bulk and tanker | 1100 | 0 | 346 | 0 | 353 | 0 | 1800 |
| | Bulk and tank container | 1100 | 0 | 346 | 0 | 0 | 308 | 1755 |
| | Packaging and road | 0 | 1352 | 0 | 50 | 0 | 308 | 1711 |

Note: Unit is yuan/ton

The demonstration results show that the economy of bulk transportation of LNG carrier is obviously better than that of LNG tank container carrier from the perspective of upper transportation links alone; The economy of LNG tank container road is obviously better than that of LNG tanker; The economy of the multimodal transportation of LNG tank container by water and land is equivalent to that of the traditional transportation mode of LNG carriers by water and land.

## 4 Application Prospect of LNG Tank Container Along the Yangtze River

### 4.1 Analysis of Domestic LNG Application Prospect

At this stage, LNG is mainly used for urban gas storage and peak shaving, clean fuels for transportation and industry. In 2020, there will be about 5.71 million tons of LNG used for peak shaving by urban residents and industrial gas storage, 6.43 million tons of LNG used for transportation clean fuel, 11.43 million tons of LNG used for industrial fuel, and 7.14 million tons of LNG used for power generation.

Under the new requirements of the "double carbon" goals, natural gas, as a clean fossil energy, has large growth potential. According to the "14th Five-Year Plan" energy plan, it is estimated that China's natural gas consumption demand will reach about 450 billion $m^3$ in 2025.The domestic natural gas output will reach 230 billion $m^3$, and the LNG and pipeline gas imports will reach 120 billion $m^3$ and 100 billion $m^3$ respectively. In the future, LNG related application scenarios will include gas consumption in remote cities and towns, gas storage and peak shaving, transportation energy, industrial fuel gas and power generation, and gasification into the network. LNG has broad application prospects in China.

### 4.2 LNG Market Demand Along the Yangtze River

The Yangtze River Economic Belt accounts for more than 40% of China's GDP, with huge economic volume and strong demand for natural gas. In terms of natural gas supply, pipeline gas supply is the main supply at present, but the gas source is single and the gap of natural gas supply is gradually expanding, so it needs to be supplemented by imported LNG.

Based on the analysis of the current situation of the natural gas production, supply, storage and marketing system in various cities of the Yangtze River Economic Belt and the prediction of the future development trend, the LNG market scale in the Yangtze River basin is predicted to be 14 million tons by 2035 by using the production, transportation and marketing balance analysis method, taking full account of the planning and layout of LNG terminals along the Yangtze River, the promotion of early work, industry management and other factors. The unloading scale at each station is shown in Table 9.

**Table 9.** Statistics of unloading scale at each station

| Province | Port | LNG market scale (10000 tons/year) |
|---|---|---|
| Hubei | Wuhan Port | 200 |
| Hunan | Yueyang Port | 200 |
| Jiangxi | Jiujiang Port | 200 |
| Anhui | Wuhu Port | 200 |
| Jiangsu | Jiangyin Port | 150 |
|  | Suzhou Port | 300 |
|  | Zhenjiang Port | 150 |
| Total |  | 1400 |

### 4.3 Prediction of LNG Tank Container Market Scale Along the Yangtze River

First, based on the analysis of the current transportation mode and scale, the imported LNG will still be transported in bulk by LNG carriers in 2025, but some of them are still imported in the form of LNG tank container through container ships or Special ship for LNG tank container. In the future, with the expansion of demand and the diversification of import sources, the import form of special ships for LNG tank container will have great development opportunities, and will also usher in a period of rapid development.

Application Prospect of LNG Storage Tanks in the Yangtze River Coast        637

Secondly, at present, the main distribution mode is still LNG tanker. According to the calculation, if an LNG tanker operates at full load for 300 days per year and the economic haul distance is 300 km, the annual transfer capacity is 6000 tons. Based on the existing 12000 vehicles, it is theoretically consistent with the expected distribution demand of 64.29 million villages. However, there is still a big gap in the actual operation process. Especially when the transportation distance and the gas storage capacity in remote areas are low, and LNG tank container are more flexible than LNG tanker, which are suitable for storage and transportation and can realize multimodal transportation. LNG tank container will usher in a rapid development period.

Finally, with the help of the existing mature container logistics network, LNG tank container can avoid multiple LNG reloading during transportation, reduce intermediate operations, reduce logistics losses, improve logistics efficiency, and achieve a wider coverage and higher transportation efficiency. Therefore, LNG tank container transportation will become the main development direction of LNG storage and transportation in the future due to its obvious advantages such as economic investment, strong adaptability and simple operation.

To sum up, it is predicted that the maximum LNG tank container transfer volume in 2035 will reach 3.68 million tons, about 216000 FEUs.

### 4.4 Suggestions for the Development of LNG Tank Transportation Along the Yangtze River

1) It is suggested to establish management standards suitable for LNG tank container transportation system. It is suggested to optimize the relevant requirements for carrying LNG tank container on the whole ship, manage the dangerous goods carried by existing container ships, and appropriately relax the safety distance requirements. Adjust the storage management requirements of LNG tank container in the storage yard.
2) It is suggested to coordinate relevant competent departments to form an integrated LNG tank container standard system and promote multimodal transport of LNG tank container. In terms of LNG tank container construction, highway, railway and water have their own relevant standard systems. These standards are not different from each other in terms of technology and are easy to solve.
3) It is suggested to coordinate all competent departments to realize management integration while promoting the integration of LNG tank container standards and inspection.

## 5 Conclusion

In this paper, the development prospects of LNG tank container along the Yangtze River are systematically studied. The main conclusions are as follows:

1) The development status of LNG tank container storage and transportation in China is analyzed, and the role of LNG tank container is clarified from upstream supply, midstream transportation and downstream service of natural gas;

2) The ship types of LNG bulk carriers and LNG tank carriers in each section of the Yangtze River trunk line are defined, and the demonstration model of LNG multimodal transport organization in the Yangtze River trunk line is constructed. The demonstration results show that the economy of LNG tank container multimodal transport by water and land is equivalent to that of the traditional water and land transport mode of LNG carriers;
3) The study clarified that LNG will have a broad application prospect in the fields of gas consumption in remote cities and towns, gas storage and peak shaving, transportation energy, industrial gas and power generation, and gasification into the network in the future. It is predicted that the LNG market in the Yangtze River basin will reach 14 million tons in 2035, of which the maximum LNG tank container transfer volume can reach 3.68 million tons, about 216000 FEUs.

The limitations of this paper are as follows: First, the economy of LNG tank container storage and transportation is based on the demonstration of offshore LNG direct delivery, and the more complex situation of offshore cargo source transshipment has not been fully considered; Second, at present, the domestic regulations on the management of the sea transportation of LNG bulk into the river and the special ship for LNG tank containers have not been clarified, and changes in policy may have a significant impact on LNG transportation costs.

## References

1. Sahed, A., Mékidiche, M., Kahoui, H.: Forecasting natural gas prices using nonlinear autoregressive neural network. Int. J. Math. Sci. Comput. (IJMSC) **6**, 37–46 (2020)
2. Yang, L., Han, J., Wang, N., et al.: Development situations of LNG terminals in China. Oil Gas Storage Transp. **335**(11), 1148–1153 (2016). (in Chinese)
3. Ye, J., Yang, Y., Shen, G.: Strategy study on optimizing energy structure in china under low carbon economy. Shanxi Coking Coal Sci. Technol. **181**(06), 54–56 (2010). (in Chinese)
4. Meng, F., Ma, L., Wang, X.: An approach on the evaluation of LNG tank container transportation safety. Int. J. Eng. Manag. Res. **09**(5), 44–53 (2019)
5. Peng, Y.A., Ycy, A., Zjw, B., et al.: A prototype test of dynamic boil-off gas in liquefied natural gas tank containers - ScienceDirect. Appl. Thermal Eng. **05** (2020)
6. Xiong, L.: Quantitative risk assessment on inland waterway transport of LNG containers. Chongqing University (2021). (in Chinese)
7. Wahid, F., Ghazali, R., Fayaz, M., et al.: Statistical features based approach (SFBA) for hourly energy consumption prediction using neural network. Int. J. Inf. Technol. Comput. Sci. (IJITCS) **5**, 23–30 (2017)
8. Gu, H., He, P., Mei, S., Xie, J., et al.: Simulation on different proportions of coal and natural gas co-combustion in a rotary lime kiln. Int. J. Intell. Syst. Appl. (IJISA) **4**, 17–24 (2011)
9. Naim, I., Mahara, T.: Comparative analysis of univariate forecasting techniques for industrial natural gas consumption. Int. J. Image Graph. Sig. Process. (IJIGSP) (10), 33–44 (2018)
10. Tsougranis, E.-L., Wu, D.: Dual reutilization of lng cryogenic energy and thermal waste energy with organic rankine cycle in marine. Appl. Energy Procedia **142**, 1401–1406 (2017)
11. Kumar, S., Kwon, H.T., Choi, K.H., et al.: Current status and future projections of LNG demand and supplies: a global prospective. Energy Policy **39**(7), 4097–4104 (2011)
12. Tian, Y.Z., Liu, Y.: Difficulties and solutions for bulk transportation of LNG in tank container. Ship Boat **171**(01), 19–25 (2018). (in Chinese)

13. Shan, T., Zhang, C., Duan, P., et al.: Design of the integrated structure system for the pipeline and maintenance ladder inside LNG storage tank. Oil Gas Storage Transp. **358**(10), 1180–1185 (2018). (in Chinese)
14. Gao, Z., Chang, X., Zhao, S., et al.: Economic analysis of the main supply mode and the liquid distribution of LNG tank container. Int. Petrol. Econ. **30**(05), 66–73 (2022). (in Chinese)
15. Gao, Z.: Analysis of LNG tank container value chain and its development suggestions. World Petrol. Ind. **209**(02), 17–22+39 (2022). (in Chinese)
16. Zhang, G., Haiting, A.N., Liu, Y., et al.: Economic argumentation model and system design for refitting bulk carrier into aquaculture engineering ship. Fishery Modern. **45**(02), 1–5 (2018)
17. Lan, H., Zhang, P., Transportation, S.O.: Transportation organization and ship type demonstration of container lines. Logist. Technol. **35**(07), 21–23 (2016). (in Chinese)
18. Wang, S.: Discussion on current situation and prospect of LNG tank business. Pop. Standard. **311**(18), 134–135 (2019). (in Chinese)
19. Liang, Y., Zhou, S., Wang, Z., et al.: Development status and prospect of LNG tank containers. Int. Petrol. Econ. **27**(06), 65–74 (2019). (in Chinese)

# Multi-depot Open Electric Truck Routing Problem with Dynamic Discharging

Xue Yang[1] and Ning Chen[1,2(✉)]

[1] Smart Port and Logistics Laboratory, School of Transportation and Logistics Engineering, Wuhan University of Technology, Wuhan 430063, China
319203@whut.edu.cn

[2] Sanya Science and Education Innovation Park of Wuhan University of Technology, Sanya 572000, China

**Abstract.** Aiming at the problems of electric trucks in actual logistics distribution, such as short driving range, long charging time, and shortage of charging stations. The open electric vehicle routing problem for the joint distribution of multiple depots is proposed, considering the factors of load dynamics affecting discharging (DD-MDOEVRPTW). And the mixed integer programming model of this problem is established with the minimum total cost as the objective function. Based on the idea of destruction and repair in the large-scale neighborhood search algorithm, the local search operation is added and the hybrid genetic algorithm with two-layer encoding is devised to solve the problem. Finally, the model and algorithm are tested and numerically analyzed using arithmetic examples to verify their effectiveness. The results show that the DD-MDOEVRPTW is more in line with the actual operation situation of logistics enterprises. Moreover, the increase of charging function in distribution centers and the joint distribution of multiple depots can effectively increase the efficiency of enterprises.

**Keywords:** Dynamic discharging · Multi-depot · Electric truck routing problem · Hybrid genetic algorithm

## 1 Introduction

In the context of carbon neutrality, the pressure on the freight industry to reduce carbon emissions has increased dramatically. To reduce carbon emissions, the European Union will increase the construction of infrastructure related to new energy vehicles and add one million charging stations in the EU by 2025 [1]. Electric trucks, which are pollution-free and environmentally friendly, are widely used in many scenarios of logistics and distribution, especially in the 'last mile' distribution, which plays an important role. Although electric trucks have many advantages, problems such as the limited driving range, long charging times, and inadequate charging stations are still difficult points for logistics operations.

The multi-depot vehicle routing problem (MDVRP) was first proposed by Tillman (1969) in which there are multiple depots and vehicles must depart from and return to the same depot [2]. The literature has conducted various extended studies on the MDVRP problem, including the problem of the time window, backhaul, pickup and delivery, multi-depot location, as well as the intermediate depot can be used as a refueling station. MDVRP is an NP-hard problem, and the previous solution of MDVRP tends to transform this problem into a single-center vehicle routing problem. By adding a virtual warehouse center, an improved ant colony algorithm was proposed to solve the MDVRP problem [3]. Subsequently, researchers have also done various extended studies on MDVRP. The MDVRP problem that can be refueled at distribution centers was studied. A three-stage approach based on adaptive memory and taboo search was proposed and solved faster [4]. This study expands the research depth and scope of MDVRP and lays a foundation for the subsequent expansion. An improved multi-objective genetic algorithm is proposed to study the multi-depot vehicle routing problem with time windows, and the effectiveness of the proposed algorithm is verified based on real traffic data [5]. The EVRP problem with multiple depots, time windows, and nonlinear charging constraints was studied and solved by introducing a genetic algorithm with two-layer encoding and multiple crossover methods [6]. A multi-depot electric vehicle routing optimization problem considering two-dimensional loading with distance minimization as the objective was studied. A variable neighborhood search algorithm to solve the electric vehicle routing sub-problem and a space-saving heuristic algorithm to solve the box loading problem were designed [7]. The location routing problem of electric vehicles in multiple depots with time windows was studied, and two battery charging strategies were considered [8].

It can be seen from the above research that the objective function is generally the lowest total cost and is solved using a heuristic algorithm. With the popularity of the concept of low carbon and environmental protection, countries are vigorously promoting the use of electric vehicles. The application of electric vehicles in the field of logistics and transportation is also developing rapidly. In summary, there has been some research on electric vehicle routing problems, but most of the research focuses on the electric vehicle routing problem of the single depot, and there is little research combining the open multi-depot electric vehicle routing problem and the dynamic battery power consumption rate. In this paper, the dynamic influence of vehicle load, speed, road slope, upwind area, and other factors on the battery power consumption of electric vehicles is considered. The dynamic power consumption rate model is established to study the DD-MDOEVRPTW.

The main contributions of the article are as follows: (1) For the first time, the dynamic power consumption rate is combined with MDOEVRPTW to design the power consumption model, which provides a better distribution route solution for logistic operation. (2) The vehicles are allowed to charge in depots and charging stations, which effectively alleviates the imperfect status of public charging stations and provides a better charging plan. (3) Design a two-dimensional coding hybrid genetic algorithm, combined with the idea of destruction and repair in a large-scale neighborhood search algorithm, enhance the local search ability of the genetic algorithm and get the solution fast.

The rest of the paper is structured as follows: Sect. 2 describes the DD-MDOEVRPTW and presents the mathematical model. The design of the hybrid genetic algorithm is presented in Sect. 3. Section 4 describes the numerical experimental study and discussion the results in detail. Section 5 concludes with the text and discusses future research directions.

## 2 Problem Description and Mathematical Formulation

### 2.1 Problems' Description and Mathematical Notation

DD-MDOEVRPTW has multiple depots and multiple homogeneous electric trucks to jointly provide distribution service for customers with known demand, location, and time windows. However, the range of electric trucks is limited by the capacity of their batteries. Therefore, on the way to distribution, electric trucks sometimes need to make a detour to visit a nearby charging station or depot charging. After the distribution is completed, it will return to the neighboring depot.

This paper establishes the dynamic power consumption rate model and studies the DD-MDOEVRPTW. With the fixed cost, transportation cost, charging cost, and penalty cost of violating the time window, an electric truck routing optimization model with a minimum total cost is established to solve the distribution routing plan and charging plan. The notation is described in Table 1.

**Table 1.** Mathematical notation

| Notation | Notation description |
| --- | --- |
| Sets | |
| $O$ | Set of depots: $O = \{1, 2, 3, \cdots, o\}$ |
| $N$ | Set of customers: $N = \{1, 2, 3, \cdots, n\}$ |
| $K$ | Set of the number of vehicles: $K = \{1, 2, 3, \cdots, k\}$ |
| $F$ | Set of public charging stations: $F = \{1, 2, 3, \cdots, f\}$ |
| $F'$ | Set of depots and public charging stations: $F' = \{F \cup O\}$ |
| $V$ | Set of depots, customers and public charging stations: $V = O \cup N \cup F$ |
| Parameters | |
| $d_{ij}$ | Distance from node $i$ to node $j$ |
| $m_c$ | Vehicle weight |
| $p_i$ | The demand of customer node $i$ |
| $C$ | Maximum loading capacity of the vehicle |
| $Q$ | Maximum battery capacity of the vehicle |
| $e_i$ | The earliest service time of customer node $i$, $i \in N$ |
| $l_i$ | The latest service time of customer node $i$, $i \in N$ |
| $t_{ij}$ | The time for the vehicle to travel from node $i$ to node $j$, $i, j \in V$ |
| $v$ | Speed |

(*continued*)

**Table 1.** (*continued*)

| Notation | Notation description |
|---|---|
| Parameters | |
| $\partial$ | Penalty function coefficients for violation of vehicle capacity |
| $\beta_1$ | Unit penalty cost for vehicles arriving early at customer nodes |
| $\beta_2$ | Unit penalty cost for vehicles arriving later than the customer node |
| $C_0$ | Fixed cost per unit of vehicle |
| $C_1$ | Distribution cost per unit distance |
| $C_2$ | Charging cost per unit of power |
| Decision variables: | |
| $x_{ijk}$ | 1 if $arc(i,j)$ is traversed by a vehicle; 0 otherwise |
| $y_{ik}$ | 1 if a vehicle recharges at charging station $i$; 0 otherwise |
| $p_{ik}$ | The load capacity of vehicle $k$ at point $i$ |
| $q_i$ | The power of the vehicle at node $i$ |
| $q_{ik1}$ | Remaining power of vehicle $k$ when it reaches node $i$ |
| $q_{ik2}$ | Remaining power of vehicle $k$ when it leaves node $i$ |
| $w_{ik}$ | Vehicle $k$ replenished at charging station node $i$ |
| $t_{i1}$ | Time of vehicle arrival at node $i, i \in V$ |
| $t_{i2}$ | Time of vehicle leaving node $i, i \in V$ |
| $tw_i$ | Vehicle waiting time at customer node $i, i \in V$ |
| $tf_i$ | Service time of the vehicle at node $i, i \in V$ |

## 2.2 Electric Truck Power Consumption Model

The main factors affecting discharging rate of electric trucks are as follows: (1) Total vehicle weight. (2) Road slope. Driving uphill requires more power than driving on flat or downhill roads. (3) Travel speed, which affects rolling resistance and air resistance. (4) Electrical energy conversion efficiency [9].

During the driving of the electric truck, the vehicle needs to overcome air resistance, rotational resistance, and gravity and convert mechanical energy into electrical energy, which in turn is converted into battery energy [10]. The power consumption model per unit mileage was established on the $arc(i,j)$ as follows [11].

$$P_{ij} = (C_d \cdot \rho A v^2 / 2 + (m_c + p_{jk}) \cdot g \cdot \theta) \cdot v_{ij} \tag{1}$$

$$E_{ij} = \varphi \cdot \phi (C_d \cdot \rho A v^2 / 2 + (m_c + p_{jk}) \cdot g \cdot \theta) \tag{2}$$

$$\theta = C_r \cos(\alpha_{ij}) + \sin(\alpha_{ij}), \varphi \cdot \phi = \begin{cases} \varphi^d \cdot \phi^d \ P_{ij} \geq 0KW \\ \varphi^r \cdot \phi^r \ P_{ij} < 0KW \end{cases} \quad (3)$$

$P_{ij}$ is the mechanical power, $E_{ij}$ is the power consumption per unit mileage, $C_d$ is the air resistance coefficient, $C_r$ is the rolling friction coefficient, $\rho$ is the air density, $A$ is the windward area of the vehicle, $m_c$ is the vehicle deadweight, $p_{jk}$ is the load capacity of the vehicle at point $j$, $g$ is the acceleration of gravity, $\alpha_{ij}$ is the road slope, $\varphi^d$ is the regression coefficient of motor mode battery efficiency, $\varphi^r$ is the regression coefficient of generator mode battery efficiency, $\phi^d$ is the motor mode electrical efficiency regression coefficient, $\phi^r$ is the regression coefficient of generator mode electrical efficiency.

## 2.3 Mathematical Models for DD-MDOEVRPTW

Based on the analysis of the problem and the construction of the dynamic power consumption rate model for electric trucks, the DD-MDOEVRPTW model is constructed.

$$\min Z = (C_0 \sum_{k \in K} x_{ok} + C_1 \sum_{k \in K} \sum_{i \in V} \sum_{\substack{j \in V \\ j \neq i}} d_{ij} x_{ijk} + C_2 \sum_{k \in K} \sum_{i \in \{F \cup O\}} y_{ik} w_{ik})$$

$$+ \beta_1 \sum_{i \in N} \max(e_i - t_i, 0) + \beta_2 \sum_{i \in N} \max(t_i - l_i, 0) \quad (4)$$

$$\sum_{j \in N, i \neq j} x_{ijk} = 1, \forall i \in N, k \in K \quad (5)$$

$$\sum_{j \in V, i \neq j} x_{ijk} \leq 1, \forall i \in D, \forall k \in K \quad (6)$$

$$\sum_{i \in V, i \neq j} x_{jik} - \sum_{i \in V, i \neq j} x_{ijk} = 0, \forall j \in \{V \setminus O\}, k \in K \quad (7)$$

$$0 \leq p_{ok} \leq C, \forall o \in O, \forall k \in K \quad (8)$$

$$0 \leq p_{jk} \leq p_{ik} - p_j x_{ijk} + C(1 - x_{ijk}), \forall i \in V, \forall j \in \{N \setminus O\}, i \neq j, k \in K \quad (9)$$

$$q_{ik2} = Q, \forall i \in O, \forall k \in K \quad (10)$$

$$0 \leq q_{jk1} \leq q_{ik1} + w_{ik} - E_{ij} d_{ij} x_{ijk} + Q(1 - x_{ijk}), \forall i, j \in V, i \neq j, k \in K \quad (11)$$

$$0.2Q \leq q_{jk1} \leq Q, \forall j \in V, \forall k \in K \quad (12)$$

$$q_{ik1} + w_{ik} = Q, \forall i \in F', \forall k \in K \quad (13)$$

$$q_{ik1} = q_{ik2}, \forall i \in N, \forall k \in K \quad (14)$$

$$t_{o2} = 0, \forall o \in O \tag{15}$$

$$tw_i = \max[0, (e_i - t_{i1})], \forall i \in N \tag{16}$$

$$t_{i2} = t_{i1} + tf_i + tw_i, \forall i \in V \tag{17}$$

$$t_{ij} = d_{ij}/v, \forall i, j \in V \tag{18}$$

$$t_{j1} = \sum_{i \in V} \sum_{j \in V, i \neq j} x_{ijk}(t_{i2} + t_{ij}), \forall k \in K \tag{19}$$

$$x_{ijk} \in \{0; 1\}, \forall i, j \in V, \forall k \in K \tag{20}$$

$$y_{ik} \in \{0; 1\}, \forall i \in V, \forall k \in K \tag{21}$$

The objective function is to minimizes the total cost, which includes fixed cost, delivery cost, charging cost and penalty cost of violating the time window, as is shown in Eq. (4). Constraint (5) enforce that each point of customer must be visited only once by one vehicle. Constraint (6) indicates that each vehicle is used only once. Constraint (7) ensures the flow conservation of electric trucks in and out of customer points and charging stations, and the vehicle can return to the nearby distribution center after the delivery. The constraints (8)–(9) is the load constraint of vehicle capacity. Constraint (8) indicates that the maximum loading capacity of an electric truck is $C$ when it starts from the depot. Constraint (9) represents the loading capacity change of electric trucks during distribution. The constraints (10)–(14) refers to the power constraint of electric trucks. Constraint (10) indicates that the electric truck is in full charge when it leaves the depot. Constraint (11) represents the power level when the electric truck reaches each node under the influence of the dynamic discharging. Constraint (12) indicates that considering some other external factors, such as weather temperature, etc., the remaining power when the vehicle reaches any node is not less than 20% of the full power. Constraint (13) indicates the charging electricity of electric trucks at the charging node. Constraint (14) represents that the electricity of the vehicle unchanged before and after visiting the customer node. The constraints (15)–(19) are the time constraint. Constraint (15) indicates that the moment when the vehicle departs depots is 0. Constraint (16) means that if the vehicle arrives at the customer point in advance, the waiting time $tw_i$ is $e_i - t_{i1}$; otherwise it is 0. Constraint (17) means that the time when the vehicle leaves the node is equal to the time when the vehicle arrives at the point plus its service time, charging time, and waiting time at the point. Constraint (18) is the time required for a vehicle to travel from node $i$ to node $j$. . Constraint (19) indicates that the time for a vehicle to arrive at the node $j$ is equal to its time to leave the previous node $i$ plus the time it takes to travel from node $i$ to node $j$. . Constraints (20)–(21) indicate the domain of the decision variables.

## 3 Hybrid Genetic Algorithm for DD-MDOEVRPTW

The DD-MDOEVRPTW is a variant of the vehicle routing problem and an NP-hard problem. In this paper, we consider the dynamic power consumption rate problem under multiple influencing factors, which increases the difficulty of solving it. The genetic algorithm has a strong ability to find the best, but the local search ability is poor, so this paper designs a hybrid genetic algorithm to solve the problem. Drawing on the idea of destruction and repair in the large-scale neighborhood search algorithm, the local search operation is added to enhance the local search ability of the genetic algorithm.

### 3.1 Coding Design

The classical multi-depot vehicle routing problem is solved by generally using a one-dimensional integer encoding to display the index of customers to be visited in a sequential connection [12]. In contrast, for DD-MDOEVRPTW with a dynamic discharging rate, there are additional constraints of depot selection and vehicle battery capacity limitation. To solve this problem, this paper designs a two-dimensional coding method. The first-dimensional coding indicates the index of customer points to be visited and the second-dimensional coding indicates the vehicle number, as shown in Fig. 1, which can visually represent a solution to the problem.

| Index   | 2 | 4 | 9 | 8 | 7 | 6 | 5 | 3 | 1 |
|---------|---|---|---|---|---|---|---|---|---|
| Vehicle | 1 | 2 | 2 | 1 | 2 | 1 | 2 | 1 | 1 |

**Fig. 1.** Schematic diagram of chromosome coding

### 3.2 Population Initialization

Based on the above chromosome coding method, charging stations and depots are added to obtain the initial chromosome. The steps are as follows. Step 1: Randomly arrange all customer points to get the first-dimensional encoding of the chromosome. Step 2: Randomly select vehicles for distribution to get the second-dimensional encoding of the chromosome. Step 3: Consider the load constraint of the vehicle, if $\sum_{i=1}^{n} p_i \leq C$ and $\sum_{i=1}^{n+1} p_i > C$, and insert the nearest depot to the customer point after the nth chromosome. Repeat Step 3 until all customer points are visited. Step 4: Consider the power constraint of the vehicle, when the power is insufficient, insert the nearest charging station or depot to the customer point, and repeat Step 4 until all customer points are traversed. Step 5: If neither Step 3 nor Step 4 is satisfied, insert the nearest depot and repeat Step 5 until all customer points are traversed. Step 6: If all customer points are served, then output the feasible solution, otherwise return to Step 1 for customer points that cannot be accessed due to not satisfying the vehicle load or power constraints until all customer points are served.

### 3.3 Constraint Processing and Fitness Functions

According to the DD-MDOEVRPTW studied in this paper, the vehicle battery range constraint, rated load limit and time window constraint are treated in the form of a penalty function, and the objective function obtained is as Eq. (22).

$$\min Z = (C_0 \sum_{k \in K} x_{ok} + C_1 \sum_{k \in K} \sum_{i \in V} \sum_{\substack{j \in V \\ j \neq i}} d_{ij} x_{ijk} + C_2 \sum_{k \in K} \sum_{i \in \{F \cup O\}} y_{ik} w_{ik}) + \beta_1 \sum_{i \in N} \max(e_i - t_i, 0)$$

$$+ \beta_2 \sum_{i \in N} \max(t_i - l_i, 0) + F_1 \max(0.2Q - q_i, 0) + F_2 \max(\sum_{i \in N} p_i - C, 0) \tag{22}$$

To satisfy the above three constraints, $F_1$ and $F_2$ are taken as infinite positive numbers to make the objective function value of chromosomes that do not meet the constraints infinite. The inverse of the above objective function is taken as the fitness function, which is $fit(i) = 1/\min(Z)$.

### 3.4 Genetic Manipulation

In genetic operators, the selection operation combines roulette selects strategy and elite selection strategy. The crossover operation adopts the order crossover method, and the mutation operation selects the two-point mutation. Finally, the algorithm ends when it runs to a set number of iterations.

### 3.5 Insert Local Search

The local search operation first selects a customer at random from the original set of customers and removes a certain number of customers in turn according to the relevance size as in Eq. (23) and (24), and then use the repair operator to reinsert the removed customers back into the corrupted solution.

$$R(i, j) = 1/(d'_{ij} + V_{ij}) \tag{23}$$

$$d'_{ij} = d_{ij}/\max(d_{ij}) \tag{24}$$

$R(i, j)$ represents correlation; $d'_{ij}$ and $d_{ij}$ are the normalized value; $d_{ij}$ represents the Euclidean distance between point $i$ and point $j$; $V_{ij}$ represents whether point $i$ and point $j$ are on the same route, if the value is 0, otherwise it is 1.

## 4 Numerical Experiments and Analysis

In this paper, data from the VRP website [13], and the MDVRPTW cases in the public standard case library are modified to design the DD-MDOEVRPTW examples that are applicable to the study. The values of each parameter of the dynamic power consumption are set as in Table 2 [14].

**Table 2.** Power consumption model parameter values

| Parameter symbols | Parameter meaning | Values |
| --- | --- | --- |
| $g$ | Gravitational acceleration (m/s$^2$) | 9.81 |
| $C_d$ | Air resistance coefficient | 0.7 |
| $C_r$ | Rolling constant friction coefficient | 0.01 |
| $A$ | Vehicle windward area | 3.912 |
| $\rho$ | Air Density (kg/m$^3$) | 1.205 |
| $\phi^d$ | Efficiency parameters (motor mode) | 1.184692 |
| $\phi^r$ | Efficiency parameters (generator mode) | 0.846055 |
| $\varphi^d$ | Discharge efficiency parameters | 1.112434 |
| $\varphi^r$ | Recovery efficiency parameters | 0.928465 |
| $\alpha$ | Road slope | [0°, 19°] |

**Table 3.** Parameters related to energy consumption coefficient of electric trucks

| $m_c$ | $C$ | $Q$ | $C_0$ | $C_1$ | $C_2$ | $\beta_1$ | $\beta_2$ |
| --- | --- | --- | --- | --- | --- | --- | --- |
| 3760 | 1335 | 80 | 100 | 1 | 3 | 10 | 15 |

A brand of a purely electric truck is selected as the distribution vehicle in this paper, and the vehicle-related parameters are shown in Table 3 [15]. The cost of each parameter in the distribution process is set in Table 3. The hybrid algorithm is written by MATLAB (R2019a) under Windows 10 operating system, 8 GB memory, and a 2.5 GHz Intel Core i5 processor for a model solution.

The hybrid genetic algorithm is used to solve the model, and four examples with the number of customers 0–50, 50–100, 100–150, and 150–200 in the basic examples are selected for verification. Table 4 records the average results after 10 runs for each case, where the vehicle can be charged at the depot. Table 5 records the results of the cases when the depot doesn't have a charging function. Each column from left to right in the table shows the case number, the number of customer points, the number of depots, the number of public charging stations, the maximum number of available vehicles, the total cost, the fixed cost, the penalty cost, the charging cost, and the distribution cost. It can be seen from the results that with the increase of the number of customer points, the time window penalty cost, charging cost and distribution cost increase significantly. The total cost can be reduced by increasing the charging function of the depot, in which the penalty cost and charging cost are significantly reduced. Compared with the single depot in the benchmark example, the multi-depots can reduce the total cost, especially the penalty cost. Figure 2 shows the power consumption rate of each vehicle arriving at the customer points, each symbol representing a route, with the first row of depots having a charging function and the second row of depots not. It can be seen from the figure that the power consumption rate decreases as the vehicle load decreases.

**Table 4.** Results of solving the algorithm

| Instance | N | O | F | K | T | F | P | C | D |
|---|---|---|---|---|---|---|---|---|---|
| pr01 | 45 | 4 | 4 | 10 | 8017.741 | 700 | 1571.989 | 2557.486 | 3188.266 |
| pr02 | 92 | 4 | 4 | 10 | 21156.48 | 1000 | 3531.682 | 8652.585 | 7972.212 |
| pr03 | 137 | 6 | 6 | 15 | 55470.39 | 1500 | 6752.328 | 20392.99 | 21392.99 |
| pr04 | 184 | 6 | 6 | 25 | 99863.18 | 2500 | 7972.148 | 47400.82 | 41990.21 |
| Average | 115 | 5 | 5 | 15 | 46151.95 | 1450 | 5082.037 | 20125.97 | 18635.92 |
| Median | 115 | 5 | 5 | 15 | 38313.44 | 1250 | 5501.915 | 14772.79 | 14682.6 |

**Table 5.** Results of solving the algorithm

| Instance | N | O | F | K | T | F | P | C | D |
|---|---|---|---|---|---|---|---|---|---|
| pr01 | 45 | 4 | 4 | 10 | 8477.15 | 700 | 1584.73 | 2860.97 | 3131.46 |
| pr02 | 92 | 4 | 4 | 10 | 22666.29 | 1000 | 3946.37 | 9084.99 | 8634.92 |
| pr03 | 137 | 6 | 6 | 15 | 61113.16 | 1500 | 8024.35 | 28885.30 | 24703.52 |
| pr04 | 184 | 6 | 6 | 25 | 116918.85 | 2500 | 8714.96 | 59967.06 | 46936.83 |
| Average | 115 | 5 | 5 | 15 | 52293.9 | 1400 | 4567.6 | 25424.6 | 83406.7 |
| Median | 115 | 5 | 5 | 15 | 41889.72 | 1250 | 9970.72 | 18985.15 | 16669.22 |

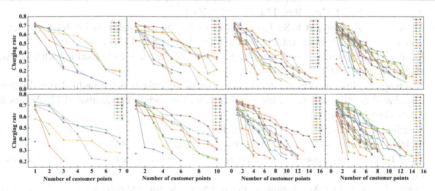

**Fig. 2.** The power consumption rate

## 5 Conclusion and Outlook

In this paper, we propose the DD-MDOEVRPTW and innovatively combines the three major problems of electric vehicle routing problem, open multi-depot joint distribution, and dynamic power consumption rate. The research related to the vehicle routing problem is extended. The conclusions are as follows:

1) The vehicle power consumption rate has a linear relationship with the vehicle load and decreases with the decrease of the vehicle load. Considering the dynamic influence of vehicle load on the power consumption rate can provide a basis for enterprises to reasonably plan the charging problem.

2) Considering the joint distribution of multiple depots can better meet the time window constraints of customers and lower logistics costs than single distribution centers.
3) The addition of the charging function in the depot can shorten the detour mileage of vehicles and reduce the impact of the shortage of charging stations on enterprise operations.
4) In terms of algorithm design, the hybrid genetic algorithm of two-dimensional coding design has a good effect on solving small, medium and large scale examples, and can quickly find the solution to the problem.

**Acknowledgment.** This work is supported by Major Science and Technology Project of Hainan Province (ZDKJ2020012) and Open Fund Project (2020KF0051).

# References

1. European Commission. The European Green Deal in a nutshell, 11 December 2019. https://ec.europa.eu/transport/themes/urban/urban-mobility/urban-mobility-package_en
2. Tillman, F.A.: The multiple terminal delivery problem with probabilistic demands. Transp. Sci. **3**(3), 192–204 (1969)
3. Yu, B., Yang, Z.Z., Xie, J.X.: A parallel improved ant colony optimization for multi-depot vehicle routing problem. J. Oper. Res. Soc. **62**(1), 183–188 (2011)
4. Crevier, B., Cordeau, J.F., Laporte, G.: The multi-depot vehicle routing problem with inter-depot routes. Eur. J. Oper. Res. **176**, 756–773 (2007)
5. Bi, X., Han, Z., Tang, W.K.S.: Evolutionary multi-objective optimization for multi-depot vehicle routing in logistics. Int. J. Comput. Intell. Syst. **10**(1), 1337–1344 (2017)
6. Karakatic, S.: Optimizing nonlinear charging times of electric vehicle routing with a genetic algorithm. Expert Syst. Appl. **164**, 114039 (2021)
7. Xiaoning, Z., Rui, Y., Zhaoci, H., et al.: Logistic optimization for multi depots loading capacitated electric vehicle routing problem from low carbon perspective. IEEE Access **8**, 31934–31947 (2020)
8. Mauricio, G.E., Paz, J.C., Escobar, J.W.: The multi-depot electric vehicle location routing problem with time windows. Int. J. Ind. Eng. Comput. **9**, 123–136 (2017)
9. Yuan, L.: Analysis of energy consumption and the sensitivity based on battery electric bus. Mech. Eng. Des. **1**, 1–7 (2013)
10. Goeke, D., Schneider, M.: Routing a mixed fleet of electric and conventional vehicles. Eur. J. Oper. Res. **245**(1), 81–99 (2015)
11. Van Keulen, T., de Jager, B., Serrarens, A., Steinbuch, M.: Optimal energy management in hybrid electric trucks using route information. Oil Gas Sci. Technol. **65**(1), 103–113 (2010)
12. Li, J., Li, Y., Pardalos, P.M.: Multi-depot vehicle routing problem with time windows under shared depot resources. J. Comb. Optim. **31**(2), 515–532 (2016)
13. VRP web: VRP instances (Multiple depot VRP with time windows), November 2006. http://www.bernabe.dorronsoro.es/vrp/
14. Murakami, K.: A new model and approach to electric and diesel-powered vehicle routing. Transp. Res. Part E Logist. Transp. Rev. **107**, 23–37 (2017)
15. Ministry of Industry and Information Technology of People's Republic of China. Catalog of new energy vehicle models exempted from vehicle purchase tax (Fifty-Sixth Batch), 12July 2022. https://www.miit.gov.cn/

# Analysis on the Selection of Logistics Distribution Mode of JD Mall in the Sinking Market

Weihui Du[1], Xiaoyu Zhang[2(✉)], Saipeng Xing[3], and Can Fang[4]

[1] School of Logistics, Wuhan Technology and Business University, Wuhan 430065, China
[2] College of Humanities, Yangtze University, Hubei 434023, China
zzxy_0317@163.com
[3] School of Management, Wuhan Technology and Business University, Wuhan 430065, China
[4] Department of Logistics and Operation Management,
Cardiff University, Cardiff CF10 3XQ, UK

**Abstract.** The development of e-commerce industry in China has gone through nearly 20 years. E-commerce has penetrated into people's daily activities, and the number of e-commerce platform users is increasing day by day. But now, the number of e-commerce platform users in the first and second tier cities has become saturated. E-commerce enterprises must change their thinking and find new footholds. At this time, the sinking market with huge user potential has become the basic market development idea. For the development of e-commerce enterprises in the sinking market, it is very important to have a suitable logistics distribution mode. Based on the analysis of different logistics distribution modes, combined with the characteristics of JD Mall and in the sinking market, Through Delphi method and fuzzy analytic hierarchy process, this paper constructs the evaluation index system of distribution mode selection, and finally helps JD Mall better select the logistics distribution mode in the sinking market.

**Keywords:** Logistics distribution mode · JD Mall · Sinking market

## 1 Introduction

In China, the development of e-commerce originated in the 1990s, and now it has gone through 20 years. Compared with the developed countries, although China's e-commerce industry developed late, it has a strong development momentum. Especially after stepping into the 21st century, the rapid development and popularization of the Internet in China, gradually penetrated into every corner of people's lives. As shown in Fig. 1. As of December 2020, the number of Internet users in China has reached 988 million, which is more than half of China's total population. The Internet penetration rate reached 70.4% by the end of 2020, an increase of 5.9% points over the end of 19 years. From these data, it is not difficult to find that the Internet is widely used in China, with a high penetration rate. In the future, with the continuous development of e-commerce, the scale of online transactions will continue to increase.

Although e-commerce continues to develop, but in general, urban consumption, especially in the first and second tier cities, has entered a stable period. In this case, all kinds of e-commerce platforms are forced to start looking for new development positions. The sinking market with huge users to be developed has been favored by many e-commerce enterprises. In 2019, the number of active users of JD increased by 27 million, More than 70% come from sinking markets. It is not difficult to see that JD Mall has great development potential in the sinking market [1].

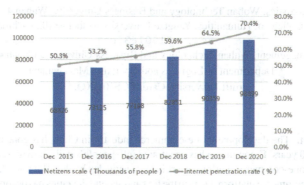

**Fig. 1.** China's Internet penetration rate and scale in 2020-1

With the development of e-commerce, people pay more attention to e-commerce logistics distribution. In order to keep up with the pace of e-commerce, logistics distribution is also keeping pace with the times and improving constantly. Various distribution modes emerge as the times require, such as the famous third-party logistics distribution, cloud distribution mode and so on. For the development of JD in the sinking market, a suitable logistics distribution mode is very important. The choice of logistics distribution mode is a key step for JD Mall to develop in the sinking market.

## 2 Overview of Relevant Theories

### 2.1 Sinking Market

The concept of sinking market first appeared in the literature in 2004. Generally speaking, the sinking market includes cities, towns and rural areas with less developed economy and huge consumption potential. China has a vast territory with many prefecture level cities, counties, towns and villages, but there are only 79 first and second tier cities, with administrative area accounting for about 5% of the total area. More than this, the population of the areas below the third line is also huge, with a population of about one billion. This shows the development potential of the sinking market.

The sinking market has three obvious characteristics. Firstly, the consumption structure is diversified, and the consumption preference of market groups presents a diversified situation; the second is large scale, large quantity, large population base and large group size; the third is that the social attribute is clear, the interpersonal relationship in the sinking market is relatively close, and the consumer's decision-making is greatly influenced by the society and acquaintances.

## 2.2 The Sinking Market E-commerce Theory

In recent years, many e-commerce platforms have developed in the sinking market, and relevant research has also increased. Pan Helin [2] pointed out that the popularity of e-commerce in the sinking market is getting higher and higher, and it is gradually extended to small areas, towns, etc. in addition, the development level has also been improved. According to the e-commerce market situation in 2018, Zhu Xiaopei [3] concluded that the consumption of e-commerce users tends to be more rational. Through the questionnaire survey, concludes that when choosing e-commerce platform, users shift their focus from price to quality, especially for logistics and service quality [4].

## 2.3 Related Theories of E-commerce Logistics Distribution

Logistics distribution has always been an important part of the development of e-commerce, which is related to the final transaction of e-commerce and the credibility of e-commerce platform. At present, the development of e-commerce in the sinking market is facing the problems of insufficient infrastructure level and poor logistics distribution.

Therefore, for e-commerce enterprises, it is essential to have efficient and reasonable logistics distribution. Sherriff T.K thinks that keeping the low efficiency of distribution service can greatly increase the core competitiveness of enterprises [5]. Wang xiao and Gao Heyun attach great importance to distribution efficiency [6]. Having an efficient logistics distribution mode can improve customer satisfaction and trust, it will also have an important impact on the development of e-commerce enterprises. After analyzing the online review information of e-commerce logistics, Hong W found that service attitude is very important to customer satisfaction in logistics service [7].

# 3 Analysis on the Current Situation of JD Mall in the Sinking Market

## 3.1 The Reason Why JD Mall Entered the Sinking Market

JD Mall chooses to enter the sinking market, which is not only the general trend of the development of national e-commerce, but also the decision made after multi-dimensional investigation. The main reasons can be summarized as follows.

The First and second tier cities tend to be saturated: at present, the development of China's e-commerce industry has been more than 20 years, and concentrated in the development of the first and second tier cities. Therefore, the user coverage rate of the e-commerce platform in the first and second tier cities has become saturated, and the potential for tapping is not great. The disappearance of dividends makes JD Mall shift its focus to the sinking market where the Internet coverage is not high enough.

Huge market demand: According to many survey data, in the sinking market, although the level of economic development cannot catch up with the first and second tier cities and the income gap of residents, there are many residents in the sinking market, and the pressure of housing loans is small, so there is more funds for flexible use. So various county-level cities and towns are among the engines to promote the rapid development of JD Mall.

There are many blank areas and broad market: although the network environment in counties, villages and other areas with sinking market is continuously improving, there is still a big gap compared with the first and second tier cities. At present, there is a lack of large-scale shopping malls, and the business circle is short of goods, but the shopping demand is large. Therefore, consumers show great enthusiasm for the entry of e-commerce platform, so as to meet the needs of life [8–10].

### 3.2 Analysis on the Existing Logistics Distribution Mode of JD Mall

At present, there are various types of e-commerce platforms in China, and there are great differences in the operation mode and development level of the platforms. Therefore, the conditions to be considered when choosing the logistics distribution mode are also different. At present, JD Mall mainly chooses four logistics modes, which are outsourcing distribution mode, logistics alliance mode, crowd sourcing distribution mode and self-operated logistics mode. These four distribution modes have their own advantages and disadvantages as well as the suitable environment. The service experience is slightly different, and the operation mode and profit amount are also different.

Outsourcing distribution mode: the outsourcing distribution mode of logistics, that is, the third-party logistics distribution, this mode is that JD Mall outsources the logistics to the third-party logistics, and JD does not directly participate in the logistics distribution process.

In this mode, JD can focus on the main business, reduce other business, reduce the risk degree and cost of JD operation, increase the core competitiveness of the enterprise, and help to play the long board and avoid the short board. In addition, outsourcing distribution can reduce distribution costs, platform users can also enjoy better and more professional logistics services without additional costs.

Logistics Alliance Distribution Mode: Logistics alliance mode refers to the formation of logistics alliance based on e-commerce platform and cooperation with other enterprises in the supply chain. With win-win as the common goal, we signed agreements, cooperated to share resources and information, and comprehensively used the resources of various enterprises for logistics distribution. Under the mode of logistics alliance, JD cooperates with other logistics enterprises to complement each other, make full use of all logistics resources.

Crowdsourcing distribution mode: crowdsourcing distribution mode is a new distribution mode. It refers to the continuous development of modern social sharing concept, outsourcing the distribution task that should be undertaken by professional delivery personnel to the public in a paid and resource way. This process needs to be used in the Internet platform. The crowdsourcing distribution mode effectively reduces the distribution cost of JD, some of the logistics equipment required for distribution are provided by the distribution personnel in crowdsourcing, and JD is not required to provide them. In addition, crowdsourcing distribution makes use of the resources in the society, improves the efficiency of resource integration, provides employment opportunities for many people in society, alleviates the employment pressure to a certain extent, and is conducive to social stability and economic development [9–11].

Self operated distribution mode: As a result, the self-supporting logistics service providers should build their own logistics service platform to meet their own demand. Under the mode of self operated distribution, the service quality of logistics distribution can be guaranteed and the efficiency is high, it is safe and timely. Personnel can also be better managed in place, can better grasp the control of distribution arrangements. In addition, the self-operated distribution mode improves the brand effect of JD, improves the recognition degree of customers for JD Mall, and also improves the market competitiveness of JD.

### 3.3 Analysis of Problems Existing in Logistics Distribution Mode

Although JD Mall has gained some development in the sinking market, and the above logistics distribution mode has its own advantages, there are still some problems in practical application.

#### 3.3.1 Outsourcing Distribution Mode

In the mode of outsourcing distribution, outsourcing the distribution service leads to the failure of effective guarantee of distribution service quality, which increases the risk of inadequate service. Moreover, it will form a whole after cooperation with logistics enterprises. Some logistics enterprises with inadequate service have a joint impact on JD.

#### 3.3.2 Logistics Alliance Distribution Mode

In the choice of logistics alliance mode, due to the different management methods between JD Mall and other enterprises, it is difficult to coordinate in place, and the service level is uneven, so it is not easy to get the minimum guarantee. At the same time, JD Mall needs to share information with multiple enterprises in the distribution process, this has led to the leakage of customer information and other issues, leading to the impact of consumers on JD Mall.

#### 3.3.3 Crowdsourcing Distribution Mode

There are certain risks in the selection of crowdsourcing distribution mode, which brings some security risks to JD. Due to the participation of many social personnel in the distribution process, it is difficult and difficult to restrict the management of these personnel, which leads to some people with ulterior motives to seek private interests. Moreover, the information leakage risk of this mode is too high and the security risks are serious, some customers are losing out.

#### 3.3.4 Self Operated Distribution Mode

There are some inevitable defects in the self-supporting distribution mode. The self-supporting logistics distribution of JD leads to a large cost in the early stage and high maintenance cost in the later stage, which increases the capital turnover risk and operation risk of JD. At the same time, it has high requirements for various professional logistics facilities and equipment as well as a large number of logistics professionals, which increases the operation difficulty of JD Mall.

## 4 Analysis on the Logistics Distribution Model Selection of JD Mall in the Sinking Market

### 4.1 Establishment of Evaluation Index System for Logistics Distribution Mode Selection

The choice of logistics distribution mode is a complex and difficult work, which is not only determined by some reasons. Different logistics distribution modes have their own advantages and disadvantages. JD Mall needs to select the best logistics distribution mode according to its own characteristics and external environment with a comprehensive perspective. The factors that affect the choice of logistics distribution mode can be summarized into three categories: enterprise scale strength, logistics service requirements and sinking market environment.

According to the above three factors, the evaluation index system of logistics distribution mode of JD Mall can be constructed, as shown in Fig. 2. It is divided into four layers, namely, sub layer, criterion and scheme. The target layer is the optimal distribution mode $U$ of JD Mall, and the criterion layer is the enterprise scale strength $A$, logistics service requirement $B$ and sinking market environment $C$. The sub criteria layer refers to the influencing factors in the criteria layer. The enterprise scale strength $A$ specifically includes enterprise scale $A_1$, facilities and equipment level $A_2$, informatization level $A_3$; logistics service requirement $B$ specifically includes service quality $B_1$, delivery speed $B_2$, commodity safety $B_3$; sinking market environment c specifically includes regional economic level $C_1$, policy support $C_2$, and market dispersion $C_3$. The scheme layer is set to $D$, Among them, $D_1$ is outsourcing distribution mode, $D_2$ is logistics alliance mode,

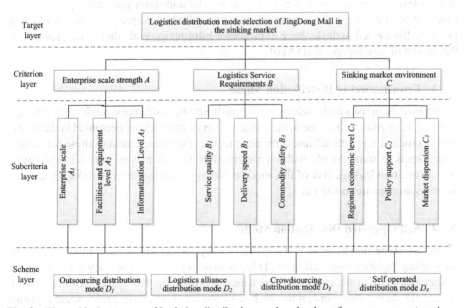

**Fig. 2.** Hierarchical structure of logistics distribution mode selection of e-commerce enterprises in the sinking Market

$D_3$ is crowdsourcing distribution mode, $D_4$ is self-supporting distribution mode. The specific hierarchical structure is shown in Fig. 2.

### 4.1.1 Enterprise Scale

The first is the size of the enterprise. The size of enterprises will have a great impact on their choice of logistics distribution mode. Large e-commerce enterprises have higher requirements for distribution services, and these enterprises also have the ability to build their own logistics. The second is the level of facilities and equipment. The level of enterprise logistics facilities also restricts the choice of distribution mode, the third is the level of informatization. In the modern society with the rapid development of information technology, the level of enterprise information will affect the decision-making of enterprise to a great extent [12–14].

### 4.1.2 Logistics Service Requirements

The first is the quality of service. Logistics and distribution is the last link of e-commerce, and it is also a key link. Therefore, service attitude is very important, which is directly related to customers' perception of the platform. The second is delivery speed. The speed of delivery will greatly affect customers' shopping experience. The third is the safety of goods. This is related to the quality of delivery. Customers not only need delivery speed, more need to receive goods are complete and safe.

### 4.1.3 Sinking Environment Market

The first is the level of regional economy. The level of local economy determines the tendency of consumers for distribution demand, and the demand of consumers in economically developed areas will be more diversified. The second is policy support. If JD Mall chooses the logistics distribution mode supported by national policy, it will be more beneficial for long-term development. The third is market dispersion. The dispersion degree of residents in the sinking market area is also an important factor affecting the decision-making of logistics distribution mode [15, 16].

**Table 1.** The influencing factor system of distribution mode selection in the sinking market of JD Mall

| Purpose of evaluation | Primary influencing factors | Secondary influencing factors |
|---|---|---|
| Logistics distribution mode selection of JD Mall in the sinking Market | Enterprise scale strength $A$ | Enterprise scale $A_1$<br>Facility level $A_2$<br>Informatization level $A_3$ |
| | Logistics service requirements $B$ | Service quality $B_1$<br>Delivery speed $B_2$<br>Commodity safety $B_3$ |
| | Sinking market environment $C$ | Regional economic level $C_1$<br>Policy support $C_2$<br>Market dispersion $C_3$ |

To sum up, by the analysis mentioned above, the following Table 1 can be obtained (Tables 2 and 3).

### 4.2 Index Weight Determination and Consistency Testing at All Levels

According to the data in the above table, the results are shown in Table 4. When the consistency ratio Cr is less than 0.1, the consistency test is passed, and the judgment matrix does not need to be adjusted again.

**Table 2.** Judgment matrix of target layer U

| Indicator $U$ | A | B | C |
|---|---|---|---|
| A | 1 | 1/3 | 1/2 |
| B | 3 | 1 | 4 |
| C | 2 | 1/4 | 1 |

**Table 3.** Judgment matrix of hierarchy

| A | $A_1$ | $A_2$ | $A_3$ | B | $B_1$ | $B_2$ | $B_3$ | C | $C_1$ | $C_2$ | $C_3$ |
|---|---|---|---|---|---|---|---|---|---|---|---|
| $A_1$ | 1 | 1/3 | 1/4 | $B_1$ | 1 | 1/4 | 1/3 | $C_1$ | 1 | 1/5 | 1/2 |
| $A_2$ | 3 | 1 | 2 | $B_2$ | 4 | 1 | 3 | $C_2$ | 5 | 1 | 3 |
| $A_3$ | 4 | 1/2 | 1 | $B_3$ | 3 | 1/3 | 1 | $C_3$ | 2 | 1/3 | 1 |

**Table 4.** Test results

| Indicator | $W_i$ | $\lambda_{max}$ | $C_I$ | $C_R$ | Consistency judgment |
|---|---|---|---|---|---|
| X | [0.151 0.630 0.218] | 3.108 | 0.054 | 0.093 | agreement |
| A | [0.124 0.517 0.359] | 3.108 | 0.054 | 0.093 | agreement |
| B | [0.117 0.614 0.268] | 3.074 | 0.037 | 0.064 | agreement |
| C | [0.122 0.648 0.230] | 3.004 | 0.002 | 0.003 | agreement |

According to the above data, we can get the comprehensive weight and ranking of each factor in the index level, as shown in Table 5.

It can be seen from the above table that the index weights of the standard layer are 0.124 for the enterprise scale strength, 0.517 for the logistics service requirements and 0.093 for the sinking market environment.

The weight of enterprise scale strength a includes enterprise scale $A_1$, facilities and equipment level $A_2$ and informatization level $A_3$, which are 0.124, 0.517 and 0.359, respectively. The weight of service quality $B_1$, delivery speed $B_2$ and commodity safety $B_3$ in logistics service requirement $B$ are 0.117, 0.614 and 0.268 respectively. The weights of regional economic level $C_1$, policy support $C_2$ and market dispersion $C_3$ included in sinking market environment $C$ are 0.122, 0.648 and 0.230 respectively.

Table 5. Comprehensive weight and ranking of each element at index level

| Criterion layer | Individual weight | Subcriteria layer | Individual weight | Comprehensive weight | Importance ranking |
|---|---|---|---|---|---|
| Enterprise scale Strength $A$ | 0.151 | Enterprise scale $A_1$ | 0.124 | 0.019 | 9 |
| | | Facility level $A_2$ | 0.517 | 0.078 | 4 |
| | | Informatization level $A_3$ | 0.359 | 0.054 | 6 |
| Logistics service requirements $B$ | 0.630 | Service quality $B_1$ | 0.117 | 0.074 | 5 |
| | | Delivery speed $B_2$ | 0.614 | 0.387 | 1 |
| | | Commodity safety $B_3$ | 0.268 | 0.169 | 2 |
| Sinking market environment $C$ | 0.219 | Regional economic level $C_1$ | 0.122 | 0.027 | 8 |
| | | Policy support $C_2$ | 0.648 | 0.142 | 3 |
| | | Market dispersion $C_3$ | 0.230 | 0.050 | 7 |

It can be seen from the importance ranking of the comprehensive weight shown in the chart that the impact of the selection of delivery mode is mainly considered from the aspect of logistics service requirements, among which the distribution speed and commodity safety have the most prominent impact on the selection of distribution mode, and their weights are 0.387 and 0.169 respectively.

### 4.3 Construction of Fuzzy Hierarchy Model for Logistics Distribution Mode of JD Mall in Sinking Market

To determine the importance of the scheme layer, Delphi method is still used (20 relevant employees of JD Mall are invited to fill in the questionnaire), and the four logistics distribution modes are compared with each other under the premise of considering the sub criteria layer factors, and then the score is made based on the actual situation of JD Mall. The evaluation results are divided into four grades: excellent, good, medium and poor, based on the statistical results of employee feedback, the fuzzy evaluation matrix R of sub criteria level and scheme level is established, as shown in Table 6.

Firstly, the single factor fuzzy evaluation of outsourcing distribution is calculated Enterprise scale strength:

$$B_{U1} = W_{U1} * R_{U1} = (0.124, 0.517, 0.359) * \begin{bmatrix} 0.05 & 0.25 & 0.55 & 0.15 \\ 0.1 & 0.5 & 0.3 & 0.1 \\ 0.2 & 0.4 & 0.4 & 0 \end{bmatrix}$$
$$= (0.1297, 0.4331, 0.3669, 0.0703) \tag{1}$$

Logistics service requirements: (0.1326, 0.3628, 0.4038, 0.1).
Sinking market environment: (0.1641, 0.346, 0.3777, 0.1122).

The comprehensive evaluation matrix of outsourcing distribution mode can be obtained:

$$R_{os} = \begin{bmatrix} 0.1297 & 0.4331 & 0.3669 & 0.0703 \\ 0.1326 & 0.3628 & 0.4038 & 0.1 \\ 0.1641 & 0.346 & 0.3377 & 0.1122 \end{bmatrix} \tag{2}$$

**Table 6.** Fuzzy evaluation matrix

| Primary indicators | Secondary indicators | Outsourcing distribution mode ||||  Distribution mode of Logistics Alliance |||| Crowdsourcing distribution mode |||| Selfoperated distribution mode ||||
|---|---|---|---|---|---|---|---|---|---|---|---|---|---|---|---|---|---|
| | | excellnt | good | mesne | poor | excellnt | good | mesne | poor | excellnt | good | mesne | poor | excellnt | good | mesne | poor |
| Enterprise scale strength | Enterprise scale | 1 | 5 | 11 | 3 | 2 | 8 | 7 | 3 | 3 | 12 | 4 | 1 | 1 | 6 | 13 | 0 |
| | Facilities and equipment level | 2 | 10 | 6 | 2 | 2 | 12 | 4 | 2 | 2 | 14 | 4 | 0 | 0 | 8 | 8 | 4 |
| | Informatization level | 4 | 8 | 8 | 0 | 2 | 9 | 6 | 3 | 2 | 6 | 8 | 4 | 2 | 10 | 6 | 2 |
| Logistics service requirements | Service quality | 3 | 4 | 11 | 2 | 2 | 12 | 4 | 2 | 0 | 6 | 10 | 4 | 4 | 10 | 3 | 3 |
| | Delivery speed | 2 | 8 | 8 | 2 | 2 | 4 | 12 | 2 | 0 | 9 | 9 | 2 | 4 | 6 | 8 | 2 |
| | Product safety | 4 | 7 | 7 | 2 | 4 | 6 | 9 | 1 | 2 | 6 | 8 | 4 | 4 | 6 | 10 | 0 |
| Sinking market environment | Regional economic level | 0 | 6 | 10 | 4 | 4 | 15 | 1 | 0 | 2 | 4 | 12 | 2 | 0 | 8 | 10 | 2 |
| | Policy support | 4 | 6 | 8 | 2 | 2 | 6 | 10 | 2 | 2 | 8 | 6 | 4 | 3 | 8 | 9 | 0 |
| | Informatization Market dispersionlevel | 3 | 10 | 5 | 2 | 2 | 4 | 12 | 2 | 2 | 6 | 6 | 6 | 2 | 6 | 8 | 4 |

It can be calculated according to the formula.

$$B_{OS} = W_U * R_{os} = (0.151, 0.630, 0.218) * \begin{bmatrix} 0.1297 & 0.4331 & 0.3669 & 0.0703 \\ 0.1326 & 0.3628 & 0.4038 & 0.1 \\ 0.1641 & 0.346 & 0.3377 & 0.1122 \end{bmatrix}$$
$$= (0.1389, 0.3694, 0.3834, 0.098) \tag{3}$$

The other logistics modes can be evaluated in the same way.

$$B_{LA} = (0.1194, 0.3233, 0.4633, 0.0924)$$

$$B_{CS} = (0.0547, 0.4058, 0.3929, 0.1449)$$

$$B_{SO} = (0.1584, 0.5178, 0.4002, 0.0835)$$

We assume that the rating levels are excellent, good, medium and poor, and their corresponding scores are *100, 80, 60* and *40*, it can be derived $V^T = [10, 8, 6, 4]$. According to the formula $D = B*V^T$, the comprehensive evaluation value of various distribution modes can be obtained through calculation, as follows:

$$D_1 = (0.1389, 0.3694, 0.3834, 0.098) * (10, 8, 6, 4)^T = 7.0358$$

$$D_2 = (0.1194, 0.3233, 0.4633, 0.0924) * (10, 8, 6, 4)^T = 6.9298$$

$$D_3 = (0.0547, 0.4058, 0.3929, 0.1449) * (10, 8, 6, 4)^T = 6.5504$$

$$D_4 = (0.1584, 0.5178, 0.4002, 0.0835) * (10, 8, 6, 4)^T = 8.4616$$

### 4.4 Final Evaluation Results and Analysis

According to the above calculation results, Table 7 below can be obtained.

**Table 7.** Comprehensive analysis of JD Mall's declining market logistics distribution method choosing strategy

|  | Logistics distribution mode $D_1$ | Logistics distribution mode $D_2$ | Logistics distribution mode $D_3$ | Logistics distribution mode $D_4$ |
|---|---|---|---|---|
| Logistics distribution mode | Outsourcing distribution mode | Distribution mode of Logistics Alliance | Crowdsourcing Distribution mode | Self operated distribution mode |
| Comprehensive evaluation value | 7.0358 | 6.9298 | 6.5504 | 8.4616 |

### 4.4.1 Analysis of Main Factors Affecting the Evaluation Results

After analyzing the weight of the influencing factors, we can draw the following conclusions: among the primary indicators, the weight of logistics service requirements is the largest among the main factors affecting the logistics distribution mode of JD Mall in the sinking market, indicating that the impact is also the greatest. Therefore, JD Mall should pay attention to the influence of logistics service, especially logistics distribution, when developing the sinking market, fully consider the advantages and disadvantages of various logistics distribution. With the development of economy and the progress of society, residents in the sinking market will pay more attention to the quality of distribution service when shopping on the e-commerce platform. Therefore, when JD Mall enters the sinking market, it is necessary to consider not only the "quantity" of distribution, but also the "quality" of distribution.

### 4.4.2 Selection and Analysis of Evaluation Results

According to the above calculation results, the comprehensive weight of outsourcing distribution mode is 7.0358, that of logistics alliance distribution mode is 6.9298, that of crowdsourcing distribution mode is 6.5504, and that of self operated distribution mode is 8.4616.After comparing the data, it is concluded that: for the logistics distribution of JD Mall in the sinking market, among the four logistics distribution modes, the self-supporting distribution mode is the most suitable, and the self-supporting logistics distribution mode is also in line with the actual situation of JD Mall in the sinking market, so the conclusion of this paper is scientific and practical.

## 5 Conclusion

After analyzing the characteristics of the sinking market and the development environment of JD Mall in the sinking market, this paper constructs the evaluation index system for the selection of logistics distribution mode in the sinking market of JD Mall. On this basis, the fuzzy analytic hierarchy process and Delphi method are used for analysis and calculation. Finally, the most appropriate logistics distribution mode is selected. The main conclusions are as follows.

1) Great opportunities and challenges for e-commerce platforms to enter the sinking market. In addition, the competition in the same industry is also very fierce.
2) It analyzes the four main existing logistics distribution modes (outsourcing distribution mode, logistics alliance distribution mode, crowdsourcing distribution mode and self-operated distribution mode). JD Mall should make decisions according to the actual situation and its own needs when choosing the logistics distribution mode in the sinking market.
3) When establishing the evaluation index of logistics distribution mode selection, it is concluded that there are three main factors that affect the selection of logistics distribution mode of JD Mall in the sinking market, which are the enterprise scale strength, logistics service requirements, sinking market environment and their subordinate secondary indicators.
4) The final conclusion of this paper is that JD Mall should choose its own logistics mode when entering the sinking market, and combining with examples, it shows that the final conclusion of this paper is scientific and practical.

**Acknowledgment.** This project is supported by Humanities and Social Sciences Research Planning Project of Ministry of Education. Research on the "double-edged sword" effect of leadership empowerment on employees and team creativity from the perspective of digital transformation (22YJA630097).

# References

1. Li, J.: Research on the competitive strategy of JD group logistics company. Beijing University of Posts and Telecommunications, Beijing (2019)
2. Pan, H.: It's time to correct the name of the sinking market. Global Times 2019–10–26 (007)
3. Zhu, X.: Taking stock of the 2018 e-commerce market: traffic sinks to consumer grading. Shanghai Informatization **02**, 37–40 (2019)
4. Echo: Eview released 2019 online shopping report: consumption in the sinking market is being upgraded and accelerated. China Optical (05), 44–45 (2019)
5. Sherriff, T.K.: Kerry logistics-paving the new silk road. Asian Case Res. J. **23**(01), 153–191 (2019)
6. Wang, X., Gao, H.: Research on the vehicle path problem in urban distribution system. Logist. Eng. Manag. **41**(04), 85–87 (2019)
7. Hong, W., Zheng, C., Wu, L., et al.: Analyzing the relationship between consumer satisfaction and fresh e-commerce logistics service using text mining techniques. Sustainability **11** (2019)
8. Kai: Research on the selection of logistics and distribution mode of Yang e-commerce enterprises in the sinking market. Tianjin University of Technology, Tianjin (2020)
9. Su, D., Cen, W.: Research on the selection of logistics distribution mode based on fuzzy hierarchical analysis. Pop. Sci. Technol. **22**(09), 139–142 (2020)
10. Xu, F.: Research on Jingxi sinking market value creation. Shandong University of Finance and Economics, JiNan (2021)
11. Zhou, Y.: 2010–2019: This decade of consumerism. People's Forum **03**, 52–57 (2020)
12. Wang, H.: The growth of users is slowing down. The e-commerce platform is driving down the market. China Bus. Daily (11) (2021)
13. Niu, B.: Analysis of factors influencing the choice of enterprise logistics mode. Market Forum **09**, 58–61 (2020)
14. Li, J.: Research on the competitive strategy of jd group logistics company. Beijing University of Posts and Telecommunications (2019)
15. Gruauskas, V., Gimauskien, E., Navickas, V.: Forecasting accuracy influence on logistics clusters activities: the case of the food industry. J. Clean. Prod. **240**, 118225 (2019)
16. Akeb, H., Moncef, B., Durand, B.: Building a collaborative solution in dense urban city settings to enhance parcel delivery: an effective crowd model in Paris. Transp. Res. Part E Logist. Transp. Rev. (2018)
17. Fulzele, V., Shankar, R., Choudhary, D.: A model for the selection of transportation modes in the context of sustainable freight transportation. Ind. Manag. Data Syst. (2019)
18. Li, F., Fan, Z.P., Cao, B.B., et al.: A method for selecting enterprise's logistics operation mode based on Ballou model. Math. Probl. Eng. **2019**(4), 1–9 (2019)
19. Xu, X., Li, Y., Tang, R.: Simulation optimization of discrete logistics processes: a case study on logistics of an e-commerce enterprise in Shanghai. Discrete Dyn. Nat. Soc. (1) (2019)
20. Ramaekers, K.: Using an integrated order picking-vehicle routing problem to study the impact of delivery time windows in e-commerce. Eur. Transp. Res. Rev. **10**(2), 224–245 (2018)

# Research on Cruise Emergency Organization Based on Improved AHP-PCE Method

Long Zhang[1] and Zhengxie Li[2(✉)]

[1] School Logistics, Wuhan Technology and Business University, Wuhan 430065, China
[2] Philippine Christian University Center for International Education, 1004 Manila, Philippines
435840200@qq.com

**Abstract.** With the prosperity and growth of China's tourism economy, cruise tourism is gradually entering the Chinese market. The cruise safety management has become a hot issue. On the one hand, due to the complex structure of the cruise ship itself and the concentration of personnel, once an accident occurs, it will cause serious consequences; on the other hand, due to the fact that China's cruise ship emergency management is in the initial stage, and the emergency response organization system for major cruise emergencies is still not perfect, which leads to the difficulty of accident rescue work and affects the emergency rescue ability. This paper mainly analyzes the characteristics related to major cruise ship emergencies, discovers the relevant factors affecting the emergency organization capacity of major cruise ship emergencies from the internal factors of organization management capacity, uses the improved fuzzy comprehensive evaluation and hierarchical analysis method to construct the emergency organization evaluation model of major cruise ship emergencies, and improves the emergency management capacity of major cruise ship emergencies by enhancing the emergency organization capacity.

**Keywords:** Emergency management · Organizational system · AHP-PCE

# 1 Introduction

In recent years, with the rapid development of China's economy, the pace of people's lives is getting faster and faster, and the pressure of life is increasing, people prefer to be able to relax during leisure travel, cruise tourism has become the current people's first choice of a combination of leisure, entertainment, tourism and holiday travel relaxation way, people are also increasingly concerned about the safety of cruise tourism. The high growth of cruise ship traffic and complex cruise personnel structure will inevitably put forward higher requirements for cruise safety issues, however, cruise tourism related supporting infrastructure is not perfect, cruise emergency management organization system is relatively backward, in recent years cruise major emergencies occur from time to time, cruise emergency management capacity aspects are facing a severe test [1]. The emergency management capability will certainly restrict the orderly development of the cruise industry to a certain extent. Cruise ship emergency management lacks a complete emergency organization system, and the emergency coordination mechanism is not yet perfect. It is urgent to improve the cruise ship emergency organization system, enhance the emergency organization capacity of cruise ship major emergencies, strengthen the level of cruise ship emergency management, and maximize the level of cruise ship emergency management capacity [2].

## 1.1 Characteristics of Major Cruise Emergencies

### 1.1.1 Definition of Major Cruise Emergencies

The so-called major emergencies are mainly based on the degree of harm of cruise transport emergencies to divide, different degrees of social harm and different loss of life and property, the level of cruise transport emergencies are also different, this paper mainly divided into four levels of major cruise emergencies particularly significant, significant, large, general, respectively, with I, II, III and IV to indicate, and according to the size of different levels were used red, orange, yellow and blue to represent different levels of cruise emergencies, in this paper, major cruise emergencies are generally referred to I and II level [3].

### 1.1.2 Classification and Characteristics of Major Cruise Emergencies

This article is mainly based on the provisions of China's Ship Traffic Accident Statistics Code and IMO MSC Circular No. 953 "Investigation Reports on Marine Accidents and Incidents", and combined with the statistical data reports on marine traffic safety accidents in China in recent years, this article classifies major cruise ship emergencies into the following nine types: major cruise ship fires and explosions, major cruise ship public health incidents, major cruise ship terrorist attacks, major cruise ship mass incidents, major cruise ship natural disaster incidents, major cruise ship supply shortages, major cruise ship collisions and touchdowns, major cruise ship groundings, and extensive damage to cruise ship hulls and serious equipment failures.

This paper summaries and concludes the main features of major cruise ship emergencies based on the classification of cruise ship emergencies and taking into account the relevant characteristics of the emergencies [4].

(1) Serendipity: major cruise ship emergencies often occur suddenly and by chance, and cannot be foreseen, while they do not occur with any certain regularity.
(2) Public: Due to the rapid development of modern communication technology and media, when a major cruise ship emergency occurs, it will cause a high degree of public concern within a short period of time, and whether the government can efficiently, quickly, safely and reasonably handle a major cruise ship emergency will directly affect the credibility of the government, as well as people's concern about the safety of cruise travel.
(3) Hazards: Due to the increasingly rapid development of modern cruise ships and the increasing tonnage and cost of cruise ships, when an emergency occurs on a cruise ship, it may not only pose a serious threat to the lives and property of the passengers and crew on board at the time, but will also be a long-term threat to the ecological environment and the local economic development of the area where the incident occurred, with follow-up treatment often taking decades to recover [5].

### 1.1.3 Optimizing Emergency Response Capabilities for Major Cruise Ship Emergencies

As we can see from the IMO cruise ship emergency statistics, with the continuous development of science and technology, the safety of the cruise ship industry has improved and the number of major cruise ship emergencies has gradually decreased in recent years, but from the point of view of the damage caused by the accident, the damage and loss caused by major cruise ship emergencies has gradually increased, which has brought a serious threat to the safety of passengers' lives and property and huge losses [6]. We should not only be prepared to prevent accidents before they occur, but we should also be prepared to respond to major cruise ship emergencies through a comprehensive emergency response system and improved emergency response efficiency to reduce the damage caused by major cruise ship emergencies. It is therefore imperative to improve the level of emergency response capability for major cruise ship emergencies [7].

The types of major cruise ship emergencies can be divided into the following nine main categories, and according to the IMO 2012–2022 Cruise Ship Major Incidents by Type of Incident Statistical Table, the proportion of each type of incident is shown in Table 1.

**Table 1.** Major cruise accidents by event type in 2012–2022

| Type of accident | Accident ratio (Unit: %) |
| --- | --- |
| Fire and explosion | 15% |
| Public health events | 4% |
| Terrorist attacks | 5% |
| Group events | 10% |
| Natural disaster events | 8% |
| Missing supplies | 4% |
| Collision and touch | 18% |
| Stranded | 25% |
| Extensive hull damage and serious equipment failure | 11% |

From Table 1, the proportion of cruise ship accidents by type, it can be concluded that the most frequent cruise ship accidents are cruise ship groundings, more often cruise ship collisions and touch-and-go accidents, followed by other accidents such as fire and explosion accidents, extensive hull damage and serious equipment failure [8].

## 2 Cruise Ship Emergency Organization System for Major Emergencies

### 2.1 Cruise Ship Emergency Management Capability for Major Emergencies

In the process of developing emergency plans for major cruise ship emergencies, we should pay full attention to the rationality and practicality of the emergency plan, whether it can play a good rescue guidance role after the accident, whether it is operable, and whether it can maximize the reduction of accident losses [9]. It is very important to have a complete and reasonable emergency organization structure, which is a guiding principle for the implementation of emergency rescue, and a complete organization structure will have clear provisions for mutual coordination between departments, division of labour, areas of responsibility and authority in the rescue process [10].

#### 2.1.1 The Relationship Between Emergency Response Organizations and Emergency Management

Emergency management refers to the activities of government departments and other public institutions in the whole process of emergencies, through the establishment of the necessary emergency response mechanisms and the adoption of a series of emergency rescue management tools, and thus reduce the harm of accidents to promote the harmonious development of society; emergency organization is an extremely important part of emergency management, organization as a carrier of management, when the organization expands and ages, it will inevitably lead to The internal structural friction coefficient of the organization gradually increases and the complexity of the management increases, resulting in internal friction between the organizations and a gradual decline in energy, resulting in poor information transfer, weakened departmental functions and weakened response capabilities, thus affecting the level of emergency management capabilities [11].

#### 2.1.2 Definition of Organizational Structure

Organizational structure is the division of labour, the grouping and the coordination between the various departments. Organizational structure is a model that reflects the order of each department in the organizational structure, its spatial location, the state of dispersion, the contact details of the relevant departments and the interrelationship between the departments. It is a systematic "framework" for the entire emergency response plan [12].

#### 2.1.3 Composition of Organizational Structure

The emergency response structure of a cruise ship is a complex structure and system consisting of many departments with clear hierarchy, clear functions, smooth information transmission and functional coupling [13]; it mainly consists of four parts: the emergency function management body, the emergency response hierarchy, the emergency response department structure and the emergency response authority structure:

(1) Emergency response functions are the various rescue efforts and the proportions and relationships required to achieve emergency response objectives.
(2) The emergency response hierarchy refers mainly to the composition of the emergency response management hierarchy and the number of people managed by each emergency management department [14].
(3) The emergency department structure refers primarily to the composition of the various emergency response management departments.
(4) The structure of emergency powers refers to the interrelationship and clear division of powers and responsibilities between the various levels and departments of emergency response [15].

### 2.2 Analysis of Factors Affecting the Emergency Response System for Major Cruise Ship Emergencies

As the cruise ship emergency plan is still a blank in China, the cruise ship emergency organization structure system is not yet perfect, this paper draws on the organization structure of the maritime emergency plan and combines the characteristics of foreign emergency organization structure system to develop Fig. 1 cruise ship emergency organization structure chart.

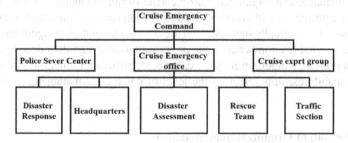

**Fig. 1.** Emergency organization structure of major cruise emergencies

#### 2.2.1 Complexity of Emergency Response Organizational Functions for Major Cruise Ship Emergencies

The cruise ship emergency response organization is the actual implementer of emergency management. Any organizational system is composed of a collection of elements formed by the combination of various closely related and closely associated relevant factors, and there is a certain coupling relationship between all relevant influencing factors [16]. However, the more departments there are in the cruise ship emergency function, the easier it is to reduce the degree of aggregation of the emergency organization system, resulting in a scattered emergency organization system and weakened functions; in the cruise ship emergency management organization system, the degree of coupling of functions between various departments reflects the level of emergency rescue between departments, mainly including whether the emergency departments will cooperate tacitly and whether the division of work between departments is clear. In the process of emergency rescue, if the division of labour is not clear and the departments do not cooperate tacitly, then it will inevitably lead to panic and disorder in emergency rescue and affect the normal start of emergency rescue work [17].

### 2.2.2 Cruise Ship Major Emergency Response Organization Information Conversion Degree

According to the management theory of information theory, the most important supporting factor in management decision-making is accurate information, and the transmission of information accompanies the entire emergency management process [18]. Accurate information is the cornerstone of scientific management. When emergencies occur, how to accurately and timely and smoothly transmit information related to cruise ship emergencies is very important to enhance the efficiency of emergency rescue and improve the level of emergency management, due to the suddenness of cruise ship emergencies, when emergencies occur, we need to immediately obtain relevant and accurate information to deal with emergencies in a timely and rapid manner; in the case of major cruise ship emergencies When a major cruise ship emergency occurs, we must handle the transformation of information correctly and reasonably [19].

### 2.2.3 Complexity of the Organizational Structure for Major Cruise Ship Emergencies

One of the most significant features of the emergency organization system for major cruise ship emergencies is the complex structure of the emergency organization, the main features of which are the complex levels of the emergency management organization system, the many departments, distinct upper, lower levels and the many managers, all of which will affect the decision-making, the main influencing factors of which include the status of the management level, respectively the level of education of the managers and whether the managers have rich experience in emergency management, rich experience in emergency management is very important for the help of emergency rescue, which can well help the managers in formulating a reasonable emergency rescue strategy [20].

## 3 Evaluation Study of Cruise Ship Emergency Response Organizations for Major Emergencies

### 3.1 Index Selection Criteria

After analyzing and understanding the factors affecting the emergency organization of major cruise ship emergencies, the indicators in the evaluation system can be designed and constructed. However, it is also necessary to follow the principles of scientifically, purposefulness, systematization and certain practicality to construct an evaluation index system for the capacity of cruise ship emergency response organizations.

## 3.2 Construction of Index System

Before establishing the indicators, the authors analyse the information related to the emergency management of major cruise ship emergencies in depth, taking into account historical data and through field research. In establishing the index system, the author used on-site research and expert consultation to determine the evaluation index system, following the principles of purposefulness, comprehensiveness and feasibility. The specific index system is shown in Table 2.

**Table 2.** Evaluation index system

| The first indicators | The second indicators | The third indicators |
| --- | --- | --- |
| Organizational functional complexity | Functional coupling degree | Whether interdepartmental cooperation is tacit |
| | | Whether the division of labor of departments is clear |
| | Degree of functional polymerization | Whether there is strong cohesion between departments |
| | | Whether there is harmony between departments |
| | Hierarchical functional span | Whether the department structure is reasonable |
| | | Whether the functional division is clear |
| Degree of information conversion | Information feedback | smoothness of information transmission |
| | | Timeliness of information feedback |
| | Communication network | The complexity of the communication network |
| | Information channel | Whether the information channel single |
| | | Whether the information is accurate or not |
| Organizational structure complexity | Management level | Educational background of managers |
| | | Management experience level |

(*continued*)

**Table 2.** (*continued*)

| The first indicators | The second indicators | The third indicators |
|---|---|---|
| | Management range | Whether there are too many managers |
| | | Whether the distribution of authority is reasonable |
| | Rank relation | The upper and lower levels are clear |
| | | Whether the relationship between superiors and subordinates is harmonious |

## 3.3 Fuzzy Hierarchical Analysis

### 3.3.1 Determination of Indicator Weights

The following steps were used to analyse the factors influencing the organizational capacity of cruise ships to respond to major emergencies using hierarchical analysis:

(1) Hierarchical modeling, After defining the problem, a hierarchical model is built from the top down, based on the logical relationship between the factors. The highest level is called the target level, i.e. the problem to be solved. Evaluation index structure is shown in Table 3:

**Table 3.** Evaluation index structure table

| Target layer | First-level indicators | Second-level indicators | Third-level indicators |
|---|---|---|---|
| Emergency Organization Capability for Major Emergencies of Cruise Ships A | Organizational functional complexity $B_1$ | Functional coupling degree $C_1$ | Whether interdepartmental cooperation is tacit $D_1$ |
| | | | Whether the division of labor of departments is clear $D_2$ |
| | | Degree of functional polymerization $C_2$ | Whether there is strong cohesion between departments $D_3$ |

(*continued*)

**Table 3.** (*continued*)

| Target layer | First-level indicators | Second-level indicators | Third-level indicators |
|---|---|---|---|
| | | | Whether there is harmony between departments $D_4$ |
| | | Hierarchical functional span $C_3$ | Whether the department structure is reasonable $D_5$ |
| | | | Whether the functional division is clear $D_6$ |
| | Degree of information conversion $B_2$ | Information feedback $C_4$ | smoothness of information transmission $D_7$ |
| | | | Timeliness of information feedback $D_8$ |
| | | Communication network $C_5$ | The complexity of the communication network $D_9$ |
| | | Information channel $C_6$ | Whether the information channel single $D_{10}$ |
| | | | Whether the information is accurate or not $D_{11}$ |
| | Organizational structure complexity $B_3$ | Management level $C_7$ | Educational background of managers $D_{12}$ |
| | | | Management experience level $D_{13}$ |
| | | Management range $C_8$ | Whether there are too many managers $D_{14}$ |
| | | | Whether the distribution of authority is reasonable $D_{15}$ |

(*continued*)

**Table 3.** (*continued*)

| Target layer | First-level indicators | Second-level indicators | Third-level indicators |
|---|---|---|---|
| | | Rank relation $C_9$ | The upper and lower levels are clear $D_{16}$ |
| | | | Whether the relationship between superiors and subordinates is harmonious $D_{17}$ |

(2) Construction of judgment matrix: Any system analysis has a certain amount of basic information, and the information of the hierarchical analysis method is people's judgement on the relative importance of each element of each level in relation to the previous level. These judgement data are expressed in matrix form as a judgement matrix. The method of determining the relative importance of each factor is shown in Table 4:

**Table 4.** Relationship between relative importance of various factors

| Scale | Meaning | Scale | Meaning |
|---|---|---|---|
| 1 | bi and bj are equally important | 9 | $b_i$ is more important than $b_j$ |
| 3 | $b_i$ is slightly more important than $b_j$ | 2,4,6,8 | The degree of importance is between the above cardinal numbers |
| 5 | $b_i$ is obviously more important than $b_j$ | reciprocal | $b_{ij} = 1/b_{ij}$ |
| 7 | $b_i$ is more important than $b_j$ | | |

Two-by-two comparison method is used for each layer of indicators to obtain the judgment matrix R.

$$R = \begin{bmatrix} b_{11} & b_{12} & .. & b_{1n} \\ b_{21} & b_{22} & .. & b_{2n} \\ . & . & ... & \\ . & . & ... & \\ b_{n1} & b_{n2} & .. & b_{nn} \end{bmatrix} \quad (1)$$

Use the summation method to calculate the weight of each index, and the specific calculation is as follows:

$$r_i = \frac{\sum_{j=1}^{n} b_{ij}, \ i = 1, 2, 3 \cdots n}{\sum_{i=1}^{n}\sum_{j=1}^{n} b_{ij}, \ i = 1, 2, 3 \cdots n} \qquad (2)$$

The $r = (r_1, r_2, r_3 \cdots, r_n)^T$ is the maximum eigenvector sought, so the weight of each indicator factor is obtained, which is that eigenvector.

(3) Consistency check

Is the judgment matrix of R, Calculate the normalized eigenvector Q and the maximum eigenvalue max that satisfies $RQ = Q\lambda_{max}$, $Q = (r_1, r_2, r_3 \cdots, r_n)$.

We usually need to check the consistency of the judgement matrix to avoid, for example, errors in the logic of assignment of factors when comparing two. The usual consistency check indicators are:

$$CI = \frac{\lambda_{max} - n}{n - 1} \qquad (3)$$

Typically, $CI > 0, \lambda_{max} > n$, Smaller CI means better consistency. In practice, we usually compare the CI with the average random consistency index $RI$. See Table 5 for $RI$ values:

**Table 5.** RI value table

| Matrix order | 1 | 2 | 3 | 4 | 5 | 6 | 7 | 8 | 9 |
|---|---|---|---|---|---|---|---|---|---|
| RI | 0.00 | 0.00 | 0.58 | 0.90 | 1.12 | 1.24 | 1.32 | 1.41 | 1.45 |

The first-order and second-order matrices are always consistent, and when the judgment matrix order is greater than 2, the random consistency ratio of the judgment matrix needs to be calculated according to Table 5. When RI < 0.10, we consider that the judgment matrix has satisfactory consistency, otherwise, we need to readjust the judgment matrix for recalculation.

$$CR = \frac{CI}{RI} \qquad (4)$$

### 3.3.2 Fuzzy Hierarchical Evaluation Steps

(1) Establish the index system set. The evaluation index system of emergency organization capability of cruise ship major emergencies is shown in Table 7 in Sect. 4.1. The overall target is recorded as A, and the set of A includes three elements $B_1$, $B_2$ and $B_3$, that is, $A = \{B_1, B_2, B_3\}$. And each secondary target can be further divided into different sets of $C_1$, $C_2$, $C_3$, $C_4$, $C_5$, etc.

(2) Establishing a fuzzy evaluation set. It is difficult to describe the evaluation of the emergency management capability of a cruise ship, and the only way to determine the strengths and weaknesses of the emergency management capability is to use a rubric. The emergency management capability of cruise ships can be divided into four levels: "good", "better", "average" and "poor". "This means that the evaluation set V = {good, better, fair, poor} = ($V_1$, $V_2$, $V_3$, $V_4$).

(3) Determination of indicator system weights. Determine the weight value of each indicator using hierarchical analysis.

(4) Determine the affiliation. Determination of affiliation. Qualitative analysis of each tertiary indicator is carried out by means of expert consultation and questionnaires, and the evaluation level of each indicator is also given. Based on the results of the survey, a fuzzy relationship matrix T between the evaluation indicators and the evaluation set is established, so as to obtain the fuzzy relationship from A to V.

$$t_{kj} = \frac{\text{The K tertiary index is evaluated as the number of people in the level}}{\text{Total number of surveys}} \quad (5)$$

$$T_k = \begin{bmatrix} t_{11} & t_{12} & \cdots & t_{1n} \\ t_{21} & t_{22} & \cdots & t_{2n} \\ \vdots & \vdots & & \vdots \\ t_{n1} & t_{n2} & \cdots & t_{nn} \end{bmatrix} \quad (6)$$

(5) Fuzzy Comprehensive Evaluation

1) The third level of fuzzy comprehensive evaluation is based on the secondary indicator weights Rij and the tertiary indicator affiliation judgment matrix $T_k$.

$$A_{ij} = R_{ij} T_k \quad (7)$$

2) Based on the evaluation results of the first step, the second level indicators are evaluated by fuzzy synthesis.

$$A_i = R_i A_{ij} \quad (8)$$

3) Based on the evaluation results of the second step, a fuzzy integrated evaluation of the first level indicators was carried out.

$$A = R A_i \quad (9)$$

where A is denoted as the cruise ship major emergency response organization comprehensive evaluation results affiliation matrix. The fuzzy comprehensive evaluation results are denoted by K, then:

$$K = AV^T \quad (10)$$

On the final result of K, different grades of the evaluation object can be obtained.

## 4 Cruise Ship Major Emergency Response Organization Evaluation

### 4.1 Calculation of Indicator Weights

This chapter is mainly to evaluate the emergency response organization of major cruise ships. The weights of first-level indicators are shown in Tables 6 and 7.

Table 6. Judgment matrix and weights of first-level indicators

| A | Organizational functional complexity | Degree of information conversion | Organizational structure complexity | weight |
|---|---|---|---|---|
| Organizational functional complexity | 1.00 | 5.00 | 3.00 | 0.53 |
| Degree of information conversion | 0.20 | 1.00 | 0.50 | 0.15 |
| Organizational structure complexity | 0.34 | 2.00 | 1.00 | 0.32 |

Therefore, the maximum eigenvalue $\lambda_{max} = 5.006$, $CR = 0.019 < 0.1$, $CR = 0.025 < 0.1$ conformity testing.

Table 7. $A_1$ judgment matrix and weight

| $A_1$ | Functional coupling degree | Degree of functional polymerization | Hierarchical functional span | Weights |
|---|---|---|---|---|
| Functional coupling degree | 1.00 | 2.00 | 0.34 | 0.31 |
| Degree of functional polymerization | 0.50 | 1.00 | 0.250 | 0.17 |
| Hierarchical functional span | 3.00 | 4.00 | 1.00 | 0.52 |

The weights of the secondary indicators are shown in Tables 8, 9 and 10.

Largest eigenvalue is $\lambda_{max} = 5.006$, $CR = 0.019 < 0.1$, conformity testing.

Table 8. $A_2$ judgment matrix and weights

| $A_2$ | Information feedback | Communication network | Information channel | Weights |
|---|---|---|---|---|
| Information feedback | 1.00 | 2.00 | 0.34 | 0.32 |
| Communication network | 0.50 | 1.00 | 0.20 | 0.17 |
| Information channel | 3.00 | 5.00 | 1.00 | 0.52 |

Largest eigenvalue is $\lambda_{max} = 4.923$, $CR = 0.017 < 0.1$, conformity testing.

Table 9. $A_3$ judgment matrix and weights

| $A_3$ | Management level | Management range | Rank relation | Weights |
|---|---|---|---|---|
| Management level | 1.00 | 5.00 | 6.00 | 0.51 |
| Management range | 0.20 | 1.00 | 3.00 | 0.31 |
| Rank relation | 0.17 | 0.34 | 1.00 | 0.18 |

Largest eigenvalue is $\lambda_{max} = 4.865$, $CR = 0.015 < 0.1$, conformity testing.

Through the calculation of the weight of the first-level indicators, we can determine the comprehensive weight of the second-level indicators. The Table 10 is the comprehensive weight table of the second-level indicators:

Table 10. Comprehensive weight table of secondary indicators

|  | Organizational functional complexity$B_1$ | Degree of information conversion$B_2$ | Organizational structure complexity$B_3$ | Comprehensive weight |
|---|---|---|---|---|
|  | 0.532 | 0.145 | 0.313 |  |
| Function coupling degree $C_1$ | 0.310 |  |  | 0.103 |
| Functional degree of polymerization $C_2$ | 0.169 |  |  | 0.056 |

(*continued*)

**Table 10.** (*continued*)

|  | Organizational functional complexity B$_1$ | Degree of information conversion B$_2$ | Organizational structure complexity B$_3$ | Comprehensive weight |
|---|---|---|---|---|
|  | 0.532 | 0.145 | 0.313 |  |
| Hierarchical functions span C$_3$ | 0.521 |  |  | 0.173 |
| Information feedback C$_4$ |  | 0.323 |  | 0.107 |
| Communication network C$_5$ |  | 0.165 |  | 0.055 |
| Information channel C$_6$ |  | 0.521 |  | 0.173 |
| Management level C$_7$ |  |  | 0.511 | 0.170 |
| Management amplitude C$_8$ |  |  | 0.308 | 0.102 |
| The grade is C$_9$ |  |  | 0.181 | 0.061 |

## 4.2 Fuzzy Evaluation of Emergency Organization for Cruise Major Emergencies

In this section, based on the calculation of the first-level indicators and second-level indicators in Sect. 4.1, combined with the scoring results of the questionnaire feedback, the distribution of each indicator is statistically obtained.

According to the formula (5) in Sect. 4, the three-level indicators are normalized to obtain the affiliation matrix AI, and the major emergencies of the cruise ship are analyzed according to the formula (6) ~ formula (10). The emergency organization conducts a fuzzy comprehensive evaluation. The evaluation process is shown in Table 11.

**Table 11.** Fuzzy comprehensive evaluation process of major cruise accidents

| Secondary index evaluation results | First level index evaluation results | Evaluation result |
|---|---|---|
| (0.16,0.16,0.35,0.33) | (0.18,0.22,0.30,0.30) | (0.18,0.25,0.32,0.25) |
| (0.10,0.24,0.35,0.31) |  |  |
| (0.30,0.27,0.23,0.20) |  |  |
| (0.16,0.20,0.35,0.29) | (0.19,0.17,0.33,0.31) | (0.18,0.25,0.32,0.25) |
| (0.27,0.20,0.27,0.26) |  |  |

(*continued*)

**Table 11.** (*continued*)

| Secondary index evaluation results | First level index evaluation results | Evaluation result |
|---|---|---|
| (0.14,0.13,0.40,0.33) | | |
| (0.16,0.30,0.29,0.25) | (0.19,0.29,0.30,0.12) | (0.18,0.25,0.32,0.25) |
| (0.27,0.35,0.20,0.18) | | |
| (0.14,0.13,0.40,0.33) | | |

Assuming that the comment level $V = (V_1, V_2, V_3, V_4) = \{poor, general, better, good\}$, and taking the values $\{5, 6, 7, 8\}$ respectively, it can be concluded that the emergency The organizational fuzzy comprehensive evaluation result $K = 6.32$. It can be seen from this that the emergency organization system for major emergencies of the cruise ship is between average and good, and it can be considered that the emergency organization system is basically perfect.

## 5 Conclusion

Cruise tourism is an indispensable form of travel in modern tourism, and cruise safety issues have become the most concerned aspects of passengers, cruise companies, and cruise management departments. The occurrence of various large and small cruise accidents not only brought huge losses to the lives and property of passengers, but also made people afraid of cruise tourism, a modern way of travel, and dare not try it. Whether the emergency management of major cruise emergencies is perfect and reasonable is an important support for maintaining the rapid development of the cruise economy.

This paper mainly focuses on the relevant research on the emergency organization of major emergencies of cruise ships. Through the investigation and analysis of relevant factors affecting the emergency organization of major emergencies of cruise ships, the relevant indicators of emergency organization are put forward, and the evaluation model of emergency organization for major emergencies of cruise ships is constructed at the same time.

## References

1. Wang, C., Guo, J., Kuo, M.: The building of social resilience in Sichuan after the Wenchuan earthquake: a perspective of the social-government interactions. Saf. Sci. (C) (2020)
2. Harrison, S., Johnson, P.: Challenges in the adoption of crisis crowd sourcing and social media in Canadian emergency management. Gov. Inf. Q. **8**(3), 3–4 (2019)
3. Zaw, T.N., Lim, S.: The military's role in disaster management and response during the 2015 Myanmar floods: a social network approach. Int. J. Disaster Risk Reduct. **18**(10), 6–8 (2017)
4. Espinoza, A.E., Osorio-Parraguez, P., Quiroga, E.P.: Preventing mental health risks in volunteers in disaster contexts: the case of the Villarrica Volcano eruption, Chile. Int. J. Disaster Risk Reduct. **21**(13), 12–14 (2018)

5. Whittaker, J., McLennan, B., Handmer, J.: A review of informal volunteerism in emergencies and disasters: definition, opportunities and challenges. Int. J. Disaster Risk Reduct. 15(8), 6–7 (2015)
6. Bundy, J., Pfarrer, M.D., Short, C.E., Coombs, W.T.: Crises and crisis management: integration, interpretation, and research development. J. Manag. 11(6), 31–33 (2017)
7. Bisri, M.B.F., Beniya, S.: Analyzing the national disaster response framework and inter-organizational network of the 2015 Nepal/Gorkha earthquake. Procedia Eng. 21(11), 12–14 (2016)
8. Jiang, P., Wang, Y., Liu, C., Hu, Y., Xie, J.: Evaluating critical factors influencing the reliability of emergency logistics systems using multiple-attribute decision making. Symmetry 13(4), 6–7 (2019)
9. Liu, D., Li, Y., Fang, S., Zhang, Y.: Influencing factors for emergency evacuation capability of rural households to flood hazards in western mountainous regions of Henan province, China. Int. J. Disaster Risk Reduct. 8(3), 16–17 (2017)
10. Eisenman, D.P., Long, A., Setodji, C., Hickey, S., Gelberg, L.: Differences in individual-level terrorism preparedness in Los Angeles County. Am. J. Prev. Med. 10(2), 6–7 (2012)
11. Hearst, M.A., Scholkopf, B., et al.: Trends and controversies-support vector machines. IEEE Intell. Syst. 16(5), 24–34 (2005)
12. Paciarotti, C., Cesaroni, A., Bevilacqua, M.: The management of spontaneous volunteers: a successful model from a flood emergency in Italy. Int. J. Disaster Risk Reduct. 11(3), 12–14 (2018)
13. Yue, H., Zhu, T.L.: Improvement of evaluation method of elderly family medical product design based on AHP. Math. Probl. Eng. 11(8), 6–7 (2022)
14. Dai, X., Wu, X.: Safety and stability evaluation of the uranium tailings impoundment dam: based on the improved AHP-cloud model. J. Radiat. Res. Appl. Sci. 15(1), 21–31 (2022)
15. Niloufar, V., Mehdi, G.: Preference of hybrid steel frame with exclusive seismic performance using the analytic hierarchy process. J. Earthquake Eng. 26(10), 12–14 (2022)
16. Jiang, L., Li, Y., Jiang, C.: Employment competitiveness of college students based on improved AHP. Eurasia J. Math. Sci. Technol. 13(8), 3–4 (2017)
17. Zhang, J., Xiao, B.: Application of analytic hierarchy process in abnormal vibration of Marine machinery. J. Phys: Conf. Ser. 17(6), 11–12 (2021)
18. Han, B., Ming, Z.: Comprehensive risk assessment of transmission lines affected by multi-meteorological disasters based on fuzzy analytic hierarchy process. Int. J. Electric. Power Energy Syst. 13(3), 6–7 (2021)
19. Dhanapal, S., Panneer, D.D., Sarit, M.: Composite techniques of structural equation modeling and analytic hierarchy process for information technology vendor selection. Int. J. Inf. Technol. Decis. 20(4), 11–15 (2021)
20. Nie, H.: Fuzzy evaluation model of the teaching quality in colleges and universities based on analytic hierarchy process. Basic Clin. Pharmacol. Toxicol. 127(3), 189–190 (2020)

# Fresh Agricultural Products Supplier Evaluation and Selection for Community Group Purchases Based on AHP and Entropy Weight VIKOR Model

Gong Feng[1], Jingjing Cao[1(✉)], Qian Liu[1], and Radouani Yassine[1,2]

[1] School of Transportation and Logistics Engineering, Wuhan University of Technology, Wuhan 430072, China
bettycao@whut.edu.cn
[2] International University of Rabat, 999055 Rabat, Morocco

**Abstract.** The outbreak and persistence of COVID-19 and the lack of effective management of the agricultural product supply chain have made the green and safe supply of fresh agricultural products an increasingly prominent issue. As a derivative of the new retail model, the community group purchases platforms have become the main spots for Chinese residents to consume fresh agricultural products. Therefore, the selection of safe and reliable suppliers of fresh agricultural products has become an inescapable focal issue in the development of community group purchases platforms. This paper proposes a three-stage evaluation index system for sustainable supplier selection of fresh agricultural products from the perspective of safety and sustainability, taking into account several aspects of suppliers such as product quality, service, development, technology and green. Then combine AHP with the entropy-weighted VIKOR method to calculate the combined weight of each indicator and prioritize candidate suppliers. Meanwhile, an numerical example is presented to verify the effectiveness of the proposed evaluation model. This paper aims to provide a reference for relevant community group purchases platforms and suppliers to form their own core competitiveness for better development.

**Keywords:** Fresh agricultural products · Supplier evaluation and selection · Community group purchases platforms · AHP · Entropy Weight · VIKOR

## 1 Introduction

Recently, with the development of the national economy and the increase in consumer income, ecological pollution has intensified [1]. And COVID-19 is still active. Consumer's expectations for fresh agricultural product have shifted these days [1]. During COVID-19, developing community group purchasing platforms played a unique role in safeguarding people's livelihoods, as traffic control and community lockdown pose strong impact on the public's normal life. Consumer consumption habits and patterns of reliance persist in today's post-epidemic environment, providing a chance for community group purchasing platforms that focus on fresh agricultural product to grow fast in China. The selection of fresh food suppliers is essential for a community group purchases platform's procurement system, which affects the quality of the products and the company's procurement costs [3]. And directly determines a community group purchases platform's market competitiveness [4].

Research on suppliers of fresh agricultural products in China started late and hasn't formed a mature theoretical and methodological system yet [5]. Now, homogenization of major community group purchases platform is serious. This has led to problems such as the poor competitiveness of community group purchases platforms and the low quality of fresh agricultural products provided [6, 7]. Therefore, there is an urgent demand to establish an agricultural products supplier selection and evaluation system to reduce the procurement costs of community group purchases platforms effectively and guarantee the quality of fresh agricultural products. This paper proposes the AHP and Entropy-weight VIKOR method to construct a new evaluation system for the selection of suppliers for community group purchases platforms, the contributions and innovations of this paper are as follows:

1) We established a supplier evaluation index system in five aspects: quality criteria, service criteria, technical criteria, development criteria and green criteria, and construct an indicator system including 15 indicators.
2) We proposed an integrated hybrid approach of AHP and Entropy weights VIKOR to select suppliers for community group purchases platforms. The weights of the indicators consider both subjective and objective factors; the VIKOR judging method reduces the tendency of negative indicators' being overlooked, maximizes group benefits and minimizes individual regret.
3) We concentrated on community group purchases platforms, which has recently emerged and is a derivative of the new retail model. This paper can fill the gap in the research on the selection of suppliers for fresh products in community group purchases platforms to a certain extent and has certain significance for relevant enterprises.

The paper is organized as follows. In Sect. 2 the indicators related to the selection of community group purchases platforms suppliers are described, and the evaluation system is determined. The data processing and calculation process of the AHP-Entropy VIKOR method is described in Sect. 2.4, following by a numerical application and case study of a community group purchases platform in Sect. 3. Recommendations on the management issues involved are provided in Sect. 4. In Sect. 5, the relevant research on green suppliers in China and abroad is reviewed. Finally, in Sect. 6 concluding remarks and future research directions are presented.

### 1.1 Evaluation System of Green Suppliers of Fresh Agricultural Products

Considering the background of the post-epidemic era, the relevant characteristics of fresh agricultural products supply chain and the traditional evaluation factors of suppliers, the evaluation indicators of green suppliers of fresh agricultural products for community group purchase platform are determined, and an evaluation system composed of target layer, constraint major element layer, constraint sub-element layer and program layer is constructed from top to bottom, as shown in Fig. 1.

In order to display the evaluation indicators for the selection of community group purchases platform suppliers better, this paper starts from five aspects: quality criteria, service criteria, technology criteria, development criteria and green criteria, and further decomposes them into 15 sub-elements, aiming to make a comprehensive, specific and objective evaluation of fresh agricultural products suppliers of community group purchases platforms.

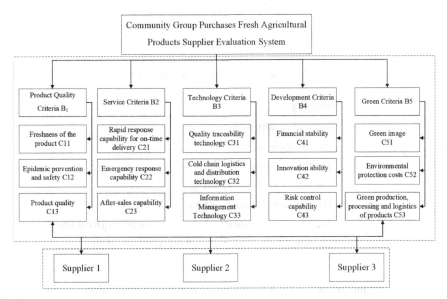

**Fig. 1.** Community group purchases fresh agricultural products supplier evaluation system

## 2 Methodology

In this work, a novel three-stage approach is proposed for the evaluation and selection of suppliers for Community Group Purchases platforms. In the first stage, the subjective weights of each criterion and indicator are determined based on the scoring of relevant criteria and indicators by professional evaluators using the AHP method. In the second stage, the objective weights of each indicator are determined by entropy weighting method based on the supplier's raw data. Then the subjective and objective weights are combined to derive a composite weight. Lastly, the VIKOR method was used to prioritize the community group Purchases platform suppliers.

### 2.1 Hierarchical Analysis to Determine Subjective Weights

Hierarchical analysis (AHP) provides a simple decision-making method for complex decision-making problems with multiple objectives, multiple criteria or unstructured characteristics by mathematizing the decision-making process with less quantitative information based on an in-depth analysis of the nature of complex decision-making problems, influencing factors and their intrinsic relationships. The basic steps are as follows.

1) Construction of the judgment matrix. The importance levels between the two indicators are shown in the Table 1.

$$A = (a_{ij})_{n \times m} = \begin{bmatrix} a_{11} & \cdots & a_{1m} \\ \vdots & \ddots & \vdots \\ a_{n1} & \cdots & a_{nm} \end{bmatrix} \tag{1}$$

**Table 1.** Importance level and scale value

| Scale | Meaning |
|---|---|
| 1 | Both are equally important |
| 3 | The former is slightly strong |
| 5 | The former is strong |
| 7 | The former is obviously strong |
| 9 | The former is absolutely strong |
| 2, 4, 6, 8 | The effect of the former is between above two adjacent levels |
| 1, 1/2, ..., 1/9 | The ratio of influence is exactly the reciprocal of aij |

2) Solve the judgment matrix A to obtain the maximum eigenvalue.

$$\lambda_{max} = \frac{1}{n} \sum_{i=1}^{n} \frac{(Aw)_i}{w_j}, i = 1, 2, ...n, w = (w_0, w_1, w_2, w_3, w_4) \tag{2}$$

3) Calculate consistency indicator CI, random consistency indicator RI and consistency ratio CR for consistency testing.

$$CI = \frac{\lambda_{max} - n}{n - 1}, CR = \frac{CI}{RI}. \tag{3}$$

When $CR = 0$, there is perfect consistency; when $CR < 0.1$ consider A to be a satisfactory consistency matrix. CI、CR is given in the (3), RI is obtained by looking up the Table 2 according to the order of the matrix; when $CR < 0.1$, the eigenvector sought is the weight value of each criterion layer factor, otherwise reconstruct the judgment matrix A.

## 2.2 Entropy Weighting Method to Determine Objective Weights

The entropy weight method is an objectively determined weight method. In information theory, the amount of information provided by data can be reflected by entropy. When

**Table 2.** Average random consistency index

| n  | 3    | 4    | 5    | 6    | 7    | 8    | 9    |
|----|------|------|------|------|------|------|------|
| RI | 0.52 | 0.89 | 1.12 | 1.26 | 1.36 | 1.41 | 1.46 |

the data of each plan of an indicator differs greatly, the entropy is low and the amount of information provided is large, while the opposite is true when the data difference is small. The basic steps are as follows.

1) According to the raw data of each indicator, an evaluation matrix of type can be constructed based on the m standardized indicator values of the given n evaluation objects, and the matrix is as follows.
2) Normalize each element in the evaluation matrix. The evaluation indexes are divided into two categories, one is the benefit index and the other is the cost index. The larger the benefit index value, the better, and the smaller the cost index value, the better.

$$X = \begin{bmatrix} x_{11} & \cdots & x_{1n} \\ \vdots & \ddots & \vdots \\ x_{m1} & \cdots & x_{mn} \end{bmatrix} \quad (4)$$

$$y_{ij} = \begin{cases} \frac{x_{ij} - \min(x_i)}{\max(x_i) - \min(x_i)}, & \text{benefit} - \text{type indicators} \\ \frac{\max(x_i) - x_{ij}}{\max(x_i) - \min(x_i)}, & \text{cost} - \text{type indicators} \end{cases} \quad (5)$$

$max(x_i)$, $min(x_i)$ are the maximum and minimum values of the ith indicator respectively.

The standardized raw data can be formed into a new evaluation matrix of type $Y = (y_{ij})_{m \times n}$ and the elements of the matrix are the standardized data.

1) Calculate the entropy value for indicator.

$$E_i = -\frac{\sum_{j=1}^{n} P_{ij} \ln P_{ij}}{\ln(n)}, \quad P_{ij} = \frac{y_{ij}}{\sum_{j=1}^{n} y_{ij}} \quad (6)$$

2) Calculate the weights from the calculated entropy value.

$$w_i = \frac{1 - E_i}{m - \sum_{i=1}^{m} E_i} \quad (7)$$

### 2.3 Calculate the Composite Weights

Different weighting methods may result in different final optimal solution. The subjective weighting method can reflect the decision maker's intention, but the evaluation results are more subjective; while the objective weighting method is more theoretical, but doesn't consider the decision maker's intention. Thus, we use a combination of subjective and objective weighting methods to determine the weights.

Assuming that the indicator weight determined by the hierarchical analysis method is $w_i$ and the entropy weighting method determines a weight of $\Theta_i$, the composite weight is $\lambda_i$. A common formula for a combined subjective-objective approach to calculate composite weight is:

$$\lambda_i = \frac{w_i \times \Theta_i}{\sum_{i=1}^{m} w_i \times \Theta_i} \quad i = 1, 2, 3, \ldots, m \tag{8}$$

Song et al. improved the formulae commonly used for the combined subjective-objective assignment method [8]. And one of them can be described as follows:

$$\lambda_i = \frac{w_i + \Theta_i}{\sum_{i=1}^{m} w_i + \Theta_i} \quad i = 1, 2, 3, \ldots, m \tag{9}$$

$$\sum_{i=1}^{m} (w_i + \Theta_i) = \sum_{i=1}^{m} w_i + \sum_{i=1}^{m} \Theta_i = 2 \tag{10}$$

Calculate Eq. 10 to get Eq. 11

$$\lambda_i = 0.5 \times w_i + 0.5 \times \Theta_i \tag{11}$$

## 2.4 VIKOR Method Judging Model

VIKOR is a method for optimizing compromise solutions in multi-attribute decision making. The basic idea of VIKOR is to define the ideal solution and the negative ideal solution first and then prioritize the alternatives according to how close they are to the ideal solution in terms of their evaluated values. The basic steps are as follows.

1) Calculate the positive and negative ideal solutions $y_i^+, y_i^-$

$$(y_i^+, y_i^-) = \begin{cases} (max_j y_i^+, min_j y_i^+), \forall i, \text{ Benefit type} \\ (min_j y_i^+, max_j y_i^+), \forall i, \text{ Cost Type} \end{cases} \tag{12}$$

2) Calculate utility value and individual regret value $R_i$.

$$S_i = \sum_{j=1}^{n} \frac{\lambda_j(y_j^+ - y_{ij})}{y_j^+ - y_j^-}, \quad R_i = \max_j \left\{ \frac{\lambda_j(y_j^+ - y_{ij})}{y_j^+ - y_j^-} \right\} \tag{13}$$

where

$y_j^+$ denotes positive ideal value, $y_j^+ = \max_j\{y_{ij}\}$ \hfill (14)

$y_j^-$ denotes negative ideal value, $y_j^- = \min_j\{y_{ij}\}$ \hfill (15)

3) Calculate the compromise value $Q_i$

$$Q_i = \varepsilon \frac{S_i - S^-}{S^+ - S^-} + (1 - \varepsilon) \frac{R_i - R^-}{R^+ - R^-} \tag{16}$$

where ε is compounding factor, and

$$S^+ = \max_i\{S_i\}, \ S^- = \min_i\{S_i\} \quad (17)$$

$$R^+ = \max_i\{R_i\}, \ R^- = \min_i\{R_i\} \quad (18)$$

Ranking of alternative suppliers.

Firstly, the ranking is obtained by increasing the value of $Q_i$: $A^{(1)}$, $A^{(2)}$, $A^{(2)}$, $A^{(3)}$, ..., $A^{(i)}$. If $A^{(1)}$ is the scheme ranked first according to the value of $Q_i$ and satisfies the following conditions.

$$Q(A^{(2)}) - Q(A^{(1)}) \geq \frac{1}{n-1}.$$

Scheme $A^{(1)}$ is still optimal according to $S_i R_i$ ordering.

Then $A^{(1)}$ is the most stable optimal solution in the decision process. If the two conditions above do not hold simultaneously, there are two situations to obtain a compromise solution.

If condition 1 is not satisfied, the compromise is $A^{(1)}$, ..., $A^{(i)}$, $A^{(i)}$ which is the maximised i-value determined by $Q(A^{(1)}) < \frac{1}{n-1}$.

If condition 2 is not satisfied, the compromise solution is $A^{(1)}$, $A^{(2)}$.

## 3 Empirical Research

A community group Purchases platform has three fresh agricultural products suppliers which are evaluated to determine the best supplier. Based on the actual situation of the three suppliers we have compiled data on the performance of the three suppliers in various indicators, which is presented in Table 3.

Table 3. Supplier raw data

| Item | | Supplier 1 | Supplier 2 | Supplier 3 | Indicator Type |
|---|---|---|---|---|---|
| Criterion | Indicators | | | | |
| Quality | C11 | 80 | 95 | 85 | Benefit type |
| | C12 | 80 | 95 | 85 | Benefit type |
| | C13 | 80 | 90 | 85 | Benefit type |

(*continued*)

### 3.1 Determination of Indicator Weights

After consulting a large number of relevant documents and consulting relevant experts and scholars, according to the basic principles of AHP, we constructed the judgment

**Table 3.** (*continued*)

| Item | | Supplier 1 | Supplier 2 | Supplier 3 | Indicator Type |
|---|---|---|---|---|---|
| Criterion | Indicators | | | | |
| Services | C21 | 85 | 80 | 85 | Benefit type |
| | C22 | 90 | 75 | 70 | Benefit type |
| | C23 | 85 | 80 | 90 | Benefit type |
| Technology | C31 | 75 | 95 | 80 | Benefit type |
| | C32 | 80 | 90 | 85 | Benefit type |
| | C33 | 85 | 95 | 75 | Benefit type |
| Development | C41 | 90 | 85 | 95 | Benefit type |
| | C42 | 70 | 95 | 85 | Benefit type |
| | C43 | 85 | 90 | 95 | Benefit type |
| Green | C51 | 85 | 80 | 90 | Benefit type |
| | C52 | 85 | 75 | 70 | Cost type |
| | C53 | 80 | 90 | 85 | Benefit type |

matrix O-B in Table 4. For the target layer corresponding to the criterion layer and the judgment matrix *B1-C, B2-C, B3-C, B4-C* and *B5-C* for the criterion layer corresponding to the indicator layer (Table 5). Then we used Excel to calculate the weights and conduct consistency checks. Finally, we calculate the subjective weight of the target layer corresponding to the index layer in In the same way, the judgement matrix B2-C, B3-C, B4-C, B5-C is constructed and the related metrics can be calculated. Then the hierarchical indicators is sorted and ranked as Table 6.

**Table 4.** Judgement matrix O-B

| Judgement Matrix O-B and Consistency Test | | | | | | | |
|---|---|---|---|---|---|---|---|
| Criterion | $B_1$ | $B_2$ | $B_3$ | $B_4$ | $B_5$ | Wi | $\lambda_{max}$ |
| $B_1$ | 1 | 3 | 5 | 5 | 5 | 0.5127 | 5.0040 |
| $B_2$ | 1/3 | 1 | 2 | 2 | 2 | 0.1907 | |
| $B_3$ | 1/5 | 1/2 | 1 | 1 | 1 | 0.0989 | |
| $B_4$ | 1/5 | 1/2 | 1 | 1 | 1 | 0.9889 | |
| $B_5$ | 1/5 | 1/2 | 1 | 1 | 1 | 0.0989 | |
| CI = 0.0010 CR = 0.009 RI = 1.12 | | | | | | | |

In the same way, the judgement matrix B2-C, B3-C, B4-C, B5-C is constructed and the related metrics can be calculated. Then the hierarchical indicators is sorted and ranked as Table 6.

**Table 5.** Judgement matrix B1-C

| Judgement Matrix $B_1$-C and Consistency Test | | | | | |
|---|---|---|---|---|---|
| Quality criterion | $C_{11}$ | $C_{12}$ | $C_{13}$ | Wi | $\lambda_{max}$ |
| $C_{11}$ | 1 | 1/3 | 4 | 0.2499 | 3.0092 |
| $C_{12}$ | 3 | 1 | 9 | 0.6813 | |
| $C_{13}$ | 1/4 | 1/9 | 1 | 0.0688 | |
| CI = 0.0046 CR = 0.0088 RI = 0.52 | | | | | |

**Table 6.** Composite ranking of hierarchical indicators

| Indicator | Quality | Services | Technology | Development | Green | W |
|---|---|---|---|---|---|---|
| | 0.5127 | 0.1907 | 0.0989 | 0.0989 | 0.0989 | |
| $C_{11}$ | 0.2499 | – | – | – | – | 0.1282 |
| $C_{12}$ | 0.6813 | – | – | – | – | 0.3493 |
| $C_{13}$ | 0.0688 | – | – | – | – | 0.0353 |
| $C_{21}$ | – | 0.2297 | – | – | – | 0.0438 |
| $C_{22}$ | – | 0.6483 | – | – | – | 0.1236 |
| $C_{23}$ | – | 0.1220 | - | – | – | 0.0233 |
| $C_{31}$ | – | – | 0.6370 | – | – | 0.0630 |
| $C_{32}$ | – | – | 0.1047 | – | – | 0.0104 |
| $C_{33}$ | – | – | 0.2583 | – | – | 0.0255 |
| $C_{41}$ | - | – | – | 0.6483 | – | 0.0641 |
| $C_{42}$ | - | – | – | 0.1220 | – | 0.0121 |
| $C_{43}$ | - | – | – | 0.2297 | - | 0.0227 |
| $C_{51}$ | - | – | – | – | 0.1095 | 0.0108 |
| $C_{52}$ | - | – | – | – | 0.3090 | 0.0306 |
| $C_{53}$ | - | – | – | – | 0.5816 | 0.0575 |

The entropy and entropy weights of each evaluation indicator are calculated according to (4) to (7), where environmental protection expenditure is a cost-based indicator and the rest of the indicators are benefit-based indicators. The entropy weights for each indicator were calculated as shown in Table 7.

Based on the weights of each indicator derived from the hierarchical analysis and the entropy weighting method, the combined weights were calculated according to (11) in Tables 7 and 8.

**Table 7.** Entropy weighting calculation results

| Criterion | Indicator | Entropy value $e_i$ | Coefficient of variation $\partial_i$ | Entropy weights $w_i$ |
|---|---|---|---|---|
| Quality | C11 | 0.5119 | 0.4881 | 0.0741 |
|  | C12 | 0.5119 | 0.4881 | 0.0741 |
|  | C13 | 0.5794 | 0.4206 | 0.0639 |
| Services | C21 | 0.6309 | 0.3691 | 0.0561 |
|  | C22 | 0.4555 | 0.5445 | 0.0827 |
|  | C23 | 0.5794 | 0.4206 | 0.0639 |
| Technology | C31 | 0.4555 | 0.5445 | 0.0827 |
|  | C32 | 0.5794 | 0.4206 | 0.0639 |
|  | C33 | 0.5794 | 0.4206 | 0.0639 |
| Development | C41 | 0.5794 | 0.4206 | 0.0639 |
|  | C42 | 0.6022 | 0.3978 | 0.0604 |
|  | C43 | 0.5794 | 0.4206 | 0.0639 |
| Green | C51 | 0.5794 | 0.4206 | 0.0639 |
|  | C52 | 0.6126 | 0.3874 | 0.0588 |
|  | C53 | 0.5794 | 0.4206 | 0.0639 |

**Table 8.** Composite weight calculation results

| Serial number | Indicator | Subjective weights $\theta_i$ | Entropy weights $w_i$ | Composite weights $\lambda i$ |
|---|---|---|---|---|
| 1 | C11 | 0.1282 | 0.0741 | 0.1011 |
| 2 | C12 | 0.3493 | 0.0741 | 0.2117 |
| 3 | C13 | 0.0353 | 0.0639 | 0.0496 |
| 4 | C21 | 0.0438 | 0.0561 | 0.0499 |
| 5 | C22 | 0.1236 | 0.0827 | 0.1032 |
| 6 | C23 | 0.0233 | 0.0639 | 0.0436 |
| 7 | C31 | 0.0630 | 0.0827 | 0.0728 |
| 8 | C32 | 0.0104 | 0.0639 | 0.0371 |

(*continued*)

## 3.2 VIKOR Method to Determine the Optimal Option

The positive and negative ideal solutions for each indicator can be derived from Eqs. (4), (5) and (12).

$y_i^+ = (1, 1, 1, 1, 1, 1, 1, 1, 1, 1, 1, 1, 0, 1); y_i^- = (0, 0, 0, 0, 0, 0, 0, 0, 0, 0, 0, 0, 1, 0)$

Fresh Agricultural Products Supplier Evaluation and Selection 691

**Table 8.** (*continued*)

| Serial number | Indicator | Subjective weights $\theta_i$ | Entropy weights $w_i$ | Composite weights $\lambda i$ |
|---|---|---|---|---|
| 9 | C33 | 0.0255 | 0.0639 | 0.0447 |
| 10 | C41 | 0.0641 | 0.0639 | 0.0640 |
| 11 | C42 | 0.0121 | 0.0604 | 0.0362 |
| 12 | C43 | 0.0227 | 0.0639 | 0.0433 |
| 13 | C51 | 0.0108 | 0.0639 | 0.0374 |
| 14 | C52 | 0.0306 | 0.0588 | 0.0447 |

$\lambda = $ (0.1011, 0.2117, 0.0496, 0.0499, 0.1032, 0.0436, 0.0728, 0.0371, 0.0447, 0.0640, 0.0362, 0.0433, 0.0374, 0.0447, 0.0607)

The S, R and Q values for each of the 3 suppliers were calculated and ranked according to Eqs. (13), and the results are shown in Table 9.

**Table 9.** S, R, Q for the three suppliers

| Suppliers | S | | R | | Q | |
|---|---|---|---|---|---|---|
| | Value | Ranking | Value | Ranking | Value | Ranking |
| Supplier 1 | 0.7074 | 3 | 0.2117 | 3 | 1 | 3 |
| Supplier 2 | 0.3236 | 1 | 0.0774 | 1 | 0 | 1 |
| Supplier 3 | 0.5439 | 2 | 0.1411 | 2 | 0.5244 | 2 |

Firstly, the ranking is obtained by decreasing value: Supplier 2, Supplier 3, Supplier 1. This is the option ranked first according to the value of Q and satisfies the following conditions.

1) $Q(A^{(2)}) - Q(A^{(1)}) = 0.5244 \geq 0.5$
2) Option $A^{(1)}$ is still the best option according to the S and R ranking.

So the supplier selection ranking is: Supplier 2 > Supplier 3 > Supplier 1.

## 4 Analysis and Suggestions

Through hierarchical analysis, an evaluation system consisting of a target layer, a constraint major element layer, a constraint sub-element layer and a program layer was constructed. By the case studied in Sect. 4, we found that in the supplier selection process of the community group purchases platforms, indicators such as product freshness, epidemic prevention and safety, emergency response capability and cold chain logistics and distribution technology have a greater composite weight.

This also reflects consumer preference for quality and safety of fresh agricultural products in the post-epidemic era. By theoretical analysis, it is clear that whether fresh agricultural products carry or spread COVID-19 during processing and transportation, and whether the freshness of the products can be guaranteed are crucial. It affects the competitiveness of the community group purchases platforms directly. If community group purchases platforms want to obtain further development and better foothold in the market, they should take measures from the following aspects.

1) Combine subjective judgement and objective data to make supplier selections. The issue of supplier selection contains both qualitative and quantitative elements. Community group purchases platforms should be both subjective and objective to improve the accuracy of comprehensive evaluation when making decisions on supplier selection. The research results in this paper show that the epidemic prevention and safety, freshness of the products, the emergency response capability and financial stability of the suppliers are given greater weight. This reflects that in the special era, consumers' preference for the safety and quality of fresh agricultural products, and enterprises should focus on these influencing factors when determining supplier evaluation indicators.
2) Construct a scientific and perfect supplier evaluation system. Determining reasonable supplier evaluation indicators and building a scientific supplier evaluation system can largely reduce enterprise procurement, logistics costs and the potential risks in business management, while ensuring the quality of products and increasing the profit point of enterprises. When determining evaluation indicators, enterprises can consider traditional supplier evaluation and selection indicators such as development and service, but also need to consider the needs of the enterprise's own development and changes in the social environment.

## 5 Related Work

Research on supplier selection focuses on the determination of supplier evaluation indicators and the establishment of the system and the selection of evaluation methods. In this section we will first review relevant research on fresh products supply chains, and then we will review the current status of research on evaluation indicators and systems and evaluation methods at home and abroad. Previous research on supplier evaluation and selection systems has mainly focused on social performance indicators, economic performance indicators and environmental performance indicators. In recent years, under the influence of epidemics and other related factors, the relevant indicators for supplier evaluation should be adjusted and improved. At the same time, with more and more indicators included in the supplier evaluation index system, the supplier evaluation methods are becoming more and more complex.

### 5.1 Determination of Indicator Weights

Companies must use green supplier selection strategies to respond to market pressures [4]. Mousak hani et al. consider the selection of green suppliers based on environmental competence to be one of the most important issues facing companies [9]. Parkouhi et al. argue that green supplier selection is one of the core issues in green supply chain management and contains a large number of qualitative and quantitative factors as an

MCDM issue [10]. At present, China has not yet developed a mature theoretical and methodological system for the establishment of an evaluation system for the selection of fresh produce suppliers [11]. Community group purchases and other e-commerce platforms have formed a new supply chain operation model for fresh products e-commerce, and how suppliers are evaluated and selected has become extremely important [3]. As an innovative new retail model, community group purchases are booming. However, as a new e-commerce model, the community group purchases platforms have problems such as serious homogeneity and poor product quality [6]. In terms of bringing convenience to consumers, the social e-commerce model has great advantages and potential, but there are still major problems in terms of product quality and product traceability [8].

## 5.2 Evaluation Criteria

Lo et al. identified supplier evaluation indicators as green performance, product quality, etc., based on the attributes of products purchased by consumers [4]. Mousak hani et al. divided the evaluation criteria of cost, quality, delivery, technology, environmental competence, organization and green image into green and general criteria for the evaluation of green suppliers [9]. Zhang et al. propose environmental performance, economic performance and social performance as integrated evaluation criteria for sustainable suppliers [12]. Rong et al. proposed evaluation indicators based on the impact of the epidemic such as epidemic safety and quality traceability techniques [2]. Incorporating environmental factors into supplier selection leads to higher margins and greater customer satisfaction [13]. Freshness of the product is an important factor in the quality of fresh agricultural products [14].

## 5.3 Evaluation Methods

Hou et al. used the entropy weighting method to obtain the weights of each evaluation index, and then used the improved TOPSIS method to select the appropriate green supplier [15]. Chandra et al. proposed a comprehensive evaluation method for integrated green suppliers with BWM and VIKOR [16]. Yuan et al. proposed an integrated evaluation method combining SWOT strategic analysis and entropy-weighted fuzzy synthesis evaluation [17]. Wang et al. used hierarchical analysis to select a low-carbon supplier of fresh produce [11]. Agarwal et al. adopted AHP method for supplier selection in dynamic environment [18]. Zhou et al. proposed an evaluation research method combining intuitionistic fuzzy sets and TOPSIS method [19]. As supplier selection is a Multi-Criteria Decision Making problem, Sahai et al. usaged Data Envelopment Analysis to measure supplier performance [20]. Liao et al. used the AHP and TOPSIS composite method to determine the weight of each index for the evaluation and selection of green suppliers for paper manufacturing enterprises [21]. Singh et al. proposed a new hybrid approach by using logarithmic fuzzy preference programming (LFPP) and artificial neural network (ANN) to generate requirement prioritization of the washing machine supplier [22].

## 6 Conclusion and Future Work

This paper utilized the AHP-Entropy VIKOR method to evaluate suppliers quantitatively by introducing five criteria: quality, service, technology, development and green, and refining them into 15 secondary indicators. When selecting suppliers, enterprises can refer to the evaluation system established in this paper and calculate the performance of suppliers by combining the composite weights of the calculated secondary indicators; the calculated weights reflect that fresh products suppliers should mainly focus on freshness, epidemic prevention, safety and other product quality indicators. They should also focus on their own financial stability and information technology level. The improvement of these aspects will allow them to have their own unique advantages and increase their competitiveness. Meanwhile, the community group purchases platform that chooses such suppliers has stronger market competitiveness. The three suppliers analyzed in this paper each have their own strengths in product quality, technology or low carbon green, and the relevant indicators meet certain standards, but none of the suppliers has consider all the five aspects of quality, service, technology, development and green, which reflects the current situation of most suppliers to a certain extent. The final selected supplier 2 has a higher priority due to the high quality of the fresh products provided and its technical capabilities, but there is still room for improvement in service guidelines.

In future research, we will investigate the development of fresh agricultural product suppliers on community group purchases platforms in greater depth, so as to identify more accurate and comprehensive evaluation indicators and establish a more detailed and complete evaluation system. In terms of evaluation methods, we will seek more scientific and reasonable evaluation models. For example, we will consider the complexity of objective things and the fuzziness of human thinking, and conduct a multi-attribute decision-making process in a fuzzy environment.

**Acknowledgment.** This project is supported by the National Natural Science Foundation, China (No. 61502360).

## References

1. Du, Y., Zhang, D., Zou, Y.: Sustainable supplier evaluation and selection of fresh agricultural products based on IFAHP-TODIM model. Math. Probl. Eng. 1–15 (2020)
2. Rong, L., Wang, L., Liu, P.: Supermarket fresh food suppliers evaluation and selection with multigranularity unbalanced hesitant fuzzy linguistic information based on prospect theory and evidential theory. Int. J. Intell. Syst. **37**(3), 1931–1971 (2021)
3. Dan, H., Jing, M.: The research on the factors of purchase intention for fresh agricultural products in an E-Commerce environment. IOP Conf. Ser. Earth Environ. Sci. **100**, 0123173 (2017)
4. Lo, H.-W., Liou, J.J.H., Wang, H.-S., et al.: An integrated model for solving problems in green supplier selection and order allocation. J. Clean. Prod. **190**, 339–352 (2018)
5. Deng, Y.: Selection of cold chain logistics suppliers for fresh agricultural products based on AHP. Logist. Technol. **04**, 91–93 (2017). (in Chinese)
6. Li, M., Fang, L.: Business model analysis and development prospect of community Group-buying platform. Econ. Res. Guid. **03**, 83–86 (2022). (in Chinese)

7. Cai, G.: Research on the current situation of agricultural product supply chain considering fresh-keeping efforts in the context of social e-commerce. China Logist. Procurement **12**, 95–97 (2022). (in Chinese)
8. Song, H., Wang, Z.: Weighing between objective weight and subjective weight. Tech. Econ. Manag. Res. **03**, 62 (2003). (in Chinese)
9. Mousakhani, S., Nazari-Shirkouhi, S., Bozorgi-Amiri, A.: A novel interval type-2 fuzzy evaluation model based group decision analysis for green supplier selection problems: A case study of battery industry. J. Clean. Product. **168**, 205–218 (2017)
10. Parkouhi, S.V., Ghadikolaei, A.S.: A resilience approach for supplier selection: using fuzzy analytic network process and grey VIKOR techniques. J. Clean. Product. **161**, 431–451 (2017)
11. Wang, Y., Zhang, P., Chen, X.: Empirical analysis of low-carbon fresh agricultural product supplier evaluation. Bus. Econ. Res. **05**, 127–130 (2018). (in Chinese)
12. Jing, Z., Dong, Y., Qiang, L., et al.: Research on sustainable supplier selection based on the rough DEMATEL and FVIKOR methods. Sustainability **13**(1), 88 (2020)
13. Verma, M., Prem, P.R., Ren, P., et al.: Green supplier selection with a multiple criteria decision-making method based on thermodynamic features. Environ. Dev. Sustain. 1–33 (2022)
14. Ma, X., Wang, S., Islam, S.M.N., et al.: Coordinating a three-echelon fresh agricultural products supply chain considering freshness-keeping effort with asymmetric information. Appl. Math. Model. **67**, 337–356 (2019)
15. Bin, H., Wang, Y.: Supplier evaluation and selection based on improved TOPSIS method in green supply chain. J. Hunan Univ. Technol. **28**(02), 81–86 (2014)
16. Garg, C.P., Sharma, A.: Sustainable outsourcing partner selection and evaluation using an integrated BWM–VIKOR framework. Environ. Dev. Sustain. **22**(2), 1529–1557 (2020)
17. Ying, Y., Yi, Z., Long, D., et al.: Research on large supermarket fresh food supplier evaluation and selection based on SWOT-entropy weight fuzzy comprehensive model. In: 2015 International Conference on Advanced Mechatronic Systems, pp. 15–19. IEEE (2015)
18. Prince, A., Manjari, S., Vaibhav, M., et al.: Supplier selection in dynamic environment using analytic hierarchy process. Int. J. Inf. Eng. Electron. Bus. (IJIEEB) **6**(4), 20–26 (2014)
19. Zhou, Q., Wang, Q., Chen, L.: Research on green supplier selection based on intuitive fuzzy set-TOPSIS. J. Syst. Sci. **01**, 94–98 (2017). (in Chinese)
20. Manjari, S., Prince, A., Vaibhav, M., et al.: Supplier selection through application of DEA. Int. J. Eng. Manuf. (IJEM) **4**(1), 1–9 (2014)
21. Liao, J., Zhang, R.: Research on the evaluation and selection of green suppliers in papermaking enterprises based on AHP-TOPSIS method. Logist. Eng. Manag. **02**, 91–93+84 (2020). (in Chinese)
22. Yash, V.S., Bijendra, K., Satish, C.: A hybrid approach for requirements prioritization using LFPP and ANN. Int. J. Intell. Syst. Appl. (IJISA) **11**(1), 13–23 (2019)

# Logistics of Fresh Cold Chain Analysis of Joint Distribution Paths in Wuhan

Weihui Du[1], Donglin Rong[2(✉)], Saipeng Xing[3], and Jiawei Sun[4]

[1] School of Logistics, Wuhan Technology and Business University, Wuhan 430065, China
[2] National Engineering Research Center of Geographic Information System, China University of Geosciences, Wuhan 430074, China
249454030@qq.com
[3] School of Management, Wuhan Technology and Business University, Wuhan 430065, China
[4] Technical University of Munich Asia Campus, Technical University of Munich (TUM), Singapore, Singapore

**Abstract.** In the process of transportation and circulation of fresh products, the loss rate is an important reason for the high cost of logistics and the decline of product quality. Cold chain joint distribution of fresh products is an important method and means to reduce the loss rate. It is of great significance to study cold chain joint distribution for fresh cold chain logistics. The research content of this article is a common distribution route optimization fresh cold chain logistics in Wuhan city. First of all, the general situation of fresh cold chain logistics in our country is summarized, and the related concepts and connotations of cold chain logistics are deeply interpreted. Secondly, the existing problems of fresh cold chain logistics in Wuhan are analyzed and discussed. Finally, taking Wuhan Baishazhou cold chain logistics distribution center as an example, this paper uses the mileage saving method to study the route optimization of cold chain logistics distribution, so as to provide solutions to the high loss and high cost problems of fresh cold chain logistics joint distribution.

**Keywords:** Cold chain logistics · Joint distribution · Path optimization · Mileage saving method

## 1 Introduction

Compared with other countries such as the United States and Japan, China's fresh food cold chain logistics started a little later, so it is relatively backward in terms of cold chain transportation equipment and transportation skills, and there are more problems in the whole cold chain logistics transportation. At this stage, most of the food in China is still in accordance with the traditional distribution method of room temperature logistics to carry out the transportation of fresh food, the high cost, difficulty and strict time requirements of distribution have become important problems faced in the process of fresh cold chain logistics transportation. In cold chain logistics, choosing the optimal distribution route can save a lot of resources, reduce costs and speed up the circulation of fresh food, thus increasing the company's revenue [1]. For the company, optimizing the common distribution route in fresh food cold chain logistics greatly reduces the transport time of fresh food, reduces the loss of fresh food, reduces the costs consumed in transit, and improves the sales efficiency and customer satisfaction of new products. For the customer, the optimization of the common distribution route of the fresh food cold chain logistics will result in a reduction in the cost of the fresh food purchased by the customer, making the customer satisfied with our service and thus making the customer more inclined to choose and purchase fresh food cold chain products. The optimization of the cold chain logistics distribution route is the key to achieving full process optimization. The study of supply chain optimization for cold chain logistics companies has contributed to their development, effectively reducing costs, saving money as well as increasing revenue [2].

## 2 Relevant Conceptions

### 2.1 Theory of Cold Chain Logistics

Cold chain logistics, also known as low temperature logistics, is a way of freezing, storing, transporting, storing and transporting goods. The temperature of each of his logistics links is kept at a specific temperature through the use of various refrigeration technologies and refrigeration equipment, so that the environment they are in is maintained at a specific temperature all the time, which can not only protect the quality of food, but also reduce unnecessary losses and reduce the consumption of logistics costs [3, 4].

Since the 21st century, with the rapid development of science and economy, people's standard of living has gradually begun to improve. People are not just chasing after filling their stomachs, but are beginning to pursue a high quality of life, and fresh food is increasingly being enjoyed by most people as an indispensable delicacy on their tables. Due to the special nature of fresh food, the number of goods that need to be delivered through cold chain logistics is also increasing year by year. The requirements of cold chain logistics are relatively high, so if we want to promote the rapid development of cold chain logistics, we need to vigorously develop our refrigeration technology. In addition, relevant policies from the national government are needed to support this [5].

### 2.2 Joint Distribution

Joint Distribution is the combination of a number of companies in an area in order to improve the efficiency of the logistics of the company. The company delivers the fresh food to a professional third party logistics company, which organises the information and arranges the transport, so that the fresh food is transported by different companies to different sorting centers, and then delivers the different fresh food products to the corresponding customers according to their needs. The main objective of joint distribution is to rationalise distribution routes, save costs, reduce loss of goods and increase customer satisfaction [7].

## 2.3 Vehicle Distribution Routing Issues

The logistics and distribution vehicle route optimization problem refers to the use of limited vehicles or resources to meet the needs of our known customers, which in a nutshell means using minimal resources to maximize our customers' needs. In practical terms: we need to make good use of each vehicle, taking their on-board rated weight, maximum mileage and other issues into account, and then develop a distribution route with the lowest cost [8].

## 3 Analysis of the Current Situation and Problems of Common Distribution of Fresh Food Cold Chain Logistics in Wuhan

### 3.1 Fresh Cold Chain Logistics Development in Wuhan

At the algorithm level, vehicle routing problem (VRP) can be abstractly solved in digraph, undirected graph, connected graph and network graph. The cities and warehouses arriving in the logistics distribution route are represented by points, and the connection between the points represents the water, land and air routes between the two cities, which can clearly reflect the connection between each point and point. The research models of vehicle routing problem (VRP) mainly include mathematical model and network graph model. The mathematical model has the advantages of large capacity, high flexibility and strong versatility than network graph model, so the mathematical model is often used [9, 10].

#### 3.1.1 Growth of Cold Chain Market Demand in Wuhan

Information shows that the current frozen food market in China, the annual growth rate is about 10%. Wuhan cold storage development potential is great, Wuhan is located in the "nine provinces through" the geographical location, with radiation around the provinces and the role of the local. Wuhan consumes 800,000 tons of frozen products every year, of which the demand for meat, seafood and other frozen products is about 350,000 tons, of which the demand for freshwater fish is about 300,000 tons. There are also markets for vegetables and fruit, all at a low temperature of 0–5 °C. Tens of thousands of tonnes of fresh food are shipped out every day from cold stores across Wuhan. According to statistics, about 50% of the food consumed by Wuhan residents on a daily basis belongs to fresh products. With that in mind, in order to meet the daily needs of the residents, it is necessary to ensure that 1 million tonnes of fresh food can be stored to meet the rapid growth in demand.

The people of Wuhan are constantly improving their quality of life and pursuing a more nutritious and healthy lifestyle in parallel with technological progress. As a result, people also have higher demands for the quality and variety of fresh produce. The rapid growth of demand in the cold chain market has prompted a more complete system of fresh cold chain logistics in Wuhan. As the demand and quality of fresh food is getting higher and higher among Wuhan residents, this has put forward high requirements for the distribution of fresh food in Wuhan, prompting the development of the cold chain market in Wuhan.

### 3.1.2 Wuhan Cold Chain Logistics Infrastructure

The Baishazhou Cold Chain Project is the largest cold storage project in Central China, covering an area of 65,000 tonnes and covering an area of approximately 1,000 kms, which is no less than that of the Xudong Cold Storage. Its capacity has reached the point where it can drive the storage and distribution of fresh food products in various regions such as South China and North China. The development of the Baishazhou cold chain project has greatly eased the existing storage capacity of Wuhan from the source and promoted the rapid development of the fresh food cold chain logistics industry in Wuhan and even in various regions of Hubei. However, the perishable nature of fresh food has led to difficulties in the distribution process, high distribution costs and difficulties in ensuring the quality of goods. Therefore, this aspect of cold chain logistics and distribution in Wuhan is not yet developed enough [11].

## 3.2 Wuhan Cold Chain Logistics Infrastructure Improvement

### 3.2.1 Low Level of Distribution Information

Cold chain products are complex and varied, with each product having different characteristics. Each customer is independent and distributed in different locations. The demand of customers fluctuates greatly and is not stable enough, so there is no way to achieve centralised distribution without pulling through and sharing information with each other. The fresh food cold chain in Wuhan still adopts the traditional distribution method, where the distribution center is in individual contact with the customer and rarely collects the customer's information for map route analysis. Therefore, the level of information technology in the whole distribution process is low, and a mechanism or system needs to be set up to pull through information and then carry out comprehensive processing [12, 13].

### 3.2.2 Unreasonable Capacity Arrangement

Due to the large number of customer types, different customer demands and low level of information technology, the distribution center is unable to obtain effective and accurate distribution data and can only make deliveries based on past experience. Some need to be delivered in time because of the time, then arrange for special vehicles to deliver, while others have a low loading rate of delivery vehicles due to the different time, often resulting in vehicles being idle but not used, and low loading rates leading to wasted costs.

### 3.2.3 Poorly Arranged Distribution Routes

Wuhan City Cold Chain Center has not formed a complete distribution logistics system, relying on past guesses to develop distribution routes, no mechanism to map out the number and location of customer needs, how to effectively use vehicles to make deliveries, and maximize the use of the vehicle's capacity as much as possible. It is still stuck in customer-specific routes, planning routes at will without optimizing them.

### 3.2.4 High Distribution Cost

Distribution costs are too high mainly because of the following three aspects: first, the arrangement of vehicles is unreasonable, resulting in low personnel utilization, personnel efficiency is not up to standard, wasting great manpower costs; second, the distribution route is unreasonable, resulting in the vehicle driving distance repeated or multiple trips, the vehicle return journey becomes longer; third, the vehicle cost is high, various unreasonable applications of vehicles, resulting in the vehicle fuel costs, maintenance costs are too high.

## 4 Application of the Mileage Saving Method

### 4.1 The Principle of the Mileage Saving Method

The mileage saving method is one of the most efficient heuristic algorithms designed for the problem of uncertainty in the number of distribution vehicles. Its basic principle is: according to the two sides of the triangle and must be larger than the third side, if the distribution center distributes goods to two distributors or customers, the distance assigned to one car for every two customers is greater than the distance assigned by one car in order of precedence. As shown in Fig. 1 to distribute goods from distribution center P to customers A and B, two cars are needed and the car transport distance is 2PA + 2PB, but if the goods of customers A and B are loaded on the same car in accordance with the order of departure from distribution center P, arrival at customer A, unloading the goods of customer A, and then going to customer B, unloading the goods of customer B and returning, only one car is needed and the car transport distance is PA + PB + AB, then according to what we said earlier the sum of the two sides of the triangle is greater than the third side, get PA + PB > AB, saving mileage for PA + PB-AB [14–17].

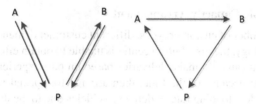

**Fig. 1.** The principle of the mileage saving

The basic idea of the mileage saving method is that, subject to some qualitative constraints, such as the weight and volume of the vehicle, the first vehicle of the distribution unit delivers the goods of the specified maximum weight and volume in the order of the customer's location under the optimal route, while the second vehicle is loaded in the same way as before. In this way, the customer's needs are all met and the route of distribution is optimized thereby achieving cost savings.

## 4.2 Conditions of Application of the Mileage Saving Method

On this basis, combined with the basic idea of the mileage saving method, several basic conditions for distribution route optimization using the mileage saving method have been summarized [18–21].

1) The location of the distribution center is known.
2) The location and demand of customers are known.
3) The distribution center has sufficient vehicles for distribution.
4) The delivery time can meet the customer's demand.
5) The distance between the distribution center and the customer is known.
6) One customer can only be delivered by one vehicle.
7) One vehicle can deliver to more than one customer.

## 4.3 Model Construction for the Mileage Saving Method

According to the data in the above table, the results are shown in Table 4. When the consistency ratio $Cr$ is less than $0.1$, the consistency test is passed, and the judgment matrix does not need to be adjusted again.

### 4.3.1 Model Description

When a distribution center carries out logistics distribution to two or more customers, it is important to establish logistics distribution routes according to the specific needs of customers, so as to achieve cost savings and reduce losses. Based on the Baishazhou logistics center in Wuhan, a customer-oriented logistics and distribution route model is constructed.

Assume that the Baishazhou fresh food distribution center is P, which simultaneously delivers and transports to various fresh food customers, each with a demand $g_i(i = 1, 2, 3\ n)$, and that the Baishazhou logistics center has m existing vehicles of 4 tons and 2.5 tons respectively. We must design the shortest route for distribution while meeting the requirements of different customers to achieve cost savings, improve efficiency and enhance customer satisfaction.

### 4.3.2 Related Variables

Several relevant parameters are explained as follows.

$n$: indicates the number of customers.

$m$: indicates the number of vehicles carrying out distribution transport.

$g_i$: the quantity of fresh food demanded by the customer.

$d_{ij}$: indicates the distance between two customers ($i, j = 1, 2......n$, where $i$ or $j$ is equal to $0$, the point is the Baishazhou fresh produce distribution center, such as $dpj$, which indicates the distance from distribution center $P$ to customer j).

$Q$: indicates the maximum carrying capacity of the distribution vehicle.

$D$: indicates the maximum number of miles that the distribution vehicle can travel.

$n_s$: indicates the number of fruit shops to be reached by the distribution vehicle for distribution ($s = 1, 2......n$, when $n = 0$, it means that the vehicle is not involved in the distribution).

$x_{ijs}$: the value of x is *0* or *1*, *1* means that the delivery vehicle *s* is travelling from customer *i* to customer *j*, *0* means that the delivery vehicle *s* is travelling from customer j to customer i.

$y_{is}$: the value of y is either *0* or *1*, with 1 indicating that the goods of customer *i* are delivered by *s* and *0* indicating that the goods of customer *i* are not delivered by *s*.

### 4.3.3 Model Building

If the number of miles (*S*) of distribution completed by the distribution center to each customer is taken as the minimum objective function, a model is developed based on the relevant variables as follows [22–24].

$$minS = \sum_{i=0}^{n} \sum_{j=0}^{n} \sum_{x=1}^{m} d_{ij} x_{ijs} \tag{1}$$

$$\sum_{s=1}^{m} y_{is} = \begin{cases} 1, i = 1, 2...n \\ m, i = 0 \end{cases} \tag{2}$$

$$\sum_{i=0}^{n} g_i y_{is} \leq Q, \quad s = 1, 2 \ldots m \tag{3}$$

$$\sum_{i=0}^{n} \sum_{j=0}^{n} d_{ij} x_{ijs} \leq D \tag{4}$$

$$\sum_{i=1}^{n} x_{ijs} = y_{is}, \quad j = 0, 1, 2 \ldots n \tag{5}$$

In the model established above, formula (1) indicates that the transport vehicles required to complete the entire distribution task are m vehicles. Equation (2) indicates that the demand of any one vehicle to deliver customers cannot exceed the rated weight of the truck. The formula (3) indicates that the distance travelled by any one vehicle cannot exceed its maximum mileage. Equation (4) means that there is one and only one delivery vehicle reaching any one customer during a delivery. Equation (5) means that one and only one vehicle leaves the fresh food shop during a delivery.

## 5 Design of a Common Distribution Route Optimization Scheme

### 5.1 Data Selection

When using the mileage saving method for route optimization, the first step is to know the quantity of goods demanded by the fresh produce shop. Since the goods sold by the fresh food shops are all fresh and have high requirements for freshness, the frequency of stocking for the fresh food shops is daily. From Table 1 and Table 2, we can see that this paper selects 10 fresh food shops to divide their annual demand evenly as a day's demand for data processing. 10 tons of goods are delivered each day, and according to the mode of direct delivery of each fresh food shop, a total of 10 refrigerated trucks are needed, and a total of 206 km is needed for the round trip.

## 5.2 Draw a Simplified Road Map

In order to provide a more visual representation of the role of the mileage saving method in path optimization, and also for better calculation, a geographical map of the fresh food shops and the distribution center was created based on the geographical location of each fresh food shop.

**Table 1.** Table of Demand for Fresh Food Stores

| Order number | Fresh food shop | volume of demand(t) |
|---|---|---|
| 1 | $A_1$ | 0.8 |
| 2 | $A_2$ | 1.4 |
| 3 | $A_3$ | 0.9 |
| 4 | $A_4$ | 0.7 |
| 5 | $A_5$ | 1.4 |
| 6 | $A_6$ | 1.1 |
| 7 | $A_7$ | 0.9 |
| 8 | $A_8$ | 0.7 |
| 9 | $A_9$ | 0.6 |
| 10 | $A_{10}$ | 1.5 |
| 11 | sum | 10 |

**Table 2.** Table of Distance between Fresh Food Store and Distribution Center

| Order number | Fresh food shop | Distance from distribution center (km) |
|---|---|---|
| 1 | $A_1$ | 12 |
| 2 | $A_2$ | 11 |
| 3 | $A_3$ | 9 |
| 4 | $A_4$ | 12 |
| 5 | $A_5$ | 10 |
| 6 | $A_6$ | 16 |
| 7 | $A_7$ | 6 |
| 8 | $A_8$ | 6 |

(*continued*)

**Table 2.** (*continued*)

| Order number | Fresh food shop | Distance from distribution center (km) |
|---|---|---|
| 9 | A9 | 12 |
| 10 | A10 | 9 |
| 11 | sum | 103 |

As can be seen from Fig. 2, $P$ denotes the Baishazhou fresh produce distribution center, $A_1$, $A_2$, $A_3$, $A_4$, $A_5$, $A_6$, $A_7$, $A_8$, $A_9$ and $A_{10}$ denote the locations of the individual fresh produce shops, $PA_1$, $PA_2$, $PA_3$, $PA_4$, $PA_5$, $PA_6$, $PA_7$, $PA_8$, $PA_9$ and $PA_{10}$ denote the distances between the distribution center and the fresh produce shops, and then based on their locations to produce A simplified route distribution diagram, from which the distances between the units can be clearly understood.

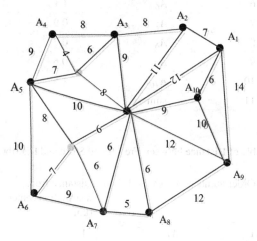

**Fig. 2.** Road map

### 5.3 Optimal Design of Paths

Based on the data we collected we started to calculate the mileage savings between customers using the mileage savings method where the sum of the two sides of a triangle is greater than the third side, using the principle that the mileage savings equals ($PA + PB - AB$), to create a mileage savings table, as shown in Table 3.

The mileage saving method is used for route optimization calculations, and a mileage saving ranking table is produced in descending order of mileage saved, as shown in Table 4.

**Table 3.** Save Odometer

|   | $A_1$ |   |   |   |   |   |   |   |   |
|---|---|---|---|---|---|---|---|---|---|
| $A_2$ | 16 | $A_2$ |   |   |   |   |   |   |   |
| $A_3$ | 6 | 12 | $A_3$ |   |   |   |   |   |   |
| $A_4$ | 1 | 7 | 13 | $A_4$ |   |   |   |   |   |
| $A_5$ | 0 | 0 | 2 | 13 | $A_5$ |   |   |   |   |
| $A_6$ | 0 | 0 | 0 | 11 | 16 | $A_6$ |   |   |   |
| $A_7$ | 0 | 0 | 0 | 0 | 0 | 13 | $A_7$ |   |   |
| $A_8$ | 0 | 0 | 0 | 0 | 0 | 8 | 7 | $A_8$ |   |
| $A_9$ | 10 | 2 | 0 | 0 | 0 | 2 | 1 | 6 | $A_9$ |
| $A_{10}$ | 15 | 7 | 0 | 0 | 0 | 0 | 0 | 0 | 11 |

From Table 4, it can be seen that A1A2 has the largest amount of mileage savings, based on the principle of the mileage saving method so the optimal distribution route can be designed, as shown in Table 5: the first distribution route is $P$-$A_1$-$A_2$-$A_{10}$-$P$, saving 31 km. The second distribution route is $P$-$A_4$-$A_5$-$A_6$-$A_8$-$P$, saving 37 km, the third distribution route is $P$-$A_7$-$A_9$-$P$, saving 1 km, and the fourth distribution route is $P$-$A_3$-$P$.

**Table 4.** Saving Sorting Table

| Order number | Connection point | Saving mileage | Order number | Connection point | Saving mileage |
|---|---|---|---|---|---|
| 1 | $A_1A_2$ | 16 | 12 | $A_2A_{10}$ | 7 |
| 2 | $A_5A_6$ | 16 | 13 | $A_2A_4$ | 7 |
| 3 | $A_1A_{10}$ | 15 | 14 | $A_7A_8$ | 7 |
| 4 | $A_3A_4$ | 13 | 15 | $A_1A_3$ | 6 |
| 5 | $A_4A_5$ | 13 | 16 | $A_8A_9$ | 6 |
| 6 | $A_6A_7$ | 13 | 17 | $A_3A_5$ | 2 |
| 7 | $A_2A_3$ | 12 | 18 | $A_2A_9$ | 2 |
| 8 | $A_4A_6$ | 11 | 19 | $A_6A_9$ | 2 |
| 9 | $A_9A_{10}$ | 11 | 20 | $A_1A_4$ | 1 |
| 10 | $A_1A_9$ | 10 | 21 | $A_7A_9$ | 1 |
| 11 | $A_6A_8$ | 8 | 22 | $A_1A_5$ | 0 |

Table 5. Path comparison table

| Optimal path | Saving mileage | Tonnage |
| --- | --- | --- |
| $P$-$A_1$-$A_2$-$A_{10}$-$P$ | 31 | 3.7 |
| $P$-$A_4$-$A_5$-$A_6$-$A_8$-$P$ | 32 | 3.9 |
| $P$-$A_7$-$A_9$-$P$ | 1 | 1.5 |
| $P$-$A_3$-$P$ | 0 | 1.5 |

## 6 Conclusion

In recent years, China's logistics industry has developed rapidly, but due to the small market share and low profits of various logistics companies, many enterprises have focused on initiatives to improve efficiency and save costs, including Deppon Express in 2021, which has started various actions to increase efficiency and reduce costs.

This paper examines the path optimization of cold chain fresh produce common distribution in Wuhan by reviewing relevant research theories at home and abroad to understand the methods of common distribution path optimization in logistics at home and abroad. The paper adopts the mileage saving method, and through the principle of mileage saving method, ten fresh food shops are selected to study the path optimisation of the distribution route of Baishazhou distribution center in Wuhan city, which achieves cost saving, loss reduction and efficiency improvement.

**Acknowledgment.** This project is supported by Humanities and Social Sciences Research Planning Project of Ministry of Education. Research on the "double-edged sword" effect of leadership empowerment on employees and team creativity from the perspective of digital transformation. (22YJA630097).

## References

1. Yao, Y.: Research on Optimization of Regional Agricultural Products Cold Chain Logistics Distribution, p. 08. China Agricultural Press, Beijing (2020)
2. Ting, P.-H.: An efficient and guaranteed cold-chain logistics for temperature-sensitive foods: applications of RFID and sensor networks. MECS Press IJIEEB **10**(05) (2013)
3. Gu, W., Lu, C.: Research on path planning based on comparative particle swarm and mileage saving algorithm. China Automotive **08**, 36–39 (2019)
4. Wang, H., Wang, X.: Optimization of the distribution path of Zhongbai supermarket based on the mileage saving method. Logist. Technol. **36**(03), 84–87+157 (2017)
5. Yan, Y., Lai, S.: Distribution route optimization based on mileage saving method. Hunan: J. Hunan Ind. Vocat. Technol. Coll. **17**(01), 29–32 (2017)
6. Li, B., Yang, L.Y., Mo, L.: Optimization analysis of express logistics distribution network in Yizhou city. Coastal Enterprises Technol. **01**, 16–19 (2018)
7. Sun, S., Sun, J., Ji, S.: Research on the distribution scheme of Jilin Chunguang Dairy based on the mileage saving method. Logist. Sci. Technol. **41**(06), 43–45 (2018)

8. Guo, Y., Li, J.: Research on the optimization of the same city distribution path of fruit products considering customer classification–take Shandong XX Company as an example. Rural Econ. Technol. **31**(02), 69–72 (2020)
9. Fan, H.: Study on the optimization of "fruit and vegetable butler" distribution system in Shijiazhuang. Hebei University of Science and Technology, Wuhan (2018)
10. Lu, X.H.: Research on optimization of drug distribution path of GK Guilin Pharmaceutical Distribution Company. Guilin University of Technology, Guilin (2021)
11. Jing, Y.: Research on the improvement strategy of logistics capacity of Lanzhou Best Express. Lanzhou Jiaotong University, Lanzhou (2019)
12. Youssef, B., Belkora, M.-J.: Distribution strategies toward nanostores in emerging markets: the Valencia case. Interfaces **47**(6), 505–517 (2017)
13. Lin, S., Yang, J., et al.: Low carbon path optimization of logistics distribution based on mileage saving method. Logist. Eng. Manag. **41**(04), 80–82 (2019)
14. Kong, L., Heng, L., Luo, H., et al.: Sustainable performance of just-in-time (JIT) management in time-dependent batch delivery scheduling of precast construction. J. Clean. Product. 193684–193701 (2018)
15. Chen, J., Shi, J.: A multi-compartment vehicle routing problem with time windows for urban distribution-a comparison study on particle swarm optimization algorithms. Comput. Ind. Eng. 13395–133106 (2019)
16. Liu, B.: Research on dynamic vehicle routing optimization method for cold chain logistics distribution with time window. Beijing Jiaotong University, Beijing (2018)
17. Raut, R.-D., Gardas, B.-B., Narwane, V.-S., et al.: Improvement in the food losses in fruits and vegetable supply chain - a perspective of cold third-party logistics approach. Oper. Res. Perspect. (2019)
18. Wan, Y.: Research on the optimization of fresh agricultural products distribution path from the perspective of low carbon. Kunming University of Technology, Kunming (2021)
19. Rezaei, N., Ebrahimnejad, S., Moosavi, A., et al.: A green vehicle routing problem with time windows considering the heterogeneous fleet of vehicles: two metaheuristic algorithms. Eur. J. Ind. Eng. **13**(4), 507–510 (2019)
20. Li, J., Liu, M., Liu, P.: Route optimization of cold chain logistics vehicles with multiple models of fresh agricultural products. J. China Agric. Univ. (07) (2021)
21. Liu, G., Hu, J., Yang, Y., et al.: Vehicle routing problem in cold chain logistics: a joint distribution model with carbon trading mechanisms. Resourc. Conserv. Recycling (2020)
22. Grigorios, D., Konstantakopoulos, Sotiris, P., et al.: Vehicle routing problem and related algorithms for logistics distribution: a literature review and classification. Oper. Res. (2020)
23. Sheng, H.: Research on the vehicle routing problem of fresh e-commerce logistics distribution. University of Electronic Science and Technology, Chengdu (2019)
24. Christian, F.: A decision support system to investigate food losses in e-grocery deliveries. Comput. Ind. Eng. (2018)

# Optimization of Logistics Distribution Route Based on Ant Colony Algorithm – Taking Nantian Logistics as an Example

Zhong Zheng[1,2(✉)], Shan Liu[1], and Xiaoying Zhou[3]

[1] School of Transportation, Nanning University, Guangxi 530200, China
zhengzhong2007@163.com
[2] King Mongkut's Institute of Technology Ladkrabang, Bangkok 10520, Thailand
[3] Guangxi Wuzhou Communications Co., Ltd., Nanning 530200, China

**Abstract.** The continuous development of social economy and the people's growing yearning for a better life have promoted the rapid development of e-commerce, and urban distribution business has become increasingly important. Urban distribution is also faced with relatively complex task requirements, such as short time window, strict requirements, complex road conditions, high frequency of tasks, etc. Choosing the right path has become the key issue to improve distribution efficiency. Aiming at the problems encountered in the distribution of urban distribution routes, the paper uses ant colony algorithm to analyze and optimize the distribution routes, and uses MATLAB programming to solve the optimized distribution scheme. By comparing the data before and after optimization, it is found that the logistics transportation cost is effectively reduced, and the logistics transportation resources are saved, which proves the feasibility of the algorithm, and provides a certain reference for improving the efficiency of urban distribution and reducing the distribution cost.

**Keywords:** Ant colony algorithm · Path optimization · Intra-city distribution

## 1 Introduction

The logistics distribution in cities is faced with unique and complex distribution environment and requirements. The distribution volume is small and the frequency is high. The order distribution demand is personalized, there are many freight points, and the distribution of transportation networks in the service area is uneven. The distribution road conditions are usually complex. How to take into account the time, economic cost and quality of urban distribution has become a concern of more and more theoretical research and enterprise practice. The improvement of transportation path requires reasonable planning [1, 2].

The problem of route distribution has attracted the attention of many scholars. Dai Tingting and others proposed an improved ant colony optimization algorithm where the ant colony algorithm is calculated at a "20 × 20" grid environment, and the search efficiency has been greatly improved [3]; Luyuan et al. proposed an improved ant colony algorithm. The improved ant colony algorithm has better search ability and improved convergence speed [7]; Liu Ziyu, Zhao Lixia and others improved the ant colony algorithm by using the piecewise function to adjust the convergence speed of the algorithm, which has the advantage of effectively reducing the path length, jumping out of the local optimum and accelerating the convergence speed [5]; Ge Dayun started from the multimodal transport network under a single task, combined with ant colony algorithm to design the model solution method, and completed his goal with the minimum cost by integrating various advantages of multimodal transport [6]; Li Shuangshuang et al. proposed a new pheromone update algorithm and a global tendency pheromone distribution model, used the grid method to build a two-dimensional plane space with obstacles and carried out simulation experiments, and the results verified the effectiveness of the model [7]; Li, Sidi et al. proposed a knowledge-based hybrid ant colony algorithm to improve the overall satisfaction of all tourism groups, improve the performance of the algorithm, and avoid the generation of local optimal solutions [8].

Aiming at the problems of the above ant colony algorithm in path planning, this paper takes Nantian Logistics Company as an example, combines the specific situation of the enterprise, and proposes an ant colony algorithm based on practical applicability. By going to the enterprise to investigate the needs of customers, obtain actual data, and build a model based on data processing. Finally, apply the model and algorithm to the actual situation of the enterprise, and then obtain the optimized distribution plan through MATLAB programming.

## 2 Problem Description

### 2.1 Description of Problems Faced by Enterprises

As a city distribution company, Nantian Logistics is under great pressure every day, and often due to the untimely vehicle arrangement, some urgent goods cannot be delivered to customers in time, which affects the work process and satisfaction of customers. The business volume is small and the frequency is high. The current distribution path and sequence mainly depend on manual judgment; It is unable to maximize the role of vehicles. Some lines pass repeatedly, but the goods are not arranged on the same vehicle, resulting in the waste of vehicle resources, which also reflects the low level of management.

### 2.2 Modeling and Optimization of Nantian Logistics Distribution Path

The logistics center in this study is used for this project with a total of 4 transport vehicles; The maximum transport capacity of each vehicle is *2500* kg; The goods distribution center number is 0, and the customer number is *1–16*; The earliest departure time of the transport vehicle is *9:00* in the morning; Each customer has the same service time. The quantity requirements and geographical coordinates of each customer's goods are shown in Table 1.

**Table 1.** Customer parameters

| Customer No | North latitude | East longitude | Number | Weight | Volume |
|---|---|---|---|---|---|
| 0 | 22°55′01.46″ | 108°21′35.77″ | 0 | 0 | 0 |
| 1 | 22°51′44.75″ | 108°18′02.95″ | 21 | 367.5 | 0.84 |
| 2 | 22°51′39.20″ | 108°15′37.33″ | 3 | 296 | 1.52 |
| 3 | 22°49′06.46″ | 108°18′24.67″ | 30 | 240 | 2.638 |
| 4 | 22°52′24.49″ | 108°24′09.54″ | 10 | 200 | 0.545 |
| 5 | 22°52′33.75″ | 108°24′04.76″ | 39 | 235.58 | 0.897 |
| 6 | 22°52′35.76″ | 108°25′05.98″ | 10 | 400 | 2.4 |
| 7 | 22°45′30.74″ | 108°26′46.36″ | 6 | 125.4 | 0.24 |
| 8 | 22°47′58.38″ | 108°16′45.43″ | 3 | 44 | 0.066 |
| 9 | 22°47′39.73″ | 108°18′35.03″ | 1 | 11 | 0.0165 |
| 10 | 22°51′02.41″ | 108°18′13.21″ | 42 | 735 | 1.68 |
| 11 | 22°52′26.46″ | 108°16′26.32″ | 19 | 325 | 0.546 |
| 12 | 22°47′57.05″ | 108°22′03.50″ | 5 | 50 | 0.38 |
| 13 | 22°49′05.98″ | 108°26′51.23″ | 23 | 460 | 0.69 |
| 14 | 22°43′57.06″ | 108°22′03.52″ | 41 | 818 | 1.23 |
| 15 | 22°47′57.08″ | 108°31′48.04″ | 36 | 718 | 1.08 |
| 16 | 22°50′40.13″ | 108°18′21.08″ | 53 | 1058 | 1.59 |

Data source: based on the relevant data of the arrival of Nantian Logistics Group in January.

## 3 Model Establishment

Taking 16 customer enterprises served by Nantian Logistics as the object of this study, the content of distribution work mainly belongs to the intra-city distribution in Nanning. The steps to build the model are as follows.

### 3.1 Determine Coordinate Points

First of all, according to the data obtained in this practice, take multiple transportation destinations as an example, obtain the longitude and latitude according to the specific location of the address, and then convert them into coordinate points through the longitude and latitude and XY conversion software.

After obtaining the coordinates of *12* transportation points, because the algorithm mainly calculates the distance between points, the origin is not set. The coordinate points use the map to query the distance between them (unit: *km*), and the distance matrix shown in Fig. 1 is generated. In the figure, *1–16* corresponds to the coordinate points of *16* transportation points, of which 0 is the headquarters of Nantian Logistics Group.

|    | 0   | 1    | 2    | 3    | 4    | 5    | 6    | 7    | 8    | 9    | 10   | 11   | 12   | 13   | 14   | 15   | 16   |
|----|-----|------|------|------|------|------|------|------|------|------|------|------|------|------|------|------|------|
| 0  | 0   | 9.8  | 11.8 | 12.2 | 6.6  | 6.3  | 7.7  | 19.8 | 15.4 | 15.1 | 9.4  | 10   | 13.2 | 15.4 | 19.8 | 10   | 6.5  |
| 1  | 9.8 | 0    | 4    | 4.7  | 10.7 | 10.7 | 13.4 | 18.8 | 7    | 6.7  | 1.2  | 3    | 9.7  | 14.5 | 27.5 | 2.6  | 3.6  |
| 2  | 11.8| 4    | 0    | 6.7  | 14.7 | 14.6 | 17.4 | 22.2 | 5.5  | 6.6  | 5    | 2.1  | 12.9 | 19.2 | 31   | 5.7  | 5.4  |
| 3  | 12.2| 4.7  | 6.7  | 0    | 11.5 | 11.6 | 14.2 | 15.7 | 4.3  | 3.4  | 3.5  | 7.1  | 6.5  | 13.6 | 24.5 | 2.1  | 8.2  |
| 4  | 6.6 | 10.7 | 14.7 | 11.5 | 0    | 0.5  | 1.8  | 13.5 | 15.1 | 15   | 10.5 | 13.2 | 9    | 7.9  | 20.6 | 10.2 | 10.7 |
| 5  | 6.3 | 10.7 | 14.6 | 11.6 | 0.5  | 0    | 1.7  | 13.9 | 15.3 | 15.1 | 10.5 | 13.1 | 9.1  | 8.2  | 21   | 10.3 | 10.6 |
| 6  | 7.7 | 13.4 | 17.4 | 14.2 | 1.8  | 1.7  | 0    | 13.5 | 17.9 | 17.7 | 13.2 | 15.9 | 10.2 | 7.8  | 19.8 | 13   | 13.2 |
| 7  | 19.8| 18.8 | 22.2 | 15.7 | 13.5 | 13.9 | 13.5 | 0    | 17.8 | 16.5 | 17.8 | 22   | 9.2  | 6    | 8.8  | 16.5 | 21.1 |
| 8  | 15.4| 7    | 5.5  | 4.3  | 15.1 | 15.3 | 17.9 | 17.8 | 0    | 3    | 6.1  | 8.3  | 9    | 16.6 | 26.5 | 5.3  | 10.6 |
| 9  | 15.1| 6.7  | 6.6  | 3.4  | 15   | 15.1 | 17.7 | 16.5 | 3    | 0    | 5.8  | 7.9  | 8    | 16.7 | 26.7 | 5.1  | 10.1 |
| 10 | 9.4 | 1.2  | 5    | 3.5  | 10.5 | 10.5 | 13.2 | 17.8 | 6.1  | 5.8  | 0    | 4.1  | 8.7  | 14.6 | 26.5 | 1.4  | 4.7  |
| 11 | 10  | 3    | 2.1  | 7.1  | 13.5 | 13.1 | 15.9 | 22   | 8.3  | 7.9  | 4.1  | 0    | 12.7 | 18.5 | 30.6 | 5.5  | 3.3  |
| 12 | 13.2| 9.7  | 12.9 | 6.5  | 9    | 9.1  | 10.2 | 9.2  | 9    | 8    | 8.7  | 12.7 | 0    | 7.6  | 18.1 | 7.3  | 12.4 |
| 13 | 14.5| 14.5 | 19.2 | 13.6 | 7.9  | 8.2  | 7.8  | 6    | 16.6 | 16.7 | 14.6 | 18.5 | 7.6  | 0    | 12.8 | 13.7 | 16.9 |
| 14 | 19.8| 27.5 | 31   | 24.5 | 20.6 | 21   | 19.8 | 8.8  | 26.5 | 26.7 | 26.5 | 30.6 | 18.1 | 12.8 | 0    | 25.3 | 29.5 |
| 15 | 10  | 2.6  | 5.7  | 2.1  | 10.2 | 10.3 | 13   | 16.5 | 5.3  | 5.1  | 1.4  | 5.5  | 7.3  | 13.7 | 25.3 | 0    | 6.1  |
| 16 | 6.5 | 3.6  | 5.4  | 8.2  | 10.7 | 10.6 | 13.2 | 21.1 | 10.6 | 10.1 | 4.7  | 3.3  | 12.4 | 16.9 | 29.5 | 6.1  | 0    |

**Fig. 1.** Transportation point distance matrix

The logistics delivery vehicle needs to start from point A and deliver the goods to customers *1–12*. For distribution companies, they also need to consider the requirements of distribution vehicle routes and costs. On the premise of meeting the constraints of loading weight and route, it is necessary to arrange the optimal transportation order and select the optimal transportation path, so as to achieve good results in distribution efficiency and cost control [9, 10]. According to the specific situation of Nantian Logistics, this paper uses ant colony algorithm to analyze this problem.

### 3.2 Assumptions

The construction of the model determines the following six assumptions:

(1) The distance between two points is the shortest. Set it as the shortest actual distance;
(2) The traffic condition is good and there is no congestion;
(3) The geographical coordinates of each distribution center are known, and there will be no shortage of vehicles [11];
(4) Knowing the needs of each customer, the goods can be mixed under the condition of meeting the on-board conditions;
(5) The truck is limited in weight, starting from the distribution center and finally returning to the starting point;
(6) The total distance required by each distribution plan will not exceed the maximum distance traveled by the distribution vehicles per tank of oil [12–15].

### 3.3 Parameter Setting

Set the number of ants participating in the algorithm model to $m$,

Based on Ma Ning's research on the optimal path of ant colony algorithm [5], the update and transfer probability of information elements are set, and the calculation formula is shown in (1) and (2).

$$\tau_{ij}(t+1) = \rho \tau_{ij}(t) + \Delta \tau_{ij} \tag{1}$$

$$P_{ij}^k(t) = \begin{cases} \dfrac{\tau_{ij}^\alpha \eta_{ij}^\beta}{\sum_{s \notin tabu_k} \tau_{is}^\alpha \eta_{sj}^\beta}, & j \notin tabu_k \\ 0, & otherwise \end{cases} \qquad (2)$$

Explanation of symbol expression: $\Delta\tau_{ij}$ represents the difference of information elements; $\tau_{ij}$ refers to the value of path $(i, j)$ information element at time t; $\rho$ Indicates the retention of information elements on the path; η Ij is the transfer probability and expectation from location $i$ to location $j$; $\alpha$ Indicates the information elements accumulated by ants; $\beta$ Expectations for transfer; Tabuk is all the nodes that the ants walk through [13, 15].

At the beginning of the model, the starting point of each ant is a randomly selected city. By maintaining the memory vector of a path, record the city points that the ant passes by in turn. When constructing each path, the ant perceives the pheromone concentration on a path and continues to walk along that path according to the pheromone concentration [14, 15]. Ant colony algorithm has been proved to achieve good results in various path selection calculations [16–19].

## 4 Model Solution and Result Analysis

### 4.1 Model Construction

#### 4.1.1 Model and Parameters

It is assumed that the delivery address and the customer's requirements are known. In the whole model, the number of ants, that is, the transportation vehicles in the freight center is $m$, and $d_{ij}$ represents the distance between customer $i$ and customer $j$; The pheromone concentration of each path is the same at the initial time point and between the points; Set pheromone factor $\alpha$ The value of is $1$; Set heuristics factor $\beta$ The value of is $\underline{3}$, set the pheromone volatilization factor $\rho$ The value of is 0.85; The number of customers is n (set is N), the vertex set in the network is D (including distribution center and customer set), the distribution center can be assigned a value of 1, and the distribution vehicle set is $K$. $N$, $D$ and $K$ are the contents of (3) respectively.

$$N = \{2, 3, \ldots, n\}$$

$$D = \{1\} \cup N$$

$$K = \{1, 2, \ldots, m\} \qquad (3)$$

The departure time of the distribution vehicles in the cargo distribution center is $T0$, the quantity of goods transported is $Q$, and the customer set served by the transport vehicle $k$ is expressed as $Sk$.

The decision of variables is shown in (4) and (5).

$$X_{ijk} = \begin{cases} 1, & k(i \to j) \\ 0, & otherwise \end{cases} \qquad (4)$$

$$Y_{ik} = \begin{cases} 1, k(i) \\ 0, otherwise \end{cases} \tag{5}$$

The mathematical representation of the model is shown in (6) to (12).

$$\min \sum_{k \in K} \sum_{i \in D} \sum_{j \in D} d_{ij} X_{ijk} \tag{6}$$

$$\sum_{i \in D} q_i Y_{ik} \le Q_k \tag{7}$$

$$\sum_{k \in K} \sum_{i \in D} X_{ijk} = 1 \tag{8}$$

$$\sum_{i=N} X_{ojk} = \sum_{i \in N} X_{iok} = 1, \forall j \in k \tag{9}$$

$$\sum_{i \in D} X_{ijk} = Y_{jk}, \forall j \in N, \forall k \in K \tag{10}$$

$$\sum_{i,j \in S_k} X_{ijk} \le |S_k| - 1, \forall k \in K \tag{11}$$

$$X = X_{ik} \in D \tag{12}$$

Among then, Eq. (6) is the path constraint (the shortest distance); Eq. (7) is the constraint on the carrying capacity of the distribution vehicle (less than the maximum carrying capacity of the vehicle); (8) Is the constraint on the number of delivery and service times (1 time); Eq. (9) is the constraint of the distribution vehicle path (cannot take the circular transportation route, and must return to the distribution center after distribution); Eq. (10) is the constraint of the delivery order (first delivery $i$ and then customer $j$); Eqs. (11) and (12) are path constraints (each distribution point and path are interconnected).

### 4.1.2 Update Pheromone

The expression of $P_{ijk}$ for the probability of selecting the next customer after completing the delivery of goods from the previous customer is shown below.

$$P_{ij}^k = \begin{cases} \dfrac{[\tau_{ij}(t)]^\alpha [\eta_{ij}]^\beta}{\sum_{s=allow_k} [\tau_{ij}(t)]^\alpha [\eta_{ij}]^\beta} j = allow_k \\ 0, otherwise \end{cases} \tag{13}$$

$$\tau_{ij}(t+1) = \rho \tau_{ij}(t) + \Delta \tau_{ij}(t, t+1) \tag{14}$$

$$\Delta \tau_{ij}(t, t+1) = \sum_{k=1}^{m} \Delta \tau_{ij}^k(t, t+1) \tag{15}$$

$$\Delta \tau_{ij}^k = \begin{cases} \dfrac{q}{l_k}, k(t, t+1) \to (i,j) \\ 0, otherwise \end{cases} \tag{16}$$

In formula (13), allow is the set of the next service object selected by vehicle k after completing the service of the previous customer at time $t$; $\tau_{ij}$ (t) represents the information elements from customer object i to customer object j at time t (the initial value is set as a constant); $\eta_{ij}$ (t) represents the visibility value from customer object $i$ to customer object $j$ at time $t$; $\alpha$ and $\beta$. It represents the degree to which pheromones and heuristic factors affect path selection. If the collection of the next customer is empty, it means that an iteration cycle has been completed.

The next iteration cycle starts from $t + 1$ and dynamically updates the information elements according to the information elements remaining in the previous cycle. In formula (14) $\rho$ (value $0-1$) indicates the persistence of information elements; Eq. (15) represents the information element increment of path $i$ to $j$ from time $t$ to time $t + 1$, $m$ is the distribution vehicle; In formula (16), $Q$ is the total amount of information elements released in an iteration cycle; $lk$ is the length of the path that the transport vehicle k passes in an iteration cycle.

### 4.1.3 Algorithm Steps

The algorithm mainly has the following four steps.

(1) Initialize various parameters to determine the distance between Nantian logistics distribution center and each customer, as well as the distance between customers.
(2) Set the distribution vehicle to depart from the cargo center ($i = 1$); According to the conditions of (11), after completing the demand service for the $j^{th}$ customer, find the next customer who needs service, and repeat this; After many searches, it provides distribution services for all customers in the system; Finally, return to Nantian Logistics Hub.
(3) Select the next customer through formula (13), find the next customer meeting the shortest path from the previous customer service path set, and redistribute the delivery vehicles according to the customer's service needs.
(4) Then update the information elements according to formula (8), and return to (2) to start again; Until the maximum number of iteration cycles is reached.

## 4.2 Model Solution Results

After using Matlab to optimize the model, the operation results not only delivered the goods to the customers in time, but also met the requirements of the model in terms of time, path selection and minimum cost.

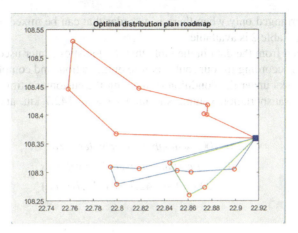

**Fig. 2.** Route map of optimal distribution scheme

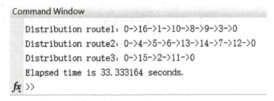

**Fig. 3.** Optimal route allocation diagram

## 5  Result Analysis

It can be seen from Fig. 2 and Fig. 3 that the transportation plan only needs three vehicles, which are divided into three routes for transportation, and all the points have been completed and the distribution task has been completed on time.

**Table 2.** Original data

| Vehicle No. | 1 | 2 | 3 | 4 |
|---|---|---|---|---|
| Driving route | 0 → 4 → 5 → 6 → 0 | 0 → 12 → 13 → 7 → 14 → 0 | 0 → 2 → 1 → 11 → 16 → 0 | 0 → 10 → 15 → 3 → 9 → 8 → 0 |
| Transport volume/t | 835.58 | 1453.4 | 2046.5 | 1748 |
| Mileage/km | 16.5 | 62.9 | 28.6 | 34.7 |

Data source: compiled according to relevant data of daily work of Nantian Logistics Group

The vehicle allocation of Nantian Logistics is mainly manual. When receiving the transportation request from the customer, it will first print out the list to be delivered, and then put it neatly on the table, while the list with closer location will be arranged for delivery together. If there is an order that is far away, the delivery time will be delayed, and it will be arranged only when there is an order that can be mixed, so the original data assumed in Table 2 is available.

It can be seen from the data in the table that the logistics center used four vehicles to deliver goods according to four routes before optimization, and completed all goods transportation tasks under the condition of meeting the customer's time requirements. The completed transportation volume and mileage were 142.7 km, and the cost was shown in (17).

$$\begin{aligned} Cost &= 1000 * number\ of\ vehicles\ used \\ &+ total\ distance\ traveled\ by\ vehicles \\ &= 1000 * 4 + 142.7 = 4142.7 (yuan) \end{aligned} \qquad (17)$$

**Table 3.** Optimized data

| Vehicle No. | 1 | 2 | 3 |
|---|---|---|---|
| Driving route | 0 → 16 → 1 → 10 → 8 → 9 → 3 → 0 | 0 → 4 → 5 → 6 → 13 → 14 → 7 → 12 → 0 | 0 → 15 → 2 → 11 → 0 |
| Transport volume/t | 2455.5 | 2288.98 | 1339 |
| Mileage/km | 42.2 | 60.6 | 27.8 |

It can be seen from the analysis results in Table 3 that the number of vehicles in some optimized schemes has been reduced from 4 to 3, and the corresponding routes have also been changed to 3. All customers' freight tasks have been completed within the time required by customers. Comparing the two algorithms, we can see that although the traditional transportation routes also deliver goods, this is not the optimal solution in practice, and the improved arrangement is also more labor-saving.

$$\begin{aligned} Cost &= 1000 * number\ of\ vehicles\ used \\ &+ total\ distance\ traveled\ by\ vehicles \\ &= 1000 * 3 + 130.6 = 3130.6 (yuan) \end{aligned} \qquad (18)$$

From the comparison between Table 2 and Table 3, it can be seen that the improved algorithm saves *12.1* km of total vehicle mileage; At the same time, *1* vehicle was built for use; Compared with the original data, it saved 1012.1 yuan, as shown in (17) and (18).

To sum up, an optimization model is established for the distribution of Nantian Logistics in the same city; Based on ant colony algorithm and combined with the actual situation of Nantian Logistics, it is more suitable for practical application. Through this case study, it can be seen that ant colony algorithm has effectively saved transportation resources, reduced logistics and transportation costs, and made great progress towards low cost and high efficiency in intra-city distribution.

## 6 Conclusion

In the context of the continuous development of e-commerce, customers' demand for online and offline single offline distribution is increasing, and urban distribution has become a key link to improve consumer satisfaction. At the same time, due to the constraints of urban congestion, large orders, small distribution orders and high frequency, the selection of appropriate routes has become an important issue of urban distribution.

Aiming at the increasingly important urban distribution, the paper selects the typical distribution enterprise Nantian Logistics Group as the research object, and proposes a path planning scheme based on the data collection and analysis of urban distribution routes, and uses ant colony algorithm to analyze the path optimization. The practice of enterprises shows that the optimized scheme has effectively reduced the transportation distance, saved the logistics cost and significantly improved the logistics efficiency.

**Acknowledgments.** This paper is supported by: Young and middle-aged teachers basic ability improvement project of Guangxi University "Research on Key Technologies of Urban and Rural Logistics Integration Operation Based on Big Data and Cloud Computing" (2023KY1862).

## References

1. Tang, P., Tian, H.: Research on the layout of national economic mobilization logistics centers. Int. J. Mod. Educ. Comput. Sci. **11**, 44–50 (2010)
2. Liu, Z.: Application and research of two-population ant colony algorithm based on heuristic reinforcement learning. Shanghai University of Engineering and Technology, Shanghai (2020). (in Chinese)
3. Dai, T., Liu, X., Hu, Y.: Application of improved ant colony algorithm in path planning. J. Jiamusi Univ. (Nat. Sci. Ed.) **40**(01), 123–125 (2022). (in Chinese)
4. Land, M., Gao, H., Cui, Y.: Application of improved ant colony algorithm in express delivery routing. Comput. Technol. Dev. **31**(11), 15–20 (2021). (in Chinese)
5. Liu, Z., Zhao, L., Xue, J., et al.: Research on improved ant colony algorithm for vehicle routing problem. J. Hebei Univ. Sci. Technol. **43**(01), 80–89 (2022). (in Chinese)
6. Ge, D.: Optimal path selection of multimodal transport based on Ant Colony Algorithm. J. Phys. Conf. Ser. **2083**(3) (2021)
7. Li, S., Zhao, G., Yue, W.: Research on path planning for mobile robot based on improved ant colony algorithm. J. Phys. Conf. Ser. **2278**, P012005 (2021)
8. Li, S., Luo, T., Wang, L., Xing, L., Ren, T.: Tourism route optimization based on improved knowledge ant colony algorithm. Complex Intell. Syst. (2022). (prepublish)
9. Wang, Y.: Research on optimization method of air logistics distribution path based on ant colony algorithm. Inf. Technol. **11**, 76–80 (2021). (in Chinese)
10. Zheng, X., Qi, Q.: Optimization and application of low carbon cold chain logistics distribution model based on improved ant colony algorithm. Preserv. Process. **22**(03), 83–90 (2022). (in Chinese)
11. Lei, J., Sun, Y., Zhu, H.: Application of improved ant colony algorithm in vehicle path planning with time window. Comput. Integr. Manuf. Syst. 1–15, 15 April 2022. http://kns.cnki.net/kcms/detail/11.5946.TP.20211228.1340.008.html
12. Wang, S., Fan, S., Wang, J., Shen, R.: Research on logistics distribution route optimization based on ant colony algorithm – taking Yuzhong District of Chongqing as an example. Econ. Trade Pract. **13**, 31–32 (2017). (in Chinese)

13. Chen, J., Liu, G.: Research on cold chain logistics path optimization based on ant colony algorithm. Transp. Technol. Econ. **23**(05), 38–44 (2021). (in Chinese)
14. Zhu, Y., You, X., Liu, S.: Improved ant colony algorithm based on heuristic mechanism. Inf. Control **48**(03), 265–271 (2019). (in Chinese)
15. Ma, N.: Logistics optimal path based on clustering ant colony algorithm. Sci. Technol. Eng. **20**(31), 12911–12915 (2020). (in Chinese)
16. Zhang, W.: Application of an improved ant colony algorithm in coastal tourism route optimization. J. Coastal Res. **98**(Special), 84–87 (2019)
17. Du, P., Hu, H.: Optimization of tourism route planning algorithm for forest wetland based on GIS. J. Discret. Math. Sci. Cryptogr. **21**(2), 283–288 (2018)
18. Qian, X., Zhong, X.: Optimal individualized multimedia tourism route planning based on ant colony algorithms and large data hidden mining. Multimedia Tools Appl. **78**(15), 22099–22108 (2019)
19. Zhou, X.: Research on the optimization of tourism traffic routes based on ant colony algorithm. Revista de la Facultad de Ingenieria **32**(3), 819–827 (2017)

# Optimization Research of Port Yard Overturning Operation Based on Simulation Technology

Qian Lin[1(✉)], Yang Yan[1], Ximei Luo[1], Lingxue Yang[1], Qingfeng Chen[1], Wenhui Li[1], and Jiawei Sun[2]

[1] School of Logistics, Wuhan Technology and Business University, Wuhan 430065, China
linqian@wtbu.edu.cn
[2] Technical University of Munich Asia Campus, Technical University of Munich (TUM), Singapore, Singapore

**Abstract.** The operation of the world transport supply chain is affected by many factors, including the COVID-19 epidemic. China's production is in a relatively stable state, which to some extent alleviates the contradiction between supply and demand in the international market. With the increase of container transportation in China, the utilization rate of container transportation is getting higher and higher. The overturning rate will have a significant impact on the utilization rate and storage efficiency of containers. Based on the theory and practical experience of container port operation, this paper studies Huangshi New Port based on rail and water transport simulation technology. By optimizing the container turnover operation of the port yard and improving the container turnover technology, the operation cost of the port can be reduced, the operation efficiency of the port can be improved to a certain extent, and the competitiveness of the port can be enhanced.

**Keywords:** Simulation platform of Rail-water Intermodal · Simulation technology · Box-turning operation

## 1 Introduction

According to the 2021 Global Port Development Report released by the Shanghai International Shipping Scientific Research Center, the container throughput of the world's top 100 seaports shows a good trend of growth on the whole, among which 84 ports have a positive growth year-on-year in 2020, and 41 ports have a growth rate of 10% higher than that in 2020 (see Table 1). At present, container transport plays an important role in the world freight business [1]. Port is an important hub of cargo transportation, and its operation efficiency is directly related to the competitiveness of cargo transportation in the port, even related to the overall development of the region. Container yard is a buffer zone for port containers [2–4]. The efficiency of a container yard not only determines the port throughput, but also determines the speed of cargo operation [5].

**Table 1.** TOP20 domestic ports in terms of cargo throughput in the world in 2021 (unit: 10, 000 TEU)

| Range | Country | Name of port | In 2021, | In 2020, | Year-on-year growth |
|---|---|---|---|---|---|
| 1 (1) | China | Zhoushan, Ningbo | 122405 | 117240 | 4.4 |
| 2 (2) | China | Shanghai | 76970 | 71104 | 8.2 |
| 3 (3) | China | Tangshan | 72240 | 70260 | 2.8 |
| 4 (4) | China | Qingdao | 63029 | 60459 | 4.3 |
| 5 (5) | China | Guangzhou | 62367 | 61239 | 1.8 |
| 7 (7) | China | Suzhou | 56590 | 55408 | 2.1 |
| 9 (10) | China | Rizhao | 54117 | 49615 | 9.1 |
| 10 (9) | China | Tianjin | 52954 | 50290 | 5.3 |
| 13 (13) | China | Yantai | 42337 | 39935 | 6.0 |
| 14 (18) | China | Taizhou | 35291 | 30111 | 17.2 |
| 15 (24) | China | Jiangyin | 33757 | 24705 | 36.6 |
| 16 (15) | China | Dalian | 31553 | 33401 | −5.5 |
| 17 (17) | China | Huanghua | 31134 | 30125 | 3.3 |
| 18 (16) | China | Nantong | 30851 | 31015 | −0.5 |
| 20 (20) | China | Shenzhen | 27838 | 26506 | 5.0 |

Source: Shanghai International Shipping Center Prospective Industry Research Institute (Figures outside brackets are 2021 ranges, figures inside brackets are 2020 ranks)

## 2 Theoretical Review

### 2.1 Box-Turning Operation and Its Impact

Port container overturning is an important part of container ship turnover, including container loading, lifting and moving. Port container turnover affects the loading and unloading efficiency, especially in loading and unloading operations [6–8]. The overturning problem refers to that an overturning sequence can be found under the given shellfish position structure, initial container position state and container loading sequence, and the minimum number of overturning is required [9]. Reducing the number of overturning is helpful to improve the efficiency of overturning and thus reduce port handling costs [10].

### 2.2 Simulation Platform for Container Rail-Water Intermodal

Based on the needs of the Yangtze River Economic Belt and regional economic development, the National Multimodal Transportation Project of Huangshi New Port uses 3D virtual technology to establish the real scene and complex decision-making environment, and makes the planning scheme of container Rail-water Intermodal [11–13]. On this basis, this paper focuses on how to arrange cargo space, operation equipment scheduling and transportation path optimization under different transportation modes [14–17]. The 3D virtual simulation experiment platform was designed with B/S architecture and adopted 3D simulation modeling technology. According to the real situation of the storage yard, Maya, 3DMax and other simulation software were used to model the whole storage yard and simulate the incoming and outgoing storage environment [18, 19].

## 3 Case Analysis

### 3.1 Introduction to Huangshi Newport

Huangshi New Port in Yangxin County, Huangshi City, Hubei Province, supported by the Ministry of Transport, is the core port area of "East Hubei Combined Port" planned and constructed by Hubei Province. It is a national first-class water transport port among 28 major inland river ports in China, and there are only 2 ports of this class in Hubei Province.

**Table 2.** Port cargo and container throughput of Hubei Province in 2021

Unit of measurement: ten thousand tons, ten thousand TEU

| Port | Throughput of cargo | | | Foreign trade cargo throughput | | | Throughput of container | | |
|---|---|---|---|---|---|---|---|---|---|
| | Since the beginning of the year The cumulative | This month, | Year-on-year growth rate (%) | Since the beginning of the year The cumulative | This month, | Year-on-year growth rate (%) | Since the beginning of the year The cumulative | This month, | Year-on-year growth rate (%) |
| **Hubei Province Total** | **48, 831** | **5, 111** | **28.6** | **1, 787** | **146** | **−2.4** | **284** | **23** | **24.2** |
| Jiayu | 1, 225 | 118 | 18.8 | | | | | | |
| Wuhan | 11, 679 | 1, 503 | 10.8 | 964 | 73 | −7.7 | 248 | 20 | 26.1 |
| Huangzhou | 1, 007 | 238 | 15.6 | | | | | | |
| Ezhou | 2, 176 | 198 | 49.7 | | | | | | |
| Huangshi | 4, 992 | 495 | 5.9 | 731 | 67 | 6.2 | 4 | … | −28.0 |
| Xiangyang | 4 | … | −48.3 | | | | | | |
| Jingzhou | 4, 375 | 410 | 23.0 | 42 | 2 | 9.5 | 15 | 1 | 21.5 |
| Yichang | 11, 470 | 981 | 41.3 | 50 | 4 | −16.6 | 15 | 1 | 20.6 |
| Qianjiang | 104 | 18 | −0.7 | | | | | | |
| Tianmen | | | | | | | | | |
| Hanchuan | 9 | 1 | 31.8 | | | | | | |
| Other Ports in Hubei Province | 11, 790 | 1, 150 | 55.7 | | | | 2 | … | 44.3 |

It is located in the source Wei mouth town board, belongs to the key area of the Yangtze river economic belt, due to the superior g geographical position, is rare in the middle reach of Yangtze river of deep water port, even during the dry season, the port in front of the lowest water level also can continue to achieve 6 m.

In 2021, the cargo throughput of Huangshi New Port increased by 1.8% compared to this time of year, among which the throughput of scattered and miscellaneous items reached 1932.70, 000t, an increase of 4% compared to this time of year, and the container throughput reached 44, 060TEU, a decrease of 28% compared with the same period in previous years. The transshipment efficiency of container throughput at Huangshi New Port is lower than that at other ports along the Yangtze River (See Table 2).

Before the construction of the port, a broad reserve site has been planned. The overall planning construction land is nearly 20 km$^2$. The surrounding area has a good industrial base. The planned construction includes the container dock area, the large bulk cargo dock area and the bulk cargo dock area. It plans to build a total of 69 production berths, 46 berths will be designed in the near term and 23 berths in the long term. More than 500 million yuan will be invested in the first phase of the project. The target is to build nine 5, 000-ton class berths, which can also rely on 10, 000-ton ships, and the annual throughput can reach 10 million tons. After the port is fully completed, the annual throughput of the storage yard can reach "a large port of 100 million tons and one million Tank box". In the future, it will be built into the largest foreign trade transfer hub port in the middle reaches of the Yangtze River and an excellent site selection base for the multi-modal transport of public iron-water goods and distribution. The final planning target of this port is to be the modern comprehensive port area (integrated transport hub of water and land) closest to the sea in Hubei Province, which will become a powerful engine for the transformation and leap-forward development of Huangshi area and play an important connection role along the riverbank It can also radiate the coastal areas to the east. It is a very rare comprehensive logistics base.

**Table 3.** Main device configurations

| Type of equipment | Device Name | The number of | Main technical parameters |
|---|---|---|---|
| Bank bridge | Container handling bridge | 12 | The rated lifting weight under the hook is 45t and the rated lifting weight under the spreader is 35t Lifting height: 20 m on the rail surface |
| Dragon door crane | Container gantry crane | 10 | Fixed lifting weight: 35t under the spreader Lifting speed: full load 20 m/min, no load 30 m/min |

(*continued*)

**Table 3.** (*continued*)

| Type of equipment | Device Name | The number of | Main technical parameters |
|---|---|---|---|
| Front side crane | Front container crane | 2 | Wheelbase: 6000–6500 mm Maximum torque/RPM: $\geq$1770 N * m/2100 rpm |
| High stacking machine | Empty container stacking machine | 12 | Maximum lifting height: > 15000 mm Minimum height clearance: > 350 mm |
| Inner set card | Container truck | 20 | Maximum load: 9000 kg Loading container size: 40 ft. |

Through investigation, it is found that Huangshi New Port has built 6 professional container berths at the present stage, with the whole coastline of the berths being 606 m long. Each berth Each berth has two 2 shore bridges, 12 container unloading bridges, 10 gantry cranes, 2 front cranes, 12 empty container strollers and 20 container trucks. The port yard has been divided into 6 yard areas with a total area of 79, 200 m$^2$, among which there are 4 heavy container yard areas. Each yard area is equipped with 404TEU of 10 800 m$^2$, 2 yard bridges and 2 empty container yard areas. Each yard area is equipped with 138TEU of 9 000 m$^2$ and 2 flow machines (see Table 3).

### 3.2 There Are Problems in the Box-Turning Operation of Huangshi Newport

#### 3.2.1 Current Status of Storage Space of Huangshi Nerport Yard

The current situation of Huangshigang storage yard can be analyzed from two aspects.

(1) Storage yard space resource allocation method: According to the values of "number, area, space, row and layer" of the storage yard, the theoretical number of the storage yard is calculated through corresponding values. Storage yard space resources are generally allocated in two steps, that is, the amount of working containers in each container area of the storage yard is allocated reasonably and equally, and then the containers on site are assigned to the container area.

(2) Container stacking arrangement: 1) Empty containers and heavy containers are stacked in a reasonable area; 2) Empty containers are stacked according to intact containers, damaged containers, dirty containers, self-owned containers and rented containers; 3) Empty and intact boxes are stacked according to the size, type and bracket of the boxes; 4) Heavy boxes are divided into the arrival boxes, delivery boxes, transit boxes and auxiliary boxes; 5) The arrival boxes should be stacked separately by different owners; 6) TThe sending boxes should be stacked separately considering the name of the ship, the number of the ship and the bill of lading number; 7) According to the destination, according to the ship, ticket piled up, according to the carrier transit plan transit; 8)The same planned task container stack.

### 3.2.2 Reasons for Overturning Operation in Huangshi Newport

The overturning operation will increase the operating cost of the dock. First of all, import containers should follow the "first in, first out" mode. That is, the first unloaded import containers should be transferred out first. The later unloaded containers cannot be stacked on top of the first unloaded import containers. However, in fact, when unloaded import containers have not been shipped out, in order to improve the utilization rate of the port terminal, the newly unloaded containers will be placed on top of the original containers, resulting in the export of imported containers turning over. And in the process of stacking import containers, because cannot know in advance the owner of the container pick up the order, so cannot be stacked in a certain order. In addition, there are some objective factors, such as special containers (dangerous goods, etc.) with special places, if the special places are occupied, there will be a turning operation; Secondly, improper port organization and scheduling will also lead to container overturning operation. For example, following the principle of "card collection priority", in the process of card collection and container pick-up, the order of cargo owner is inconsistent with the order of container stacking in the storage yard, resulting in the operation of turning over the container. In this way, the running distance and running time of the field bridge vehicles in different stacks at the same bay position can be shortened. However, it will lead to unnecessary repeated operations, which will not only increase the cost of turning over the box, but also prolong the working time, resulting in a large amount of waste of waiting time for the collection card and increase the time cost. At the same time, it will also lead to the decrease of the utilization rate of the yard bridge, but also produce problems such as yard congestion.

## 4 Simulation and Optimization of Simulation Platform for Container Rail-Water Intermodal

### 4.1 Theoretical Analysis

The root cause of container overturning problem is that the order of container pickup is inconsistent with the stacking order of port yard. The main reasons are: first, the import container storage management is not good; Second, the lack of collection card to the station related data; Thirdly, the container flipping strategy is unreasonable. Therefore, it is necessary to find appropriate ways to reduce container overturning rate from the following aspects.

(1) Centralized supervision of containers at the port. Through centralized management of stacking containers, the port terminal adopts a more scientific stacking management mode to maximize the utilization rate of the port terminal and reduce the rate of overturning. Port container management has positive influence on container quality control and overall planning.

(2) Centralized stacking of the same container owner import box set. This method requires that all ports, dock and shipping companies cooperate closely during transportation, find out the owners with a large number of imported containers in the information center, and arrange unified stacking and reasonable configuration for them, so that no overturning operation occurs when the same owner takes up the containers, that is, zero overturning of containers is realized.

(3) Under the same delivery list, import containers can be exchanged accordingly. Multiple import containers in the same bill of lading, pick up in no particular order. As a result, containers can be sorted at the time of delivery, thus reducing the effect of turning over.
(4) Import containers by appointment. This method requires the consignee or the driver to make an appointment with the port terminal before picking up the goods at the terminal. Within a certain period of time, the port can limit the dock collection card to avoid the congestion caused by the collection card entering the wharf. At the same time, it can also obtain some collection card arrival situation in advance, so as to make reasonable arrangements for the gantry crane scheduling and unloading. In order to ensure that the operators arrive at the scheduled time, the port has also developed a set of reward and punishment mechanism to ensure that the operators can take out the containers at the scheduled time.

In addition, in order to effectively improve the operation of the port terminal, we need to adopt modern information technology means to improve the port operation mode. Container turning operation is the most common problem encountered when containers are picked up, which is affected by three factors, namely, the storage status of containers in the stacking area, the order of container picking by the dispatcher and the storage location when unloading. The number of container overturning can be reduced by making a reasonable overturning strategy, optimizing the unloading location and adjusting properly according to the cargo position order.

### 4.2 Simulation Experiment of Container Rail-Water Intermodal Platform

Container yard operation (see Fig. 1) has obvious randomness and dynamics, it belongs to a multi-level queuing complex system. Therefore, in this paper, the container port yard operating system model is constructed by simulation of the hot metal combined transport platform, reasonable analysis is made of the established simulation model, and feasible optimization suggestions are put forward.

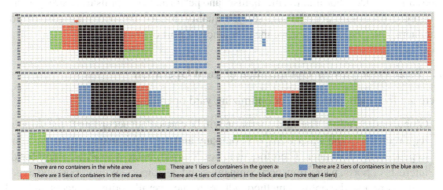

**Fig. 1.** Plane diagram of container yard

The box-turning operation process was optimized through several operation experiments, and the optimal box-turning effect was finally achieved according to the experimental results. On the premise of reducing the loss, the following optimization conclusions were drawn:

(1) Box turning strategy. At present, the extensive use of the collection card reservation port mode and GPS technology enables the port to obtain the data of the collection card arrival, thus it saves a certain amount of time for the container turning rate and the best way of unloading. Proper placement of the inlet box can greatly reduce the second and more than two turnover rate. After determining the initial status of the stacking area, if the order is not changed, it is inevitable to turn over the box operation, but the number of turning over the box can be reduced as far as possible.

(2) Adjust the container picking sequence. By adjusting the order of containers to match the storage location of containers as far as possible, it can reduce the turnover cost of containers and save the time of picking up containers. At the same time, it can also offset the waiting time of early arrival collection cards, but it cannot completely avoid the waiting time of collection cards at the port. Therefore, it is necessary to make a reasonable balance between the two to solve the problem of turning over boxes.

## 5 Countermeasures and Suggestions

### 5.1 Basic Idea of Yard Operation Optimization

The scientific outlook on development should be thoroughly implemented, the concept of resource conservation should be combined with the basic national policy of responding to environmental protection, the policy of "two-oriented" port construction of the Ministry of Transport should be implemented into daily work, the interests of shipping companies, shippers and cargo terminals should be taken into account, and the breakthrough should be to reduce the frequency and rate of overturning containers. By integrating modern information technology, supported by scientific management methods, the container stacking strategy and turning strategy are constantly optimized, and measures such as differential service, coordination mechanism and performance appraisal are strengthened to improve economic benefits and service quality, reduce the operating cost of port containers, and gradually improve the efficiency of port operations.

### 5.2 Basic Principles for Optimization of Yard Operation

There are four main principles for optimizing yard operation.

(1) Global optimization. The yard operation will be carried out according to the shipping requirements of the water transport company and the requirements of the customer to pick up the container. Therefore, it is necessary to make overall planning for the interests of ports, water transport companies and customers. With import and export as the core, use the optimal method, adhere to the overall optimal concept and seek the lowest overall turnover rate under the constraint of the requirements of ports, water transport companies and customers.

(2) Standardized management. From the entry to the exit of the container, the process sequence chain is complex, involving many positions, and requiring high degree of cooperation between different positions. Therefore, the post requirements, operation contents, operation procedures and assessment indicators of field crane drivers, yard planners and central control dispatchers should be clearly defined, so as to efficiently guide each post to reduce the number and rate of overturning under the premise of ensuring efficiency.

(3) Timely information. Obtaining accurate, comprehensive and detailed container information is the guarantee of fundamentally reducing the overturning rate. Therefore, coordination between relevant departments should be strengthened to ensure that necessary information can be obtained, updated and shared in a timely manner, avoid repeated information processing, and provide a reliable basis for scientific and reasonable management of port and wharf stacking.

(4) Minimize the movement. When the overturning operation cannot be avoided, the reasonable overturning operation can be reduced and the overturning speed can be improved. The minimization of action can be achieved through the inner turning box, the nearest turning box, the lowest falling box and other methods.

To sum up, this paper draws the port operation process, as shown in Fig. 2.

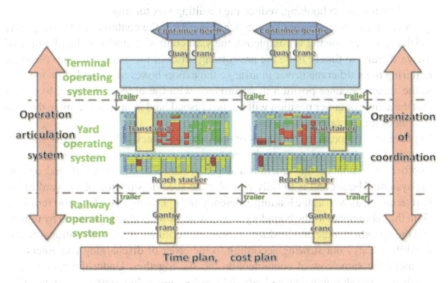

**Fig. 2.** Port operation flow

## 5.3 Reducing the Turning Rate

There are six main ways to reduce the overturning speed.

(1) Improve the system management rules. The shipping agent should consider the booked capacity and the actual carrying capacity of each ship in a balanced way

to ensure the balanced allocation of the space of the ships attached to the wharf, avoid the phenomenon of cabin bursting caused by blind acceptance of booking, and reduce the resulting overturning of the container; Secondly, the shipping agent should check the authenticity of shipper, cargo and other information when accepting booking, prevent false booking, strengthen the performance control after booking, and provide the wharf company with detailed freight information.

(2) Make stacking plans scientifically. The shipping company should strengthen the information communication with the affiliated port agent and the affiliated dock, improve the reliability of liner operation, try to follow the published shipping schedule for scheduled operation, minimize the disruption of the storage yard plan of the affiliated dock due to liner delay, and avoid unnecessary overturning. At the same time, the shipping company should also provide detailed information such as ship, container, arrival time, export date and subsequent docking port to the docking dock in a timely manner, and ensure the reliability and accuracy of the information, so that the dock can make reasonable yard plans with reliable and accurate information.

(3) Strengthen the connection with the front water transport chain. In shipping business, the booked capacity and the actual carrying capacity of each ship should be considered in a balanced manner to ensure the balanced allocation of the space of the ships attached to each port and dock, so as to avoid the phenomenon of "cabin bursting" caused by too much booking, reduce the resulting box turning.

(4) Strengthen the arrangement of container areas. After the container gathering port is finished and the stowing is completed, the yard controller shall use the idle time of the yard crane to timely issue the pre-turning order of the same container position, requiring the yard crane driver to arrange the export boxes to be loaded on the ship in the same container position in accordance with the order of loading, to ensure that the same container position will not be crushed, and in principle realize the zero turning of the container on the ship.

(5) Improve the comprehensive quality of relevant personnel. Firstly, improve the efficiency of yard control personnel communication and exchange, and constantly improve and enhance the function and performance of the port terminal operating system; The second is to fully consider the specific situation of the yard of the port terminal, as well as the loading efficiency and operating cost of the port terminal. With the cabin board as the boundary, reasonable stowing plan is prepared by region, and the operation route and loading sequence are optimized to ensure rationality and flexibility. Carry out stowing according to the order of discharging containers, try to match the same row of containers on the site together; Under the condition of confirming the shipping date and schedule, as far as possible, start stowing when the clearance condition is good; Only by understanding the true intentions of the stow crew can ship controllers better implement stow plans. At the beginning of loading, container collection, delivery and loading are carried out according to serial number. In fact, due to the different management modes of each port and wharf, accidents often occur in bridge crane, yard crane and collection card, etc., so ship control personnel should timely adjust their working ideas. Therefore, ship control personnel need to strengthen the communication with stowage personnel, flexible use of container interchange rules, to reduce unnecessary overturning operations.

(6) Reasonable implementation of box-turning operations. It is difficult for any port terminal to do zero turn over. When it is really necessary to turn over containers, yard controllers should rely on the function module of turning over optimization of port terminal operating system based on the information of export container ship name, loading date, unloading port, weight and so on, and aim at the minimum amount of turning over containers Through scientific calculation to determine the best order and stacking position of turning over boxes, to avoid the unreasonable operation caused by the double or serial turning over boxes.

### 5.4 Specific Methods to Reduce Overturning Speed

The specific measures to reduce the overturning speed can be embodied in six aspects.

(1) The Tabu algorithm is used to optimize the integrated scheme of storage position-set card scheduling. There are many operation targets and a large number of uncertain factors in container terminal yard. It is suggested to make a model plan before unloading the ship and work in order, so that the yard can classify the imported boxes within a limited time. The Tabu search algorithm can be used to pack the boxes. Under the premise that the stacking area is allowed, another Tabu search algorithm can be implemented to obtain the optimal scheduling scheme. Based on the calculation results of the model, container mixing can be reduced and overturning caused by rearrangement after unloading can be effectively reduced or avoided.

(2) Design multiple preferential and penalty strategies to reduce container stop time. The dispatch should take the departure time of containers as the core to allocate the container space. First, pre-register the container departure date of the shipping company before the ship arrives at the port, so as to facilitate the dispatch to make the container space plan. Second, strengthen communication with shipping companies and carriers to obtain more reliable departure information. After unloading the imported containers, the dock shall promptly notify the consignee, freight station or carrier to pick up the containers as soon as possible to avoid long-term container stacking; The third is to formulate preferential policies to encourage off-peak pick-up. It is suggested to offer convenience and preferential treatment to the pick-up customers during off-peak hours, and charge certain port clearing fees to the pick-up customers during peak hours, so as to ease the traffic in the evacuation port area and avoid excessive mechanical load in the storage yard. Fourth, for the customers of free stacking period timeout, develop a step charging strategy, the longer the period, the higher the fee. Design multiple strategies to guide customers to pick up containers as early as possible, flexible compression container yard parking time, better improve the yard turnover efficiency. Practice has proved that after the Port of Los Angeles raised the stacking fees for overdue containers, the stacking time of containers in the yard was reduced by 20% on average, and the stacking capacity of the yard was significantly improved.

(3) Use differential evolution algorithm to plan the differential stacking of containers. Depot planning directly affects assembly and shipment. Differential evolution algorithm can be used to guide the classification of customers. One is to open up exclusive container area for big customers of the wharf, and sign cooperation agreements with shipping companies or carriers; The second is to set up container groups for small

and medium-sized customers, and stack them in a centralized manner according to the number of the container groups. To guide customers to pick up the boxes without the designated container number and in accordance with the recommended operation order, on the one hand, it can alleviate the problems of time and cost, on the other hand, it can shorten the waiting time of picking up the boxes, so that customers have a better experience of picking up the boxes; Third, to prevent the overturning caused by mixing, it is recommended to separate the heavy containers stacked for a long time from the heavy containers stacked for a short time. This kind of containers should also be distinguished from other heavy containers and boxes to be checked after the inspection. For the cases that must be turned over, it is suggested to use differential evolution algorithm to determine the reasonable blocking position of the case from a macro perspective.

(4) Promote the appointment of suitcases. Information technology makes it possible to make an appointment to pick up a suitcase. At the beginning, a series of preferential policies can be adopted to encourage large customers to use the appointment to pick up a suitcase service. When the customer gradually adapts, and then carry out heavy container appointment, which requires yard machinery to constantly improve the operating capacity. Therefore, the port should change its thinking, provide customers with suggestions on the pick-up time and order, constantly improve the booking pick-up mode, and finally form the pick-up mode according to the suggested time period, to minimize or avoid "be on hand" and other uncertain factors.

(5) The container management right shall be transferred from the shipping company to the unified operation of the terminal. At present, most container management rights belong to shipping companies. Dalian Port container Terminal obtains container management rights through reform and innovation, and the turnover rate can be basic Keep it at around 5%. The container management right of the terminal is conducive to the balance between the utilization rate of the discharging yard and the overturning rate. It can also improve the quality of the container and the flexibility of the unified deployment of the inbound container emphasis, so as to rationally allocate the yard resources and reduce the overturning volume.

(6) Building a modern smart port by using the new generation of information technology. At present, there are some uncertain factors in the actual operation of Huangshi New Port, such as the arrival time of ships cannot be accurately predicted. With the continuous development of modern smart port equipment, more intelligent systems should be considered in the later stage to make up for this problem. The modern smart port uses the new generation of information technology to gradually integrate the port related business and management innovation deeply, make the port more intensive, efficient, convenient, safe and green, innovate the port development mode, and realize the scientific and sustainable development of the port.

## 6 Conclusion

In this paper, the overturning operation of container yard is simulated and analyzed by the simulation technology of container Rail-water Intermodal, and a number of optimization methods are proposed:

(1) Compared with the traditional method, this method analyzes the problems and causes of yard returning with the help of the simulation optimization model. The integrated model of storage position-set card scheduling is helpful to solve the optimization task allocation problem of ship unloading.
(2) Design multiple preferential and penalty strategies to solve the problem of the time of compressed containers at the port. This method has good reusability and interoperability, which is conducive to promoting customer interaction and reducing the time of compressed containers, and can well meet the requirements of port managers.
(3) Experiments show that the differential evolution algorithm can improve the solving accuracy of the simulation optimization model by planning the differential stacking of containers. At the same time, the classification of customers can be guided to further optimize the task allocation of each node, better improve the system management rules, facilitate the scientific development of stacking plans, strengthen the connection with the front-end water transport chain.
(4) The container management right is transferred from the shipping company to the unified operation of the terminal. By strengthening the arrangement of container areas, improving the comprehensive quality of relevant personnel and reasonably implementing the overturning operation, the overturning rate of the container terminal yard can be effectively reduced, the utilization rate of port resources can be improved, the loading and unloading costs can be reduced, and the competitiveness of the port can be enhanced.

The above suggestions put forward corresponding improvement suggestions for the current problem of the port turnover efficiency of Huangshi Newport. The next research should be based on the concept of smart port proposed by the Institute of Water Transport Science of the Ministry of Transport in 2022, and gradually use more intelligent systems and Internet of things technology to support the port information modernization. At present, the commercialization of 5G, the Internet of Things, big data and the development of artificial intelligence provide more and more underlying technical support for the automation and intelligence of ports.

By September 2022, Jiangsu Province of China currently has the largest number of smart port patent applications in China, with a cumulative number of smart port patent applications reaching 930. Shanghai, Shandong and Guangdong have all applied for more than 500 smart port patents. Figure 3 shows the top ten provinces in the number of smart port patent applications in China, where Hubei is in the NO.8.

On the basis of the underlying technology, AI identification technology identifies port gate and container code, etc., and relies on unmanned transportation technology and network layer to realize low-delay port communication and high-bandwidth video transmission. These technologies enable the linkage development of various sectors of smart ports. It is suggested that Huangshi Newport Dry Port should set up intelligent production management system, equipment control system and intelligent monitoring/remote

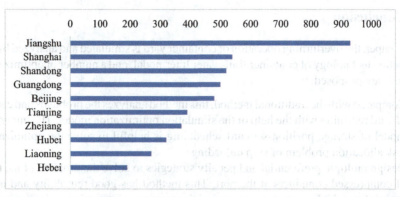

**Fig. 3.** TOP10 applications for smart port patents in provinces (municipalities and autonomous regions) in China by September 2022 (Unit: item) Data source: Intelligent Bud Foresight Industry Research Institute

control system of large equipment according to the actual development status and product and cargo characteristics, combine the frontier intelligent loading and unloading equipment, storage yard intelligent loading and unloading equipment and horizontal intelligent loading and unloading equipment of each port inside the port, and coordinate the subdivided shore bridge system, storage yard system and fleet management system. It will better realize the intelligent operation and operation of the whole process from the overall management, monitoring and overall control of the port, so as to improve the transportation safety of the terminal and improve the cargo efficiency of the terminal.

**Acknowledgment.** This research is supported by Wuhan University of Business and Technology's scientific research project: Modern Logistics Park Planning and Design Research (No. A2017005).

# References

1. Kizilay, D., Eliiyi, D.T.: A comprehensive review of quay crane scheduling, yard operations and integrations thereof in container terminals. Flex. Serv. Manuf. J. **33**(1), 1–42 (2021)
2. Yan, Y.: Development status and trend of container terminal operating system. Containerization **25**(1), 19–23 (2014). (in Chinese)
3. Cai, P.: Review of the development of container transport in China's ports in 2019 and outlook for 2020[J]. Containerization **31**(2), 16–19 (2020). (in Chinese)
4. Huangshi New Port officially opened Phase I with an annual throughput of 10 million tons [EB/OL]. http://news.cnhubei.com/xw/hb/hs/201509/t3402605.shtml. Accessed 30 Sept 2015. (in Chinese)
5. Stojaković, M., Twrdy, E.: Determining the optimal number of yard trucks in smaller container terminals. Eur. Transp. Res. Rev. **13**(1), 1–12 (2021)
6. Ma, S.: Research on the problem of container turnover in container terminals. East China Normal University, Shanghai (2021). (in Chinese)
7. Zhang, Y., Han, H.: Reducing the turnover rate of container terminals. Containerization **9**, 107–111 (2013). (in Chinese)

8. Yu, X., Ding, L.: Research on optimization of container turnover operation at the wharf. J. Logist. Sci. Technol. **12**, 19 (2013). (in Chinese)
9. Zheng, S.: Research on the model and algorithm of container picking operation in container yard based on the optimization of container turnover path. South China University of Technology, Guangzhou (2018). (in Chinese)
10. Li, H.: Collaborative optimization model of import container picking order and turnover strategy based on partial container information. Dalian Maritime University, Dalian (2017). (in Chinese)
11. Otti, E.E., Okorie, E.C., Bulus, S.M.: Analysis and numerical simulation of deterministic mathematical model of pediculosis capitis. Int. J. Eng. Manuf. (IJEM) **13**(1), 1–13 (2023)
12. Mei, S., Theo, N., Tao, Z., Xin, Z., Tianbao, Q.: Simulation model to determine ratios between quay, yard and intra-terminal transfer equipment in an integrated container handling system. J. Int. Logist. Trade **19**(1), 2–19 (2021)
13. Luo, S., Zhan, Y., Guo, Y.: Research on the teaching and training methods of matrix theory in engineering practice. Inf. Commun. (12), 137–139 (2020). (in Chinese)
14. Zhang, Q., Yang, S., Zeng, Q., Yu, T.: Storage pricing model of container yards under fluctuating demand. Appl. Econ. **52**(39), 4223–4235 (2020)
15. He, J., Xiao, X., Yu, H., et al.: Dynamic yard allocation for automated container terminal. Ann. Oper. Res. 1–22 (2022)
16. Yu, M., Zhang, Y.: Multi-agent-based fuzzy dispatching for trucks at container terminal. Int. J. Intell. Syst. Appl. (IJISA) **2**(2), 41–47 (2010)
17. Zhang, Y., Tang, G., Yu, X., et al.: The impact of the length of container yard on the efficiency of terminal operation. Water Transp. Eng. **11**, 94–98 (2016). (in Chinese)
18. Aveshgar, N., Huynh, N.: Integrated quay crane and yard truck scheduling for unloading inbound containers. Int. J. Prod. Econ. **159**, 168–177 (2015)
19. Cahyono, R.T., Kenaka, S.P., Jayawardhana, B.: Simultaneous allocation and scheduling of quay cranes, yard cranes, and trucks in dynamical integrated container terminal operations. IEEE Trans. Intell. Transp. Syst. **23**(7), 8564–8578 (2022)

# A Review of Epidemic Prediction and Control from a POM Perspective

Jing Wang[1], Yanbing Xiong[1], Qi Cai[1], Ying Wang[2], Lijing Du[1,3,4](✉), and Kevin Xiong[5]

[1] School of Safety Science and Emergency Management, Wuhan University of Technology, Wuhan 430070, China
dulijing@whut.edu.cn
[2] School of Business, Wuchang University of Technology, Wuhan 430223, China
[3] School of Management, Wuhan University of Technology, Wuhan 430072, China
[4] Research Institute of Digital Governance and Management Decision Innovation, Wuhan University of Technology, Wuhan 430072, China
[5] Information Technology Consulting Services, Ontario Limited, Markham, ON 1750351, Canada

**Abstract.** Infectious disease outbreaks have occurred many times in the past decades. They have had a tremendous impact on global health and the economy. Based on the learnings from these outbreaks, this study reviews the POM literature related to the epidemic and discusses what POM can help to address the challenges of the pandemic. This research divides the epidemic forecasting model into three types, namely compartmental model, statistical model and hybrid model. These models can be used to study the role of measures such as lockdowns and medical supplies in epidemic control. In addition, the pandemic has caused supply chain disruptions and led to a decline in the production level of manufacturers. It is important to ensure the supply of medical materials. Therefore, the relevant literature on the production of medical materials should be analyzed. Based on the comprehensive analysis of these articles, the future research directions are proposed. The results of this study are expected to be helpful for epidemic control research.

**Keywords:** Epidemic simulation and prediction · Epidemiological model · Control measures · Medical materials production · POM model

## 1 Introduction

Infectious disease outbreaks have been an immense challenge for humanity. The COVID-19 broke out in December 2019 and spread rapidly across the globe within a few dozen days. It has become a crucial public health problem worldwide, disrupting the lives of tens of millions of people in many countries. Although the outbreak is now under preliminary control, it is still spreading globally and continues to produce new mutant strains with increasing transmission. Moreover, some people are infected more than once. The world is still facing a severe challenge.

Almost all nations have imposed some kind of restrictions to some extent to control the spread of the COVID-19 [1]. Since the outbreak of the epidemic, many scientists have tried to make contributions to the prevention of the epidemic in various disciplines such as medicine, economics, sociology, etc. This paper analyzes the relevant literature on epidemic prevention and control from the perspective of Production and Operations Management (POM) to help mitigate the epidemic.

The contributions of this review are as follows. First, in order to predict and control the development of the epidemic, scholars have established various models to simulate the evolution of the epidemic. In view of this, the research investigates the epidemiological models and classify them. Second, non-pharmacological interventions taken to mitigate the epidemic are discussed in this study. Third, the production of medical materials is an important factor in controlling the epidemic, and how to secure their production is also the concern of this paper. Finally, possible future research directions are suggested in this research.

## 1.1 Selection of Journals

The search engine used for this study was the Web of Science. In the stage of article selection, articles using the following keywords in the topic of the articles: "pandemic", "infection", "epidemic", "COVID-19", "outbreak", "infectious", "influenza" or "coronavirus" were selected. The scope of the search was limited to major POM journals. In the initial screening process, 476 research papers were found. The title and abstract of each article were read carefully and irrelevant articles were excluded. Finally, 37 articles were obtained.

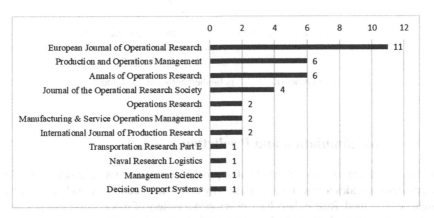

**Fig. 1.** Distribution of articles across journals

Figure 1 shows the distribution of published articles across 11 journals. It is observed that European Journal of Operational Research has published the highest number of research papers. Apart from European Journal of Operational Research, Production and Operations Management and Annals of Operations Research also represents most publications in this domain. These 23 papers of 37 papers, published in three journals, represent about 60% of all papers. Although many journals have published articles related to epidemics, their focus is completely different from epidemic control. They focus on finance, supply chain and some other issues. Therefore, only articles related to epidemic control are considered in this paper.

## 1.2 Research Trend

The time period of articles selected for this research was from 2014 to 2023. Figure 2 shows the number of these articles based on the year of the publication. It can be seen from the figure that the curve of the number of papers from 2014 to 2019 is flat, with a sudden increase in 2020. The reason is the outbreak of the COVID-19 in 2020, which brought a huge shock to the world and posed a challenge to the research on epidemic control. The number of related articles continue to increase in 2021 and 2022. By January 8, 2023, the number of papers published has reached 8. It clearly indicates a growing interest of POM researchers in the prevention and control of the epidemic.

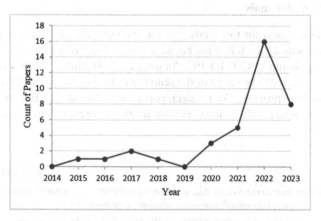

**Fig. 2.** Number of articles published between 2014 and 2023

## 2 Epidemic Simulation and Prediction

Epidemic outbreak is a complex diffusion process occurring among people. Modeling this process provides a way to understand why and how infections spread and how they might be prevented. Researchers have developed a variety of models over many years to study the evolution of epidemic. In this study, these models are divided into three types, which are the compartmental model, the statistical model and the hybrid model. The application of these models in epidemic prediction is described below.

### 2.1 Compartmental Model

The compartmental model is currently the most common infectious disease dynamics model. This model divides the population into compartments of susceptible (S), exposed

(E), infected (I), treated (T), recovered (R), deceased (D), etc. according to epidemiological status, and describes the dynamics of continuity between compartments using ordinary differential equations. The compartmental model is more convenient compared to other models and can better fit the epidemiological trend of infectious diseases.

Researchers use mathematical methods to study differential equations to simulate the epidemiological process of infectious diseases in different situations and predict the epidemic spread rate, mortality, cases of infection, etc. The classic SEIR model is mostly used to simulate the epidemic, and some scholars expand this model by considering hospitalized people, asymptomatic infected people and other groups to increase the accuracy of model simulation [1–5]. In addition, some scholars have considered the factors that affect the spread of the epidemic from multiple perspectives. For example, Liu and Zhang [6] predict the number of infections during the spread of an outbreak based on a time-discrete SEIR model to forecast the demand for healthcare resources. Kumar et al. [7] propose an extended SEIR model to study the measures to prevent the spread of infection and predict possible pandemic dynamics. Table 1 lists the articles mentioned above.

**Table 1.** Characteristics of compartmental model

| Article | Model | Goal | Country |
| --- | --- | --- | --- |
| Lu and Borgonovo (2023) | Extended SEIR model | Modeling COVID-19 epidemic, sensitivity analysis of the model | Italy, USA |
| Perakis et al. (2022) | Multiwave SEIRD model | Forecasting COVID-19 cases | USA |
| Liu et al. (2021) | SEIDR model | Simulating the spread of COVID-19 | USA, UK |
| Kumar et al. (2021) | Extended SEIR model | Forecasting possible pandemic dynamics | USA, seven European countries |
| Büyüktahtakn et al. (2018) | SITR-FB model | Simulating the Ebola disease transmission | West Africa |
| Liu and Zhang (2016) | time-discrete SEIR model | Predicting the trajectory of an epidemic diffusion | |
| He and Liu (2015) | Modified SEIR model | Forecasting medical demand | China |

## 2.2 Statistical Model

Statistical model is generally driven by historical data. It grasps the key factors of epidemic transmission. Compared with the compartmental model, statistical model has fewer parameters and more accurately shows the changing patterns of the epidemic.

Table 2 lists the statistical models used to forecast the epidemic spread. It can be seen from the table that the prediction models used by scholars include renewal equation approach, Stochastic Fractal Search algorithm, machine learning, etc., which are mainly used to forecast hospital admissions, ICU admission, mortality, newly infected people and other data [8–15]. These models take into account transmission networks and stochastic factors and allow for more detailed simulation of epidemic process.

**Table 2.** Characteristics of statistical model

| Article | Model | Goal | Country |
|---|---|---|---|
| Chang et al. (2023) | Renewal equation approach | Modeling local coronavirus outbreaks | USA |
| Bekker et al. (2023) | Statistical model | Modeling COVID-19 hospital admissions and occupancy | Netherlands |
| Lotfi et al. (2022) | Regression-based robust optimization approach | Predicting COVID-19 epidemic | Iran |
| Saadatmand et al. (2022) | Machine learning | Predicting ICU admission, mortality, and length of stay | Iran |
| Sbrana (2022) | Novel state-space approach | Predicting the number of newly infected people | USA |
| Khalilpourazari and Doulabi (2022) | Novel hybrid reinforcement learning-based algorithm | Predicting COVID-19 outbreak | Canada |
| Khalilpourazari and Doulabi (2021) | Stochastic Fractal Search algorithm | Predicting the number of symptomatic, asymptomatic, life-threatening, recovered, and death cases | Canada |
| Liu et al. (2020) | Mixed-integer non-linear programming model | Controlling the H1N1 outbreak | China |

## 2.3 Hybrid Model

With the deepening of research, scholars use more complex models to simulate the epidemic. Nikolopoulos et al. [16] present statistical, epidemiological, machine- and deep-learning models, and a new hybrid forecasting method based on nearest neighbors and clustering to forecast COVID-19 growth rates in different countries. Evgeniou et al. [17] propose a hybrid model that combines the standard SEIR model with the equivalent standard machine learning classification model to simulate isolation and exit policies in France. Taylor et al. [18] formulate a compartmental and statistical model to empirically compare the interval and distributional prediction combination methods of cumulative mortality of COVID-19. Furthermore, some scholars combine these models with logistics models to predict the material demand and use logistics models to distribute these materials to control the epidemic. According to the above research, these models have different characteristics and scope of application. Hybrid model can study more complex problems and predict the spread of the epidemic more accurately.

## 3 Epidemic Control Measures

Epidemic prevention and control measures include pharmacological interventions and non-pharmacological interventions. In the field of POM, pharmacological interventions mainly refer to vaccination, while non-pharmacological interventions include social distance, stay-at-home orders, lockdowns, etc. [19–28]. This research has investigated and analyzed the literature related to these control measures to understand the main methods of epidemic control. Table 3 shows these control measures and lists the models used in these papers. It can be seen from the table that medical materials such as ventilators and nucleic acid tests are effective for epidemic control [19, 20, 24, 29]. But how to ensure the supply of medical materials is a problem that needs to be studied.

Although intervention measures can effectively contain the spread of infectious diseases, they impose substantial direct and indirect costs on societies [30]. Therefore, some scholars study the relationship between intervention measures and economy to reduce the economic losses caused by them. Birge et al. [31] propose a spatial epidemic model illustrating population mobility to restrict economic activity in different neighborhoods of a city at different levels. Their study indicates the potential to limit the economic costs of unemployment while containing the spread of a pandemic. Eryarsoy et al. [30] formulate a mathematical model to manage the lost lives during an epidemic through controlling intervention levels. Their findings demonstrate that when the projected economic costs of the epidemic are large and the illness severity is low, a no-intervention strategy may be preferable. Chen et al. [32] provide an efficient social distancing policy that minimizes the total risks of disease transmission and economic volatility.

**Table 3.** Epidemic control measures

| Article | Control Measures | Model | Country |
| --- | --- | --- | --- |
| Abdin et al. (2023) | Testing capacities, control strategies | Epidemiological compartmental model | France |
| Rezapour et al. (2023) | Movement restrictions, social distancing, proactive testing | Multi-scale reaction-diffusion model | USA |

(*continued*)

**Table 3.** (*continued*)

| Article | Control Measures | Model | Country |
|---|---|---|---|
| Hosseini et al. (2023) | Behavioral changes in the population, vaccination | Statistical model | Iran |
| Wang (2022) | Stay-at-home orders | Statistical model | USA |
| Li et al. (2022) | Restriction on mass gatherings, closure of schools, stay-at-home | DELPHI model | USA, UK, Russia |
| Chen et al. (2022) | Medical resources, social distancing policy | Modified SEIR model | China |
| Ertem et al. (2022) | Early social distancing measures | Age-structured compartmental simulation model | USA |
| Biswas et al. (2022) | Lockdowns, curfews | Mixed Integer Non-Linear Programming epidemic model | France |
| Baveja et al. (2020) | Travel restrictions | Pandemic-management service value chain | |
| Kumar et al. (2021) | Social media | SEIR-V model | |
| Mehrotra et al. (2020) | Ventilators | stochastic optimization model | USA |

## 4 Medical Materials Production in an Epidemic

Medical materials play an important role in controlling the epidemic. However, due to the rapid outbreak of the epidemic, the demand for medical supplies has skyrocketed. It is critical to produce sufficient medical supplies to meet the demand in a timely manner. Many scholars have studied the production of medical materials in this special situation. Li et al. [33] formulate a risk-averse two-stage stochastic programming model to study the comprehensive production planning problem under uncertain demand and explore the impact of different risk preferences on the production of three types of masks. Sun et al. [34] analyze the construction of vaccine at-risk production capacity under the integrated and outsourcing mode and make recommendations for improving at-risk production capacity under both modes. Angelus et al. [35] discuss the large-scale production of new vaccines under two yield improvement strategies.

Table 4 lists the articles on the production of medical materials under the epidemic. Current research mainly focuses on vaccine production. Some scholars have conducted the incentive program for vaccine manufacturers, some have studied the at-risk production capacity building of vaccines, and some have investigated the large-scale production of new vaccines [34–38]. However, there are few studies on the production of medical materials other than vaccines. Only a few scholars have studied the production of masks and health equipment [33, 39].

**Table 4.** Literature reviews of medical materials production

| Article | Type of Material | Model | Outbreak |
| --- | --- | --- | --- |
| Li et al. (2023) | Mask | Risk-averse two-stage stochastic programming model | COVID-19 |
| Sun et al. (2022) | Vaccine | Signaling game model, incomplete contract model | COVID-19 |
| Angelus et al. (2022) | Vaccine | Stochastic, multiperiod, sequential-decision model | COVID-19 |
| Soltanisehat et al. (2022) | health equipment | Mixed-integer linear programming model, Monte Carlo simulation | Epidemic |
| Arifoglu & Tang (2022) | Vaccine | Backward induction | Influenza |
| Jansen & Ozaltin (2017) | Vaccine | Cournot competition model | Influenza |
| Chick et al. (2017) | Vaccine | Public procurement model | Influenza |

## 5 Conclusions and Future Research Direction

Based on the review of articles related to epidemic control in the past decade, this study mainly considers the review of epidemic simulation and prediction, epidemic prevention and control measures and medical material production. The models used for epidemic simulation and prediction are divided into three types, namely compartmental model, statistical model and hybrid model. Compartmental model is the main model for exploring the epidemic evolution at present. With the deepening of research, the use of statistical model and hybrid model is increasing. The epidemic prevention and control measures mainly include government measures such as social distancing policy, lockdown and timely supply of medical materials. As for the production of medical materials, the current research mainly focuses on vaccine production. There is little research related to medical materials such as protective clothing and ventilators.

Based on our findings, the following recommendations are made for future research. First, the epidemic is now frequent, and predicting of the epidemic wave is a major challenge. Researchers should explore how to accurately predict the next possible pandemic and propose countermeasures. Second, control measures can be effective in reducing the spread of epidemic, but they can also lead to problems such as economic stagnation and social unrest. Researchers could explore how to balance the relationship between the two. Third, the supply of medical materials plays a key role in epidemic prevention and control, but currently relevant research mainly focuses on vaccine production. Future research can focus more on the production of other medical materials. Future research is considered to be combined with other disciplines to investigate epidemic control from multiple perspectives compared to just the POM perspective.

**Acknowledgment.** This work was supported by the National Natural Science Foundation of China (72104190); and the Humanities and Social Sciences Fund of Ministry of Education (20YJC630018).

# References

1. Liu, Y., Srivastava, S., Huang, Z., et al.: Pandemic model with data-driven phase detection, a study using COVID-19 data. J. Oper. Res. Soc. (2021). https://doi.org/10.1080/01605682.2021.1982652
2. Lu, X., Borgonovo, E.: Global sensitivity analysis in epidemiological modeling global sensitivity analysis in epidemiological modeling. Eur. J. Oper. Res. **304**(1), 9–24 (2023)
3. Perakis, G., Singhvi, D., Lami, O.S., et al.: COVID-19: a multipeak SIR-based model for learning waves. Product. Oper. Manag. **32**, 13681 (2022). https://doi.org/10.1111/poms.13681
4. Büyüktahtakın, I.E., des-Bordes, E., Kıbış, E.Y.: A new epidemics–logistics model: insights into controlling the Ebola virus disease in West Africa. Eur. J. Oper. Res. **265**(3), 1046–1063 (2018)
5. He, Y., Liu, N.: Methodology of emergency medical logistics for public health emergencies. Transport. Res. E-Log **79**, 178–200 (2015)
6. Liu, M., Zhang, Ding.: A dynamic logistics model for medical resources allocation in an epidemic control with demand forecast updating. J. Oper. Res. Soc. **67**(6), 841–852 (2016)
7. Kumar, A., Choi, T.-M., Wamba, S.F., Gupta, S., Tan, K.H.: Infection vulnerability stratification risk modelling of COVID-19 data: a deterministic SEIR epidemic model analysis. Ann. Oper. Res. (2021). https://doi.org/10.1007/s10479-021-04091-3
8. Chang J.T., Kaplan, E.H.: Modeling local coronavirus outbreaks. Eur. J. Oper. Res. **304**(1), 57–68 (2023)
9. Bekker, R., Broek, M., Koole, G.: Modeling COVID-19 hospital admissions and occupancy in the Netherlands. Eur. J. Oper. Res. **304**(1), 207–218 (2023)
10. Lotfi, R., Kheiri, K., Sadeghi, A., et al.: An extended robust mathematical model to project the course of COVID-19 epidemic in Iran. Ann. Oper. Res. (2022). https://doi.org/10.1007/s10479-021-04490-6
11. Saadatmand, S., Salimifard, K., Mohammadi, R., et al.: Using machine learning in prediction of ICU admission, mortality, and length of stay in the early stage of admission of COVID-19 patients. Ann. Oper. Res. (2022). https://doi.org/10.1007/s10479-022-04984-x
12. Sbrana, G.: Modelling intermittent time series and forecasting COVID-19 spread in the USA. J. Oper. Res. Soc. (2022). https://doi.org/10.1080/01605682.2022.2055499
13. Khalilpourazari, S., Doulabi, H.H.: Designing a hybrid reinforcement learning based algorithm with application in prediction of the COVID-19 pandemic in Quebec. Ann. Oper. Res. **312**(2), 1261–1305 (2022)
14. Khalilpourazari, S., Doulabi, H.H.: Robust modelling and prediction of the COVID-19 pandemic in Canada. Int. J. Prod. Res. (2021). https://doi.org/10.1080/00207543.2021.1936261
15. Liu, M., Xu, X., Cao, J., et al.: Integrated planning for public health emergencies: a modified model for controlling H1N1 pandemic. J. Oper. Res. Soc. **71**(5), 748–761 (2020)

16. Nikolopoulos, K., Punia, S., Schafers, A., et al.: Forecasting and planning during a pandemic: COVID-19 growth rates, supply chain disruptions, and governmental decisions. Eur. J. Oper. Res. **290**(1), 99–115 (2021)
17. Evgeniou, T., Fekom, M., Ovchinnikov, A., et al.: Pandemic lockdown, isolation, and exit policies based on machine learning predictions. Prod. Oper. Manag. (2022). https://doi.org/10.1111/poms.13726
18. Taylor, J.W., Taylor, K.S.: Combining probabilistic forecasts of COVID-19 mortality in the United States. Eur. J. Oper. Res. **304**(1), 25–41 (2023)
19. Abdin, A.F., Fang, Y.-P., Caunhye, A., et al.: An optimization model for planning testing and control strategies to limit the spread of a pandemic-The case of COVID-19. Eur. J. Oper. Res. **304**(1), 308–324 (2023)
20. Rezapour, S., Baghaian, A., Naderi, N., et al.: Infection transmission and prevention in metropolises with heterogeneous and dynamic populations. Eur. J. Oper. Res. **304**(1), 113–138 (2023)
21. Hosseini-Motlagh, S.-M., Samani, M.R.G., Homaei, S.: Design of control strategies to help prevent the spread of COVID-19 pandemic. Eur. J. Oper. Res. **304**(1), 219–238 (2023)
22. Wang, G.: Stay at home to stay safe: effectiveness of stay-at-home orders in containing the COVID-19 pandemic. Prod. Oper. Manag. **31**(5), 2289–2305 (2022)
23. Li, M.L., Bouardi, H.T., Lami, O.S., et al.: Forecasting COVID-19 and analyzing the effect of government interventions. Oper. Res. (2022). https://doi.org/10.1287/opre.2022.2306
24. Chen, Z., Kong, G.: Hospital admission, facility-based isolation, and social distancing: an SEIR model with constrained medical resources. Prod. Oper. Manag. (2022). https://doi.org/10.1111/poms.13702
25. Ertem, Z., Araz, O.M., Cruz-Aponte, M.: A decision analytic approach for social distancing policies during early stages of COVID-19 pandemic. Decis. Support Syst. **161**, 113630 (2022)
26. Biswas, D., Alfandari, L.: Designing an optimal sequence of non-pharmaceutical interventions for controlling COVID-19. Eur. J. Oper. Res. **303**(3), 1372–1391 (2022)
27. Baveja, A., Kapoor, A., Melamed, B.: Stopping Covid-19: a pandemic-management service value chain approach. Ann. Oper. Res. **289**(2), 173–184 (2020)
28. Kumar, S., Xu, C., Ghildayal, N., et al.: Social media effectiveness as a humanitarian response to mitigate influenza epidemic and COVID-19 pandemic. Ann. Oper. Res. **319**, 823–851 (2021)
29. Mehrotra, S., Rahimian, H., Barah, M., et al.: A model of supply-chain decisions for resource sharing with an application to ventilator allocation to combat COVID-19. Nav. Res. Logist. **67**(5), 303–320 (2020)
30. Eryarsoy, E., Shahmanzari, M., Tanrisever, F.: Models for government intervention during a pandemic. Eur. J. Oper. Res. **304**(1), 69–83 (2023)
31. Birge, J.R., Candogan, O., Feng, Y.: Controlling epidemic spread: reducing economic losses with targeted closures. Manage. Sci. **68**(5), 3175–3195 (2022)
32. Chen, K., Pun, C.S., Wong, H.Y.: Efficient social distancing during the COVID-19 pandemic: integrating economic and public health considerations. Eur. J. Oper. Res. **304**(1), 84–98 (2023)
33. Li, Y., Saldanha-da-Gama, F., Liu, M., et al.: A risk-averse two-stage stochastic programming model for a joint multi-item capacitated line balancing and lot-sizing problem. Eur. J. Oper. Res. **304**(1), 353–365 (2023)
34. Sun, H., Toyasaki, F., Sigala, I.F.: Incentivizing at-risk production capacity building for COVID-19 vaccines. Prod. Oper. Manage. **32**(5), 1550–1566 (2022). https://doi.org/10.1111/poms.13652
35. Angelus, A., Ozer, O.: On the large-scale production of a new vaccine. Prod. Oper. Manag. **31**(7), 3043–3060 (2022)
36. Arifoglu, K., Tang, C.S.: A two-sided incentive program for coordinating the influenza vaccine supply chain. M&Som Manuf. Serv. Oper. Manage. **24**(1), 235–255 (2022)

37. Jansen, M.C., Ozaltin, O.Y.: Note on cournot competition under yield uncertainty. M&Som Manuf. Serv. Oper. Manage. **19**(2), 305–308 (2017)
38. Chick, S.E., Hasija, S., Nasiry, J.: Information elicitation and influenza vaccine production. Oper. Res. **65**(1), 75–96 (2017)
39. Soltanisehat, L., Ghorbani-Renani, N., Gonzalez, A.D., et al.: Assessing production fulfillment time risk: application to pandemic-related health equipment. Int. J. Prod. Res. (2022). https://doi.org/10.1080/00207543.2022.2036381

# Application of SVM and BP Neural Network Classification in Capability Evaluation of Cross-border Supply Chain Cooperative Suppliers

Lei Zhang[1(✉)] and Jintian Tian[2]

[1] School of Logistics, Wuhan Technology and Business University, Wuhan 430065, China
495997665@qq.com
[2] School of Postgraduate Studies, PSB Academy, Singapore 039594, Singapore

**Abstract.** With the trade-driven expansion of the Belt and Road Initiative, overseas engineering projects are increasing. The success of overseas engineering projects is marked by the progress and benefit, and to achieve this goal, the supplier's ability of the supply chain supporting the project is the key to the progress and benefit of the project. On the basis of literature analysis and many years of foreign engineering practice, combined with China's reality, this paper puts forward the subcontractor competency evaluation index. That is, the initial index set of supplier competency evaluation system is constructed based on 6 main factors and 24 sub-factors including supplier credit strength, quality assurance, supply ability, service level, technical level and price level. On the basis of careful analysis of qualitative and quantitative factors, support vector machine and BP neural network method are applied to compare, analyze and evaluate the capability of engineering project suppliers, and explore the effective solution of quantitative and qualitative coordination correlation between evaluation indicators.

**Keywords:** BP neural network · Support vector machine (SVM) · Competency

## 1 Introduction

Since the Belt and Road Initiative was put forward in 2013, its goal is to establish a free and open economic system for countries along the Belt and Road through a reasonable mix of production factors to make them flow smoothly, and promote the common development of countries and regions along the Belt and Road through economic means. The engineering projects in the countries along the Belt and Road are the cornerstone of the complete chain of the Belt and Road, and the safe and smooth flow of the material supply chain is the guarantee of the smooth progress of the engineering projects. The competence of the cooperative supplier that undertakes the supply chain is the guarantee of the safety and smooth of the supply chain. Therefore, the competency evaluation of cross-border suppliers becomes more and more important.

Competency theory originated from McClelland's landmark paper Measuring Competency, not Intelligence, published in 1973 [1]. Since then, this theory has been gradually developed and improved. In 1980, McClelland proposed the concept of "competency model" and defined it as "the key abilities needed to complete the work", which included a series of knowledge, skills and attitudes. In 1982, the book Competent Manager published by Rechard E Boyatzis regarded competency as a stable internal characteristic of a person, which is knowledge, cognition, behavior, skill, motivation, trait, etc. [2]. At present, in western countries, competency research results have been widely applied to recruitment, training, selection, performance management and other human resource management practices. Based on the strategic significance of organizational development, a strong human resource system has been established, so as to comprehensively improve the competitiveness of the organization. The manager's competency model has been widely applied in major organizations. Chinese research on manager competency started relatively late, but some important achievements have been made in recent years. For example, Gu Qinxuan explored the effectiveness of shared leadership from the perspective of shared leadership and participation in a safe atmosphere [3]. Qi Yue developed the competency model of middle managers in banks [4]. Feng Xufang built a competency structure model based on the practical teaching of teachers in higher vocational colleges [5]. Wang Haixia and Tang Zhisong studied the competence of teachers' core literacy education [6]. Tan Qiwei analyzed the construction method and application of the competency model [7]. Wu Qianying established a competency model for optimizing the civil servant training system [8]. Zhang Qianxue established a competency model based on human resource performance management [9]. Kang Fei and Zhang Shuibo used text analysis to analyze the research status of project manager competency at home and abroad [10]. Zhang Shuibo et al. constructed the competency evaluation index system of construction project managers and established the evaluation model [11]. He Qizong analyzed the research progress of Chinese college teachers' competence [12]. Hou Yanhui et al. conducted a competency study on multi-project human resource allocation of engineering construction enterprises [13]. Zhao Berg analyzed the competency structure of management teachers in ethnic universities based on grounded theory [14]. Chen Zhixia and Guo Jinyuan established the structural model of graduate competency and predicted it [15]. Yang Yuekun and Lu Nan built a "trinity" evaluation system for innovative scientific and technological talents based on exploratory and confirmatory factor analysis [16]. Although domestic and foreign scholars have conducted a lot of research on the competency of managers, the research on the competency of cooperative suppliers has not received extensive attention. At the same time, there is a lack of research results on how to apply the competency model to evaluate the competency of cooperative suppliers.

Based on literature analysis and expert opinions, this paper analyzes the feature vector and structural dimension of supplier capability, and establishes an evaluation system based on project cooperative supplier capability.This is a complex and multi-factor issue, and it needs to screen out the important influencing factors among the many influencing factors as the evaluation criteria. Some of these factors can be quantitatively analyzed, such as using data envelopment analysis to evaluate; Many other factors, however, are difficult to be evaluated by quantitative analysis, and can only be evaluated by qualitative analysis, among which analytic hierarchy process (AHP) is a more commonly used method.

This paper aims to establish the cross-border supplier capability evaluation system, and based on the theory of BP neural network and support vector machine, to evaluate the cross-border supplier capability evaluation.

## 2 Introduction to Evaluation Model

### 2.1 BP Neural Network Model

In 1986, Rumelhard and Mc Clelland proposed BP neural network, which has one input layer, one output layer and several hidden layers. It is a typical multi-layer forward type, and the neurons in the same layer are not connected, and the mode between layers is fully connected [17].

Neural network is based on the learning and memory function of network, which enables the neural network to learn the characteristics of various categories of sample information, to treat the identification samples by comparing the characteristic information mastered with the input vector, and finally to determine the categories of samples to be identified. Therefore, it obtains decision information through the learning mechanism, which is an abstract, simulation and simplified human brain information processing model. The key is the mathematical model of neuron, the way of network connection and the way of neural network learning. BP neural network is a widely used neural network model.

BP neural network has two typical learning processes of forward propagation and back propagation. In the forward propagation process, the input information is transmitted through the hidden layer to the output layer to generate the output signal, during which the connection weight of neurons is constant. By comparing the expected value with the output result, if the error between the expected value and the output result does not meet the accuracy requirements, then the error propagates backward from the output layer will constantly adjust the threshold and weight, so that the gap between the expected value and the output result is closer and closer until the accuracy requirements are met.

BP artificial neural network has been widely used in pattern recognition and regression fitting, because it can select the appropriate number of hidden elements and network hierarchy, multi-precision approximation of nonlinear functions. However, BP artificial neural network has some shortcomings. One is that the network structure needs to be specified in advance or modified by heuristic algorithm in the training process, so it is difficult to ensure the optimization of the network structure. Secondly, the method of adjusting the network weight coefficient is very limited, which is easy to get the local optimal, but difficult to get the optimal solution. Thirdly, sample data is too dominant to model performance in the training process. Fourthly, it is difficult to obtain the final result due to the large amount of high quality training data and the high dimensional input in many practical problems. Lots of engineering problems are faced with similar situations [18, 19]. Engineering practice shows that the application of neural network model is an effective way.

## 2.2 Support Vector Machines Model

Support vector machine (SVM) theory is a new machine learning method based on statistical learning theory. It has many unique advantages in solving small sample, nonlinear and high-dimensional pattern recognition problems. It has strong learning ability and good generalization performance. It has unique advantages in pattern recognition, function approximation and probability density estimation [20].

The core of support vector machine (SVM) is to construct a decision surface so that the sample points in the training set are as far away from the segmentation plane as possible and the blank areas on both sides of the segmentation plane can be maximized. If the optimal classification surface with satisfactory classification effect cannot be obtained in the original space, it needs to be transformed into a linear problem in high dimensional space through nonlinear transformation, and solved in the transformation space to obtain the optimal classification surface [21].

Support vector Machine (SVM) theory is a new general learning method developed on the basis of statistical learning theory, which has been applied in many fields and achieved good results [22–24]. The function of support vector is effectively realized by the accurate fitting of high-dimensional nonlinear system with small samples, and has good generalization performance based on the principle of minimizing structural risk. The regression function of support vector machine can be expressed as:

$$y_i[(w \cdot x_i) + b] - 1 \geq 0 \quad i = 1, \ldots n \quad (1)$$

If Mercer can be satisfied, the objective function can be converted into a high-dimensional spatial function:

$$Q(\alpha) = \sum_{i=1}^{n} \alpha_i - \frac{1}{2} \sum_{i,j=1}^{n} \alpha_i \alpha_j y_i y_j K(x_i, x_j) \quad (2)$$

The corresponding function is:

$$f(x) = \text{sgn}\{(w^* \cdot x) + b^*\} = \text{sgn}\left\{\sum_{i=1}^{n} \alpha_i^* y_i K(x_i, x_j) + b^*\right\} \quad (3)$$

## 3 Establishment of Supplier Competency Evaluation System

According to the principle that the selection of evaluation indexes should be scientific, representative, objective and systematic, CNKI database and Wanfang database were searched to select the indexes with high frequency, and the evaluation indexes in domestic and foreign literatures were also used for reference.

Six main level factors and 24 sub-level factors, including credit and strength, quality assurance, supply ability, service level, technical level and price level, are extracted as indicators of supplier competence, and the initial index set of the competency evaluation system of cooperative suppliers is constructed, as shown in Table 1.

**Table 1.** The initial index set of the competency evaluation system of cooperative suppliers

| Target layer | Layer of criterion | Indicator layer | The serial number |
|---|---|---|---|
| Supplier competency system | Credit and strength | Contract performance capability | C1 |
| | | Management system | C2 |
| | | Financial position | C3 |
| | | Reputation Brand | C4 |
| | | Level of sales | C5 |
| | | Quality of personnel | C6 |
| | | Social responsibility | C7 |
| | Quality assurance | Quality control system | C8 |
| | | Quality certification system | C9 |
| | | Production quality inspection | C10 |
| | | Ex factory quality control | C11 |
| | | Product performance level | C12 |
| | Ability to supply | Timeliness of delivery | C13 |
| | | Accuracy of delivery point | C14 |
| | | Completeness of goods | C15 |
| | Level of service | Attitude towards Service | C16 |
| | | Speed of response | C17 |
| | | Effect of customer demand | C18 |
| | Level of technology | Production technology level | C19 |
| | | Production technology level | C20 |
| | | Level of inspection technology | C21 |
| | | Level of information | C22 |
| | Level of price | Rationality of price | C23 |
| | | Price control | C24 |

## 4 Classification Experiment Design

### 4.1 Experimental Design of BP Neural Network Classification

To determine the transfer function, the number of hidden layer neurons, the number of hidden layers, etc., is the premise of network design. On this basis, the influence of momentum factor and random number selection in stochastic gradient descent on classification performance is studied respectively.

Performance evaluation index: The algorithm performance was investigated from three aspects: classification accuracy, convergence time and curve characteristics. High accuracy is the primary goal of the algorithm, while stochastic gradient descent method and additional momentum factor are used to reduce running time and improve learning efficiency, so convergence time and curve features are selected as evaluation items.

### 4.2 Experimental Design of SVM Classification

The parameters of SVM algorithm are less than that of BP algorithm. The influence of different kernel functions and sample number on classification performance is mainly investigated, and the performance of BP algorithm is combined with the number of training samples. There are four common kernel functions:

(1) Gaussian radial basis kernel function

$$K(x_i, x_j) = \exp(-\gamma \|x_i - x_j\|^2) \tag{4}$$

(2) Polynomial kernel function

$$K(x_i, x_j) = ((x_i \cdot x_j) + 1)^d \tag{5}$$

(3) Linear kernel function

$$K(x_i, x_j) = (x_i \cdot x_j) \tag{6}$$

(4) One-dimensional Fourier kernel function

$$K(x_i, x_j) = (1 - q^2) \Big/ [2(1 - 2q\cos(x_i - x_j) + q^2)] \tag{7}$$

Radial basis kernel function has a wide convergence domain in these four kinds of kernel functions, and contains only one parameter which is easy to optimize, so it is the most widely used kernel function.

### 4.3 Experimental Analysis

#### 4.3.1 Competency Evaluation Data Preparation

Samples can be used as examples for learning and training. The selection of samples is crucial to the competency evaluation results, so it should be comprehensive and typical. 160 units were selected as training samples.

In order to facilitate the comparison between BP neural network and SVM model and make them comparable, it is necessary to maintain the consistency of learning samples. Here, MATLAB neural network toolkit and SVM algorithm software package are used respectively to train and evaluate the performance of the samples. In this paper, 120 samples were randomly selected as test data and 40 samples were selected as verification data in the training samples.

### 4.3.2 BP Neural Network Evaluation Model

BP artificial neural network selection function, transfer function and transfer function of hidden layer and output layer training function, weight threshold learning function respectively use logsig, tansig, traingdx, leard to calculation.

### 4.3.3 SVM Evaluation Model

The kernel function (RBF) commonly used in SVM has few parameters and good fitting effect. As a basic function, SVM environmental assessment model is established.

After 120 test data are inputted, automatic optimal parameter selection is realized by using total error minimum programming. The model parameters were selected by cross validation method and programmed with Lib SVM library file.

### 4.3.4 Model Comparison

The 40 verification data were substituted into the model established above, and the unit environmental assessment score was output. The predicted results are compared with those of BP artificial neural network and support vector machine, most of the absolute errors of the two models can meet the accuracy requirements of competency evaluation.

In the construction of BP artificial neural network and support vector machine, although the number of hidden layer nodes determined by the former model can be adjusted, but the optimal parameters need to be tested manually repeatedly, which is a large degree of complexity. In the SVM model, the method of cross validation is adopted to realize the automatic optimization of parameter selection, which reduces the human intervention and greatly reduces the total error. Compared with the two models of BP artificial neural network and support vector machine, the absolute error curve, MSE, ME and MAPE of support vector machine are all lower than that of BP neural network, and the convergence speed of support vector machine is faster, which requires less time. Overall, therefore, support vector machine model is better than BP artificial neural network model.

When the number of random samples is large enough to reflect the basic law of samples, it can effectively reduce the running time. Too much will increase the calculation amount, while too little will reduce the prediction accuracy.

The accuracy of BP artificial neural network and support vector machine is not strictly correlated with the number of samples, but enough samples can improve the accuracy and generalization ability of BP artificial neural network. For support vector machine, when the number of samples is insufficient, the generalization ability of support vector machine is more powerful than that of BP artificial neural network.

## 5 Conclusions

Based on the establishment of BP neural network and SVM model respectively, and applied to the actual situation of competency, the conclusions can be drawn as follows:

(1) In order to ensure the comparability of the model, 160 learning samples were selected, 24 evaluation indexes were taken as input vectors, and evaluation scores were taken as output vectors. The data results showed that the accuracy of competency evaluation could meet the requirements.

(2) It can be seen from the comparison that, compared with the traditional BP neural network model, the support vector machine model can realize the automatic optimization of parameters, avoid the difficult to determine the hidden layer neurons of BP neural network, reduce the influence of human factors in the evaluation process, and convergence speed is faster and less time consuming.

(3) There are many factors affecting competency. Compared with analytic hierarchy process (AHP) or fuzzy comprehensive evaluation, both BP neural network model and SVM model reduce the influence of subjective factors to a certain extent. The reasonable selection of the number of training samples determines the final accuracy of BP neural network model and SVM model.

In short, compared with other evaluation methods, the BP neural network model and SVM model adopted in this paper for competency evaluation can greatly reduce the influence of human factors and avoid the factor problem of artificially determining index weights in the traditional qualitative evaluation methods, so that the data processing is more in line with the actual situation.

**Acknowledgment.** This paper is supported by: (1) Hubei Business Service Development Research Center Fund Project: Hubei Intelligent Logistics Development Support System Research (2020Z01). (2) Doctoral Research Foundation of Wuhan Institute of Industry and Technology: Analysis and Countermeasures of Cross-border Logistics Security of Petroleum projects in China and Kazakhstan based on "Belt and Road Initiative" (D2019004).

## References

1. McClelland, D.C.: Testing for competence rather than for intelligence. Am. Psychol. **28**(1), 1–24 (1973)
2. Boyatzis, R.E.: The Competent Manager: A Model for Effective Performance. John Wiley & Sons Inc. (1982)
3. Gu, Q.: A Study on the effectiveness of shared leadership competency from the perspective of shared leadership and participation in a safe atmosphere. J. Manag. **12**, 52–60 (2020)
4. Qi, Y.: Construction and application of financial manager competency model under the background of new regulations on capital management - a case study of P Bank. Reform Opening Up **15**, 14–22+28 (2020)
5. Feng, X.: Construction of practical teaching competence structural model of higher vocational college teachers. Vocat. Educ. Forum **1**, 107–114 (2021). (in Chinese)
6. Wang, H., Tang, Z.: Research on competence of teachers' core literacy education. Curriculum. The textbook. Teaching Methods **2**, 132–138 (2020)

7. Tan, Q.: Analysis on the construction method and application of competency model. Mod. Hosp. **12**, 46–48 (2018)
8. Wu, Q.: Optimization of civil servant training system based on competency model. Mod. Mark. (Manage. Ed.) **1**, 16–17 (2020)
9. Zhang, Q.: Construction of human resource performance management system: from the perspective of competency model. Chin. Collect. Econ. **1**, 122–123 (2019). (in Chinese)
10. Kang, F., Zhang, S.: Research on project manager competency: current situation and prospect. J. Tianjin Univ. (Social Science Edition) **15**(1), 35–40 (2013). (in Chinese)
11. Zhang, S., Kang, F., Li, X.: Competency evaluation of construction project manager based on support vector machine. China Soft Sci. **11**, 83–90 (2013). (in Chinese)
12. He, Q.: Research on teacher competency in Chinese universities: progress and thinking. Res. High. Educ. **35**(10), 39–45 (2014). (in Chinese)
13. Hou, Y., Rao, W., Hao, M.: Research on multi-project human resource allocation in engineering construction enterprises: based on competency and two-stage optimization perspective. Oper. Res. Manage. **26**(4), 192–199 (2017)
14. Zhao, B.: A Study on the competence structure of management teachers in ethnic colleges and universities: based on grounded theory. Res. Ethnic Educ. 20 **31**(6), 152–156
15. Chen, Z., Guo, J.: The construction of graduate competency structure model and its predictive effect. Degrees Grad. Educ. **7**, 55–60 (2018). (in Chinese)
16. Yang, Y., Lu, N.: Construction of evaluation model of innovative scientific and technological talents based on knowledge value. Leadersh. Sci. **1**, 98–102 (2019)
17. Wang, L., Kuang, Y.: Research of BP neural network optimizing method based on ant colony algorithm. J. New Industrial. **2**(4), 8–15 (2012)
18. Zhang, H., Xiaobo, Z.: Quantitative analysis of organizational behavior of container shipping in the upper and middle reaches of the Yangtze river based on hub-and-spoke network. J. Coast. Res. **73**, 119–125 (2015)
19. He, Q.-M., Zhang, H., Ye, Q.: An M/PH/K Queue with constant impatient time. Math. Methods Oper. Res. **18**(11), 139–168 (2018)
20. Zhang, X.: On statistical learning theory and support vector machines. Acta Automat. Sin. **26**(1), 32–42 (2000)
21. Zhang, L.: Evaluation of China's petroleum security system: based on rough set and support vector machine. Chin. Soft Sci. **11**, 13–19 (2022). (in Chinese)
22. Joseph, I., Imoize, A.L., Ojo, S., Risi, I.: Optimal call failure rates modelling with joint support vector machine and discrete wavelet transform. Int. J. Image Graph. Signal Process. **14**(4), 46–57 (2022)
23. Wiradinata, T.: Folding bicycle prospective buyer prediction model. Int. J. Inf. Eng. Electron. Bus. **13**(5), 1–8 (2021)
24. Ahmed, N., Rabbi, S., Rahman, T., Mia, R., Rahman, M.: Traffic sign detection and recognition model using support vector machine and histogram of oriented gradient. Int. J. Inf. Technol. Comput. Sci. **13**(3), 61–73 (2021)

# Research on Port Logistics Demand Forecast Based on GRA-WOA-BP Neural Network

Zhikang Pan[1] and Ning Chen[1,2(✉)]

[1] School of Transportation and Logistics Engineering, Wuhan University of Technology, Wuhan 430063, China
1772232879@qq.com
[2] Wuhan University of Technology, Sanya Science and Education Innovation Park, Sanya 572025, China

**Abstract.** Port logistics demand is a crucial part of China's coastal industrial economy. In order to achieve a balanced supply and demand for port logistics and a reasonable allocation of logistics resources, the accurate prediction of port logistics demand has a vital impact on the future development of the port and the economic development of the hinterland cities. Firstly, the grey relational analysis method is used to screen out nine principal influencing factors of the logistics demand of Haikou port to realize the dimensionality reduction behavior of the input variables, the whale optimization algorithm is also introduced to optimize the weightings and thresholds of the BP neural network model, based on which the GRA-WOA-BP neural network port logistics demand forecasting model is established. The results indicate that in port logistics demand forecasting, the GRA-WOA-BP neural network to predict the results of the prediction error is all less than the BP neural network and WOA - BP neural network model. Its prediction accuracy is of significance better than that of the traditional prediction methods, with an accuracy rate of 98.82%. The research results also have certain reference significance for the strategic planning of port logistics development.

**Keywords:** Port · Logistics demand forecasting · Grey relational analysis · Whale optimization algorithm · BP neural network

## 1 Introduction

Along with our country economy continues to develop rapidly, at the same time also led to a rapid increase in China's logistics needs, and the port as an important node in the transport network, to assume the main task of freight logistics distribution and transit, and at the same time, as the extent of our foreign trade exchange continues to expand, the progress of port logistics corresponding to the service of foreign trade is particularly important.

At present, scholars' research on port logistics mainly focuses on port logistics and regional economic development, sustainable evolution of port logistics, and prediction of port logistics demand, etc. Chen et al. used gray correlation analysis to screen the relevant indicators between port logistics and regional economy in Xiamen, and throw out policy recommendations for this [1]. Xiao et al. studied the correlation between Yangtze River shipping logistics and inland economy from qualitative and quantitative perspectives, respectively, and the results of the study have important guiding significance for inland river ports [2]. Ji constructed a vector autoregressive VAR model and established an autoregressive model between Shanghai GDP and port cargo throughput to study the logistics development of Shanghai port [3]. Xu studied the sustainable evolution of port logistics to build a port logistics carbon emission evaluation system in the context of low carbon environment and studied the emission reduction efficiency of Dalian port to propose reasonable green optimization [4]. In order to explore the collaborative development trend between port logistics and urban economy, Zha and Tu developed a neural network prediction model on port synergistic development using port logistics indicators [5]. Yang used BP neural network to establish a forecasting model for import and export logistics demand at coastal ports and compared the analysis with the commonly used forecasting model, and the results indicated that the established forecasting model has a better prediction effect [6]. Zhuang et al. used principal component analysis to determine the factors influencing port green indicators and predicted port throughput through a gray prediction model [7]. Eskafi proposed a Bayesian statistics-based model for port cargo throughput prediction, which was verified to have a strong adaptive capability and high prediction accuracy [8]. Yan et al. used exponential smoothing and grey smoothing to forecast logistics demand, respectively, while constructing a combined forecasting model to overcome the lack of accuracy of a single forecasting model [9].

Therefore, a reasonable analysis of port logistics demand is not only a significant part of the development of the port industry or the port operations, the overall design, the target direction to play a key role. And not clear forecast of port logistics demand will affect its strategic development. In the process of forecasting, choosing the appropriate forecasting method for port logistics demand forecasting analysis is an urgent problem in the development of modern logistics.

## 2 Model Approach

### 2.1 Grey Relational Analysis

Grey relational analysis is a method applicable to analyze the degree of influence of multiple variable factors on the target variable factors from a quantitative research perspective in a complex environment of variable factors, with the following main steps.

Step 1: Given the target sequence $y(k)(k = 1, 2, \cdots, m)$, reference sequence $x_1(k), x_2(k), \cdots, x_n(k)$.

Step 2: Standardize each sequence to eliminate the influence of quantity. Get the $\widetilde{y}(k)$ and $\widetilde{x}_1(k), \widetilde{x}_2(k), \cdots, \widetilde{x}_n(k)$.

Step 3: Calculate the relational coefficient.

$$r_i(k) = \frac{d_{\min} + d_{\max}}{d_i(k) + \lambda d_{\max}} \tag{1}$$

where $d_{\min}$ and $d_{\max}$ are the minimum and maximum differences between the two levels, $d_i(k) = |\widetilde{y}(k) - \widetilde{x}_i(k)|$, $\lambda$ is the resolution coefficient, generally $0 < \lambda < 1$.

Step 4: Calculate and rank the relation.

$$R_i = \frac{1}{m}\sum_{k=1}^{m} r_i(k) \qquad (2)$$

where, $0 \leq R_i \leq 1$, the closer to 1 indicates the stronger relation with the target sequence in the reference sequence.

## 2.2 GRA-WOA-BP Neural Network Model

### 2.2.1 BP Neural Network

BP neural network is a kind of the input information is passed, then in the other direction, according to the error of multi-layer feedback neural network. It is often used as a function fitting, classification and prediction aspect because of its adaptive and self-learning, parallel processing of information and strong nonlinear mapping capability. The workflow of an usual three-layer BP neural network is shown in Fig. 1.

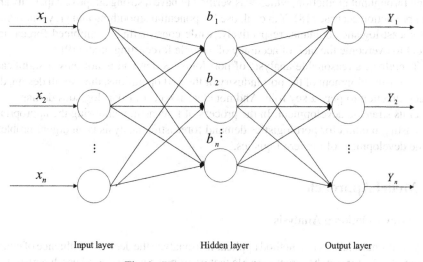

**Fig. 1.** BP neural network flow chart

### 2.2.2 Whale Optimization Algorithm

Whale optimization algorithm was first proposed by the Australian scholar Mirjalili in 2016, which simulates the predatory behavior of humpback whales, mainly including the three stage process of encircling prey, bubble network attack, and searching for prey [10].

1) Surrounding the prey

During the initial stage of predation, the individual position of each whale is unknown, it is necessary to determine the location of the current prey to update the individual whale location and then lock the envelope as follows:

$$\vec{D} = \left| \vec{C} \cdot \vec{X^*}(t) - \vec{X}(t) \right| \tag{3}$$

$$\vec{X}(t+1) = \vec{X^*}(t) - \vec{A} \cdot \vec{D} \tag{4}$$

where $\vec{X^*}(t)$ is the location of the optimal individual; $\vec{X}(t)$ is the whale's current location; $t$ is the current number of iterations; $\vec{D}$ is the distance between the prey and the individual whale; $\vec{C}$ and $\vec{A}$ are the vector of coefficients to control how the whale swims, denoted as:

$$\vec{A} = 2\vec{a} \cdot \vec{r} - \vec{a} \tag{5}$$

$$\vec{C} = 2 \cdot \vec{r} \tag{6}$$

$$\vec{a} = 2 - 2t/T_{\max} \tag{7}$$

where, $\vec{a}$ is the control parameter; $\vec{r}$ is the random vector between [0, 1]; $T_{\max}$ is the maximum number of iterations.

2) Bubble network attack (Local search)

Individual whales use a bubble net feeding strategy during predation. The whale optimization algorithm is designed with two mechanisms of shrinkage envelope and spiral update position, and the model assumes a 50% probability of each for the experiment, and the specific model is shown below:

$$\vec{X}(t+1) = \begin{cases} \vec{X^*}(t) - \vec{A} \cdot \vec{D} & p < 0.5 \\ \vec{D'} \cdot e^{bl} \cdot \cos(2\pi l) + \vec{X^*}(t) & p \geq 0.5 \end{cases} \tag{8}$$

where, $\vec{D'} = \left| \vec{X^*}(t) - \vec{X}(t) \right|$ is the distance between the whale and its prey; $b$ is a constant; $l$ is a random number between $[-1, 1]$.

3) Searching for prey (Global search)

At this stage, whales no longer update their positions according to the prey, but randomly achieve position update by replacing the prey with other whale positions within the population, which guarantees that at this stage whales can complete the global search and avoid the defect of falling into the local optimum.

$$\vec{D} = \left| \vec{C} \cdot \vec{X_{rand}} - \vec{X}(t) \right| \tag{9}$$

$$\vec{X}(t+1) = \vec{X_{rand}} - \vec{A} \cdot \vec{D} \tag{10}$$

where $\vec{X_{rand}}$ represents the position of random individuals in the current whale population.

The whale optimization algorithm has few adjustment parameters and is simple to operate, it can effectively solve the problems of BP neural network that is easy to fall into local optimum and slow convergence speed. The flow chart shown in Fig. 2.

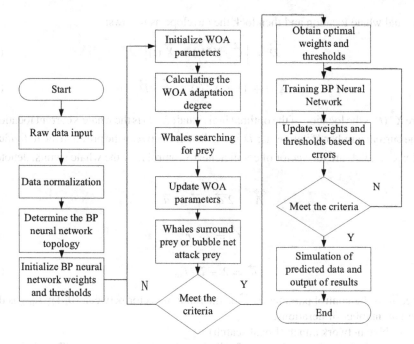

**Fig. 2.** WOA-BP neural network model flow chart

## 3 Port Logistics Demand Impact Indicator System Construction

### 3.1 Influencing Factor Indicator Selection

With the national "One Belt One Road" strategic plan and the construction of Hainan Free Trade Port, the trend of globalization of the source of goods, trade and investment facilitation, and the trend of efficient flow of goods are obviously strengthened, and Hainan Province gradually formed the pattern of "four directions and five ports", and As the largest port in Hainan Province, Haikou Port logistics development plays a key role in promoting the economic development of Hainan province the development of Haikou port logistics plays an important role in promoting the economic development of the whole Hainan Province. The foreign trade container route mainly connects with the international container route network of Southeast Asia region through Yangpu Port Container Port, or radiates to the national container routes of Europe and America region and East Asia region through Hong Kong, China, and the intensive domestic and foreign trade routes meet the huge market of goods import and export.

Port logistics demand will be affected by a variety of factors, in order to can be more comprehensive port logistics demand forecast, combined with the existing literature research angle analysis, this paper selected three perspectives of socio-economic development, logistics industry development, port logistics level development for model index factor selection [11, 12].

1) Socio-economic development. Industrial development and social and economic prosperity are inseparable, among which the three major industries of the national economy also play different degrees of influence on port logistics. Total city GDP and total retail sales of consumer goods reflect the degree of logistics demand and economic growth, while the service area of Haikou port is not only Haikou city, and the proportion of GDP of Haikou city to the province's GDP is chosen to indicate the influence of Hainan economy on Haikou port. The total investment in fixed assets also plays a significant role in port logistics.

2) Logistics industry development. The development of the logistics industry is also as a significant influencing factor for the development of logistics in Haikou port, based on the availability of data and avoiding subjective will to lead to bias in the analysis of the results. So we choose two quantitative indicators for the analysis: the number of resident population in Haikou and the proportion of the number of people in the logistics industry to the total number of employees.

3) Port logistics level development. The comprehensive strength of Haikou port's own logistics is a prerequisite for studying its logistics demand. In this paper, we comprehensively select four indicators, namely, the total value of foreign trade import and export, the number of berths above 10,000 tons, container throughput and freight turnover, as the reference factors affecting the development of port logistics level.

## 3.2 Input and Output Variables are Determined

After clarifying the factors influencing the logistics demand of Haikou port, the following factors are mainly selected as input variables in this paper: Haikou GDP ($X_1$), the first industrial output value ($X_2$), the second industry output ($X_3$), the tertiary industry output value ($X_4$), total fixed asset investment ($X_5$), total retail sales of consumer goods($X_6$) proportion of Haikou GDP to provincial GDP ($X_7$), foreign trade Import and export volume ($X_8$), berths above 10,000 tons ($X_9$), container throughput($X_{10}$), freight turnover ($X_{11}$), resident population($X_{12}$), proportion of employees in logistics industry to total employees($X_{13}$), and port cargo throughput ($Y$) are selected as output variables. Actual data's are shown in Table 1.

**Table 1.** Base data for input variables 2005–2020

| Year | $X_1$ | $X_2$ | $X_3$ | $X_4$ | $X_5$ | $X_6$ | $X_7$ |
|---|---|---|---|---|---|---|---|
| 2005 | 301.35 | 23.09 | 83.14 | 195.12 | 137.17 | 138.45 | 33.30 |
| 2006 | 350.12 | 25.94 | 102.12 | 222.06 | 158.23 | 161.35 | 33.30 |
| 2007 | 393.69 | 26.81 | 111.12 | 255.75 | 181.83 | 189.38 | 32.30 |
| 2008 | 443.18 | 31.40 | 113.28 | 298.50 | 221.43 | 234.75 | 30.40 |
| 2009 | 481.28 | 34.11 | 111.49 | 335.69 | 277.03 | 277.20 | 29.09 |
| 2010 | 595.14 | 38.19 | 142.83 | 414.13 | 349.65 | 326.94 | 28.83 |

(continued)

**Table 1.** (*continued*)

| Year | $X_1$ | $X_2$ | $X_3$ | $X_4$ | $X_5$ | $X_6$ | $X_7$ |
|---|---|---|---|---|---|---|---|
| 2011 | 713.30 | 47.71 | 177.91 | 487.68 | 404.59 | 387.18 | 28.28 |
| 2012 | 797.24 | 50.76 | 197.13 | 549.34 | 510.38 | 436.26 | 27.92 |
| 2013 | 904.64 | 58.10 | 217.03 | 629.51 | 649.33 | 505.35 | 28.75 |
| 2014 | 1091.70 | 57.07 | 217.49 | 817.15 | 821.53 | 558.47 | 31.19 |
| 2015 | 1161.96 | 57.09 | 223.67 | 881.21 | 1012.05 | 613.51 | 31.38 |
| 2016 | 1257.67 | 63.91 | 233.56 | 960.20 | 1271.73 | 673.30 | 31.10 |
| 2017 | 1390.58 | 62.51 | 252.22 | 1075.85 | 1415.50 | 742.72 | 31.16 |
| 2018 | 1510.51 | 63.21 | 276.00 | 1171.31 | 1327.74 | 787.25 | 31.30 |
| 2019 | 1671.92 | 71.18 | 275.99 | 1324.75 | 1123.27 | 823.94 | 31.49 |
| 2020 | 1791.58 | 79.88 | 269.56 | 1442.14 | 1234.47 | 835.89 | 32.40 |
| Year | $X_8$ | $X_9$ | $X_{10}$ | $X_{11}$ | $X_{12}$ | $X_{13}$ | $Y$ |
| 2005 | 12.35 | 2 | 22.03 | 406.60 | 173.73 | 5.34 | 2523.80 |
| 2006 | 28.50 | 2 | 24.48 | 443.16 | 176.68 | 4.72 | 3018.80 |
| 2007 | 35.20 | 3 | 28.6 | 530.97 | 179.45 | 5.65 | 3617.99 |
| 2008 | 45.40 | 5 | 34.64 | 555.74 | 183.50 | 5.84 | 3813.23 |
| 2009 | 38.10 | 7 | 43.52 | 465.16 | 187.85 | 6.25 | 4014.57 |
| 2010 | 39.48 | 7 | 61.34 | 564.86 | 204.62 | 6.78 | 4795.52 |
| 2011 | 39.36 | 7 | 80.8 | 594.91 | 209.73 | 6.70 | 5520.60 |
| 2012 | 42.15 | 7 | 100 | 665.66 | 214.13 | 6.93 | 6122.89 |
| 2013 | 51.40 | 7 | 116.82 | 839.76 | 217.11 | 7.67 | 7421.39 |
| 2014 | 34.00 | 7 | 134.67 | 963.54 | 220.07 | 8.36 | 7581.00 |
| 2015 | 43.40 | 6 | 127.5 | 708.73 | 222.30 | 9.38 | 8209.90 |
| 2016 | 39.20 | 17 | 140.18 | 669.68 | 224.36 | 9.33 | 8866.93 |
| 2017 | 31.10 | 25 | 163.6 | 460.17 | 227.21 | 10.49 | 10112.78 |
| 2018 | 50.88 | 25 | 184.67 | 519.42 | 230.23 | 9.90 | 10764.90 |
| 2019 | 48.12 | 25 | 197.26 | 1292.73 | 232.79 | 9.42 | 11197.56 |
| 2020 | 53.97 | 23 | 197.07 | 1497.86 | 288.66 | 9.76 | 10466.60 |

The 11 selected influencing factors were correlated with the logistics demand using grey correlation analysis, and the analysis results are shown in Table 2. In accordance with the grey correlation analysis, the influencing factor variables with a grey correlation of 0.65 or more were selected for the WOA-BP neural network model for training and testing, and $X_1, X_2, X_3, X_4, X_5, X_6, X_{10}, X_{11}, X_{13}$ were selected.

**Table 2.** Grey relation degree between logistics demand and various factors in Haikou port

| Influencing Factors | $X_1$ | $X_2$ | $X_3$ | $X_4$ | $X_5$ | $X_6$ | $X_7$ |
|---|---|---|---|---|---|---|---|
| Grey relation degree | 0.8491 | 0.8174 | 0.8429 | 0.7689 | 0.6618 | 0.8747 | 0.58 |
| Influencing Factors | $X_7$ | $X_8$ | $X_9$ | $X_{10}$ | $X_{11}$ | $X_{12}$ | $X_{13}$ |
| Grey relation degree | 0.58 | 0.6131 | 0.5848 | 0.7618 | 0.6624 | 0.6264 | 0.7433 |

## 4 Analysis of Simulation Experiments

### 4.1 Neural Network Parameter Setting

The accuracy of GRA-WOA-BP neural network model, traditional BP neural network and WOA-BP neural network to forecast port logistics demand is studied, the neural network model parameters were set to a maximum training number of 1000, a training target error of 0.00001, and a learning rate of 0.01; the WOA model parameters were set to a population size of 30 and a maximum evolutionary generation of The research examples and model runs were conducted under the software MATLAB R2019b.

### 4.2 Analysis of Model Results

The GRA-WOA-BP neural network model is used to train the logistics demand of Haikou port, and the prediction results are compared with the other two models, which are shown in Table 3.

**Table 3.** Prediction results and relative errors of each model

| Year | True Value | BP Predicted value | BP Relative Error | WOA-BP Predicted value | WOA-BP Relative Error | GRA-WOA-BP Predicted value | GRA-WOA-BP Relative Error |
|---|---|---|---|---|---|---|---|
| 2018 | 10764.90 | 9003.9 | 16.4% | 10073.9 | 6.4% | 10934.9 | 1.6% |
| 2019 | 11197.56 | 8838.56 | 21.1% | 10072.56 | 10% | 11042.56 | 1.4% |
| 2020 | 10466.60 | 8845.6 | 15.5% | 10073.6 | 3.8% | 10407.6 | 0.6% |

As can be seen from Table 3, the GRA-WOA-BP neural network model performs better than the traditional BP neural network model and the WOA-BP neural network model in predicting the logistics demand of Haikou port, and the relative error interval range of prediction is significantly smaller than the other two models. Meanwhile, to further study the prediction accuracy of the model, MAE, RMSE, MAPE and prediction accuracy are selected as the evaluation indexes in this research, and the smaller the error indexes representation model of prediction accuracy is higher, and the specific results are shown in Table 4.

**Table 4.** Comparison of prediction errors of each model

| Model | MAE | RMSE | MAPE | Accuracy rate |
| --- | --- | --- | --- | --- |
| BP | 1913.8 | 1940.29 | 17.61% | 82.36% |
| WOA-BP | 736.50 | 795.42 | 6.74% | 93.26% |
| GRA-WOA-BP | 128.22 | 137.40 | 1.18% | 98.82% |

The GRA-WOA-BP model outperforms the other two models in port logistics demand prediction, and its prediction accuracy is as high as 98.82% from Table 4. In summary, the whale optimization algorithm for BP neural network is practical and feasible for weight and threshold finding, and the GRA-WOA-BP neural network model constructed in this study is applicable to the logistics demand prediction of Haikou port, and the prediction accuracy is higher than other prediction models.

## 5 Conclusion

This research introduces the GRA-WOA-BP neural network model into the field of port logistics demand forecasting, selects the influencing factors affecting the logistics demand of Haikou port, forecasts the logistics demand of Haikou port from 2005 to 2020 by constructing the neural network model, and makes a comprehensive comparison of the forecasting results of the three models The results of the three models are compared and analyzed. Through the above study, the following main conclusions were obtained.

1) The factors affecting the development of logistics demand at Haikou port are systematically reviewed. 13 factors influencing the logistics demand of Haikou port were selected from three perspectives: socio-economic development, logistics industry development and port logistics level development, and 9 main influencing factors were clarified through grey relational analysis, among which the total retail sales of social consumer goods, GDP of Haikou city, primary industry, secondary industry and the logistics demand of Haikou port have a strong correlation relationship.

2) In port logistics demand forecasting, by introducing the whale optimization algorithm to optimize the initial weights and thresholds in the BP neural network, the relative error, mean absolute error, root mean square error and mean absolute percentage error of the forecasting model results are smaller than those of the BP neural network and WOA-BP neural network models, and the forecasting accuracy is higher. The GRA-WOA-BP neural network model can better clarify the nonlinear relationship in the complex network system, help to provide more valuable reference information for port decision makers, and enhance the rationality of port logistics demand planning, so as to provide a certain reference basis for the future port logistics industry.

**Acknowledgment.** This project is supported by Major science and technology projects in Hainan (ZDKJ2020012) and Hainan open fund project (2020KF0051).

# References

1. Meng, C., Wanshan, Z., Jing, P.: Research on the relationship between port logistics and regional economy in Xiamen city based on gray relational analysis. In: Proceedings of 2018 5th International Conference on Education Reform and Management Innovation (ERMI 2018), pp. 92–98 (2018)
2. Xiao, R., Pan, L., Xiao, H., et al.: Research of intelligent logistics and high-quality economy development for Yangtze river cold chain shipping based on carbon neutrality. J. Mar. Sci. Eng. **10**(8), 1029 (2022)
3. Ji, W.: Research on Shanghai port logistics based on VAR model. IOP Conf. Ser. Earth Env. Sci. **791**(1), 012042 (2021)
4. Jinyu, X., Mo, L.: Carbon emission assessment of port integrated logistics in low-carbon environment. Int. J. Front. Sociol. **3**(18), 55–60 (2021)
5. Zha, A., Tu, J.: Research on the prediction of port economic synergy development trend based on deep neural networks. J. Math. **2022**, 1–9 (2022)
6. Yang, D.: Logistics demand forecast model for port import and export in coastal area. J. Coast. Res. **103**(sp1), 678–681 (2020)
7. Zhuang, X., Li, W., Xu, Y.: Port planning and sustainable development based on prediction modelling of port throughput: a case study of the deep-water Dongjiakou port. Sustainability **14**(7), 4276 (2022)
8. Eskafi, M., Kowsari, M., Dastgheib, A., et al.: A model for port throughput forecasting using Bayesian estimation. Marit. Econ. Logistics **23**(2), 1–21 (2021)
9. Peng, Y., Lin, Z., Zhiyun, F.: Research on logistics demand forecast of port based on combined model. J. Phys. Conf. Ser. **1168**, 032116 (2019)
10. Mirjalili, S., Lewis, A.: The whale optimization algorithm. Adv. Eng. Softw. **95**, 51–67 (2016)
11. Ma, H., Luo, X., Varatharajan, R.: Logistics demand forecasting model based on improved neural network algorithm. J. Intell. Fuzzy Syst. **40**(4), 6385–6395 (2021)
12. Ren, X., Tan, J., Qiao, Q., et al.: Demand forecast and influential factors of cold chain logistics based on a grey model. Math Biosci. Eng. **19**(8), 7669–7686 (2022)

# Evaluation and Optimization of the Port *A* Logistics Park Construction Based on Fuzzy Comprehensive Method

Xin Li[1], Xiaofen Zhou[1], Meng Wang[1], Rongrong Pang[2], Hong Jiang[1(✉)], and Yan Li[1]

[1] School of Logistics, Wuhan Technology and Business University, Wuhan 430065, China
lixin@wtbu.edu.cn, 1281361476@qq.com

[2] Philippine Christian University Center for International Education, 1004 Manila, Philippines

**Abstract.** With the rapid development of China's cities, manufacturing and commerce, logistics, as a supporting service industry, has occupied an increasingly important position in the development of the national economy. The construction of the logistics park has become a hot spot for the promotion and revitalization of regional logistics industry and economic development. The demand for the logistics industrial park has gradually expanded, but it has also been accompanied by uneven management, services, facilities and equipment, resulting in the waste of a package of resources and not playing its due role. The logistics park plays an active role in improving economic efficiency and logistics efficiency. At the same time, it has the characteristics of large investment, long development cycle, many uncertain factors and high risk in the construction of logistics park. Taking the project of A port Logistics Park as an example, this paper analyzes various risks and related factors in the construction of the logistics park, establishes a risk evaluation index system, analyzes the overall system of the logistics park, uses the fuzzy comprehensive evaluation method to evaluate the risk level of Port A Logistics Park, and puts forward corresponding optimization strategies and plans based on the evaluation results.

**Keywords:** Logistics Park Construction · Index System · Fuzzy Comprehensive Method · Evaluation and Optimization

## 1 Introduction

The National Logistics Park Development Plan (2013–2020) (hereinafter referred to as the "Plan") has been released recently, clarifying the development objectives and overall layout of the national logistics park, and drawing a "road map" for the development of the logistics park [1].

Port construction usually faces many problems and challenges, such as channel engineering, queuing phenomenon, management organization behavior, etc., which will affect port operation to varying degrees [2–4]. Port A is committed to improving the comprehensive handling capacity of containers in the port, making the container industry bigger and stronger, striving to achieve its position as the main channel in container transport in central China, and playing a positive role in the construction of Wuhan Shipping Center and port A. As one of the earliest ports opened to the outside world in Wuhan, Port A has developed into a major distribution center for foreign trade import and export goods in Hubei Province, with more than 400 flights per month, accounting for about two-thirds of the total container transport in Wuhan, and is the largest professional international container terminal in the middle and upper reaches of the Yangtze River [5].

Therefore, the quality of port logistics park project construction directly affects China's foreign trade economy. A set of index system for the construction project of Wuhan Xingang Logistics Park has been constructed, and the risk assessment method for the construction project of Port A Logistics Park has been summarized, which not only provides a reference for the development research of the construction planning of Port A Logistics Park in Wuhan in the future, but also provides a reference for the research of the construction investment project of Port A Logistics Park to a certain extent, which is conducive to the evaluation of the development level of the relevant port logistics system in theory. The logistics park has large investment scale, long construction period and many uncertain factors. The evaluation of Port A's logistics construction project and the analysis of various regions, elements and their interactions in the port logistics system are helpful to find out some problems in the development of Port A's logistics, and help the relevant departments and institutions in the construction of Port A to solve these problems, which has certain guiding significance for the construction and development of Port A [6].

## 2 Overview of Relevant Theories

### 2.1 Fuzzy Comprehensive Method

The fuzzy evaluation method is based on fuzzy mathematics. Fuzzy mathematics was born in 1965, and its founder is American automatic control expert L.A. Zadeh. This comprehensive evaluation method transforms qualitative evaluation into quantitative evaluation according to the theory of membership degree of fuzzy mathematics, that is, to make an overall evaluation of things or objects subject to multiple factors by using fuzzy mathematics. It has the characteristics of clear results and strong systematisms, can better solve fuzzy and difficult to quantify problems, and is suitable for solving various uncertain problems [7].

Due to a series of problems such as the complexity of evaluation factors, the hierarchy of evaluation objects, the fuzziness of evaluation criteria, the fuzziness or uncertainty of evaluation influencing factors, and the difficulty in quantifying qualitative indicators, it is difficult for people to accurately describe the objective reality with the absolute "either this or that", and there is often the fuzzy phenomenon of "this or that", and its description is often expressed in natural language, The biggest feature of natural language is its fuzziness, which is difficult to be measured uniformly by classical mathematical models [8]. Therefore, the fuzzy comprehensive evaluation method based on the fuzzy set can comprehensively evaluate the subordination level of the evaluated object from multiple indicators. It divides the change range of the evaluated object. On the one hand, it can take into account the hierarchy of the object, so that the fuzziness of the evaluation criteria and influencing factors can be reflected; On the other hand, human experience can be fully used in the evaluation to make the evaluation results more objective and consistent with the actual situation. Fuzzy comprehensive evaluation can combine qualitative and quantitative factors, expand the amount of information, improve the degree of evaluation, and make the evaluation conclusion credible [9].

### 2.2 Characteristics of Logistics Park Construction Project

Logistics Park refers to a place where multiple logistics facilities and different types of logistics enterprises are centrally arranged in space in areas where logistics operations are concentrated and where several modes of transportation are connected. It is also a gathering point for logistics enterprises with a certain scale and multiple service functions [10].

1) Multiplicity of construction objectives. The objectives of general construction projects mainly focus on the use value and economic value of the project service itself. Different from general construction projects, the logistics park has a public welfare attribute, which leads to the project objectives of the logistics park not only focusing on the economic benefits of the construction of the logistics park itself, but also focusing on the social effectiveness of the development and construction and the driving effect of the construction of the logistics park on the local economy. In short, the logistics park project should pay attention to both spectator benefits and macro benefits [11].

2) Diversity of construction contents. The logistics park project is different from the single objective logistics facility construction (such as warehouse) and transportation road construction projects. As a logistics park construction project at a certain level, for example, in domestic and international logistics services, the construction content of the logistics park not only corresponds to the basic logistics functional elements of logistics operators such as transportation, warehousing, packaging, distribution, loading and unloading, but also includes the development and construction of diversified logistics infrastructure and logistics service facilities such as customs declaration, inspection, catering, accommodation, maintenance, transportation, finance, insurance, so on and so forth [12].

### 2.3 Risk Characteristics of Logistics Park Construction Projects

In addition to the typical characteristics of general construction project risks, i.e., wide sources, long development cycle, dynamics, objectivity, regularity, relativity and uncertainty, the logistics park construction project risks also show the following significant characteristics:

1) Diversity and multilevel of risks. The logistics park construction project has the characteristics of large investment, long development cycle, many uncertain factors and complex types, and the cross impression between a large number of risk factors and the outside world makes the risk present multi-level.
2) One-way transitivity. In the $m$ stages of logistics park project construction, the risk impact has a one-way transmission from front to back in time / stage, and the risk factors in the previous stage will have an impact on the subsequent stages, that is, the

risk has a one-way transmission from front to back. Generally, for example, the risk factors in the $m$ stages of logistics park construction are $U_1, U_2, \cdots U_m$, and the risk impacts of each stage are $R_1, R_2, \ldots R_m$, then the one-way transmission of project construction risk can be expressed as

$$R_1 = f_1(U_1), R_2 = f_2(U_2), \cdots, R_m = f_m(U_1, U_2, \cdots U_m)$$

3) Environmental adaptability. Every logistics park construction project exists in a certain environment, and exchanges logistics, information and capital flow with the external environment. The same type of logistics park construction project risk in different project construction environment, the impact of environmental factors is also different.
4) Policy impact. Policy risks mainly include industrial development policy, land policy, investment policy, fiscal and tax policy and monetary policy, which may have an adverse impact on the investment and construction environment of the logistics park.

## 3 Analysis on Risk Factors of Port a Logistics Park Construction

### 3.1 Analysis on the Environment of Port a Logistics System

1) Shoreline resources. Port A planning area includes Hanjiang River and Yangtze River. The conditions of the Yangtze River channel are relatively superior. The Wuhan Shanghai section of the Yangtze River channel is a first-class channel with a deeper water depth than the upstream section of the Wuhan Yangtze River Bridge. 5000-ton ships can enter and leave the port all year round, ensuring the normal navigation of river sea direct ships throughout the year. The "river sea direct" route can be opened normally and the "river sea combined transport" can be realized.
2) Collecting and distributing network resources. The collection and distribution network resources of Wuhan Newport are mainly composed of three levels. The first level is the general framework composed of the planned "two rivers, four rings, four railways and nine shoots". The second level is composed of some expressways and expressways connected with each other. The third level is the traffic structure composed of the traffic in the port. The third level constitutes the collection and distribution network system of Wuhan Newport [13].

### 3.2 Analysis on Risk Factors of Logistics Park Construction

The main contents of the investment, construction and operation stages of port A logistics park project are analyzed, as shown in Fig. 1.

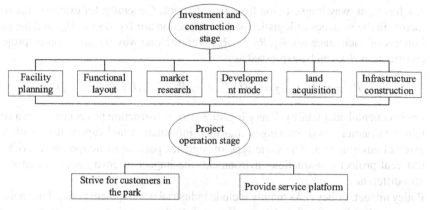

**Fig. 1.** Phase tasks of logistics park

Many risk factors are involved in the whole implementation cycle of port A logistics park project. Through the analysis of the risks and risk nature of each stage of the project, it is summarized that the main investment risks of the port A logistics park project are as follows:

Technical risks. In the two stages of port A logistics park project implementation, risks will be caused due to the uncertainty of relevant technical factors. For example, the rationality of site selection, the suitability of scale, the rationality of planning and layout, the suitability of functional design, the advanced applicability of technology, etc.

Market risks. It is also the most important risk faced by port A logistics park after its operation. Port A logistics park needs to objectively investigate and analyze the logistics demand in Central China and even the entire economic hinterland, the status of market competitors and the current market situation of logistics services, and analyze whether the potential of the future logistics market matches the scale and service level of the project [14].

Project management risks. For the port A logistics park project, the factors that affect the management risk mainly include: the rationality of the management organization and division of labor, the soundness of the quality safety system, the coordination of public relations, the selection of contracting mode and the management ability of the management personnel.

Economic and financial risks. For the port A logistics park project, the investment scale is huge, the financing channels are diverse, and the construction cycle is long. The land acquired by the project, the project infrastructure construction and the follow-up operation of the project all need a lot of funds. Therefore, fund raising is the biggest factor affecting the economic and financial risks of the project.

Environmental risks. In the investment, construction and operation stages of port A logistics park project, many factors will affect the normal implementation of the park project, such as the port climate, the national political and economic environment, changes in the microeconomic environment [15].

## 4 Evaluation of Port a Logistics Park Construction Based on Fuzzy Comprehensive Method

### 4.1 Construction of Risk Evaluation Index System for Logistics Park

According to the characteristics of port A logistics park project with multi-level risk factors and many uncertain factors, the fuzzy comprehensive evaluation method is selected as the risk evaluation method of port A logistics park construction project. At the same time, the risk grade of port A logistics park project risk and the weight of each influencing factor are determined through expert investigation.

Through the analysis of the implementation stage and risk identification of the port A logistics park project, it can be seen that the port A logistics park project is divided into two stages, and the risk factors are different in different stages. The specific evaluation index system is shown in Fig. 2.

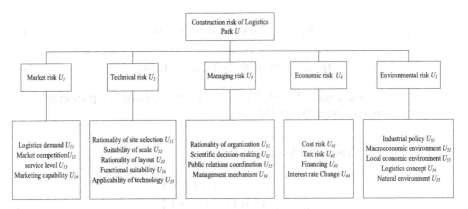

**Fig. 2.** Schematic diagram of risk assessment index system

### 4.2 Establish Indicators and Evaluation System

To make the students become the "protagonist" in the class, first of all, let the students look at the major and study from a different perspective, let the students clearly know what to learn and how to use it? Driven by the project, with the international logistics operation process as the core and team cooperation as the necessary form, students are guided to in-depth practice and form the project results in the course. Integrate the course achievements with the enterprise needs to achieve the "conversion agent" of learning style.

1) Define the main factor layer indicator set as $U: (U_1, U_2, U_3, U_4, U_5)$, and its corresponding weight set as $W: (W_1, W_2, W_3, W_4, W_5)$, where $W_i$ represents the proportion of indicator $U_i$ in $U$.

2) Define the sub factor layer indicator set as $U_i = (U_{i1}, U_{i2},\ldots, U_{ij})$, and the corresponding weight is $W: (W_{i1}, W_{i2},\ldots, W_{ij})$, where $W_{ij}$ represents the proportion of indicator $U_{ij}$ in $U_i$.

The expert investigation method is used to judge the importance of each risk factor, and the weights of the primary and secondary risk factor indicators are as follows:

$$W = (0.20, 0.20, 0.10, 0.30, 0.20) \tag{1}$$

$$W_1 = (0.30, 0.30, 0.20, 0.20) \tag{2}$$

$$W_2 = (0.30, 0.30, 0.20, 0.20) \tag{3}$$

$$W_3 = (0.20, 0.30, 0.30, 0.20) \tag{4}$$

$$W_4 = (0.30, 0.30, 0.30, 0.20) \tag{5}$$

$$W_5 = (0.20, 0.20, 0.25, 0.20, 0.15) \tag{6}$$

Define the comment set as $V: (V1, V2,\ldots, V5)$, which respectively represent low risk, low risk, medium risk, high risk and high risk.

### 4.3 Effect Fuzzy Evaluation Matrix

$$R_1 = \begin{bmatrix} 0.3 & 0.3 & 0.2 & 0.1 & 0.1 \\ 0.3 & 0.4 & 0.2 & 0.1 & 0.0 \\ 0.4 & 0.4 & 0.1 & 0.1 & 0.0 \\ 0.4 & 0.3 & 0.2 & 0.1 & 0.0 \end{bmatrix} \tag{7}$$

$$R_2 = \begin{bmatrix} 0.3 & 0.2 & 0.2 & 0.2 & 0.1 \\ 0.3 & 0.3 & 0.2 & 0.2 & 0.0 \\ 0.3 & 0.4 & 0.1 & 0.1 & 0.1 \\ 0.4 & 0.3 & 0.3 & 0.0 & 0.0 \end{bmatrix} \tag{8}$$

$$R_3 = \begin{bmatrix} 0.3 & 0.4 & 0.2 & 0.1 & 0.0 \\ 0.3 & 0.3 & 0.2 & 0.1 & 0.1 \\ 0.4 & 0.4 & 0.2 & 0.0 & 0.0 \\ 0.3 & 0.3 & 0.2 & 0.1 & 0.1 \end{bmatrix} \tag{9}$$

$$R_4 = \begin{bmatrix} 0.4 & 0.3 & 0.2 & 0.1 & 0.0 \\ 0.3 & 0.3 & 0.2 & 0.2 & 0.0 \\ 0.3 & 0.2 & 0.3 & 0.1 & 0.1 \\ 0.4 & 0.3 & 0.2 & 0.1 & 0.0 \end{bmatrix} \tag{10}$$

$$R_5 = \begin{bmatrix} 0.3 & 0.4 & 0.2 & 0.1 & 0.0 \\ 0.2 & 0.4 & 0.3 & 0.1 & 0.0 \\ 0.3 & 0.3 & 0.3 & 0.1 & 0.0 \\ 0.3 & 0.2 & 0.3 & 0.1 & 0.1 \\ 0.2 & 0.3 & 0.3 & 0.2 & 0.0 \end{bmatrix} \quad (11)$$

### 4.4 Fuzzy Evaluation Calculation

$$B_i = W_i \times R_i = (b_{i1}, b_{i2}, \ldots b_{i5})$$
$$B = (W_1, W_2, W_3, W_4, W_5) \times (B_1, B_2, B_3, B_4, B_5)^T$$
$$= (0.321, 0.311, 0.214, 0.215, 0.119, 0.031)$$

According to the principle of maximum risk subordination, the comment $V_1$ represented by 0.322 - low risk is the overall risk level of the logistics park project. As the overall risk is at a low level, the project can be implemented from the risk perspective.

## 5 Optimization of Port a Logistics Park Construction

### 5.1 Improve the Service Mode of Diversified Logistics Parks

The logistics park has continuously improved its service capacity and provided basic supporting services such as office, catering, property, parking, accommodation, industry and commerce, taxation, etc. for the settled enterprises. Some logistics parks have extended the service chain and provided value-added services such as logistics consulting, logistics finance, commodity display, facility leasing, insurance agency, etc. for the settled enterprises. The service categories are increasingly rich, becoming a new growth point of the park. By improving the standardization of logistics operations, reducing the cost of logistics services and striving to improve the level of logistics services can enterprises improve their competitiveness in the logistics market with high-quality services and reasonable service prices.

### 5.2 The Logistics Park is Upgraded to Intelligence

Increase the establishment of the information platform of Port A Logistics Park, provide basic services such as information release, goods tracking, data exchange, property management, and improve the informatization level of the park. Some parks have developed business auxiliary functions such as capacity trading, payment and settlement, financing insurance, credit management, etc. in combination with business needs to accelerate the digital development of the park. With the development of equipment technology, the integration of new generation communication technologies such as the Internet of Things, big data, cloud computing, artificial intelligence and logistics parks is the main feature of the development of logistics parks at present.

## 5.3 Actively Seek Government Support

The introduction of the national logistics development plan and policy will clarify the investment orientation and support for logistics infrastructure projects and enterprises, including logistics parks, and will be more beneficial to the development of logistics parks that meet the planning requirements or are included in the government planning. For example, preferential land tax policies, supporting infrastructure construction, and some construction funds. The active support of the government can effectively reduce the construction cost, reduce the financing difficulty, facilitate the coordination between various departments, and reduce the risk of the planning and implementation of the logistics park to a certain extent [16].

## 6 Conclusion

Logistics parks play an increasingly important role in improving the efficiency of logistics services, promoting the adjustment of industrial structure, transforming the mode of economic development, and serving the national development strategy. As a distribution center for inbound and outbound goods and a logistics platform for the procurement and distribution of large manufacturers, the logistics park occupies a core position in the entire logistics chain.

The construction of logistics park is characterized by large investment, long development cycle and many uncertain factors. Therefore, the risk of logistics park construction project is high. Taking Port-A Logistics Park project as an example, this paper analyzes various risks and related factors in the construction of the logistics park, establishes a risk evaluation index system, analyzes the overall system, evaluates the risk level of Port A Logistics Park through fuzzy comprehensive evaluation method, and puts forward corresponding optimization strategies.

**Acknowledgment.** This research was funded by Industry-education integration curriculum project of Wuhan Technology and Business University.

## References

1. Nan, L., Zhao, J., Lyu, M.: Exploration of land development intensity index of port container Logistics Park based on quantitative algorithm and pent analysis method. Pol. Marit. Res. **25**(3), 61–67 (2018)
2. He, Q.-M., Ke, J., Dong, S., Wang, X., Yuan, Z.: Simulation Modelling of State-dependent Queueing Network: impact of deepening on vessel traffic in Yangtze River Estuary. Adv. Mech. Eng. **11**(5), 1–12 (2019)
3. He, Q.-M., Zhang, H., Ye, Q.: An M/PH/K queue with constant impatient time. Math. Methods Oper. Res. **18**(11), 139–168 (2018)
4. Zhang, H., Zhao, X.: Analysis of organizational behavior of container shipping in the upper and middle reaches of the Yangtze River based on Hub-and-Spoke network. J. Coastal Res. **Sp**(73), 119–125 (2015)
5. Mainwaring, G., Olsen, T.O.: Long undersea tunnels: recognizing and overcoming the logistics of operation and construction. Engineering **4**(2), 249–253 (2018)

6. Guan, S.: Smart E-commerce logistics construction model based on big data analytics. J. Intell. Fuzzy Syst. **40**(2), 1–9 (2020)
7. Wu, Y., Zhang, T.: Risk Assessment of offshore wave-wind-solar-compressed air energy storage power plant through fuzzy comprehensive evaluation model. Energy **223**(5), 120057 (2021)
8. Zhang, H., He, Q.-M., Zhao, X.: Balancing herding and congestion in service systems: a queueing perspective. Inf. Syst. Oper. Res. **58**(3), 511–536 (2020)
9. Qian, J., Wu, J., Yao, L., et al.: Comprehensive performance evaluation of wind-solar-cchp system based on emergy analysis and multi-objective decision method. Energy **230**(190), 120779 (2021)
10. Liu, L.: Oil spill risk assessment of submarine oil pipeline based on fuzzy comprehensive evaluation and accounting (Retraction of vol 14, art no 1911, 2021). Arab. J. Geosci. **24**, 14 (2021)
11. Wang, J., Liu, S., Wang, S., et al.: Multiple indicators-based health diagnostics and prognostics for energy storage technologies using fuzzy comprehensive evaluation and improved multivariate grey model. IEEE Trans. Power Electron. **36**(11), 12309–12320 (2021)
12. Verbic, G., Keerthisinghe, C., Chapman, A.C.: A project-based cooperative approach to teaching sustainable energy systems. IEEE Trans. Educ. **60**(3), 221–228 (2017)
13. Fatima, S., Abdullah, S.: Improving teaching methodology in system analysis and design using problem based learning for ABET. Int. J. Mod. Educ. Comput. Sci. **5**(7), 60–68 (2013)
14. Kaviyarasi, R., Balasubramanian, T.: Exploring the high potential factors that affects students' academic performance. Int. J. Educ. Manag. Eng. (IJEME) **8**(6), 15–23 (2018)
15. Al-Hagery, M.A., Alzaid, M.A., Alharbi, T.S., et al.: Data mining methods for detecting the most significant factors affecting students' performance. Int. J. Inf. Technol. Comput. Sci. **12**(5), 1–13 (2020)
16. Zhiqin, L., Jianguo, F., Fang, W., Xin, D.: Study on higher education service quality based on student perception. Int. J. Educ. Manage. Eng. **2**(4), 22–27 (2012)

# Advances in Technological and Educational Approaches

# OBE Oriented Teaching Reform and Practice of Logistics Information System Under the Background of Emerging Engineering Education

Yanhui Liu[1], Jinxiang Lian[2], Xiaoguang Zhou[2(✉)], and Liang Fang[2]

[1] Department of Automation, Century College,
Beijing University of Posts and Telecommunications, Beijing 102101, China
[2] School of Automation,
Beijing University of Posts and Telecommunications, Beijing 100876, China
zxg@bupt.edu.cn

**Abstract.** The construction of Emerging Engineering Education is a major action plan to continuously deepen the reform of engineering education to respond the challenges of the new economy, meet the needs of the industry and face the future development. It has the characteristics of reflecting the feature of the times, new and rich connotation, multi-disciplinary integration, multiplex subject participation, and wide coverage. Under the background of Emerging Engineering Education construction, guided by the OBE teaching concept and taking the logistics information system course as an example, this paper has conducted in-depth research and practical innovation from the aspects of courses' content, teaching methods, assessment ways, and the analysis of the degree of reaching the curriculum objectives, which reflects the multidisciplinary cross integration curriculum design idea of focusing on students and facing industrial needs and capacity training. The research of this paper has a certain reference significance for the training of compound and applied talents and the reform of curriculum teaching in relevant majors of colleges and universities.

**Keywords:** Emerging Engineering Education · OBE · Logistics information system · reform in education

## 1 Introduction

Engineering education is an important task to promote national development [1]. For the purpose of actively responding to a new round of scientific and technological revolution and industrial reform, and promoting the reform and innovation of engineering education, the Ministry of Education of China began to actively promote the construction of Emerging Engineering Education (EEE) disciplines in 2017 [2, 3]. It is a response to China's implementation of major national strategies such as innovation driven development, "Made in China 2025", "Internet plus", "Network Power", and "the Belt and Road", and is also a strong driver of China's engineering education reform [4, 5].

One of the core concepts of China's engineering education reform is outcome based education (OBE for short), which mainly follows the principle of reverse design [6–8]. Starting from the needs (including internal and external needs) setting which is the first step; the second step is to determine the training objectives; the third step is to establish the graduation requirements; and then fix on the curriculum system according to the graduation requirements. Finally, decide the course content and the class hours of the teaching content according to the contribution of a specific course to achieving the graduation requirements [9–12].

This paper takes the logistics information system course as the research object, combined with the OBE education concept, aiming at the three major problems of teachers' teaching, students' learning and knowledge application difficulties [13]. Focusing on project teaching, our teaching practice and innovation are carried out from five aspects, namely: teaching objectives, teaching content, teaching strategies, evaluation system and ability achievement, so as to effectively improve the teaching quality of our application-oriented undergraduate professional education.

## 2 Reform Objectives and Ideas of Course Teaching

### 2.1 Ideas for Curriculum Teaching Reform

The course Logistics Information System is one of the core course of logistics specialty in colleges and universities, which has the characteristics of multiple contents, wide coverage and complex knowledge structure. The difficulties of teachers' teaching are mainly reflected in the selection of teaching materials and the design of teaching contents; The trouble s of students' learning are mainly reflected in the unclear learning objectives, low learning initiative, and lack of engineering thinking and system thinking; The difficulty in comprehensive application is mainly reflected in the weak ability of students in comprehensive application of single or multiple professional courses [14]. Considering the above problems and the OBE concept, combined with the actual needs of enterprises, this paper introduces project teaching, carries out the project team as the teaching object, takes a single semester as the teaching cycle, forms a project based through and full cycle teaching management, and carries out feedback and improvement through teaching practice.

The basic idea of curriculum construction is shown in Fig. 1. Under the guidance of OBE concept, the training objectives for college students are formulated and graduation requirements are clarified. On this basis, carry out curriculum construction. The main problems to be solved are "why to learn", "what to learn", "how to learn", "how to evaluate" and "how to improve". The contents involved include curriculum objectives, content reconstruction, teaching organization, curriculum evaluation and degree of achievement analysis. In the meantime, the quality and effect of curriculum construction need to be constantly improved through information feedback. In the course construction, the integration between schools and enterprises is also needed to achieve the transformation of learning achievements.

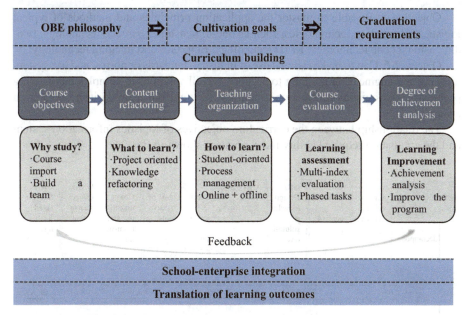

**Fig. 1.** Logistics information system course construction ideas

### 2.2 Determination of Course Teaching Objectives

Under the guidance of OBE concept, the curriculum teaching objectives are determined according to the degree of contribution of the curriculum to the graduation required ability index points [15, 16]. Those ability index points with a contribution of 50% or more are selected as the objectives of the course teaching [17, 18]. The teaching of logistics information system course has the following six specific objectives.

Objective 1: Have ability to apply mathematics, computer, engineering foundation and professional knowledge to the analysis of complex engineering problems of logistics information system;

Objective 2: Be capable to master certain methods of literature search and investigation and analysis, identify, analyze and express complex problems in the logistics information system by using relevant basic engineering knowledge and engineering technology methods, and have the ability to learn independently;

Objective 3: Have a certain awareness and ability of project management, be able to effectively carry out project communication and team cooperation, and complete project progress management, scope management, cost management, quality management and risk management;

Objective 4: Be familiar with the business and data processing process of typical logistics information systems such as order management, procurement management, warehousing management, distribution management and transportation management, and be able to use software development technology, database technology, prototype design and other auxiliary tools to design and develop typical logistics information systems;

Objective 5: Be able to master the application principles and methods of typical logistics information technologies, such as one-dimensional barcode technology, two-dimensional code technology, RFID technology, EDI, GIS and tracking technology;

Objective 6: Can to make rational use of industry standards or norms, and use system thinking and engineering thinking to propose overall solutions to complex engineering problems in the logistics information system, and have a strong team spirit and innovation awareness.

The relationship between the curriculum objectives of logistics information system and the graduation required ability index points is shown in Fig. 2.

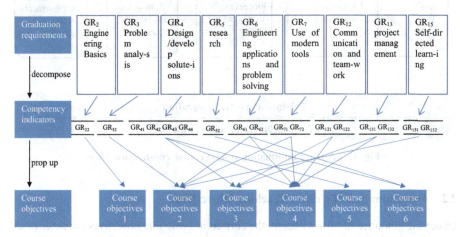

**Fig. 2.** The relationship between curriculum objectives and graduation requirements

In which, $GR_i$ is the graduation requirement, $GR_{ij}$ refers to the ability index points required for each graduation, which are described as follows:

$GR_{22}$: Be able to apply mathematics, computer, engineering foundation and professional knowledge to the analysis and interpretation of complex engineering problems of logistics information system;

$GR_{32}$: Be able to identify the main links and key factors in the logistics system, and conduct qualitative and quantitative analysis of the survey results with the help of commonly used analysis tools;

$GR_{41}$: Have a certain project management awareness and ability, be able to effectively communicate, cooperate and motivate the team, and implement quantitative assessment of the project process;

$GR_{42}$: Be familiar with the business and data processing process of typical logistics systems, such as procurement management, warehousing management, distribution management, transportation management, and carry out project planning and implementation according to specific objects;

$GR_{43}$: Master the application principles and methods of typical logistics technologies in the logistics system, such as one-dimensional barcode technology, two-dimensional code technology, RFID technology, EDI, GIS and some tracking technologies;

$GR_{44}$: Be able to assist the design and development of typical logistics information systems with the help of commonly used tools such as prototype design, database design and process drawing;

$GR_{52}$: Propose reasonable and effective solutions based on the actual needs of the enterprise and taking into account the operability, implement ability, application value and other indicators;

$GR_{61}$: Have systematic and innovative thinking, and be able to analyze and solve problems from multiple perspectives such as project management, project planning and technology development;

$GR_{62}$: In combination with the actual needs of the enterprise, considering the operability, implementability and application value of the scheme, propose an integrated logistics information system solution for complex problems;

$GR_{71}$: Use EXCEL, SPSS and other statistical analysis tools or some big data analysis tools to make statistics, analysis and prediction on complex engineering problems;

$GR_{72}$: Make scientific and rational use of network resources, and reserve some commonly used technical tools and platform resources;

$GR_{121}$: Establish communication awareness, master communication skills, and be able to communicate and report effectively with peers or the public by modern means;

$GR_{122}$: Through team activities, we will improve our ability to collate data, organize words, write documents and make PPT for summary and report.

$GR_{131}$: Be able to master the basic principles and methods of project management;

$GR_{132}$: Be able to complete the project planning and implementation under the interdisciplinary environment as a team member or leader;

$GR_{151}$: Establish the awareness of lifelong learning and explore the methods of lifelong learning;

$GR_{152}$: Develop goals and plans for independent learning to adapt to long-term career development.

## 3 Content Reconstruction and Organization Implementation of Curriculum Teaching

### 3.1 Reconstruction of Course Teaching Content

In view of the characteristics of multi-disciplinary intersection and integration emphasized in the construction of EEE courses, this paper focuses on project teaching and constructs systematic theoretical teaching content and continuous practical teaching content according to the course teaching objectives.

In the design of theoretical teaching content, the process management of the project is added, including the project application, scope management plan, schedule management plan, cost management plan, quality management plan and risk management plan; The design of practical teaching content focuses on comprehensive and designed experimental projects, including feasibility analysis, project planning, system analysis and system design. This combination of in class experiments and project teaching not only reflects the integration of teaching content design, but also reflects the project oriented, all cycle teaching management concept. It is helpful to cultivate students' systematic thinking and engineering practice ability.

## 3.2 Organization and Implementation of Teaching Process

Figure 3 shows the arrangement of key teaching links of logistics information system. The teaching process of logistics information system is carried out in the form of a project team. A total of 18 weeks from the beginning to the end of the semester are required to complete the phased tasks. The class hours include two parts: explicit class hours and implicit class hours. Teachers use explicit class hours to explain, comment, supervise and guide, and students use implicit class hours to investigate, plan, analyze and design. Through the organization and implementation of this teaching process, the participation and initiative of students majoring in logistics engineering in practical teaching have been significantly enhanced, and the quality of learning achievements has also been significantly improved, which truly reflects the student-centered education and teaching philosophy.

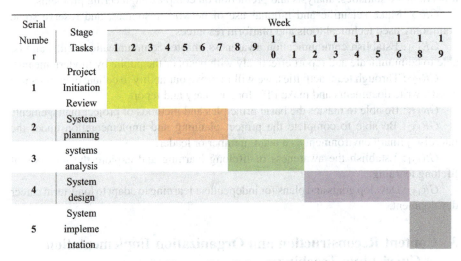

**Fig. 3.** Arrangement of key teaching links of logistics information system

## 4 Process Based Multi Indicator Comprehensive Course Evaluation and Analysis of Course Goal Achievement

The logistics information system curriculum has reformed a single outcome evaluation, built a multi indicator assessment mechanism under phased tasks with process management, and formed a process assessment file. The specific curriculum evaluation indicators are shown in Fig. 4.

The evaluation indicators mentioned above are classified according to the teaching links to obtain the supporting relationship and assessment proportion between the curriculum objectives and teaching links, as shown in Table 1.

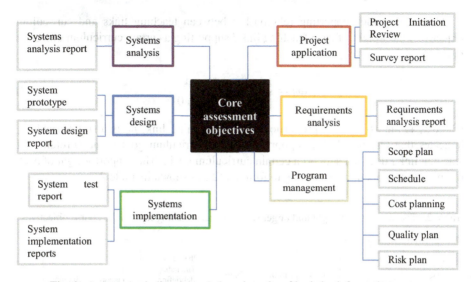

**Fig. 4.** Assessment indicators and phased results of logistics information system

**Table 1.** The supporting relationship between curriculum objectives and teaching links and the assessment ratio.

| Course objectives \ assess prop up | Classroom performance (5%) | Home-work (5%) | Experiment (15%) | Phased tasks (including debriefing, discussion and reporting) (30%) | Team collaboration (5%) | Big jobs (40%) |
|---|---|---|---|---|---|---|
| Objective 1 |  | √ | √ | √ |  | √ |
| Objective 2 |  | √ | √ | √ |  | √ |
| Objective 3 |  |  | √ | √ |  | √ |
| Objective 4 |  | √ | √ | √ |  | √ |
| Objective 5 |  | √ |  | √ |  | √ |
| Objective 6 |  | √ | √ | √ |  | √ |
| Objective 7 | √ |  |  |  | √ | √ |

According to Table 1, the supporting weight of each teaching link to the curriculum goal and its corresponding goal score are calculated, as shown in Table 2. Assume that the sum of the supporting weights of each teaching link $j$ for a course objective $i$ is 1, and the calculation formula is as follows:

$$\sum_{j=1}^{n} \omega_{ij} = 1 \tag{1}$$

Considering the supporting relationship between teaching links and curriculum objectives, the weight of each teaching link $j$ supporting a certain curriculum objective $i$ is:

$$\omega_{ij} = \begin{cases} \dfrac{x_j}{\sum x_j p_{ij}}, & p_{ij} = 1 \\ 0, & p_{ij} = 0 \end{cases} \qquad (2)$$

where $x_j$ is the assessment proportion of each teaching link, $p_{ij}$ is a 0–1 variable. 1 indicates that teaching link $j$ supports a certain curriculum goal $i$, and 0 refers that teaching link $j$ does not support a certain curriculum goal $i$. The support weight of each teaching link is obtained according to formula (2), as shown in Table 2.

**Table 2.** The supporting weight and target score of each teaching link on the course objectives.

| Course objectives \ assess weight | Attendance (5%) | Homework (5%) | Experiment (15%) | Phased tasks (including debriefing, discussion and reporting) (30%) | Team collaboration (5%) | Big jobs (40%) |
|---|---|---|---|---|---|---|
| Objective 1 |  | 0.0556 | 0.1667 | 0.3333 |  | 0.4444 |
| Objective 2 |  | 0.0556 | 0.1667 | 0.3333 |  | 0.4444 |
| Objective 3 |  |  | 0.1765 | 0.3529 |  | 0.4706 |
| Objective 4 |  | 0.0556 | 0.1667 | 0.3333 |  | 0.4444 |
| Objective 5 |  | 0.0667 |  | 0.4 |  | 0.5333 |
| Objective 6 |  | 0.0556 | 0.1667 | 0.3333 |  | 0.4444 |
| Objective 7 | 0.1 |  |  |  | 0.1 | 0.8 |
| Target score | 5 | 5 | 15 | 30 | 5 | 40 |

According to the support weight of each teaching link $j$ to a certain curriculum goal $i$ in Table 2, the degree of achievement of the curriculum goal is calculated, and the calculation formula is:

$$d_i = \sum \frac{f_j \times \omega_{ij}}{F_j} \qquad (3)$$

where $d_i$ is the degree of achievement of the course objectives, $f_j$ is the actual score value obtained by students corresponding to teaching link $j$, $F_j$ is the target score of teaching link $j$.

If the assessment scores of each teaching link of a student's logistics information system are respectively 4, 3, 11, 28, 5 and 35, according to formula (3), it can be got that the degree of achievement of each goal is 0.8555, 0.8555, 0.8706, 0.8555, 0.8800 and 0.8555, and then compare the degree of achievement of each goal with the set qualified standard value.

If it is greater than the qualified standard value, it means that the course goal is achieved, if it is less than the qualified standard value, it implies that the learning objectives of the course have not been achieved.

This evaluation method is used in our teaching practice, and the assessment results obtained are reasonable, pertinent, consistent with the actual situation, and generally accepted by teachers and students.

## 5 Conclusion

The construction of EEE courses in colleges and universities, which has attracted much attention from the education and related industries, has gradually entered the implementation stage. Compared with the traditional engineering, the "EEE" emphasizes more on the practicality, intersection and comprehensiveness of the discipline. Different from the traditional engineering talents, the new economy in the future needs high-quality composite EEE talents with strong engineering practice ability, innovation ability and international competitiveness. This forces colleges and universities to adopt reverse design to continuously promote the reform of education and teaching.

Based on the OBE teaching concept, the research analyzes and studies the teaching content, teaching methods, teaching evaluation and goal achievement of the logistics information system course, and puts forward a project based teaching through and full cycle teaching management idea. The teaching content reflects the intersection and integration of multiple disciplines. The teaching method adopts an effective process management mechanism. The teaching evaluation has designed a multi indicator comprehensive evaluation system based on projects. The goal achievement degree analysis attaches the corresponding weight to each teaching link to evaluate the achievement of the curriculum goals. The teaching reform practice of the course has provided new ideas and laid a solid foundation for the training of engineering and technical talents in the new era.

**Acknowledgment.** This paper is supported by: (1) Project of Century College of Beijing University of Posts and Telecommunications: "Research and Practice of Logistics Engineering Construction Based on CDIO Engineering Education Mode" (JSKY-1604); (2) Project of the Teaching Steering Committee of Logistics Management and Engineering Specialty in Colleges and Universities of the Ministry of Education: "Integration and Teaching Practice of Applied Undergraduate Technical Courses - Taking Logistics Information System as an Example" (JZW2022038).

## References

1. Varekan, K., Renukadevi, S.: A review of knowledge management based career exploration system in engineering education. Int. J. Mod. Educ. Comput. Sci. **8**(1), 8–15 (2016)
2. Liu, K., Chen, T.: On the governance of new engineering education. China Univ. Teach. **64**(1), 37–41 (2020). (in Chinese)
3. Hadgraft, R.G., Kolmos, A., et al.: Emerging learning environments in engineering education. Australas. J. Eng. Educ. **5**(1), 3–16 (2020)

4. Xie, B., Wu, L.: New engineering education construction integrated with regional development: logic, challenge and approach. China Higher Educ. Res. **6**, 51–56 (2021). (in Chinese)
5. Korkmaz, Ö.: The effect of project-based cooperative studio studies on the basic electronics skills of students' cooperative learning and their attitudes. Int. J. Mod. Educ. Comput. Sci. (IJMECS) **10**(5), 1–8 (2018)
6. Xu, P., Tang, Q., Liu, X.: Teaching reform of OBE oriented "Mechanical System Design" under the background of new engineering. Heilongjiang Educ. (Theory and Practice) **01**, 91–92 (2022). (in Chinese)
7. Niu, G., Wei, W., Zhu, D., Tang, Y., Zhu, L.: Teaching reform and practice of storage and processing of fruits and vegetables curriculum based on OBE concept. Curriculum Teach. Methodol. **5**(9), 20–24 (2022)
8. Zhang, X.: The application of flipped classroom in the marketing course based on OBE concept. J. Educ. Res. Policies **4**(2), 56–58 (2022)
9. Zhang, W.: Design of capstone projects under CDIO mode. Int. Conf. Comput. Sci. Comput. Intell. (CSCI) **12**, 849–852 (2019)
10. Xing, H., Xu, Z.: Construction and application of project driven one body, two wings and three stages teaching mode based on OBE concept. Int. J. Educ. **10**(3), 29–37 (2022)
11. Luo, J., Xiao, S., Chen, L.: Research on the evaluation system of practical teaching quality based on OBE concept. Comput. Informatization Mech. Syst. **5**(1), 22–24 (2022)
12. Mallikarjuna, B., Sabharwal, M., Kumar, P.: Research on the teaching reform of Python programming curriculum based on the OBE-CDIO concept. Front. Educ. Res. **5**(11), 67–70 (2022)
13. Ma, C., Zhang, H., Zhang, S.: An inquiry into the teaching reform of programmingbasic curriculum based on OBE-CDIO concept. J. Higher Educ. **01**, 89–91 (2020)
14. Dong, J., Li, Q., Peng, K., Cui, J., Lu, Y.: Research on evaluation methods for achieving curriculum objectives in the certification of engineering education. Higher Sci. Educ. **4**, 121–125 (2019). (in Chinese)
15. Li, M., Hongmin, L., Wei, Y., Kun, Q., Mei, X.: Research on quantitative evaluation of course goal achievement based on engineering education professional certification. Sci. Technol. Vis. **14**, 72–74 (2021). (in Chinese)
16. Zhang, J.: Teaching design and implementation of design mode in the course of Java programming in higher vocational colleges. J. Phys. Conf. Ser. **1856**(1), 1–7 (2021)
17. Herala, A., Knutas, A., Vanhala, E., Kasurinen, J.: Experiences from video lectures in software engineering education. Int. J. Mod. Educ. Comput. Sci. (IJMECS) **9**(5), 17–26 (2017)
18. Faizi, A., Umar, M.S.: A conceptual framework for software engineering education: project based learning approach integrated with industrial collaboration. Int. J. Educ. Manag. Eng. (IJEME) **11**(5), 46–53 (2021)

# Teaching Practice of "Three Integration" Based on Chaoxing Learning Software – Taking the Course of "Complex Variable Function and Integral Transformation" as an Example

Huiting Lu[1] and Xiaozhe Yang[2(✉)]

[1] The Faculty of General Education, Nanning University, Nanning 530299, China
[2] The Faculty of Intelligent Manufacturing, Nanning University, Nanning 530299, China
970423706@qq.com

**Abstract.** It is necessary to explore more and more effective ways to cultivate talents with all-round development. "Three integrations", that is, integrating the elements of application, innovation and moral education into the teaching of the curriculum, is a method that has been gradually recognized, accepted and valued. Taking the course of "complex variable function and integral transformation" as an example, according to the characteristics of application-oriented undergraduate students, this paper uses the information-based teaching tools in the new era, that is, relying on the Chaoxing Learning Software, combining with the teaching conditions of smart classrooms, to optimize the structure of classroom teaching, implement the "three integration", highlight the cultivation of students' application ability and innovation ability, and help students establish correct values, To solve the "pain point" problem of low classroom activity, low ideological and political integration and low professional integration of this course. Through research, we can gain experience that is worth learning and promoting, and better serve the training of professional talents.

**Keywords:** Complex function and integral transformation · Innovative teaching · Chaoxing learning APP

## 1 Introduction

It is a highly demanding work involving a wide range of factors for colleges and universities to cultivate high-quality compound talents with a solid foundation [1–3]. Complex variable function and integral transformation is a public basic course (or public compulsory course) for science and engineering majors in colleges and universities. It is a prerequisite course for power electronic technology, signal and system, automatic control principle, signal analysis and image processing, intelligent control and other courses. The complex variable function has its specific object theory, function operation skills, theory and calculation method, which provides an important analytical method for mathematics and is a further extension of the real variable function. The integral transformation is widely used in other disciplines and engineering fields, especially in physics, automation technology, communication technology, etc. [4]. The course focuses on cultivating students' abstract thinking ability, spatial imagination, logical reasoning ability and scientific and technological ability, and is another important mathematics course after advanced mathematics [5–7]. However, the actual teaching hours of the complex variable function and integral transformation course in colleges and universities are compressed. Most colleges and universities have 48 class hours of this course, and some colleges and universities have compressed it to 32 class hours. However, the subject itself has abstract knowledge, various concepts, and strong theoretical nature. The calculation of many knowledge is relatively complex. Therefore, many engineering students have great difficulties in learning complex variable functions and integral transformations. If they want to truly master the subject and flexibly use knowledge to solve practical problems, it is a challenging task, which puts forward reform and innovation requirements for the teaching of the course [8].

At the same time, the construction of new engineering courses and the social demand for talents also make the school education work face new challenges [9–11]. In combination with the requirements of contemporary education for students' ideology and morality and overall quality, the continuous improvement of the curriculum is also an urgent task to be completed [12–14]. Teaching reform has therefore become a topic of increasing concern in the education sector [15–17]. In order to solve the "pain point" problem of low class activity, low ideological and political integration and low professional integration in the course of "Complex Variable Function and Integral Transformation", this paper explores the common problems encountered in the teaching of complex variable function and integral transformation [18, 19]. Through the implementation of "three integration" teaching based on the Chaoxing learning software, it explains the way to solve the problem, and shows some teaching results obtained in teaching practice, to represents the effectiveness of the teaching reform.

## 2 Three Major Problems in Course Teaching

In teaching practice, the problems commonly existing in classroom teaching mainly include the following three aspects.

### 2.1 Lack of Activity in Large Class Teaching

With the increase in the number of engineering majors and classes offered, it is difficult for the construction of the teaching team to follow up with the demand of teaching in the short term. The teaching of public courses in many schools has to be carried out in the form of large classes, which refer as a centralized teaching with a large number of students. This is the case with Nanning University, where the number of classes in which the complex variable function and integral transformation courses are taught exceeds 75. Large class teaching restricts the form of classroom teaching activities and makes it difficult to carry out effective discussions with the participation of teachers and students. Most of the teaching methods and means, such as group discussion teaching method, role playing and debate teaching method, are also difficult to operate in large class classes, which leads to the lack of innovation in teaching activities, The participation of students is not high, and the teachers give lectures in the form of full talk, which makes the classroom atmosphere dull. This situation needs to be solved urgently. Under the existing teaching resources, innovative teaching is also imminent if we want to achieve the innovation of large class teaching activities and improve the enthusiasm of students to participate in classroom activities.

## 2.2 Low Integration of Teaching Content and Ideological and Political Education

For many years, the integration of the teaching of college mathematics public foundation and ideological and political education elements is relatively low, mainly due to the following reasons:

(1) The teaching content of mathematics courses has the characteristics of strong theoretical and abstract content, and it is difficult to extract the elements that significantly include "moral education";
(2) The teaching methods used by teachers when introducing ideological and political elements are inappropriate and easy to copy mechanically, resulting in insufficient guidance of students' values and the lack of integration of ideological and political elements;
(3) The teaching team has a superficial understanding of the connotation of the ideological and political construction of the curriculum, does not combine the requirements of the new era, has few new cases, and lacks novelty and attractive highlights.

Taking the complex variable function and integral transformation course of Nanning University as an example, as the school's model course of ideological and political education, certain teaching results have been achieved in the early stage, but the integration of ideological and political education is still not enough. Some ideological and political elements are only "cut in" added to the teaching content in a relatively rigid way, and have not achieved the effect of moistening moral education. Therefore, it is possible to achieve better results only by deepening theoretical research, improving teaching methods, strengthening moral education construction and summarizing practical experience.

## 2.3 Mathematics Teaching and Professional Application Derailed

Under the background of professional certification of engineering education, it is required that the curriculum objectives be combined with the graduation indicators of the talent training program, which requires that the university mathematics curriculum be constructed in combination with the professional needs. Taking Nanning University as an example, complex variable function and integral transformation are offered to students majoring in electrical engineering and automation, communication engineering and robotics engineering. These majors focus on the knowledge points of the course, which is integral transformation, while the demand for conformal mapping is small. Therefore, the class schedule should focus on the second part of integral transformation, and combine the following professional knowledge to carry out teaching design. For engineering students, the tedious proof and deduction process in most mathematics courses can be omitted, but the teaching of basic concepts, basic theories and basic methods must be strengthened. Mathematical reasoning depends entirely on basic concepts. The basic concepts are not clear, and many contents are not understood at all, and basic methods cannot be mastered, let alone correct application. Because the LBL (Lesson-based Learning) model currently used is a knowledge-based teaching concept, which ignores the connection between mathematical knowledge and practical problems to a certain extent. In addition, most of the mathematics teachers are researchers in the direction of mathematics, and there are more theoretical studies, while the understanding of professional knowledge is not deep enough. For them, it is difficult to design the teaching content according to the professional situation, which leads to the derailment of mathematics teaching and professional application. Therefore, we need to think about how to realize the intersection and integration of public basic courses and professional basic courses through teaching reform. While consolidating students' mathematical foundation, we should also strengthen students' mathematical application ability and break the barriers between traditional disciplines.

## 3 Practice of Innovative Teaching

Faced with the "pain point" problem of low class activity, low ideological and political integration and low professional integration in the course of Complex Variable Function and Integral Transformation, the teaching team carried out innovative teaching practice in the teaching process of this course.

### 3.1 Innovative Thinking of "Three Integration"

The course teaching uses Chaoxing Learning APP, an online learning software, to complete the moral education process from three different perspectives through the innovative idea of "three integration", i.e., integrating application elements, innovation elements and ideological and political elements.

(1) The integration of application elements in the curriculum requires teaching to break the barriers between basic courses and professional courses, and organically integrate professional knowledge into the original knowledge;
(2) The integration of innovative elements in the curriculum requires the innovation of teaching content, which is based on and higher than the textbook, and is in line with the cutting-edge knowledge. It also requires the innovation of teaching mode to change the boring traditional classroom and make the classroom appear a new atmosphere;
(3) The integration of ideological and political elements in the curriculum requires the implementation of moral education objectives in the whole process of teaching, and the "salt" of moral education into the "soup" of the curriculum.

The innovative measures of "three integration" must be built on the premise of the original teaching objectives of the course. It is not necessary to mechanically copy the knowledge points in the professional courses and turn them into professional courses because of the need to integrate the application elements. Instead, it is necessary to explain the teaching contents according to the professional characteristics and cultivate the students' thinking of applying knowledge on the basis of the original knowledge; The emphasis on the integration of innovative elements should not deviate from the teaching content of the basic course, making the class become a paper report, and the teaching content is too difficult to exceed the requirements of the syllabus; Nor can it become a moral education classroom and deviate from the teaching objectives because of the consideration of ideological and political elements.

In short, the "three integration" is inseparable, and can be generative integration, embedded integration, or migration integration at any stage of the teaching process. They are an organic combination of mutual influence and promotion. Under the teaching idea of "three integration" based on Chaoxing Learning, the teaching team adhere to the teaching concept of "OBE", process the teaching content, cultivate and guide students with multiple teaching methods, improve teaching evaluation, assist the implementation of teaching methods, ensure teaching effects, and create efficient classroom. Figure 1 shows the basic idea of the project plan.

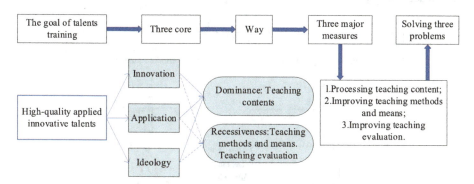

**Fig. 1.** Basic ideas of teaching practice

## 3.2 Processing Teaching Content

Complex variable function and integral transformation is a highly theoretical and abstract subject for engineering students in colleges and universities, and it is also a common mathematical tool to solve engineering problems. This requires that the course needs to integrate teaching content. Math teachers should communicate with professional teachers to learn, and put knowledge points into application elements to meet the requirements of engineering construction. In other words, the teaching content is based on the "service professional curriculum", adding the relevant professional application content in the corresponding knowledge points, expanding the breadth of content and digging the depth of knowledge, infiltrating professional ideas, not rigidly bound to textbook knowledge, solving the "pain point" problem of students' mathematical knowledge and professional technology application derailment. At the same time, through the integration with professional knowledge, the teaching content is reconstructed, and the ideological and political elements of the curriculum are added. Let's take a teaching section to show

how to integrate the "three elements" into teaching through the treatment of teaching content.

### 3.2.1 Add Innovative Elements

Take the Fourier change in the course as an example to see the integration of innovation elements. First, it is needed to explore the physical meaning of Fourier transform:

Example 1: Find the Fourier transform of rectangular pulse function $f(t) = \begin{cases} 1, & |t| \leq 0.5 \\ 0, & |t| > 0.5 \end{cases}$ and its amplitude spectrum and phase spectrum.

On the platform of Chaoxing Learning, teachers and students work together to solve the answer, and then compare with an example in the previous lesson: the function expression $f(t) = \begin{cases} 1, & |t| \leq 0.5 \\ 0, & |t| > 0.5 \end{cases}$ of a function with a period of 2 in a period is to find its Fourier series, as well as its amplitude spectrum and phase spectrum. Figure 2 is drawn accordingly.

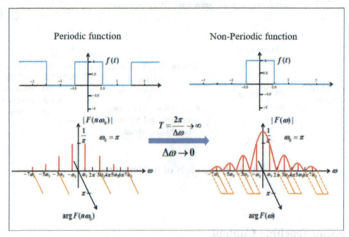

**Fig. 2.** Comparison between the spectrum of periodic function and that of non-periodic function

Integrate the original knowledge and methods with the current situation to form a new understanding and cognition. From concrete to special, help students better understand and master the physical meaning of Fourier transform, cultivate students' limit thinking in mathematics, and develop innovative thinking ability.

Then the continuous spectrum generation of a simple rectangular pulse signal function is shown in a dynamic diagram, as shown in the Fig. 3, so as to obtain the physical meaning of Fourier transform - the aperiodic function is composed of infinitely continuous harmonic waves of multiple frequencies.

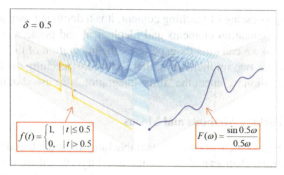

**Fig. 3.** Time domain and frequency domain of rectangular pulse signal function

The characteristics of innovation are reflected through the integration, expansion and extension of knowledge through the above methods.

### 3.2.2 Add Application Elements

According to the physical meaning of Fourier transform, the signal in the time domain can be converted into the frequency domain through this integral transform, so as to obtain the picture of amplitude and frequency correlation, as well as the picture of phase and frequency correlation. Although the signal is relatively intuitive in the time domain, it is not perfect enough, and it is difficult to study and process some problems, so the frequency domain is needed, which makes Fourier transform have a very broad role in signal processing. Here, we only study the Fourier transform of continuous aperiodic signals, while in engineering, Fourier transform has several forms shown in Table 1.

**Table 1.** Form of Fourier transform in engineering

| Time function | Frequency function |
| --- | --- |
| Continuous and aperiodic | Aperiodic and continuous |
| Continuous and periodic | Aperiodic and discrete |
| Discrete and aperiodic | Periodic and continuous |
| Discrete and periodic | Period and Discrete |

Expand knowledge, point out topics, arouse students' attention to learning, and pave the way for subsequent learning of relevant professional knowledge.

### 3.2.3 Add Ideological Elements

Add 5G-related news to the lecture. For example, China achieved full coverage of 5G at the 2022 Winter Olympics, reflecting China's leading position in 5G technology, and building students' national pride.

Through the processing of teaching content, it is a dominant way to integrate innovation elements, application elements and ideological and political elements into the teaching process, so we can directly experience the integration of these three elements. In addition, there are two implicit ways, that is, through teaching methods and means, and teaching evaluation to reflect the "three integration" discussed in this article.

### 3.3 Improve Teaching Methods and Means

Because the teaching content of complex variable function and integral transformation is difficult, the concept is abstract and the calculation is large, the efficiency of the classroom is very low before the teaching reform. In this teaching practice, under the teaching conditions of the smart classroom, the teaching mode shown in Fig. 4 is adopted by using digital teaching equipment and combining with Chaoxing Learning Communication software.

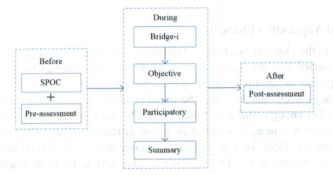

**Fig. 4.** Teaching mode of the course

SPOC teaching relies on the Chaoxing Learning Connect platform to effectively integrate the existing curriculum resources and teaching resources, put learning videos, learning materials, courseware documents and other resources in the Learning Connect resource library, provide students with an online platform to acquire knowledge, and guide students how to learn and what to learn. For example, teachers can carry out "task driven" teaching by using the "group task" function of Learning Pass to cultivate students' collaborative spirit, and use the "person selection" function of Learning Pass to carry out "discussion method" teaching, improve students' courage, improve students' communication and communication ability, and achieve the integration of "ideological and political elements"; Use the function of "classroom activities" to carry out "practice method" and "discussion method" teaching. The teacher throws questions on the platform, displays students' learning results, and encourages students to actively apply knowledge to solve problems, that is, "integrate application elements". At the same time, in the ideological collision with others, cultivate innovation ability, that is, integrate "innovation elements"; This "three integration" teaching method and means based on Chaoxing learning has greatly improved the participation and efficiency of classroom activities, and truly let students have a sense of gain.

The reform of teaching mode makes teaching activities dynamic and innovative. Large-class teaching with a large number of students can also have a highly participatory classroom, so that every student, even the students sitting in the last few rows, can obtain better learning results and make the classroom active. At the same time, the relevant data of the study conducted on the Chaoxing platform can be effectively saved, providing teachers with the original materials of the teaching process, and helping teachers better obtain the teaching feedback information.

### 3.4 Improve Teaching Evaluation Methods

Teaching evaluation is mainly about the evaluation of students' learning effect. The teaching evaluation will be diversified, the objective quantitative scoring and the subjective effect evaluation will be organically combined, and the students' cognition, emotion, values and other contents will be included in it. The traditional evaluation method will be combined with the novel evaluation method, the result evaluation and the process evaluation will be combined, and the traditional evaluation method of 30% of the usual performance plus 70% of the final performance will be retained. The innovation will be made in the evaluation of the ordinary performance, making the evaluation system more inclusive It is feasible and easy for students and teachers to accept, help students complete learning tasks, and solve the problems of low teaching effect and low participation in classroom activities encountered by teachers in teaching. Among them, the result evaluation is mainly based on the students' final paper performance, while the process evaluation is reflected in the evaluation criteria of their usual performance. Relying on the learning pass platform, independent pre-class preparation, participation in in-class activities, testing and after-class review will be recorded in the Chaoxing learning pass platform, and all will be given a certain score. The class score quantifies the participation and learning status of students in each activity, and determines the class performance, See Fig. 5 for details.

The process evaluation criteria are divided into homework scores, attendance scores and classroom performance scores, in which the classroom performance scores are further subdivided. The calculation methods of class points are as follows:

(1) Classroom practice: score (2–5 points) is set according to the difficulty of the exercise, and you can get points no matter whether you participate in it right or wrong;
(2) Ask questions: students who can ask constructive questions will be given 4 points;
(3) Answer: 3 points correct, 2 points partially correct, 0 point wrong;
(4) Select the person to answer the question: no matter right or wrong, you can get points as long as you participate;
(5) Group tasks: team leader 2–5 points, other members 1–4 points (score according to actual completion);
(6) Awards in the competition: according to the situation of students' mathematics awards, the first prize is 10 points, the second prize is 8 points, the third prize is 5 points, the first prize is 4 points, the second prize is 3 points, and the third prize is 2 points;

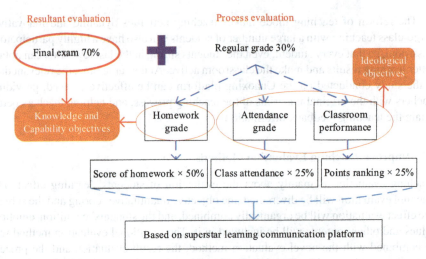

**Fig. 5.** Teaching evaluation structure

(7) Learning situation of general micro-class: self-study micro-class and score (1–5 points) according to the completion.

The results of the teaching evaluation reflect the growth of students, and systematically reflect the degree of combination of knowledge transfer and value guidance, realizing the improvement of teaching effect through scientific evaluation.

The results of teaching evaluation reflect the growth of students, and systematically reflect the combination of knowledge transfer and value guidance, which is conducive to the realization of the goal of improving teaching effectiveness through scientific evaluation.

## 4 Effect of Innovative Teaching

### 4.1 Improvement of Students' Learning Autonomy

Through the implementation of teaching reform and innovation in two demonstration classes (also known as experimental classes, or key class), students' learning autonomy and interest in learning mathematics have been greatly improved, and the "pain point" problem of low participation of students in large class teaching activities has been solved, which is mainly reflected in the active classroom atmosphere. The number of students with class score above 76 points is up to 86%. The numerical distribution automatically generated by Superstar Learning is shown in Fig. 6.

Teaching Practice of "Three Integration" Based on Chaoxing Learning Software 797

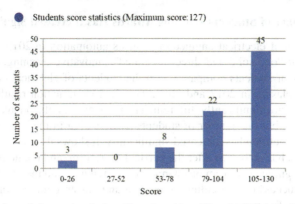

**Fig. 6.** Distribution of class points in key Class 1 and key Class 2 of Electrical Engineering and Automation of grade 2019

Two months after the end of the course in December 2020, the background of the course shows that there are still students who continue to learn the content of the course by watching the micro-class, which also reflects the goal of teaching innovation to cultivate students' lifelong learning ability. As shown in Fig. 7, there are still good learning records in February and March.

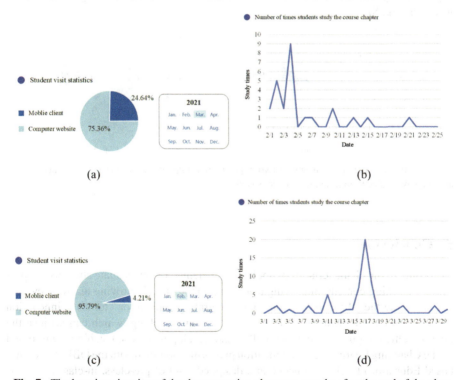

**Fig. 7.** The learning situation of the demonstration class two months after the end of the class

## 4.2 The Number of Students Winning Awards Has Increased Significantly

The four classes of electrical engineering and its automation in 2019, as the demonstration class of "one solid and three integrated" innovative teaching, have the same overall number of students compared with the students of electrical engineering and its automation classes in 2017 and 2018. The final examination results of advanced mathematics are not significantly different. The three classes of students have the same math foundation, but the number of students participating in the national college students' math competition has increased significantly, It shows that students' enthusiasm for mathematics has increased after the curriculum reform, and the number of winners has increased significantly, from one winner to two winners and then to eleven winners, reflecting the success of curriculum innovation and construction of complex function and integral transformation.

## 4.3 The Students' Learning Feedback is Constantly Praised

The questionnaire survey shows that the vast majority of students approve of innovative teaching, and more than 80% of students are well accepted (as shown in Fig. 5). The hybrid teaching launched by Chaoxing Learning Communication embodies the teaching concept of "student-centered". Students really learn knowledge from teaching, and feel the love and care of teachers. Finally, students should shape correct values from knowledge learning and realize the unity of value shaping and ability training in knowledge teaching (Fig. 8).

**Fig. 8.** Student acceptance survey of two experimental classes (light blue indicates acceptance and approval, dark blue refers to not accustomed)

## 5 Conclusion

According to the characteristics of application-oriented undergraduate students, by processing teaching content, improving teaching methods and improving teaching evaluation, promote the "three integration", i.e., integrating application elements, innovation elements and ideological and political elements into teaching, which can achieve the goal of cultivating more talents with all-round development. The teaching team, based on the Chaoxing Learning Communication platform, combined with the OBE (Outcome-based Education) teaching concept and, designed a set of pre-class, in-class and post-class teaching activities suitable for large class teaching, which effectively promoted the implementation of "three integration" teaching. On the whole, the teaching reform is relatively successful, and the research of the project has achieved more results. Teaching and learning are mutually beneficial. While students progress, teachers also get a lot of gains.

Some problems have also been encountered in the research. Because the teaching goal of "three integration" is a higher level goal, it is difficult for a few students with poor mathematical foundation to fully complete the two requirements of "application" and "innovation". Therefore, it is necessary to further consider how to help these students better master knowledge. In this teaching practice, it is a pity that due to the limited team energy and the lack of communication between the teachers of the school's interdisciplinary and inter-school, we have not further tracked the students' learning of professional knowledge after class. In the next new teaching reform practice, it is hopeful to introduce teachers of different professional courses, so that we can observe the students' learning situation and learning results on the whole front of talent cultivation.

**Acknowledgment.** This project is supported by the project of 2022 Nanning University Educational reform (2022XJJG22) References.

## References

1. Beltadze, G.N.: Game theory-basis of higher education and teaching organization. Int. J. Mod. Educ. Comput. Sci. (IJMECS) **8**(6), 41–49 (2016)
2. Singh, V., Dwivedi, S.K.: Two way question classification in higher education domain. Int. J. Mod. Educ. Comput. Sci. (IJMECS) **7**(9), 59–65 (2015)
3. Hamada, H.: Action research to enhance quality teaching. Arab World English J. Conf. Proceed. **1**, 4–12 (2019)
4. Liu, C.: Curriculum reform of complex function and integral transformation in the context of engineering education. Xueyuan **11**(18), 65–66 (2018). (in Chinese)
5. Zhang, Q., Li, H., Shi, K.: Exploration of teaching reform of complex variable function and integral transformation course for engineering majors. J. Higher Educ. **23**, 120–123+126 (2018) (in Chinese)
6. Yu, R., Chen, M., Fei, X.: Exploration of teaching reform of complex variable function and integral transformation in the context of new engineering. Educ. Modernization **7**(49), 34–36 (2020). (in Chinese)
7. Zhang, J.: On the importance of higher mathematics foundation in engineering basic education. Math. Learn. Res. **22**, 8 (2018). (in Chinese)
8. Meng, G., Zhao, H., Li, X.: The exploration of complex variable function and integral transformation in engineering teaching. For. Teach. **11**, 87–88 (2012). (in Chinese)
9. Ni, X.: Research on the integration path of general education and professional education under the OBE concept. Educ. Rev. **01**, 48–55 (2020). (in Chinese)
10. Yan, D., Xue, J., Zhang, X.: Exploring the construction path of gold courses in higher vocational colleges based on OBE concept. Educ. Teach. Forum **604**(01), 157–160 (2023). (in Chinese)
11. Lin, J.: China's new engineering construction facing the future. Tsinghua Univ. Educ. Res. **38**(02), 26–35 (2017). (in Chinese)
12. Lili, Y., Zhang, L., Fanbo, M.: Research on the cultivation of innovative talents in universities under the background of new engineering 2 Sci. Technol. Entrepreneurship Mon. **1**, 87–89 (2018). (in Chinese)

13. Zhan, J.: Learn the "melting" formula of ideological and political education in higher vocational courses. China Education News, 2022-11-29 (in Chinese)
14. Lin, X., Xu, S., Liu, Z., et al.: Analysis of curriculum teaching reform under the background of engineering education certification. J. Higher Educ. **8**(6), 136–138+142 (2022) (in Chinese)
15. Weiss, K.A., McDermott, M.A., Hand, B.: Characterising immersive argument-based inquiry learning environments in school-based education: a systematic literature review. Stud. Sci. Educ. **58**(1), 15–47 (2022)
16. Liu, Z., Fei, J., Wang, F., Deng, X.: Study on higher education service quality based on student perception. Int. J. Educ. Manag. Eng. **2**(4), 22–27 (2012)
17. Hui, C., Li, Z., Li, W., Mao, H.: Discussion on teaching pattern of cultivating engineering application talent of automation specialty. Int. J. Educ. Manag. Eng. **2**(11), 30–34 (2012)
18. Robles, A.C.M.O.: The use of educational web tools: an innovative technique in teacher education courses. Int. J. Mod. Educ. Comput. Sci. (IJMECS) **5**(2), 34–40 (2013)
19. Al-Hagery, M.A., Alzaid, M.A., Alharbi, T.S., Alhanaya, M.A.: Data mining methods for detecting the most significant factors affecting students' performance. Int. J. Inf. Technol. Comput. Sci. **12**(5), 1–13 (2020)

# Transformation and Innovation of E-Commerce Talent Training in the Era of Artificial Intelligence

Lifang Su[1] and Ke Liu[2(✉)]

[1] School of Economics and Management, Wuhan Railway Vocational College of Technology, Wuhan 430205, China
[2] School of Marxism, Wuhan Railway Vocational College of Technology, Wuhan 430205, China
591934757@qq.com

**Abstract.** Artificial intelligence technology has changed the form and structure of social business services, and the relationship between technology and labor tends to be diversified. E-commerce in the era of artificial intelligence has transformed into the integration of Internet-based digital algorithms and professional knowledge, which requires workers engaged in e-commerce to have diversified knowledge of cross-border integration. In order to meet the needs of economic and social development in the era of artificial intelligence, the cultivation of e-commerce talents should be closely combined with the market. At present, the cultivation mode of business talents in colleges and universities mainly adopts school-enterprise cooperation, combined with the national 1+X certificate system. Through the investigation of vocational education and training evaluation organization, we found that the current Chinese commercial talent training still stays in the shallow mode of single and directed cooperation, which is not conducive to the change and transformation of commercial talent in the era of artificial intelligence. Therefore, colleges and universities should change the concept of enterprise talent training, innovate and develop the school-enterprise cooperation system, improve the commercial talent training system, and realize the deep integration of production and education.

**Keywords:** Artificial intelligence · E-commerce · University-enterprise cooperation

## 1 Introduction

As a prominent form of technology after "Internet+", artificial intelligence is accompanied by "big data", "informatization" and "digital", "Economy", "platform economy", "intelligent manufacturing", "virtual world", "automation" and other keywords entered the academic and public vision. The emergence of new technologies has brought a new business model of information technology, big data technology, cloud technology, Internet of things technology and mining technology on the Internet as a platform [1]. In the intelligent era, traditional business must undergo digital transformation if it wants to develop, which requires workers to have cross-boundary integration and multi-compound knowledge structure and skill level, so as to promote the improvement of system innovation capability. In the era of intelligence, the commercial form has reshaped the new job demand of modern business, and the cultivation of commercial talents has attracted much attention [2, 3]. To cultivate talents for the country and society, colleges and universities need to adapt to the different needs of social development in different periods. At present, the training of business talents should conform to the upgrading of skills and diversified transformation in the intelligent era, clarify the goal of talent training, and constantly enrich China's high-level technical talents.

## 2 The Requirements for Business Talents in the Era of Artificial Intelligence and the Orientation of Their Training Transformation

### 2.1 Requirements for Business Talent in the Era of Artificial Intelligence

The cultivation of business talents in the intelligent era faces various reform missions. On the one hand, the integration of complex knowledge structure. The "fusion" characteristics of business talents in the era of artificial intelligence are shown in the intersection of disciplines and the fusion of knowledge, that is, traditional business knowledge is integrated into intelligent technology and new methods of skills, such as collaboration and negotiation skills, data mining and analysis technology, intelligent optimization technology, collaborative decision-making, new business model operation method, Internet of things knowledge, etc. [4]. On the other hand, mathematical literacy has become a new requirement for workers and consumers. Under the condition of digital economy, digital literacy becomes the ability that workers and consumers should possess. With the continuous improvement of digital technology, artificial intelligence technology has penetrated into various fields, and "digitalization" has been upgraded to "digital intelligence" [5]. As a result, workers increasingly need to have "dual" skills: digital and professional. Therefore, having higher intellectual literacy has become an important factor for workers to win in the job market [6].

### 2.2 Path Selection of Business Talent Training in the Era of Artificial Intelligence

In the cultivation of vocational ability, teaching, learning and practice are inseparable, especially related to the cultivation of numerically intelligent business professionals [7]. In the era of artificial intelligence, technology is developing rapidly. University teachers have high business theory knowledge, but they are not as good as enterprises closely connected with the market for new technology knowledge such as artificial intelligence and big data [8]. Based on this, it is an effective way to cultivate digital intelligent business talents by connecting with industrial reality and deepening school-enterprise cooperation.

The university-enterprise cooperation and production-education cooperative education have undergone three modes of evolution in China. The first model is a single, directed cooperation based on students' going out. This model starts from the early vocational education, which is characterized by student employment-oriented on-the-job practice [9]. Schools are active partners in cooperation, while enterprises usually do not have strong enthusiasm and initiative for cooperation because they have no initiative in the choice of internship time and interns. This model is not suitable for business majors in the intelligent era, because business majors cover loose and numerous enterprises and a certain enterprise cannot provide a large number of internship positions, resulting in low student management efficiency [10].The second mode is the platform mutual assistance and production-education docking introduced by enterprises, which is mainly reflected in the construction of training base "school in factory". However, the business in the era of artificial intelligence there is no need for large-scale production lines and hardware facilities. This mode of school-enterprise cooperation is only carried out through entrepreneurship incubation projects, which cannot realize the cultivate on of talents with multiple intelligence. Third, the mode of co-education, sharing and strategic alliance with deep integration of industry and education. This is reflected in the deep involvement of the enterprise in the talent training of the university, the integration of artificial intelligence, big data and other skills into the setting, curriculum standards and learning practices of business majors, and the absorption of opinions from enterprises and industry experts [11–13]. Schools and enterprises began to focus on the construction of "virtual full simulation, full process" business work scenes.

## 3 Influencing Factors of School-Enterprise Collaboration in Cultivating E-Commerce Talents in the Era of Artificial Intelligence

The orientation of application-oriented talents training in colleges and universities, combined with the cross-border knowledge accomplishment of e-commerce talents, requires that the training of business talents must rely on the market, industry and enterprise, and break through the single knowledge imparting of traditional education. The introduction of school-enterprise cooperative education mode is widely accepted by the commercial talents training of colleges and universities at the Age of Artificial Intelligence [14]. As new technologies such as artificial intelligence, automation and digital platforms change the skills of the service industry, skill-oriented scientific and technological growth and industrial transformation and upgrading in the intelligent era rely more on the optimization and upgrading of the labor skill structure [15]. These demands are driving the country's transformation of higher education [16]. Currently, China is actively promoting the 1+X certificate system, and the pilot vocational skills areas are oriented to 20 skills shortage areas such as modern agriculture, advanced manufacturing industry, modern service industry and strategic emerging industries, so as to promote school-enterprise cooperation and train high-quality workers and technical skills personnel [17]. The certificates related to modern business talents include: e-commerce data analysis, financial intelligence, financial robotics, logistics management, etc. These certificates are closely related to artificial intelligence technology. In order to accurately understand the influencing factors of the current school-enterprise collaborative education of e-commerce students, so as to construct an exact and effective education model of new business students, we conducted surveys on enterprises, higher vocational teachers and students through online and offline interviews and questionnaires [18, 19].

## 3.1 The Basic Situation of E-commerce Education in Higher Vocational Colleges

Enterprises are an indispensable part of the new business education in higher vocational colleges. The participation and initiative of enterprises in the school-enterprise cooperation are related to the effect of the new business education. In this research, the enterprise as an important theme research. The specific data are as follows:

### 3.1.1 Establishment Year of the Enterprise Carrying Out School-Enterprise Cooperation and Education in New Business

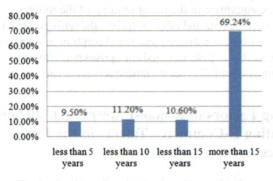

**Fig. 1.** Carrying out school-enterprise cooperation

As shown in Fig. 1, Research is carried out from the years of school-enterprise cooperative education, there are two distinct types of school-enterprise cooperation experience, one is with more than 15 years of school-enterprise cooperation experience, the other is with less than 15 years of school-enterprise cooperation experience. Specific data are as follows: Enterprises with less than five years of school-enterprise cooperation experience account for 9.5% of the respondents; Enterprises with less than ten years of experience in school-enterprise cooperation account for 11.2% of the respondents; Enterprises with less than 15 years of experience in school-enterprise cooperation account for 10.06% of the respondents; Enterprises with more than 15 years of school-enterprise cooperation experience account for 69.24% of the respondents。

According to the above data, the foundation of school-enterprise cooperation in Chinese colleges and universities is well. A large number of enterprises have established the school-enterprise cooperation model of business talent training with universities. However, we also find that most of the school-enterprise cooperation has more than 15 years of experience, and most of the school-enterprise cooperation carried out is a single superficial school-enterprise cooperation. How to use Internet technology to develop the school-enterprise cooperation mode of business talent training in the age of digital intelligence is the direction that needs to be explored.

### 3.1.2 Enterprises Participate in the Goal Work Integrated Education

Enterprise participation is the goal of cooperative education between New Business School and enterprises, among them, "provide skills standard services for the enterprise" accounted for up to 72.63%. The proportion of "reserving the skilled talents of the unit" and "expanding the influence of the unit in the industry" was 12.29% and 13.41% respectively. The proportion of profit-seeking was the smallest, only 1.68%. (As shown in Fig. 2).It can be seen that the purpose of most enterprises participating in school-enterprise collaborative education is to achieve their own sustainable development by improving their service capabilities and influence.

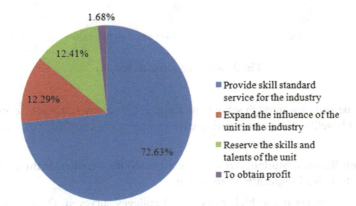

**Fig. 2.** The Goal of school-enterprise collaborative education

## 3.2 The Development of Cooperative Education between New Business Schools and Higher Vocational Colleges

### 3.2.1 Number of Teachers Who Have Received Corporate Training through School-Enterprise Cooperative Education in Recent Three Years

According to the survey data, in the school-enterprise collaborative education, in the past three years, 34.64% of teachers in higher vocational colleges have participated in enterprise training for five or more times, 6.7% for four times, and 13.41% for three times, as shown in Fig. 3.

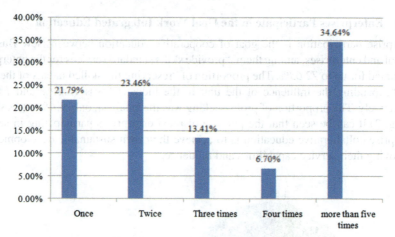

**Fig. 3.** Received corporate training

It can be seen that through school-enterprise collaborative education, teachers have significant benefits in acquiring cutting-edge technological knowledge of enterprises.

### 3.2.2 New Business Students Obtain Vocational Skills Certificates through School-Enterprise Cooperative Education

Among the students in the higher vocational colleges surveyed, the certificates they have obtained now account for 44.69% of the total, while enterprise certificates and other types of certificates account for 21.23% and 24.58% respectively.( As shown in Fig. 4)What is noteworthy is that 9.5% of students obtained skills certificates through off-campus training institutions. This reflects the need for further cooperation between higher vocational schools and enterprises in the collaborative education of new business.

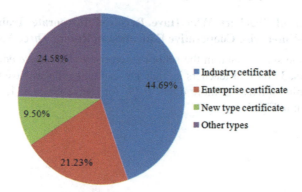

**Fig. 4.** Students obtain vocational skills certificates

### 3.2.3 Future Investment Plan of New Business School and Enterprise Collaborative Education

As shown in Fig. 5, among the surveyed schools and enterprises, 70.39% said they would increase the investment, 21.79% said they would large investment, 6.15% said they would maintain the current situation, and 1.68% said they would reduce the investment. It can be seen that the school-enterprise collaborative education model is recognized by most enterprises and vocational colleges. But at the same time, it can also be found that the enthusiasm of both sides of the new business school and enterprise collaborative education is not particularly high, which is the direction of the new business school and enterprise collaborative education needs to be improved.

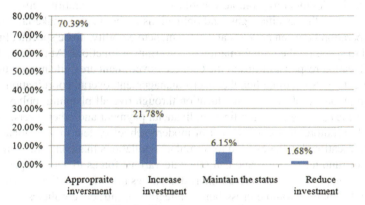

**Fig. 5.** Future investment plan

## 4 School-Enterprise Collaborative Education Strategy for E-Commerce Talents in the Era of Artificial Intelligence

### 4.1 Cultivate E-commerce Talents' Philosophical Thinking Under the Background of Artificial Intelligence

The innovation and development of colleges and universities cannot be separated from the support of enterprises in the industry and the shallow cooperation between colleges and enterprises can no longer meet the new requirements for the development of digitalized business. Under the background of the era of artificial intelligence, colleges and universities should establish the educational concept of student-centered development and cultivate business talents from various angles when training business talents. In the process of training, students' compound knowledge structure is built, including enterprise digital production, management and marketing, etc. Colleges and universities should provide students with a good learning environment, necessary network conditions, the construction of information wisdom classroom, the establishment of commercial virtual simulation training room, to provide a number of intelligent commercial training base [20]. At the same time, colleges and universities should establish the integration mode of production, education and research, build an educational ecosystem for the cultivation of e-commerce talents, promote the deep integration of colleges and industries, and make education supply and enterprise demand seamless.

### 4.2 The Systematic Construction of "Central-Local" Legislative Hierarchy of School-Enterprise Collaborative Education

To realize the integration of industry and education in vocational education, we need to overcome the disadvantages of management system and deal with the relationship between government, school and society, so as to promote the development of modern vocational education. In university-enterprise cooperation to promote vocational school, accelerate development to promote vocational education cooperation between colleges ordinance, clear the specific rights and obligations on both sides, make laws to education cooperation between colleges and more maneuverability. To be specific, first of all, we need to speed up the transformation of government functions. At present, the vocational education responsibilities of the Chinese government are quite numerous and the government needs to strengthen the macro-management, overall coordination and classification guidance of vocational education through overall planning, policy guidance and other means, as well as tax finance, financial payment and other levers. Secondly, we should promote the establishment of modern school system in vocational colleges [20, 21]. Vocational colleges need to formulate regulations that conform to the characteristics of running a school and can integrate various forces to actively participate in the management, so as to absorb all interested parties into the scientific management of the school. Finally, promote cross-border integration, introduce policies to encourage more "education-oriented enterprises" to participate in the development of vocational education, from tax incentives, direct financial subsidies, purchase services and other aspects, give qualified enterprises the legal status of "education-oriented" enterprises, vocational education organized by industrial enterprises into the vocational education system.

### 4.3 Improve the Social Training System and Realize Education and Social Equity

With the deep application and integration of the Internet in the economic field, a large number of skilled workers pour into the platform economy model. A large number of workers who rely on the platform economy, such as take-out workers, express delivery workers and ride-hailing drivers, are facing job replacement and skill crisis. Their jobs are highly mobile, highly substitutable, and repeatable. The big data algorithm logic applied by artificial intelligence technology requires certain theoretical knowledge and operation skills, and a grasp of the integrity of the production process, so as to improve the autonomy and freedom of platform economy work [22, 23]. In the context of the development of artificial intelligence, e-commerce talent training is not only a transformation within the scope of universities, but also should pay attention to the skill education and labor protection of labor under the platform economy, and establish a training system for skilled labor that is adapted to the development of intelligence and digitalization. China's 1+X certificate system needs to continue to explore and improve the skill training and vocational support system for low-end skilled labor, incorporate the platform economy labor skill education into the vocational education system, play its role as a social fairness regulator, which is conducive to ensuring educational fairness and social justice, and forming a learning society.

## 5 Conclusion

The transformation of economic structure and mode of production driven by skill-oriented scientific and technological progress needs to continuously improve the skills of workers and pay attention to the transformation and upgrading of skills of workers with alternative skills. To cope with the rapid change of industrial transformation and labor market skill demand in the smart era, the construction and transformation of higher education system are of great importance. Under the diversified differentiation pattern formed by skills in the intelligent era, higher education, as the main body of skill supply, should adhere to the training goal of compound talents with "thick foundation and wide caliber", and integrate new technology and traditional business theory. In the framework of macro system, the society and the university should form the comparative institutional advantage of the skilled society through system coordination and reform and innovation. Under the technological change with artificial intelligence as the core, higher education should be further integrated into the main economic battlefield of "artificial intelligence and industrial transformation", transform empowerment into internal motivation, serve the society, cultivate the intellectual skills of workers, and build a lifelong learning society.

## References

1. Abbasov, I.M.: Digital economy and digital education. Digit. Econ. Azerbaijan Int.Sci.-Pract. Conf. New Stage Econ. Develop. **6**(11), 5–9 (2020)
2. Nye, J.: The power we must not squander. New York Times **1**(1), 19 (2000)
3. UK Commission for Employment and Skills. Ambition 2020: world class skills and jobs for the UK (2009)
4. UNESCO Education strategy 2014–2021. UNESCO (2014)
5. Better Skills, Better Jobs, Better Lives: A strategic approach to skills policies. OECD (2012)
6. Department for Education, Department for business, innovation and skills. Rigor and responsiveness in skills 2013(4) (in Chinese)
7. Carnevale, A., Rose, S.J., Cheah, B.: The college payoff: education, occupations, lifetime earnings, p. 36. Georgetown University Center on Education and the Workforce, Washington DC (2011)
8. OECD: The Knowledge-Based Economy. Paris: Head of Publications (4) (1996)
9. Ouahbi, I., Darhmaoui, H., Kaddari, F.: Visual block-based programming for ICT training of prospective teachers in morocco. Int. J. Mod. Educ. Comput. Sci. **5**(10), 56–64 (2022)
10. Karim, M.A., Masnad, M., Ara, Y., Nandi, A.D.: A comprehensive study to investigate student performance in online education during COVID19. Int. J. Mod. Educ. Comput. Sci. **2**(3), 1–25 (2022)
11. Roy, S., Kabir, H., Ahmed, T.: Design and implementation of web-based smart class routine management system for educational institutes. Int. J. Educ. Manage. Eng. **12**(2), 38–48 (2022)
12. Rigopoulos, G.: Student satisfaction, student assessment, student feedback. Int. J. Mod. Educ. Comput. Sci. **3**(5), 1–9 (2022)

13. Karim, M.A., Masnad, M., Ara, Y., et al.: A comprehensive study to investigate student performance in online education during COVID-19. Int. J. Mod. Educ. Comput. Sci. **14**(3), 1–25 (2022)
14. Kyrpychenko, O., Pushchyna, I., Kichuk, Y., Shevchenko, N., Luchaninova, O., Koval, V.: Communicative competence development in teaching professional discourse in educational establishments. Int. J. Mod. Educ. Comput. Sci. **13**(4), 16–27 (2021)
15. Aliyev, A.G.: Technologies ensuring the sustainability of information security of the formation of the digital economy and their perspective development directions. Int. J. Inf. Eng. Electron. Bus. **5**(14), 1–14 (2022)
16. Aliyev, A.G.: Problems of regulation and prospective development of E-commerce systems in the post-coronavirus era. Int. J. Inf. Eng. Electron. Bus. **14**(6), 14–26 (2022)
17. Mohdhar, A., Shaalan, K.: The Future of e-commerce systems: 2030 and Beyond. In: Al-Emran, M., Shaalan, K. (eds.) Recent Advances in Technology Acceptance Models and Theories, pp. 311–330. Springer, Cham (2021). https://doi.org/10.1007/978-3-030-64987-6_18
18. Joma, H.N., Abdelazez, M.J., Mohammed, H.E., Ali, A.A.: Data-driven e-commerce techniques and challenges in the era of the fourth industrial revolution. Sci. J. Inform. **7**(2), 291–302 (2020)
19. Ministry of Education. Ministry of education: National adult education strategies and implementation, Addis Ababa, Ethiopia, unpublished. http://bit.ly/2JO5Job (in Chinese)
20. Faizi, J., Umar, M.S.: A conceptual framework for software engineering education: project based learning approach integrated with industrial collaboration. Int. J. Educ. Manage. Eng. **11**(5), 46–53 (2021)
21. Martin, J.G., López, C.L., Martínez, J.E.P.: Supporting the design and development of project based learning courses. In: Proceedings of IEEE Frontiers in Education Conference Proceedings, pp. 1–6. IEEE (2014)
22. Harms, S., Hastings, J.: A cross-curricular approach to fostering innovation such as virtual reality development through student-led projects. In: Proceedings of IEEE Frontiers in Education Conference, pp. 1–9. IEEE (2016)
23. Bender, W.N.: Project-based learning: differentiating instruction for the 21st Century. Corwin. **2**(3), 26–29 (2012)

# Talent Training Mode Based on the Combination of Industry-Learning-Research Under the Background of Credit System Reform

Shanyong Qin and Minwei Liu[✉]

Shandong Women's University, Jinan 250300, Shandong, China
986043935@qq.com

**Abstract.** The reform of credit system is the further deepening of educational reform. The flexible school system is conducive to the training of comprehensive talents, and promotes the reform of the talent training mode of industry-university-research cooperation education. This paper is based on combing the flexible training mode and characteristics of applied talents in the United States, Britain and Germany, and finding out the advantages of the flexible training mode of applied talents in foreign countries. Through the questionnaire survey, it is concluded that the talent cultivation mode of industry-university-research cooperation education under the credit system background is divided into two types: the mode based on the main body of industry-university-research cooperation and the mode based on the form of industry-university-research cooperation. This paper puts forward the problems that should be paid attention to in the context of the credit system reform from two dimensions: the flexible training of talents in the credit system reform and the training of talents in the industry-university-research cooperation education. This research is conducive to the innovation of talent training mode, promoting the teaching reform of credit system, and promoting the flexible training of application-oriented talents, which has strong theoretical guiding significance and practical reference value.

**Keywords:** Credit system · Industry-learning-research cooperation education · Talent training · Flexible training curriculum system

## 1 Introduction

In China, the credit system originated from Cai Yuanpei's elective system at Peking University in 1918. By 1996, nearly one-third of colleges and universities in China had implemented and experimented with the talent training mode of the credit system. With the further deepening of educational reform, a good internal and external environment was provided for the credit system reform.

With the implementation of the credit system, the curriculum has broken the boundaries of disciplines, and has been set as compulsory courses and interdisciplinary elective courses. When students choose courses independently, they realize the diversification and personalized development of training. The credit system is conducive to the talent training mode of industry-learning-research cooperation education dominated by cultivating students' strong comprehensive theoretical knowledge and solving practical problems. Under the credit system, the cultivation of talents in industry-learning-research cooperation education should not only focus on the improvement of students' learning ability and the mastery of basic knowledge of system theory, but also cultivate students' strong technological innovation ability and adaptability to the future employment market and entrepreneurship market [1–4].

Based on the research on talent cultivation of industry-learning-research cooperation education under the background of credit system reform, the paper aims to clarify the connotation of industry-learning-research cooperation education under the background of credit system reform, identify its advantages and disadvantages, and propose corresponding strategies to deepen the reform of the credit system, so as to lay a foundation for the subsequent training of comprehensive and application-oriented talents.

## 2 Literature Review

### 2.1 Research on Credit System

Brubac and John Seiler summarized the academic progress of American students and believed that American students could make their own choice of courses according to the price indicated in the credit. They could not only control the cost, but also learn the courses they liked. Mestenhuser and Breeder summarized the three credit system forms and characteristics of the United States, and believed that the American credit system includes semester, quarter and tri meter. The research of domestic scholars can be roughly divided into the following three stages [5–7]. In the first stage (1994–2000), there are many research literatures, but the research content is relatively shallow. In 1994, Chang Feng of Dalian University of Technology published his first article on the reform of the credit system in Higher Education Research. The title of the article is "On Student Work under the Credit System". In the second stage (2001–2005), there were 308 articles on the credit system, and the research content developed in depth. Compared with the first stage, the research content in this stage is more extensive, and the discussion on model, system and system is gradually increasing. In the third stage (since 2006), the number of research articles on the credit system is increasing, researchers are also increasing, and the research content is more extensive and in-depth. After several years of practice in the credit system, all colleges and universities are more experienced in dealing with the credit system, and the research literature is more rational and mature, and can explore problems from a deeper level [8, 9].

### 2.2 Research on the Cooperation Between Industry, University and Research

Azaroff L V and Boyle K (1982) applied case analysis, taking universities as research objects, to study the characteristics of industry-learning-research. Becker and Peter (2000) found that the innovative scientific research model of school enterprise cooperation has improved the innovation vitality of enterprises and is conducive to the incubation of patent achievements. Schmiedeber (2008) analyzed the practice of industry-learning-research cooperation in German manufacturing industry, and found that there was a significant complementary effect between internal R&D and external R&D cooperation. Jirjah and Kraft (2006) found that independent R&D and collaborative innovation R&D of enterprises have significant complementary and substitutive effects under different control variables. Yang and Wang et al. (2008) pointed out that the contract spirit and trust relationship between the participants in the production university research cooperation would directly affect the stable development of the alliance relationship. Cheng Qiang and Shi Linna (2016) believed that industry-learning-research cooperation has its unique characteristics. Based on the perspective of partner heterogeneity, the study found that knowledge heterogeneity and relationship heterogeneity have a positive impact [10]. Wu Jun et al. (2016) showed that government R&D subsidies and enterprises' technology absorption capacity have a steady and significant positive impact, and enterprises without the above characteristics have a steady and significant negative impact on enterprises' innovation performance [11]. Lu Yanqiu and Ye Yingping (2017) believed that network practices were the key factors affecting the performance of industry-learning-research cooperation innovation. In the cooperation of industry, university and research, the influence of network practice on cooperative innovation performance presents an inverted U-shape, and inter organizational learning plays a significant intermediary role between network practice and cooperative innovation performance [12].

### 2.3 Research Review at Home and Abroad

Scholars at home and abroad have made fruitful achievements in the research on credit system teaching and industry-learning-research cooperation. The research on the basic concept, content, type and significance of the credit system and industry-learning-research cooperation is relatively in-depth, in place and relevant, but there are still problems such as low research level, lack of professional theoretical support, less people-oriented perspective and concept, and lack of systematic and comprehensive research. The research literature on the cultivation of talents in industry-learning-research cooperation education under the credit system background is few and not in-depth [13–15]. It is urgent to study the flexible educational system, course selection system, curriculum design, and practical teaching of the cultivation of talents in industry-learning-research cooperation education under the credit system background, and propose feasible plans to promote the talent cultivation plan and training process to a new level.

## 3 Analysis of Talent Training Mode Under the Background of Foreign Credit System

### 3.1 Characteristics of Talent Training Mode Under the Credit System in the United States

The first is that the specialized courses of different majors in different colleges and departments are selected by students in this scope. The credits of compulsory courses in basic education require students to take courses from different disciplines. The second is that students still emphasize general education after choosing their major. The third is that the university does not rigidly stipulate the specific length of schooling to complete their studies. Students can choose to postpone graduation or graduate ahead of schedule according to their own interests and life plans. Elective courses account for a large proportion, leaving more choices to students, with a large degree of freedom. Fourth, students can change their major at any time after selecting their major. Fifth, the knowledge of natural sciences, social sciences and humanities and social sciences should be equally emphasized. Sixth, tutor production is an important part of the credit system. Seventh, not only pay attention to professional courses and theoretical learning in the classroom, but also emphasize the development of practical ability and practical ability. The main features are summarized in Fig. 1.

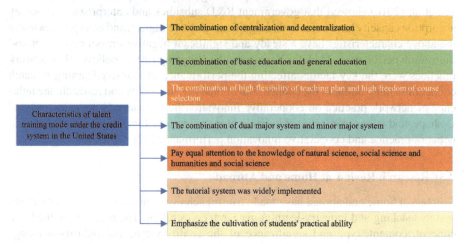

**Fig. 1.** Characteristics of talent training mode under the credit system in the United States

## 3.2 Characteristics of Talent Training Mode Under the Background of British Credit System

The first is that the duration of the bachelor's degree in British universities and colleges is usually three years. The credits of the three academic years are interlinked and progressive, which is conducive to the scientific training and comprehensive development of students. The second is that the modularization of the curriculum design makes the curriculum structure independent and examinable, and the curriculum content comprehensive and comprehensive, which can reduce the students' workload and realize the scientific and reasonable arrangement of the curriculum; The flexibility of the elective system is to provide a large number of optional courses and respect the students' right to choose independently. The thirdly is that the duties of the tutor mainly include academic guidance and spiritual care, which not only guide students in learning and academic guidance, but also care for and care for students in life. The main features are summarized in Fig. 2.

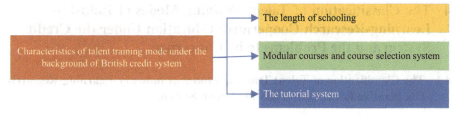

**Fig. 2.** Characteristics of talent training mode under the background of British credit system

## 3.3 Characteristics of Talent Training Mode Under the Background of German Credit System

The first is that credits are divided into two stages. The study time of the first stage is mainly for the study of basic courses, and the corresponding credits of basic courses will be obtained after passing the course study assessment; The second stage of study is the study of specialized courses. The second is that the average student must accumulate 30 credits each semester to complete university studies within the standard study period. The third is that the length of study is usually four to six academic years. Each academic year is divided into two different semesters, winter and summer. It usually takes eight to twelve semesters to get a bachelor's degree from a German university. Fourth, lectures, discussions and training are the three most common and basic teaching methods in German universities. In the lecture class, teachers teach and answer students' knowledge, and students listen and record. The discussion class allows students to participate in class discussions and interact with teachers. The training course is to train students' corresponding skills according to the content of the course. Fifthly, by recording and evaluating students' majors, grades, teaching logs, degree application records, etc. before and after going abroad, the credits obtained before and after going abroad are converted and unified to achieve mutual evaluation and recognition of credits between different schools. The main features are summarized in Fig. 3.

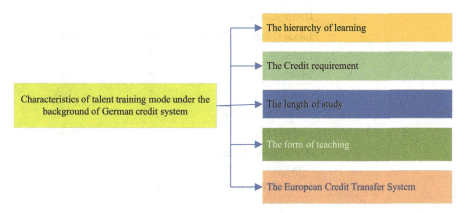

**Fig. 3.** Characteristics of talent training mode under the background of German credit system

## 4 The Classification of Talent Training Modes of Industry-Learning-Research Cooperative Education Under the Credit System and the Problems to be Concerned

### 4.1 The Classification of Talent Training Modes of Industry-Learning-Research Cooperative Education Under the Credit System

In order to more accurately understand the talent training mode of industry-learning-research cooperation education under the background of credit system reform, the project team used questionnaire survey to obtain data. In order to ensure the logic and preciseness of the questionnaire design process, the questionnaire design was carried out according to Gerbing and Anderson (1988)'s suggestions and in combination with the conclusions of the discussion with questionnaire design experts, including three stages: literature review, field interview and questionnaire prediction test. Then the questionnaire survey enters the sample data collection stage. The data collection scope of this paper involves universities, scientific research institutes, participating school enterprise cooperation enterprises and other subjects. See Table 1 for sample survey.

**Table 1.** Valid Sample Questionnaire

| Factor | Category | Valid sample | Proportion |
|---|---|---|---|
| Gender | Male | 156 | 54.17 |
|  | Female | 132 | 45.83 |
| Industry Category | Colleges and Universities | 58 | 20.14 |
|  | Scientific Research Institutes | 35 | 12.15 |
|  | Enterprise | 195 | 67.71 |
| Education | Doctor | 65 | 22.57 |
|  | Master | 51 | 17.71 |
|  | Undergraduate | 83 | 28.82 |
|  | Junior College and Below | 89 | 30.9 |
| Age | Under 28 | 28 | 9.72 |
|  | 28–45 | 163 | 56.6 |
|  | Over 45 | 97 | 33.68 |
| Personnel Classification | Senior Management | 26 | 9.03 |
|  | Middle Manager | 89 | 30.9 |
|  | Grass Roots Management Personnel | 65 | 22.57 |
|  | Scientific Researchers | 108 | 37.5 |

Through combing and summarizing the survey data, combined with the special lecture interview, the talent training mode of industry-learning-research cooperation education under the credit system background is divided into two categories: the mode based on the main body of industry-learning-research cooperation and the mode based on the form of industry-learning-research cooperation.

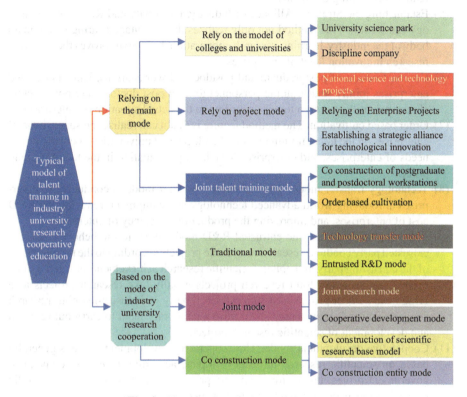

**Fig. 4.** Classification of talent training modes

The mode of relying on the main body of industry-learning-research cooperation includes the mode of relying on colleges and universities, the mode of relying on projects and the mode of jointly cultivating talents. The mode based on the form of industry-learning-research cooperation includes traditional mode, joint mode and joint construction mode. See Fig. 4 for specific classification.

(1) University Science Park. University science park is a kind of joint transformation of scientific research achievements based on the technology of universities or scientific research institutions and the capital of enterprises.

(2) Disciplinary companies. The main characteristics of discipline oriented companies are based on the advantages of disciplines, investing in scientific and technological strength, and then establishing a technology company with core technology, scientific research ability and development prospects.

(3) Relying on national science and technology projects. Rely on the scientific and technological innovation activities guided by the state and the government to carry out important key technologies, core strategies and national key projects.
(4) Relying on enterprise projects. Relying on the enterprise project mode can not only improve the efficiency of scientific research achievements transformation, but also realize the training of talents.
(5) Establishing the Strategic Alliance of Industry, University and Research Technology Innovation. This method not only realizes the advantage sharing of the main body of the industry-learning-research cooperation, but also improves the scientific research innovation level of enterprises.
(6) Co construction of postgraduate and postdoctoral workstations. Universities and enterprises jointly send training personnel to train post doctors, and post doctors choose scientific research topics according to the actual situation of enterprises.
(7) Order based cultivation. The method feature is that it can realize the supervision of enterprises on the talent training mode of colleges and universities, focusing on the needs of enterprises, and enterprises have high participation in the talent training process.
(8) Technology transfer mode. The technology transfer mode is conducive to enterprises introducing world advanced technology and equipment, reducing the R&D cost of enterprises, and improving the production efficiency of enterprises.
(9) Entrusted R&D mode. The entrusted R&D mode refers to the behavior that the entrusted party conducts research on specific projects according to the requirements of the entrusting party and obtains scientific research achievements at a given time.
(10) Joint research mode. Joint research projects usually take research projects as a bridge and subject groups as a support. Each party of industry-learning-research cooperation sends representatives to form a temporary team to carry out research and development of scientific research projects.
(11) Cooperative development mode. The cooperative development mode is generally based on scientific research projects or topics, according to market demand or government driven, universities and enterprises cooperate to complete scientific research projects or solve scientific and technological problems.
(12) Co construction of scientific research base model. The mode of co building scientific research base refers to the behavior that enterprises, scientific research institutes and universities invest a certain amount of funds, equipment and other resources to jointly build scientific research bases.
(13) Co construction entity mode. The co construction entity mode refers to the mode in which all parties of industry, university and research form a research and development entity by investing capital, manpower, equipment, technology to carry out technology development or technology management.

## 4.2 Problems that Should Be Paid Attention to in Flexible Talent Cultivation Under the Background of Credit System Reform

Members of the project team conducted in-depth interviews with universities, scientific research institutions and enterprises to understand the problems in the talent training mode of industry-learning-research cooperation education under the background of the credit system reform, and then put forward operable and feasible suggestions, and divided the problems that should be concerned about in the talent training mode of industry-learning-research cooperation education under the background of the credit system reform into two categories. See Fig. 5 for the specific classification.

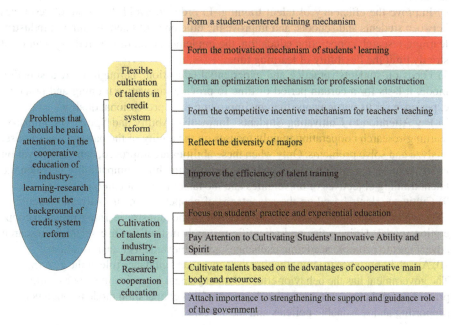

**Fig. 5.** Problems that should be paid attention to in flexible talent cultivation

1) Form a student-centered training mechanism. By implementing the credit system reform, students can choose their own courses and teachers, which fully respects the wishes of students, their interests and hobbies, and is conducive to students' personality.
2) Form the motivation mechanism of students' learning. The reform has been promoted to give students the right to choose. The school allows ordinary undergraduate students in the first and second grades to change their majors once in school, advocates students' independent design and free choice, fully arouses their enthusiasm for learning, and forms a learning motivation mechanism.
3) Form an optimization mechanism for professional construction. The reform of teaching measures plays a direct role in promoting the professional construction of the school, and also provides an opportunity for the adjustment and optimization of the professional structure, promoting the overall professional construction of the school and improving the level of teaching management.
4) Form a competitive incentive mechanism for teachers' teaching. Students choose courses and teachers independently, which has an internal competitive incentive effect on teachers' teaching. Teachers' teaching responsibilities have also been greatly enhanced, teaching energy has been invested more, updating teaching content and reforming teaching methods are more active and self-conscious than in the past.

5) Reflect the diversity of majors. The cultivation of talents in industry-learning-research cooperation education is more diversified. The cultivation of talents in industry-learning-research cooperation education is more diversified. With the development of society, economy and technology, there is an increasing demand for applied talents and compound talents..

6) Improve the efficiency of talent training. In order to avoid the waste of resources between students and schools, and improve the efficiency of talent training in industry-learning-research cooperation education, the implementation of the credit system must strictly limit the flexibility of learning time.

7) Focus on students' practice and experiential education. Through the real situation mode, it lasts for a certain period of time to provide innovative training and practical experience for its participants, thereby improving their professional quality.

8) Pay Attention to Cultivating Students' Innovative Ability and Spirit. The industry-learning-research cooperation mainly focuses on the ability of the participants to raise, analyze and solve problems. Only when these abilities are improved, can the innovation ability achieve a qualitative leap, and thus improve the high compatibility between the talent training objectives of universities and the talent needs of enterprises.

9) Cultivate talents based on the advantages of cooperative main body and resources. Enterprises create talent training bases by virtue of the industry-learning-research cooperation model. Universities and scientific research institutions use scientific research advantages to promote academic exchanges and achievements transformation.

10) Attach importance to strengthening the support and guidance role of the government. The government has the behavior effect of management, organization, coordination and promotion, and can promote the development of this cooperation mode through perfect, supporting and reasonable policies and measures.

## 5 Conclusion

The main conclusions of this study are as follows:

1) On the basis of the analysis of the talent training mode under the credit system background of the United States, Britain and Germany, this paper expounds two typical types of industry-learning-research cooperation education modes based on the principal mode of industry-learning-research cooperation and the form mode of industry-learning-research cooperation.

2) The mode of relying on the main body of industry-learning-research cooperation includes the mode of relying on colleges and universities, the mode of relying on projects and the mode of jointly cultivating talents.

3) The mode based on the form of industry-learning-research cooperation includes traditional mode, joint mode and joint construction mode.

4) The value of the flexible training mode of applied talents under the background of the credit system reform includes six aspects: forming a student-centered training mechanism, forming a dynamic mechanism for students' learning, forming an optimization mechanism for professional construction, forming a competitive incentive mechanism for teachers' teaching, the diversity of professional settings and the efficiency of talent training.

**Acknowledgment.** This project is supported by 2017 Teaching Reform Research Project of Shandong Women's University: Research on talent training mode of Industry-learning-research education mode under the background of credit reform. 2022 Jiajiayue Shandong Women's College demonstrative off-campus practice base construction project.

# References

1. Veugelers R., Cassiman B.: R&D cooperation between firms and universities. Someempirical evidence from Belgian manufacturing. Int. J. Industrial Organ. **23**(5–6), 355–379 (2005)
2. Pierre, V.: Identifying collaborative innovation capabili ties within know ledge-intensive environments: insights from the ARPANET project. Eur. J. Innov. Manag. **1**, 152–155 (2012)
3. Zhang,Y.: Challenges and countermeasures for college class management under the credit system. Henan Educ. High. Educ. **2**, 44–47 (2018). (in Chinese)
4. He, Y., Liang, F.: Research on influencing factors and mechanism of enterprise knowledge search in industry-learning-research cooperation. Sci. Sci. Manage. Sci. Technol. **03**, 12–22 (2017). (in Chinese)
5. Yu, L.: Research on the "cooperative education" model cultivating in higher vocational education. Int. J. Educ. Manage. Eng. (IJEME). **1**, 35–41 (2012). (in Chinese)
6. Wang, P., Wang, H.: A comparative study on the innovation network of textile university industry-learning-research cooperation based on knowledge graph. J. Zhejiang Sci-Tech Univ. (Social Science Edition) **48**(05), 513–525 (2022). (in Chinese)
7. Luo, S.: Research on the cultivation of innovative undergraduate talents under the credit system. Anhui University, Hefei (2010). (in Chinese)
8. Khine, P.T.T., Win, H.P.P., Naing, T.M.: Towards implementation of blended teaching approaches for higher education in Myanmar. Int. J. Educ. Manage. Eng. (IJEME), **1**(11), 19–27 (2021)
9. Hu, N., Wan, J.: Construction of higher vocational talents training program under the full credit system. Vocat. Educ. Forum **18**, 75–78 (2015). (in Chinese)
10. Cheng, Q., Shi, L.: Research on the co-evolution mechanism of industry-university-research collaborative innovation based on self-organization theory. Soft. Sci. **30**(04), 22–26 (2016). (in Chinese)
11. Wu, J., Zhang, J., Huang, D.: Research on the impact of industry-university-research cooperation on the innovation performance of strategic emerging industries. Contemp. Finan. **09**, 99–109 (2016). (in Chinese)
12. Lu, Y., Ye, Y.: The impact of network practices on innovation performance in industry-university-research cooperation. Sci. Res. Manage. **38**(03), 11–17 (2017). (in Chinese)
13. Yang, X.: Research on entrepreneurship education and innovative talent training in Chinese universities. China High. Educ. Res. **01**, 39–44 (2015). (in Chinese)
14. De Matas, S.S., Keegan, B.P.: A case study on adult and workplace learning. J. Educ. Manage. Eng. **1**, 11–19 (2020)
15. Ye, F.: Credit system realization mode and talent cultivation of industry-learning-research cooperation education. Res. High. Eng. Educ. **6**, 106–111 (2015). (in Chinese)

# Analysis of the Innovation Mechanism and Implementation Effect of College Students' Career Guidance Courses Based on Market Demand

Jingjing Ge(✉)

School of Mechanical and Electronic Engineering, Suzhou University, Suzhou 234000, China
1330629199@qq.com

**Abstract.** The negative impact of the post-epidemic era on economic development is gradually emerging, and the current employment situation is severe and complicated. As a course focused on college students' career development and employment, there are many problems in the actual teaching process, such as incoherent curriculum, tedious teaching content and lack of practical links. This paper systematically analyzes the practical significance of the innovative research of the employment guidance course for college students, the problems and difficulties in the current employment guidance course, and how to innovate the lecture form, course content and evaluation methods of the employment guidance course from the market perspective and students' needs, so as to help college students crack the employment dilemma, clarify the employment goals and achieve more satisfactory employment. And through the comparison of course effects, the significant effect of the findings of this paper on improving college students' employment readiness is clarified.

**Keywords:** Career Guidance · Innovation · Market

## 1 Introduction

Many scholars have made a lot of research results on career guidance and career development, Wang conducted a study on the innovation of career guidance model in China in 1998, Sun et al. summarized the implementation of career guidance in China during 2001–2012 and gave targeted suggestions, Qin analyzed the situation of students with employment difficulties in electronics and provided guidance suggestions [1–3]. Zhang gave an effective method to optimize career planning for undergraduate students, Liu proposed to establish a perfect career guidance model [4, 5]. Watts A G et al. compared the similarities and differences of employment policies in several countries [6]. Grubb W N emphasized the mutual influence of government and market on employment information, Stukalina, Y pointed out the importance of career management for technical universities, Kuzmin A M discussed how to strengthen the training of engineers based on a project approach [7–9]. Vaidya N M developed an effective model for predicting the

© The Author(s), under exclusive license to Springer Nature Switzerland AG 2023
Z. Hu et al. (Eds.): ICAILE 2023, LNDECT 180, pp. 822–832, 2023.
https://doi.org/10.1007/978-3-031-36115-9_73

overall performance of students, El Mrabet H proposed that schools can provide good employment programs for students with the help of technology [10, 11]. Thus, it can be seen that career guidance and career planning are important for personal growth and social development.

In the post-epidemic era, the negative impact of the epidemic on the economic development has been gradually highlighted, and the employment situation of college students is unprecedentedly severe and complicated, while employment is a two-way choice based on market demand and college students' employment values. Therefore, colleges and universities need to change their thinking mode when providing employment guidance and training to college students, actively obtain the new changes of current society and enterprises for talents, and integrate them into the whole process of employment guidance and career planning, so as to truly realize the unity of talent cultivation, scientific research and social service of colleges and universities, and accurately help college students to find high-quality employment.

## 2 The Practical Significance of the Innovation of Career Guidance Courses for College Students

### 2.1 Be Beneficial to Talent Reserve for Industrial Technological Revolution

Employment is the largest and most basic livelihood, the 20th Party Congress report pointed out: strengthen the employment priority policy, improve the employment promotion mechanism, and promote high-quality full employment. This is not only a long-term plan made by the state at the strategic level, but also a realistic need of today's rapid technological update and industrial revolution. New industrial changes and technological development have broken the boundaries of traditional industries, and more industrial forms and employment models are emerging. This is not only a new employment opportunity that contemporary college students need to seize under the severe economic situation, but also a new challenge to the employment quality and career psychology of college students. The so-called new employment mode refers to more employment positions derived from artificial intelligence, big data, Internet of Things and 5G technology, which have distinctive requirements of high knowledge and strong technical reserve, while the professional course education received by contemporary college students in the university stage focuses on basic and systematic knowledge inculcation, with strong theoretical but lack of practical ability, and innovative thinking also need to be strengthened [12].

Therefore, colleges and universities should focus on cultivating students' industry foresight and ability, through the organic connection of career guidance courses, based on the skills education of professional courses, helping students understand the latest development trend and technological innovation of the industry, the new job demands brought by them, and the accumulation of knowledge and ability required by these jobs. Only then will students have a sense of identity and belonging in their hearts, consciously and spontaneously examine the gap between themselves and their goals, reasonably set career goals, gradually refine their career plans, strive to improve their professionalism, jointly promote the development and progress of society, and become high quality labor talents.

## 2.2 Be Beneficial for College Students to Achieve Higher Quality Employment

According to the statistics of the Ministry of Education, the number of graduates in China is gradually climbing and the number is estimated to reach 11.58 million in 2023, breaking the 10 million mark again. In the past two years, the phenomenon of "slow employment" and "delayed employment" of college students has been gradually highlighted, and many scholars have analyzed and researched this phenomenon and put forward corresponding countermeasures. A search on China National Knowledge Infrastructure with the keyword of "slow employment" shows that there are as many as 637 relevant research articles and more than half of them were published after 2021, this fully indicates that the employment motivation of contemporary college students needs to be enhanced, and the establishment of correct employment values in the post-epidemic era can help graduates achieve high-quality employment [13].

This includes both the influence of college students' self-selection mentality and the influence of external environment on college students' employment behavior, which should never be ignored. The university students are limited by the barrier of information acquisition ability and the lack of social experience, so it is difficult for them to have accurate insight into the recruitment psychology of enterprises and the changes of talent demand. This requires universities to build a strong communication bridge, provide a wide range of information channels and give accurate graduation support, so that graduates have a rational and calm employment mindset based on a full understanding of the job market and prospects.

## 2.3 Be Beneficial for Efficiency Improvement of Enterprise Human Resources

In the process of recruiting talents, enterprises need to spend a lot of manpower, time and money costs, so they hope that the comprehensive skills of the graduates recruited are better matched with the positions, which can help the efficiency of talent recruitment and thus reduce human resources costs. The reality is that the "structural difficulties" in employment have existed for a long time and have become the main reason for the current difficulties in employment of college students. On the one hand, graduates facing seemingly "massive" recruitment information is still difficult to find a satisfactory job, on the other hand, enterprises have paid a lot of human cost but cannot recruit talents that match the job demand. Solving the "structural difficulties" of employment requires the joint efforts of the government, society, enterprises, universities and individual students, colleges and universities should meet the needs of enterprises in the whole process of talent training, examine the shortcomings of the current curriculum setting, teaching methods, evaluation and feedback and carry out targeted reforms in order to really improve the quality of talent training, so that graduates have a solid reserve of professional skills and good employment competitiveness [14]. Therefore, career guidance courses for college students should be the leading place for reform and innovation, thus driving other courses to gradually integrate into the system and forming a synergistic mechanism of extensive consultation, joint construction and sharing [15].

# 3 The Common Problems Existing in the Current Career Guidance Courses for College Students

## 3.1 Insufficient Attention from Colleges and the Curriculum is not Systematic

Although college students' career guidance course is included in the talent training program of most colleges and universities as a compulsory general education course, it often faces many problems in the actual teaching process, such as the course opening time is arbitrary, the teachers are not professional, and the career planning and career guidance courses are not coherent. For example, some colleges and universities put the career planning course in the first semester of freshman year, while the career guidance course is not reopened until the second semester of junior year or even the first semester of senior year, when students are already facing the key choice of choosing career and further education [16].

Therefore, colleges and universities should appropriately consider moving the career guidance course forward or in the form of short-term classes of four to six hours per semester to enhance the relevance and consistency of career guidance, especially in the sophomore year, when students have just passed the confusion stage of freshman year and have certain knowledge about the prospect and knowledge structure of their majors, but they still have many questions about how to better combine their strengths with their majors, so they need constructive guidance and advice from experienced professional teachers. Furthermore, most of the career planning and career guidance courses in colleges and universities are taught by young counselors or class teachers, many of whom are just going from college to college and are very young as career professionals themselves, their classes are more in accordance with the textbook content of theoretical knowledge, while the career guidance course is precisely a very practical and application-oriented course [17]. Therefore, colleges and universities should focus on coordinating internal and external resources, optimizing faculty allocation, and enhancing students' sense of experience and access to the course.

## 3.2 The Teaching Content is Boring and not Very Practical

Most of the current career guidance courses are still presented in the traditional classroom teaching style, and the course content is set up in a similar way, generally divided into basic parts such as employment policy explanation, employment psychological counseling, resume making, interview process and etiquette, labor relations establishment, etc. Students are not interested in facing the boring theoretical indoctrination, and the classroom effect is not ideal, and the applicability is even worse in the actual job search process. Graduates in the first step out of the ivory tower job search need to determine their own job search goals, followed by the need to establish a confident and generous job search mentality and a positive job search image, and then adequate interview preparation and appropriate interview skills. Therefore, colleges and universities can re-integrate and re-order the content of the traditional career guidance courses, and after students have been guided by the career planning course to have a preliminary understanding of their own personality, interests, hobbies and career selection factors, etc. in the early stage, combining with the current employment prospect of this profession to help graduates to reasonably choose the intended enterprises and positions, so as to effectively avoid the result of "fishing in the sea" and eventually getting nothing.

The career guidance department of colleges and universities can also regularly invite the personnel department of large-scale enterprises in the industry to open lectures or forums and other forms, from the perspective of employers to give students professional interview image and job search skills advice, which is far more than the lectures in the classroom away from the actual job search scene to bring students more sense of job search reality and a sense of self-acquisition, students will also have the motivation to self-improvement and progress. At the same time to link up with the professional power of the counseling department, in advance to teach students the skills and ways to relieve anxiety and overcome nervousness, to avoid students have experienced a number of unfavorable blow to the job search, and then to seek psychological counseling and help, this can very easily have a negative impact on the later job search process, and research results have shown that psychological education taken in modern college students' career guidance can further improve their mental health [18].

### 3.3 Large Class Teaching, Lack of Individual Guidance

Unlike ordinary general education courses that are suitable for large classes, the purpose of the career guidance course is to enable everyone to master the skills applicable to their own job search, which requires the teacher to pay attention to the subjective motivation and receptivity of each individual student in the process of teaching, in addition to the general theoretical knowledge [19]. In contrast, colleges and universities are limited by the lack of class time and faculty, and career guidance courses are generally taught in large classes. Take electronic information major, where the author once taught, for example, the standard class size is about 60 people. No matter it is simulated one-to-one interview scene or leadless group interview discussion, 60 people are obviously not suitable to carry out similar activities at the same time, including the classroom venue is also difficult to simulate the real interview environment.

Therefore, colleges and universities should consider opening career guidance courses in small classes and establishing career profiles for each student, so that each student can have a clearer understanding of their own employment advantages and shortcomings through theoretical learning and practical activities of the courses, so that they can target to enlarge their advantages and make up for their shortcomings, and gain knowledge and ability accumulation that can really benefit the future job search process.

## 4 Analysis of the Innovation Mechanism of College Students' Career Guidance Course Based on Market Demand

### 4.1 Help College Students Broaden Their Horizons and Develop a "Strong Heart" for Employment

A good employment mentality is a prerequisite for college students to achieve high quality employment. The graduates of 2023 are generally the post-00 s generation. They grew up in the two decades of China's rapid economic development. Their parents provided them with good material conditions and rich experience, which shaped them to have a strong sense of self, strong subjective initiative and strong will to realize themselves. But at the same time, most of them are the only child, and they are easy to show multiple problems such as poor teamwork, weak anti-pressure ability and weak empathy. This is bound to cause their job hunting and career selection behavior will exist double contradiction.

The author once conducted a simple test on 120 students in two classes of electronic information major, which asked what are the first factors you are most concerned about in the process of determining your ideal career? The complete statistics result of factors that affects graduates showed as Fig. 1. Among them, more than 80% of the students put salary in the first priority, followed by career prospects, company scale accounts for 6.2%, 4.9% cares about working site most, which also fully shows that contemporary college students in the mentality of choosing a career is impatient, the vision is narrow.

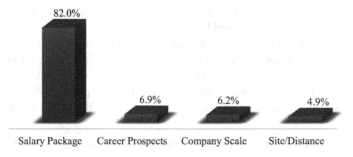

**Fig. 1.** Factors that affect graduates' choice of career

Engels said in *Dialectics of Nature*: "The development or negation of negation caused by contradiction – the spiral form of development. The same is true of career development. Spiraling means that no matter which career development path they choose, they will experience both twists and turns and progress. College career guidance courses can try to help graduates rationally view the spiraling progress of their career through the following three methods. Only when graduates have a clear judgment and rational cognition of employment goals can they obtain a "strong heart" in employment.

Method 1: Invite outstanding student representatives who have already developed in the industry to come into the class and exchange their experiences, which can increase students' sense of inclusion.

Method 2: Make full use of opportunities such as social practice, internship and training in winter and summer vacation to encourage students to enter local enterprises to assume certain work roles and experience the real world of work.

Method 3: Make good use of new media technology to provide accurate employment services for college students, promote the spirit of labor in a way that students are willing to accept [20].

## 4.2 Promote Graduates to Improve Themselves and Have a "Voice" in Career

The ultimate goal of talent recruitment is to create economic value for the enterprise. Therefore, if graduates want to stand out in the cruel job market and get the ideal job opportunities, they must "cultivate both inside and outside" and constantly improve their comprehensive quality. At present, a common problem of career guidance courses in colleges and universities is that the teaching content is divorced from reality, and it is difficult for students to acquire practical career skills through the course learning. This is caused by many factors. First, colleges and universities generally do not pay enough attention to this kind of course. There is no corresponding teaching content setting and no assessment requirements, which results in teachers ignoring the cultivation of students' practical skills when teaching. On the other hand, college teachers are far away from the real workplace environment, and they are not familiar with what practical skills the enterprise needs in addition to professional knowledge. According to a survey the author ever did on 120 students in two classes of electronic information major (there are exactly 60 students in each class), students desire to acquire a variety of career skills and the detailed data showed in Table 1 and Fig. 2.

From Table 1, we see that the skills students care about most is career goals setting, total 44 out of 120 students go to this option, followed by office skills guidance. According to Fig. 2 we can find this survey results place resume making skills as being 16.7% in all options, with 14.2% concern interpersonal relationship most, while 7.5% choose interview experience and ability and the reminder goes to labor rights related content.

**Table 1.** Statistics of different skills that students desire to acquire in career guidance courses

| Skills | Number in class1 | Number in class2 | Total number |
|---|---|---|---|
| Career goals setting | 23 | 21 | 44 |
| Office skills guidance | 11 | 14 | 25 |
| Resume making skills | 8 | 12 | 20 |
| Interpersonal relationship | 9 | 8 | 17 |
| Interview experience and ability | 4 | 5 | 9 |
| Labor rights related | 2 | 3 | 5 |

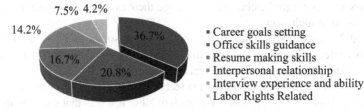

**Fig. 2.** Percentage of demand for different skills

Therefore, the career guidance courses of colleges and universities need to strengthen the in-depth cooperation with enterprises, pay attention to the recruitment points of different positions set by employers and students' demand for different skills, then make up for it in the course teaching process [21].

Method 1: Contact long-term cooperated internship units or intern employment bases to ask for the professional human resources assistance, they can provide practical office skills guidance for students to help them improve their "hard strength".

Method 2: Invite enterprises to participate in the simulated recruitment competition held by schools or colleges, creates a real environment for recruitment and interview, sets up on-site recruitment links, so that graduates can really find their own gap and find the goal of efforts through the simulated recruitment competition.

Method 3: The university coordinates with the human resources security department to hold regular lectures on issues related to the interests of graduates, such as the signing of labor contracts, the protection of labor rights and interests, and how to protect the legitimate rights and interests of graduates through legal channels in case of labor disputes [22].

These are precisely the general knowledge of the workplace that graduates must have before entering the workplace, and the knowledge reserve that universities should build up for graduates through career guidance and education. Only with solid knowledge and skills and a clear understanding of the rights and obligations of workers can graduates master the workplace "voice".

### 4.3 Whole Process and Whole Staff Participation in Talent Cultivation, to Create a Sense of Professional "Atmosphere"

Most of the career planning and career guidance courses currently offered in colleges and universities are taught by student work managers or administrative personnel, who often have very trivial and complicated daily work and find it difficult to devote a lot of time to course preparation and design, so the classroom teaching effect is not particularly satisfactory. Therefore, to strengthen the education effect of the career guidance course, we should take the optimization of the faculty as a breakthrough and establish a full staff and whole process education mechanism [23]. Graduates need to gain a detailed understanding of the current employment situation, the frontiers of development and job opportunities in the industry and advice available to them through career guidance courses, which are often sorely lacking for non-professional faculty. The external environment is changing rapidly and policies are being introduced very quickly, so it is essential to help graduates keep abreast of industry developments in real time.

Career guidance courses in colleges and universities should give due consideration to the participation of professional teachers in the preparation and teaching process. On the one hand, it can provide students with a systematic introduction to explain the industry-related enterprises, how to choose a suitable position from the many positions in the enterprise, to maximize their strengths and specialties. On the other hand, professional teachers have been developing in the industry for many years and have rich enterprise resources, which can help graduates achieve fast and high-quality employment. In addition, graduates can seek help from professional teachers when they encounter difficult problems in the job search process, so as to lay a good foundation for the next successful job application.

Furthermore, all teachers in the process of teaching professional courses should consciously guide students in the concept of employment and career selection, cultivate innovative and entrepreneurial ability, educate professional ethics and other comprehensive professional literacy education, create a sense of professional "atmosphere", help students to establish a good employment concept as early as possible, give full play to the function of ideological and political education of each course, provide the whole process of accurate guidance services, and jointly promote high-quality employment of graduates to realize their career dreams.

## 5 Implementation Effect Evaluation

In order to evaluate the effect of the research results on the promotion of the career guidance course for college students, 25 students were randomly selected from the two parallel classes of electronic information major (60 students in each standard class) that the author used to teach. A total of 50 students were evaluated and scored on their degree of employment readiness at the beginning and end of the course. The full score of the survey is 10 points, and the higher the comprehensive score, the better the preparation for employment. The course lasted for 8 weeks, and the teaching content is reasonably arranged according to the course innovation mechanism of this paper. The paired t-test method was used to sort out and analyze the scores of 50 students by SPSS software. As shown in Table 2, the results show that after continuous employment guidance and assistance for students according to the research results of this paper, the students' preparation degree of employment has been significantly improved. $t = -13.212, p < 0.001$, 95%CI was $(-1.69 \sim -1.24)$.

Table 2. Evaluation of curriculum implementation effect based on students' survey

| Group | df | $\overline{X} \pm S$ | t | p | 95%CI of the difference |
|---|---|---|---|---|---|
| Pre-test | 49 | $5.62 \pm 1.02$ | $-13.212$ | $< 0.001$ | $-1.69 \sim -1.24$ |
| Post-test | 49 | $7.08 \pm 0.85$ | | | |

## 6 Conclusion

The quality of talent training in colleges and universities directly determines whether there is a qualified and high-quality workforce to undertake the important task of future social development and progress, and the fierce job market requires graduates to change their employment willingness, improve their employment literacy and enhance their job-seeking skills in response to the market and enterprise needs. Therefore, colleges and universities should strengthen the overall coordination, improve the education effect of the employment course from the curriculum design, faculty allocation, school-enterprise contact and other multi-dimensions, innovate the teaching methods, broaden the practice platform, really create a deep, high and warm employment guidance gold course for students, so that the career and employment guidance course can escort the graduates' job search road and play the proper value of the course.

**Acknowledgements.** This project is supported by Suzhou University 2021 Community-level Party Building Innovation Secretary Project(2021sjxm011) and Suzhou University 2022 the Second Batch of School-level Platform Open Subject Project(2022ykf29)

# References

1. Lijuan, W., Muzhi, W.: Vocational guidance and career planning in China. Int. J. Adv. Couns. **20**, 27–35 (1998)
2. Sun, V.J., Yuen, M.: Career guidance and counseling for university students in China. Int. J. Adv. Couns. **34**, 202–210 (2012)
3. Qin, Y., Peng, J.: The Employment Services and Guidance of the Disadvantaged Groups of Electronics Professional College Graduates. In: Zhang, L., Zhang, C. (eds) Engineering Education and Management. Lecture Notes in Electrical Engineering, vol 111. Springer, Berlin, Heidelberg (2012). https://doi.org/10.1007/978-3-642-24823-8_114
4. Zhang, T., Shao, B.: The Research of Theory Education and Practice in University Career Planning. In: Zhang, L., Zhang, C. (eds) Engineering Education and Management. Lecture Notes in Electrical Engineering, vol 111. Springer, Berlin, Heidelberg (2012). https://doi.org/10.1007/978-3-642-24823-8_80
5. Fangfang, L.: Innovative Research on the Whole-process Model of college employment guidance in the new situation. Front. Educ. Res. **3**(9), 49–51 (2020)
6. Watts, A.G., Sultana, R.G.: Career guidance policies in 37 countries: contrasts and common themes. Int. J. Educ. Vocat. Guidance **4**, 105–122 (2004)
7. Grubb, W.N.: An occupation in harmony: the roles of markets and government in career information and guidance. Int. J. Educ. Vocat. Guidance **4**, 123–139 (2004)
8. Stukalina, Y.: Career Management in a Technical University as an Essential Factor Influencing Its Competitiveness. In: Kabashkin, I., Yatskiv, I., Prentkovskis, O. (eds) Reliability and Statistics in Transportation and Communication. RelStat 2017. Lecture Notes in Networks and Systems, vol 36, pp. 639-648, Springer, Cham (2018). https://doi.org/10.1007/978-3-319-74454-4_61
9. Kuzmin, A.M., Kunina, O.O., Fedorov, A.M., et al.: From career guidance of schoolchildren to professional training of future engineers at university of engineering and technology. Mobility for Smart Cities and Regional Development-Challenges for Higher Education. In: Proceedings of the 24th International Conference on Interactive Collaborative Learning (ICL2021), pp. 654–660 (2022)
10. Vaidya, N.M., Patel, K.K.: Learner performance and preference meter for better career guidance and holistic growth. Proc. ICT Anal. Appl. **2020**, 47–54 (2019)
11. El Mrabet, H., Ait, M.A.: IoT-school guidance: a holistic approach to vocational self-awareness & career path. Educ. Inf. Technol. **26**, 5439–5456 (2021)
12. Sun, Y., Aizhen. R.: Measures to improve the employability of college students under the new employment pattern. Vocat. Educ. Mech. Ind. (11), 16–20 (2022) (in Chinese)
13. Xing, Z.: Cultivation of college students employment values in the post epidemic era. J. Bengbu Univ. **11**(5), 94–97 (2022) (in Chinese).
14. Junhui, L., Shujiao, C., Wuyu, Z., et,al.: Solution to "Slow Employment" of college students under Covid-19 Epidemic. J. Ningbo Univ. Technol. **34**(3): 126–132 (2022) (in Chinese).

15. Sihang, C., Hui. X.: Using high quality employment guidance to help with college students employment. Chinese Social Sciences Today (2022) (in Chinese).
16. Xiaoting, L., Yanfang, Z.: Reflections on promoting precise career guidance services in colleges and universities with high quality. Youth Res. **05**, 101–105 (2022). (in Chinese).
17. Hao, D., Sun, V.J., Yuen, M.: Towards a model of career guidance and counseling for university students in China. Int. J. Adv. Couns. **37**, 155–167 (2015)
18. Lu, Y.: The application of positive psychology education in modern college students' employment guidance to enhance self-confidence and improve personality. China J. Multimed. Netw. Teach. (06), 164–167 (2022) (in Chinese).
19. David, L.T., Truța, C., Cazan, A.M., et al.: Exploring the impact of a career guidance intervention program in schools: effects on knowledge and skills as self-assessed by students. Curr. Psychol. **41**, 4545–4556 (2022)
20. Xiao, L.: Discussion on precisely college students' employment service under the background of new media. Ability Wisdom **11**, 133–135 (2022) (in Chinese).
21. Kuijpers, M.: Career guidance in collaboration between schools and work organisations. Br. J. Guid. Couns. **47**(4), 487–497 (2019)
22. Yanmei, T.: Exploration of College Student Management Based on Employment Orientation. Heilongjiang Sci. **13**(21), 61–63 (2022). (in Chinese)
23. Keller-Schneider, M., Zhong, H.F., Yeung, A.S.: Competence and challenge in professional development: teacher perceptions at different stages of career. J. Educ. Teach. **46**(1), 36–54 (2020)

# Comparative Study on the Development of Chinese and Foreign Textbooks in Nanomaterials and Technology

Yao Ding[1], Jin Wen[1], Qilai Zhou[1,2], Li Liu[1], Guanchao Yin[1(✉)], and Liqiang Mai[1(✉)]

[1] School of Materials Science and Engineering, Wuhan University of Technology, Wuhan 430070, China
{guanchao.yin,mlq518}@whut.edu.cn

[2] Faculty of Science, Shizuoka University, Shizuoka 422-8529, Japan

**Abstract.** Textbooks in colleges and universities are the carriers of knowledge that can reflect the teaching content and methods. It is also the media that can broaden students' horizon and track modern scientific development, which is also the important symbol reflected the scientific research achievements. This paper compares the textbooks of "Nanomaterials and Technologies" in China with similar textbooks abroad from three aspects. First, by comparing the proportion of backgrounds, basic and expanded knowledge, the design of knowledge system is compared to find the advantages in foreign textbooks. Second, by comparing the readability, logic and self-conscious study which are provided by textbooks, the morphology of the textbooks is compared to find the weakness in Chinese textbooks. Finally, the application of information technology in textbooks is explored to find the improvements. The purpose of this study is to pave the way for the organization of Chinese textbooks in nanomaterials and technology in future.

**Keywords:** Textbooks · Comparative study · Nanomaterials

## 1 Introduction

Textbooks, one of the main means to help teachers and students understand the knowledge, needs to be systematic researched for its development in both Chinese and foreign ones, which can pave a way for the Chinese teachers in universities and colleges to promote the qualities of both traditional course teaching and E-learning [1–7]. Nanomaterials science and technology, a newly developed major, is an undergraduate major of regular higher education, which belongs to materials major [8, 9]. However, in China, only a few universities and colleges have offered this major at present [10–12]. Besides, less work has been done on the comparative research of the written and organization of Chinese and foreign textbooks [13, 14]. More importantly, as E-learning and online courses have become a new mode in teaching instead of traditional course teaching after COVID-19, the study of the differences in the usage of information technic between Chinese and foreign textbooks is important [15–20].

Thus, under the guidance of the above weakness in researches about Chinese and foreign textbooks, through a comparative study of Chinese and foreign textbooks in the major of nanomaterials science, this work explores the close relationship between cultivation objectives of students, curriculum setting and content construction, as well as the close relationship between the training mode of students' abilities and the construction of textbooks. By statistically comparing the design of content system, analyzing the whole structure and comparing the application of information technology in these textbooks, this work will also explore the approaches for the embodiment of thoughts and teaching concepts in textbooks for advanced education, and study the role of textbooks both in supporting the traditional course teaching and E-learning. By using the statistical methodology, the main features and characteristics of excellent foreign textbooks in nanomaterials and technology are pointed out. It finds that foreign textbooks usually account for a better design structure and the abundant vividness and readability of illustrations, which can provide an openminded reading experience for students and will contribute to the self-learning. Therefore, this work can reveal the relevant shortcomings in Chinese textbooks of nanomaterials science major and pave a way for the organization and written of future textbooks.

## 2  Selection of Chinese and Foreign Textbooks

The basic information of these Chinese and foreign textbooks majored in nanomaterials and technology, *e.g.*, the contents, chief editors, publishing agency and the reasons for selecting as one of the objectives in this research work, are listed as follow.

1) Chinese textbooks.

a. *Nanomaterials and Their Preparation Techniques* (1$^{st}$ Edition), Manhong Liu, Metallurgical Industry Press.

This book was published by Metallurgical Industry Press in 2014 as a textbook for the 12$^{nd}$ Five-Year Plan of national higher education in China. The book is a basic textbook or teaching auxiliary book commonly used by students in colleges or universities majores in nanomaterials and technology. This makes the book with the certain representativeness and authority, thus been chosen as one of the objectives in this research. The book firstly introduces the background and development of nanomaterials science and then introduces the preparation and surface modification principles and methods of nanoparticles and nanomaterials, *etc*. Application principles in rubber and plastic nanomaterials, textile materials, optical materials, *etc.*, are also included in this book. Therefore, this book which has a complete knowledge system has a certain research value.

b. *Fundamentals of Nanomaterials* (2$^{nd}$ Edition), Yaojun Zhang, Chemical Industry Press.

This book was published by Chemical Industry Press in 2014 as a textbook for the 12$^{nd}$ Five-Year Plan of national higher education in China. It can be used as a textbook for basic courses of nanomaterials and technology major. The book has been widely used in lots of famous universities in China, such as Beijing University of Aeronautics and Astronautics, *etc.* Thus, this book is with the certain representativeness and authority and can be used as one of the objectives in this research. The textbook systematically introduces the basic concepts and classification of nanomaterials, *e.g.*, nano-effects, properties of nanomaterials, "top-down" and "bottom-up" preparation methods of nanomaterials, self-assembly of nanomaterials and so on. The book also includes the applications of nanotechnology in the research area as clean energy. Besides, there are two editions of the textbook (1$^{st}$ edition in 2011 and 2$^{nd}$ edition in 2015). The second edition is a bilingual edition, which has a value in researching the trend of development in Chinese textbooks.

2) Foreign textbooks.

a. *Fundamentals of Nanotechnology* (1$^{st}$ Edition), Gabor L. Hornyak, CRC Press.

This book, a typical foreign textbook majored in nanomaterials, which is widely used by many American universities (such as University of California, Los Angeles, *etc.*). Therefore, the book has the universality and representativeness among foreign textbooks in this major. The main author of this textbook, Gabor L. Hornyak, is a well-known professor in the research area of nanomaterials. Besides, CRC Cham is one of the largest science and technology publishing agencies in the world. Thus, this textbook is also authoritative in the textbooks majored in nanomaterials, which makes this book suitable for this research work. At the same time, the book covers most of the basic knowledge of nanomaterials, including the basics of nanotechnology, practical applications of nanomaterials and knowledge of nanomaterials in drugs, *etc*. The book covers a wide range of knowledge for nanomaterials and technology with a complete system, which has the value for researching.

b. *Nanotechnology: Principles and Applications* (3$^{rd}$ Edition), S. Logothetidis, Springer Cham Press.

This book is a typical textbook widely used by universities in Europe and the United States (such as Harvard University, Oxford University, *etc.*) which have the major as nanomaterials and technology. Therefore, the book has the universality and representativeness among the foreign textbooks. The main author, S. Logothetidis, is also a well-known professor in the research area of nanomaterials and technology. The publishing agency, Springer Cham, is one of the largest and most famous science and technology publishing agencies. Thus, this book is also authoritative in nanomaterials and technology. The book introduces nanostructures and nanomaterials for their applications in energy and organic electronics, knowledges of advanced nanomaterials deposition and processing methods, *etc.* It also focuses on the optical, electronic, surface and mechanical properties of nanomaterials. It provides a complete system of basic knowledges in nanomaterials and deserves to be carefully compared with Chinese textbooks.

## 3 Comparison of the Design of the Structure of Content in Chinese and Foreign Textbooks

### 3.1 Comparison of the Design of Knowledge System in Chinese and Foreign Textbooks

In order to compare and analyze the design of knowledge system of the above Chinese and foreign textbooks in nanomaterials and technology, statistical analysis of the proportion of backgrounds, basic and expanded knowledge (including applied knowledge and frontier knowledge) in the whole textbook are listed as Table 1. Detailed data processing methods are also presented below.

**Table 1.** Comparison of the proportion of backgrounds, basic and expanded knowledge in Chinese and foreign textbooks

| Textbooks | | Backgrounds | Basic knowledge | Expanded knowledge | |
|---|---|---|---|---|---|
| | | | | Applied knowledge | Frontier knowledge |
| Chinese textbooks | Nanomaterials and Their Preparation Techniques | 0.10 | 0.60 | 0.13 | 0.10 |
| | Fundamentals of Nanomaterials | 0.18 | 0.42 | 0.14 | 0.16 |
| Foreign textbooks | Fundamentals of Nanotechnology | 0.11 | 0.35 | 0.29 | 0.25 |
| | Nanotechnology: Principles and Applications | 0.10 | 0.31 | 0.38 | 0.21 |

The calculation formula for the data of each observation point in this table is shown as Formula 1.

$$\varphi_i = \sum_1^n \frac{a_{in}}{A_n} \times \omega_n \qquad (1)$$

Here, $\varphi_i$ is the proportion of backgrounds, basic or expanded knowledge, respectively. $a_{in}$ is the number of pages for each type of knowledge in each chapter. $A_n$ is the number of pages in corresponding chapters. $\omega_n$ is the weight of each chapter, $\omega_n$ equals to the proportion of chapter pages in total pages; $n$ is the number of chapters. Besides, the definition of each type of knowledge is listed as follow:

a. *Backgrounds.* It includes the introduction of backgrounds in related basic subjects, e.g., mathematics, physics, chemistry and all related natural science basic courses.
b. *Basic knowledge.* It includes the important concepts, definitions and basic theories in the book.

c. *Expanded knowledge.* It includes the comprehensive analysis and application of basic knowledge, including the case analysis, market prospects, frontier scientific researches, new technology application, *etc.*

From Table 1, it can be found that the knowledge system of Chinese and foreign textbooks in nanomaterials and technology is significantly different, which is mainly reflected in the proportion of basic, applied and frontier knowledge. The similarities and differences are as follows:

- *Common points.* Chinese and Foreign textbooks include all the three parts as background, basic and expanded knowledge in the knowledge system of whole book, which can enrich the teaching material and offer help for students to understand the knowledges in nanomaterials step by step.
- *Differences.*
  1) *Depth of the knowledge system.* The proportion of background and basic knowledge (>0.4) in domestic textbooks is higher than that in foreign textbooks, while foreign textbooks usually involve more expanded knowledge points (> 0.2). Here, more elaboration of basic knowledge can provide a deeper understanding in the nanomaterials for students when learning the relative courses with the help of textbooks.
  2) *Breadth of the knowledge system.* In foreign textbooks, the proportion of applied knowledge is slightly larger (basically > 0.29) than that of domestic textbooks (only around 0.13–0.14). At the same time, foreign authors pay more attention to the content of frontier knowledge, the proportion of frontier knowledge (> 0.2) is much higher than that of Chinese textbooks (around 0.1). In Chinese textbooks, the applied knowledge and frontier knowledge are relatively lacking, which is reflected in the lacking of foresight and timeliness.

To conclude, by comparing the design of knowledge system in both Chinese and foreign textbooks, we find that Chinese textbooks prefer to show the importance of "background" and "foundation", while foreign textbooks highlight the "applications" and "frontiers". Especially, the analysis of industry applications and market prospects in foreign textbooks is very distinctive, which is worth to be adopted in the organization of future Chinese textbooks.

## 3.2 Comparation of the Design of Content of Examples and Exercises in Chinese and Foreign Textbooks

In order to explore the differences in the presentations of the above four types of knowledge, in this part we compare the design of content of examples and exercises in Chinese and foreign textbooks in nanomaterials and technology. Specially, the proportion of memorizing/conceptual and comprehensive/analytical questions in the examples and exercises in the textbooks are statistically compared respectively in Table 2.

The calculation formula for the data of each observation point in this table is shown as Formula 2.

$$\alpha_i = \sum_1^n \frac{c_{in}}{B_n} \times \omega_n \qquad (2)$$

**Table 2.** Comparison of memorizing/conceptual and comprehensive/analytical questions in the examples and exercises in Chinese and foreign textbooks

| Textbooks | | Examples | | Exercises | |
|---|---|---|---|---|---|
| | | Memorizing /Conceptual questions | Comprehensive /Analytical questions | Memorizing/Conceptual questions | Comprehensive/Analytical questions |
| Chinese textbooks | Nanomaterials and Their Preparation Techniques | / | / | 0.40 | 0.60 |
| | Fundamentals of Nanomaterials | 0.12 | 0.71 | 0.19 | 0.66 |
| Foreign textbooks | Fundamentals of Nanotechnology | 0.09 | 0.88 | 0.08 | 0.78 |
| | Nanotechnology: Principles and Applications | 0.06 | 0.79 | 0.11 | 0.75 |

Here, $\alpha_i$ is the proportion of memorizing/conceptual and comprehensive/analytical questions in the examples and exercises in the textbooks, respectively. $c_{in}$ is the number of questions for each category in each chapter. $B_n$ is the number of examples and exercises in each chapter. $\omega_n$ is the weight of each chapter. $n$ is the number of chapters. Besides, the definition of each type of observation point is listed as follow:

a. *Memorizing/conceptual questions.* Objective questions that include an examination of concepts, nouns, definitions and other basic knowledge in the textbook.
b. *Comprehensive/analytical questions.* Including the comprehensive use and analysis of basic knowledge in the textbook, and the opening and exploring questions about the expanded knowledge.

From Table 2, it can be found that there are the following differences in the design of content of examples and exercises in Chinese and foreign textbooks in nanomaterials and technology.

1) In foreign textbooks, the proportion of memorizing/conceptual questions in examples and exercises (< 0.1) are smaller than that in Chinese textbooks (> 0.1), while comprehensive, comprehensive/analytical questions have the opposite proportion. Memorizing/conceptual questions can deepen students' understanding of basic knowledge and strengthen their ability to grasp key information in the textbook. On the other hand, comprehensive/analytical questions can guide students to find diversified problem-solving methods from lots of problems and cultivate students' comprehensive application ability of knowledge, which is valuable experience for preparing the future Chinese textbooks.
2) Foreign textbooks pay more attention to diversification of problem-solving methods for examples and exercises, which is helpful to improve the ability of students to think independently and respect students' personalized solutions to the problems involved in the examples and exercises. Through students' communication in the process of pre-class preview and after-class review, this characteristic in foreign textbook can

broaden their mind and cultivate the innovation thinking, which is very important for the improvement of students' comprehensive ability.

Therefore, adding comprehensive/analytical questions in the examples and exercises of the textbook and providing multiple problem-solving methods are helpful to improve students' pioneer thinking, which is valuable experience for reference in the compilation of future Chinese textbooks in nanomaterials and technology.

To sum up, based on the results from Table 1 and 2, it can be concluded that the similarities and differences in the design of structure of content between Chinese and foreign textbooks in nanomaterials and technology are as follow:

- *Common points.* Both Chinese and foreign textbooks can take advantages of examples and exercises to provide backgrounds, basic and expanded knowledge for the readers. The system of knowledge is complete and fruitful, which is important for the classroom teaching.
- *Differences.*
  1) For foreign textbooks, they always highlight the "applications" and "frontiers", which is different from Chinese textbooks that emphasize the "background" and "foundation". Lots of extended knowledge occupy the main part of foreign textbooks, which is important for an advance textbook in nanomaterials and technology. This point should be adopted in organizing the Chinese textbooks in future.
  2) Foreign textbooks pay more attention to diversification of problem-solving methods for examples and exercises, which is helpful to improve the ability of students to think independently. The comprehensive/analytical questions occupy the most in examples or exercises in foreign textbooks. This point is significant for a textbook which can guide the students to study independently, which should also be adopted in future Chinese textbooks.

## 4 Comparation of the Morphology of Chinese and Foreign Textbooks

In order to make a detailed comparative analysis of the morphology of Chinese and foreign textbooks in nanomaterials and technology, four objects are observed and compared from the aspects of readability, logic and self-conscious study which are provided by textbooks. Among them, the readability of the textbook is evaluated by the number of figures and tables, the proportion of color pages and art font style. Specifically, statistic results of each point are concluded in Table 3.

The calculation formula and methods for the data of each observation point in this table is as follow:

a. *Figures and Tables.* The ratio of the total number of figures and tables to the number of pages in the textbook.
b. *Color pages.* The ratio of the total number of color pages to the number of pages in the textbook.

**Table 3.** Comparison of the morphology of Chinese and foreign textbooks

| Textbooks | | Readability | | | Logic | Self-conscious study |
|---|---|---|---|---|---|---|
| | | Figures (/page) | Tables (/page) | Color page (%) | | |
| Chinese textbooks | Nanomaterials and Their Preparation Techniques | 0.89 | 0.09 | 0 | | 0.64 |
| | Fundamentals of Nanomaterials | 0.12 | 0.034 | / | There is less cohesive language between chapters and sections | 0.22 |
| Foreign textbooks | Fundamentals of Nanotechnology | 0.18 | 0.066 | 0.89 | There is lots of cohesive language between chapters and sections | 0.47 |
| | Nanotechnology: Principles and Applications | 0.14 | 0.053 | 0.83 | There is lots of cohesive language between chapters and sections | 0.43 |

c. *Self-conscious study.* This point is evaluated by Formula 3, which calculates the proportion of exercises and examples in the textbook.

$$\delta_i = \sum_1^n d_{in} \times \omega_n \qquad (3)$$

Here, $\delta_i$ is the proportion of exercises and examples in the textbook. $d_{in}$ is the number of exercises and examples in each chapter. $\omega_n$ is the weight of chapter pages. $n$ is the number of chapters.

From Table 3, it can be concluded that lots of differences are existed in the morphology of Chinese and foreign textbooks, such as the proportion of figures and tables, the use of art font words and the proportion of exercises and examples. The similarities and differences are listed as follows:

- *Common points.* Both of Chinese and foreign textbooks can use figures and tables to help the illustration of knowledge, which will provide a clear view for the readers.
- *Differences.* The advantages of foreign textbooks are reflected in the beautiful layout, the use of color pages, interesting pictures and ornamental art fonts. Besides, the statement in the book maintains a strict but humorous tone. The basic knowledge of nanomaterials can be introduced vividly in foreign textbooks by means of the combination of colorful tables and pictures. At the same time, foreign textbooks are more friendly to undergraduates in terms of writing style and problem solving. For example, the explanation is easy to be understood and the steps of problem solving are provided detailly.

The shortcomings of Chinese textbooks are the lack of vividness and readability of illustrations, which are mainly reflected in the small proportion of tables and figures in textbooks. Many illustrations in Chinese textbooks lack clarity and timeliness, and cases are dated to a long time ago, which has caused some obstacles for students to independent learning according to textbooks.

In recent years, more and more online teaching (including E-learning, online courses, *etc.*) has been applied to the education in universities. In particular, COVID-19 swept across the country and even the world since 2020, forcing all universities and colleges to delay the back to school. In order to ensure the quality of education, lots of universities have implemented online teaching. It can be predicted that online teaching will occupy an important place in future and form a complement to the traditional teaching. Facing this trend, there is a problem that needs to be seriously considered that how to serve the courses and teachers with more complete textbooks in nanomaterials and technology. What kind of textbooks should be made and how to serve the online teaching in a better way are the main problems that publishers need to consider. With the rapid development of online education, textbooks should conform to the E-learning and online courses.

From the perspective of this, a textbook should not only use papers as the carrier, but also should be equipped with various digital resources by advanced information technology. Therefore, we conduct the research part about the comparison of the application of information technology in Chinese and foreign textbooks. The results are concluded as following:

- *For the Chinese textbooks.* The textbooks are mainly traditional paper textbooks, most of which lack supporting digital resources and are only suitable for traditional offline teaching rather than online teaching.
- *For the foreign textbooks.*

  1) Almost all the professional textbooks of nanomaterials published in the past 10 years have electronic versions. Students can directly purchase authentic electronic textbooks from online platforms such as Amazon.
  2) Most of the textbooks have the exercises in the electronic version, also video for explanation of the exercises and other online teaching materials, which is convenient for students to learn the knowledges independently according to the book.

## 5 Conclusion

The main purpose of this study is to compare the Chinese textbooks in nanomaterials and technology with foreign textbooks and find a new way to improve the organization in Chinese textbooks in the related fields. Here, this study compares the books in terms of their design of structure, their morphology and the application of information technology (Fig. 1).

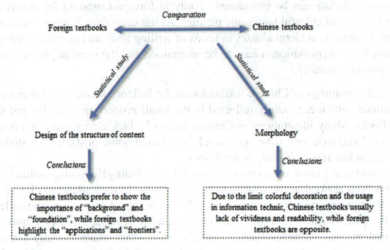

**Fig. 1.** Conclusions in this study

By comparing the structure of knowledge system in Chinese and foreign textbooks, it finds that both Chinese and foreign textbooks can contain the background knowledge, the basic knowledge and the extent knowledge while foreign textbooks have more extent knowledge to open the students' mind during learning. Besides, by statistically studying the proportion of figures, tables and color pages in the textbooks, the readability of Chinese and foreign textbooks is examined. Furthermore, the logic and the self-conscious study of the textbook are evaluated by the proportion of exercises and examples in the textbooks. It finds that the foreign textbooks usually have a large potential of art fonts, color pages and electronic materials to improve the readability and further can help the students in self-learning, as well as a larger proportion of the exercises and examples to simulate the self-motivation of students during course teaching. All these advantages this work presented in foreign books can be adopted by Chinese textbooks in future.

**Acknowledgements.** This work is supported by the Teaching Reform Project of Wuhan University of Technology (Grant No. w2022067).

## References

1. Liu Qingxian, Y., Jingyi.: Current status and innovation attempt of compiling textbooks of Chinese medicine in medical colleges in China. TMR Theory Hypothesis **2**, 478–485 (2021)

2. Rodríguez-Regueira, N., Rodríguez-Rodríguez, J.: Analysis of digital textbooks. Educ. Media Int. **59**(2), 172–187 (2022)
3. Bittar, M.: A methodological proposal for textbook analysis. Math. Enthusiast **19**(2), 307–340 (2022)
4. Pavešić, B.J., Cankar, G.: Textbooks and students' knowledge. Center Educ. Policy Stud. J. **12**(2)29–65 (2022)
5. Ramia, J.M., Soria-Aledo, V.: Textbook outcome (resultado de libro): una nueva herramienta de gestión. Cir. Esp. **100**(3), 113–114 (2022). https://doi.org/10.1016/j.ciresp.2021.06.002
6. Kyrpychenko, O., Pushchyna, I., Kichuk, Y., Shevchenko, N., Luchaninova, O., Koval, V.: Communicative competence development in teaching professional discourse in educational establishments. Int. J. Mod. Educ. Comput. Sci. **13**, 16–27 (2021)
7. Amjad, M., Akter, H.: An android based automated tool for performance evaluation of a course teacher (CTE)[J]. Int. J. Mod. Educ. Comput. Sci. **12**, 14–23 (2020)
8. Zhijuan, W., Xiaobin, Z., Wei, S., Antai, W.: Readability assessment of textbooks in low resource languages. Comput. Mater. Continua **10**, 213–225 (2019)
9. Qiyue, C., Zhongliang, N., Zhiyu, L.: Textbook outcome as a measure of surgical quality assessment and prognosis in gastric neuroendocrine carcinoma: a large multicenter sample analysis. Chin. J. Cancer Res. **4**, 433–454 (2021). (in Chinese)
10. Jiaming, Q.: A study on the representational meaning of the images in English textbooks based on visual grammar. Proceedings of Northeast Asia International Symposium on Linguistics, Literature and Teaching (2021)
11. Hongxing, J.: A brief review of root, stem and base and their application in the textbook of new senior English for China. Overseas Engl. **20**, 266–267 (2020)
12. Fei, Y.: Exploring native-speakerism in EFL textbooks in China. Overseas Engl. **14**, 260–262 (2018). (in Chinese)
13. Wei, W.: A comparative analysis of three versions of junior high school English textbook from the aspect of discourse[J]. Overseas Engl. **1**, 242–243 (2018). (in Chinese)
14. Xiaoyuan, S., Min, W.: An Eco-critical discourse analysis of texts in English textbooks based on discourse-historical approach. Overseas Engl. **17**, 221–222 (2017). (in Chinese)
15. Bakken, A.-B.: The textbook task as a genre. J. Curric. Stud. **53**(6), 729–748 (2021)
16. Benjamin Ginsberg. Thoughts on textbook writing. PS: Polit. Sci. Polit.**55**(3): 636–637 (2022)
17. Kollman, K.: The textbook road taken. PS: Polit. Sci. Polit. **55**(3): 638–640 (2022)
18. van Gijn, D., van Gijn, D., Newlands, C.: From proposal to publication-The process of writing an academic textbook. Br. J. Oral Maxillofac. Surg. **60**(10), e58 (2022)
19. Harijanto, B., Apriyani, M,E., Hamdana, E.N.: Design online learning system for Kampus Merdeka: a case study web programming course. Int. J. Mod. Educ. Comput. Sci. **13**, 1–9 (2021)
20. Long, D.T., Tung, T.T., Dung, T.T.: A facial expression recognition model using lightweight dense-connectivity neural networks for monitoring online learning activities. Int, J. Mod. Educ. Comput. Sci. **14**, 53–64 (2022)

# Practical Research on Improving Teachers' Teaching Ability by "Train, Practice and Reflect" Mode

Jing Zuo[1], Yujie Huang[1], and Yanxin Ye[2(✉)]

[1] Nanning University, Guangxi 530299, China
zuojingnancy@qq.com
[2] Guangxi Modern Polytechnic College, Guangxi 54700, China
117991414@qq.com

**Abstract.** The key to talent training quality lies in teachers' teaching level and teaching ability. As an essential way to develop the teaching ability of teachers, training has been emphasized by many colleges and universities. However, the training form of many colleges and universities nowadays is still mainly the traditional "lecturing" and "infusing" one. To test whether the mode of "train, practice and reflect" can effectively help teachers to improve their teaching ability, the teaching ability development program is designed and implemented from three aspects of training, practice and reflection, and the action research is carried out. The teachers are pre-tested and post-tested before and after the training, and the independent sample T-test is conducted by SPSS. The results show that the "train, practice and reflect" mode can significantly improve teachers' instructional design, teaching implementation, teaching evaluation, reflective ability and teaching innovation ability.

**Keywords:** Training mode · Teaching ability · Private university

## 1 Introduction

As an essential part of China's higher education system, private colleges and universities have played an essential role in popularizing higher education [1]. However, there are still many problems in classroom teaching in private application-oriented universities. Generally speaking, classroom teaching is still a one-way teaching indoctrination teaching method that lacks effective teacher-student interaction and student participation [2], of which the classroom atmosphere is dull, students are still in a passive learning state, and the concept of "student-centered, ability-oriented and output-oriented" needs to be further strengthened [3–5]. Maintaining the construction of teachers in private application-oriented universities, and significantly improving the teaching ability of teachers, has become the key to improving the teaching quality of private application-oriented undergraduate universities [6].

Many universities and scholars have carried out relevant research and practice, including research on classroom teaching quality [7, 8], teaching ability [9], the relationship between teaching ability and classroom teaching quality, etc. These help us understand the connotation of teachers' teaching ability and pinpoint the value orientation of optimizing course teaching quality. However, the following problems still lie in these studies and practices: First, the development of the teaching ability of teachers in application-oriented colleges and universities needs to be carried out fundamentally based on the standard requirements of application-oriented talent training. China's higher education can be divided into research, application, and vocational skills. And the demands and modes of talent cultivation vary in different universities. Basic theoretical knowledge is indispensable for the cultivation of application-oriented talents, with ability training as the core, and more attention must be paid to the comprehensive application of knowledge [10]. It is necessary to strengthen basic theoretical teaching, and practical application integration and improve students' ability to solve practical problems through practical projects [11]. All of these require that the teaching organization of applied teachers is mainly realized within the realm of ability [12]. However, the group of applied college teachers is influenced by the legacy of the traditional education model. They believe that the university stage is mainly for knowledge learning and theoretical research. To a conscious extent, they are more inclined to enrich academic knowledge structure [13]. A considerable number of teachers classroom teaching mode is accustomed to the traditional teaching mode, and the classroom teaching method is single, which is difficult to stimulate students' enthusiasm and initiative in learning; Second, in terms of form, the training forms for the development of teaching ability of teachers in application-oriented colleges and universities are mainly traditional "lectures" and "indoctrination", and lack the training and practice to cultivate teaching ability. Due to the large proportion of young teachers in most private application-oriented colleges and universities, the task of teacher development is heavy, and the training methods that can quickly cover a large area of teachers are mainly "invite in, go out". These training methods do play a specific role, but there are also specific problems: most of the lectures and reports are "lecture-style", only emphasizing the input of educational concepts and educational method knowledge, ignoring the cultivation and training of teaching ability and teaching skills [14]. Therefore, teachers lack the practical operation to integrate training content into actual teaching. This "lecture-style" training form will also make teachers unconsciously continue to use it in their classroom teaching, attach importance to the transmission of knowledge, and ignore the cultivation of students' abilities.

To tackle the above problems, this study designs and implements teaching ability development programs from three aspsects: training, practice and reflection, starting from the classroom teaching quality standards of applied talent training, and explores the development of teachers' teaching ability through action research. To understand the influence and role of the coupling linkage teaching ability training model of "train, practice and reflect" on teachers' ability in private colleges and universities and provide an empirical basis for scientifically improving teachers' teaching ability.

## 2 Design and Implementation of the "Train, Practice and Reflect" Coupling Linkage Teaching Ability Training Program

### 2.1 Scheme Design

Before carrying out the scheme design, the research group members sorted out the relevant literature, such as talent cultivation in application-oriented universities, curriculum construction in application-oriented universities, and classroom teaching quality in application-oriented universities. Interviews were conducted with some full-time teachers, course leaders, directors of teaching and research offices, heads of functional departments, and leaders of schools in charge of teaching in three private universities in Guangxi. And discussed in-depth issues such as "application-oriented talent training", "application-oriented classroom form" and "application-oriented classroom teaching quality standards". According to the requirements of application-oriented talent training [15], student-centered, ability and result output-oriented, the "Teaching Quality Standards for Applied Courses" was compiled from five dimensions: teaching attitude, learning objectives, instructional design, teaching implementation, and teaching effect. According to the quality standards of applied classroom teaching, combined with the goal of talent training in a private university and the needs of teachers' teaching ability development, the teacher-teaching ability development program is designed from three aspects: training, practice and reflection. Since the improvement of teaching ability is not achieved overnight and requires a specific development cycle, the entire training period is half a year when designing the program, of which the centralized training time is one week, and the half year after the intensive training will be accompanied by regular events, seminars and luncheon activities.

### 2.2 Program Implementation

The use of action research aims to solve the problem of teacher teaching capacity improvement. 200 teachers were recruited to carry out training in batches and phases, emphasizing "learner-centered, competency and outcome-output-oriented", changing the traditional training mode of only conducting lectures (lecture-based) and carrying out diversified training modes such as inquiry-based, personalized and participatory; Through assignments, tasks, projects, etc. as the carrier, let the trainees use the training content to carry out multiple teaching exercises in the training; Through unique design and specialized training on teaching reflection in training, teachers can continuously reflect and learn in practice, and constantly improve the practice of teaching reform through review and summary. At the end of the week's intensive training, the research group will also understand the situation of teachers applying what they have learned in training to their actual teaching and organize various teaching competitions, as well as seminar-like activities such as teaching luncheons and teaching salons, to provide an excellent environmental foundation for teachers' subsequent practice and reflection. After the end of half a year, the "Applied College Teacher Teaching Ability Evaluation Scale" was used to conduct pre-test and post-test before and after the program's implementation to understand the development of teachers' teaching abilities.

## 3 Research Methods

### 3.1 Survey Subjects

The research data of this paper comes from 200 teachers from a private university in Guangxi who participated in the "training, practice and reflection" coupling teaching ability training program designed by the university. Table 1 is a sample of descriptive statistical results of implementing the training plan.

Table 1. Sample descriptive statistical results

| Variable |  | Indicator |  |  |  |
|---|---|---|---|---|---|
|  |  | Before Training |  | After Training |  |
|  |  | Frequency | Percentage | Frequency | Percentage |
| Gender | man | 63 | 32.5 | 56 | 31.1 |
|  | woman | 131 | 67.5 | 124 | 68.9 |
|  | total | 194 | 100 | 180 | 100 |
| Age | 30 or less | 29 | 14.9 | 25 | 13.9 |
|  | 31–37 | 90 | 46.4 | 84 | 46.7 |
|  | 38–44 | 48 | 24.7 | 46 | 25.6 |
|  | 45–51 | 19 | 9.8 | 18 | 10 |
|  | 52 and above | 8 | 4.1 | 7 | 3.9 |
|  | total | 194 | 100 | 180 | 100 |
| Seniority | less than three years | 43 | 22.2 | 37 | 20.6 |
|  | 3–8 years | 84 | 43.3 | 81 | 45 |
|  | 9–14 years | 36 | 18.6 | 35 | 19.4 |
|  | 15–21 years | 20 | 10.3 | 17 | 9.4 |
|  | more than 22 years | 11 | 5.7 | 10 | 5.6 |
|  | total | 194 | 100 | 180 | 100 |
| Job Title | beginner and below | 27 | 13.9 | 23 | 12.8 |
|  | intermediate | 100 | 51.5 | 94 | 52.2 |
|  | deputy senior | 57 | 29.4 | 53 | 29.4 |
|  | positive advanced | 10 | 5.2 | 10 | 5.6 |
|  | total | 194 | 100 | 180 | 100 |

The training methods and approaches are designed in light of the standard process and scientific methods [16]. The questionnaire was divided into pre-training and post-training distribution, and the trainees received the questionnaire survey three days before the training, and the trainees answered the questions anonymously. A total of 194 people participated in the answers, and the questionnaire recovery rate was 97%. At the end of the half-year training, the same questionnaire was distributed again, and a total of 180 people participated in answering the questions, and the questionnaire recovery rate was 90%. The basic information of teachers before and after participation is fully displayed.

### 3.2 Scale Design

The development of the "Applied College Teacher Teaching Ability Evaluation Scale" is mainly divided into the following three processes:

First, according to Tang Yuguang's research, teaching ability includes five dimensions: instructional design, teaching implementation, teaching evaluation, teaching reflection, and teaching innovation, and combined with relevant literature, a scale is preliminarily compiled based on the understanding of the concept of this dimension. Second, the Delphi method was used to conduct multiple rounds of consultation with relevant experts and improve the scale. Third, 80 teachers from a university were selected for the pilot test, the reliability and validity of the scale were tested, and some questions and expressions of the scale were adjusted.

The resulting scale and related topics are shown in Table 2.

**Table 2.** Examples of Scale Dimensions and Their Titles

| Variable | Dimension | Questions | Examples of Topics |
|---|---|---|---|
| Teaching Ability | Instructional Design | 7 | When preparing for lessons, I can complete the instructional design relatively smoothly |
| | Teaching Implementation | 6 | I understand the diversity of students' learning and can teach them accordingly |
| | Pedagogical Evaluation | 3 | I design and conduct assessments suitable for evaluating students' learning outcomes (achievement of learning goals) |
| | Pedagogical Reflection | 3 | After teaching, I reflect on my teaching |
| | Pedagogical Innovation | 4 | I would like to explore a variety of information technology-assisted teaching activities |

## 4 Data Analysis and Discussion

Sample T-test is carried out to test whether there is a significant difference between the teaching ability and its five-dimensional sub-abilities before and after the training.

## 4.1 Reliability Test

The reliability of each dimension was analyzed using Cronbach's alpha coefficient, and the results are shown in Table 3. The reliability of the five measurements is greater than 0.7, which shows good scale reliability.

**Table 3.** Scale Reliability, Average Value and Standard Deviation of Each Dimension

| Dimension | Quantity | Average Value | Standard deviation | Cronbach's Alpha |
| --- | --- | --- | --- | --- |
| Instructional Design | 374 | 4.10 | 0.57 | 0.844 |
| Teaching Implementation | 374 | 3.92 | 0.54 | 0.774 |
| Pedagogical Evaluation | 374 | 3.88 | 0.60 | 0.722 |
| Pedagogical Reflection | 374 | 4.03 | 0.61 | 0.764 |
| Pedagogical Innovation | 374 | 3.92 | 0.56 | 0.780 |

## 4.2 Average Value, Standard Deviation and Independent Sample T-Tests

To investigate whether there is a significant difference in teaching ability before and after training, an independent sample T-test was conducted.

**Table 4.** Results of independent sample t-test in the group before and after training

| Dimension | Constituencies | Quantity | Average Value | Standard Deviation | T | Degree of Freedom |
| --- | --- | --- | --- | --- | --- | --- |
| Instructional Design | Before training | 194 | 3.96 | 0.69 | 4.974*** | 286.407 |
| | After training | 180 | 4.24 | 0.34 | | |
| Teaching Implementation | Before training | 194 | 3.71 | 0.59 | 8.551*** | 321.241 |
| | After training | 180 | 4.14 | 0.36 | | |
| Pedagogical Evaluation | Before training | 194 | 3.69 | 0.65 | 7.020*** | 347.162 |
| | After training | 180 | 4.09 | 0.46 | | |
| Pedagogical Reflection | Before training | 194 | 3.91 | 0.66 | 4.039*** | 363.710 |
| | After training | 180 | 4.16 | 0.52 | | |
| Pedagogical Innovation | Before training | 194 | 3.78 | 0.63 | 5.233*** | 341.085 |
| | After training | 180 | 4.07 | 0.43 | | |
| Total Teaching Capacity | Before training | 194 | 3.83 | 0.58 | 7.130*** | 268.654 |
| | After training | 180 | 4.16 | 0.25 | | |

Note: *** indicates significance at the 0.01 level

Table 4 shows the five dimensions of teaching ability, the mean and standard deviation of teaching ability before and after training, and the T-test results. The results show significant differences at the 0.01 level of the five dimensions of instructional design, teaching implementation, teaching evaluation, teaching reflection and teaching innovation, and the same goes with the total teaching ability at the 0.01 level. After the training, its teaching ability and its five dimensions were significantly enhanced compared with before the training.

As the data in Table 4 shows, teachers' teaching capabilities have significantly changed after training. The improvement of these capabilities is also consistent with the design idea of the "training, practice, thinking" coupling linkage teaching ability training program in this action study, discussed as follows.

(1) Instructional design ability

Gagne first proposed instructional design, which refers to planning the entire teaching system and offering specific solutions [17]. Instructional design is an essential competency for teachers. To help teachers master instructional design, a training course on instructional design is set up in this action study Help teachers master the methods of instructional design. Then the training program also sets up practical activities of teaching drills so that teachers can put instructional design to the reasonable level through the subsequent teaching drill process; In the teaching exercise, teachers were allowed to observe each other, put forward suggestions for revision and guidance, and at the same time filmed the video of the drill for the teacher. The teacher could reflect on whether the instructional design was scientific and reasonable through peer communication and a personal video review. Therefore, this training mode of training, practice and reflection helps teachers quickly understand and master the essentials of instructional design and improve their instructional design ability.

(2) Ability to implement teaching

The ability to implement teaching refers to teachers' ability to effectively organize and implement instruction according to instructional design and actual environment. Student competence can be divided into core competence and position-related competence, and position-related competence mostly depends on the effective teaching implementation of teachers. To help teachers to improve their teaching implementation ability, they are asked to carry out corresponding teaching drills in the process of participating in the training. After each teaching drill, teachers will receive suggestions and opinions from peers, and teachers will continue to carry out the next teaching drill after further improvement. Through such repeated practice, teaching skills are integrated into practical exercises, which enhances teachers' confidence in implementing teaching and strengthens teachers' skills in implementing it. Due to the repeated exercises during and after the training, most of the teachers will apply the knowledge they have learned to their own teaching for real-life teaching. And the school also cooperates with various teaching competitions to help teachers continuously polish their teaching skills and improve their teaching implementation capabilities.

(3) Teaching evaluation ability

According to the reverse design principle of instructional design, we first determine the learning objectives (learning outcomes), then decide how to evaluate whether the learning objectives have been achieved (what is the evidence of learning outcomes), and finally, design learning activities and tools. The "determine by what method to evaluate whether the goal has been achieved" refers to teaching evaluation. Teaching evaluation ability refers to the power of teachers to design scientific and reasonable teaching evaluation methods to carry out an evaluation to help students learn. In this action study, the overall design of the curriculum also reflects the concept of OBE, guiding teachers to pay attention to students' output, think about how to evaluate students' production based on students' output, and then carry out an instructional design. The guidance of these concepts runs through training, practice and reflection, so that teachers can think about how well they have achieved their goals in this teaching drill and whether they have designed reasonable evaluations based on the goals every time they do instructional design and teaching drills.

(4) Teaching reflection ability

Dewey was the first to introduce "reflection" into teaching, and teaching reflection refers to the process of teachers themselves reviewing, thinking and summarizing their teaching. Professor Ye Lan once said that a teacher who writes a lifetime lesson plan could not become a famous teacher, and if a teacher writes three years of teaching reflection, he may become a renowned teacher. The ability to reflect on teaching is also an indispensable part of a teacher's teaching ability. In this action study, the overall design of the plan is also to integrate reflection into all links. For example, after the teaching exercise, each teacher is required to fill out a self-evaluation form for this training, reflect on the advantages and shortcomings of the teaching just now, and guide teachers to think about how to improve in the future; At the same time, in peer feedback, teachers will unconsciously reflect on their entire teaching process again while listening to peer feedback; Videos were also recorded for each teacher's teaching exercise for teachers to review privately to reinforce the process of self-reflection further. After the overall training, seminars such as teaching salons and teaching luncheons are also provided, allowing teachers to put forward their confusions and problems encountered in teaching practice and constantly review and review their teaching through discussions and exchanges with peers, to improve teachers' teaching reflection ability.

(5) Ability to innovate in teaching

Teaching innovation has been a hot word in recent years. Teaching innovation ability refers to the power of teachers to start from real teaching problems and solve problems with innovative methods. Trying new teaching models provide teachers with a fresh approach and gives them tools to improve the teaching process. Teaching models give teachers conceptual as well as a practical technology to teach from. New teaching approaches can help them understand the view of the content to be taught, as well as to reflect on their ideas of learning, the learner, and their role of themselves. In this action study, the design of the training program did not specifically cover the training content of teaching innovation. Still, the final results showed that teachers' teaching innovation ability increased, and the reasons might be as follows: First, because the whole training emphasises the integration of training, practice and reflection, teachers were able to grasp ideas and methods through training, bridge the gap between learning and doing through practice, and find problems in their teaching through reflection. It is known that finding problems is very important, and this lays the foundation for teachers to innovate in education. Second, because the training process creates an atmosphere of trust and inclusiveness, it also encourages teachers to try new ways and methods to solve problems after discovering problems, which also provides environmental support for teachers to carry out teaching innovation; Third, due to the form of peer assistance in training, teachers can continue to reflect on themselves after carrying out teaching innovation, and on the other hand, they can communicate and discuss with peers, to better improve teaching. Based on the above analysis, improving teachers' teaching innovation ability is promoted.

## 5 Conclusion

Current research shows that the coupling linkage teaching ability training model of "train, practice and reflect" can significantly improve teachers' teaching ability, which is mainly reflected in the improvement of five aspects: teachers' instructional design ability, teaching implementation ability, teaching evaluation ability, teaching reflection ability and teaching innovation ability.

The coupling linkage teaching ability training model of "train, practice and reflect" can effectively help teachers improve their teaching ability, which breaks through the simple training lecture form and integrates training, practice and reflection. It allows teachers to learn through diversified learning activities, deeply involved practical training, and peer-to-peer learning mode. This also helps the institutions responsible for teacher training and teacher development in our universities to better think about the design of our training programs.

Due to the limitations of the research institute, this action plan is limited to teachers in a private university, the coverage needs to be wider, and the data needs to be more representative. Follow-up research can collaborate with more similar universities and cover a wider group of university teachers.

**Acknowledgment.** This paper is the result of the 2021 private higher education research project "Research on the improvement of teaching ability of teachers in private applied universities" (No.2021ZJY667) of the Guangxi Education Science Plan.

## References

1. Shi, Q., Zhang, C.: The internal logic of private universities' development: reconstruction and transformation paths. J. Higher Educ. Manage. **14**(04), 25–31 (2020). (in Chinese)
2. Pianta, R.C., Hamre, B.K., Allen, J.P.: Teacher-student relationships and engagement: conceptualizing, measuring, and improving the capacity of classroom interactions. Handbook of Research on Student Engagement, pp. 365–386. Springer, Boston, MA (2012). https://doi.org/10.1007/978-1-4614-2018-7_17
3. He, L.: Application of English situational teaching in classroom. Adv. Vocational Tech. Educ. **3**(2), 142–146 (2021)
4. Ahmad, N.N.N., Sulaiman, M.: Case studies in a passive learning environment: some Malaysian evidence. Account. Res. J., (2013)
5. Gu, W.: Reform and exploration of database principle course based on OBE learning outcome-oriented. In: The 4th Annual 2018 International Conference on Management Science and Engineering (MSE2018). Francis Academic Press, UK, pp. 70–75 (2018)

6. Steele, D., Zhang, R.: Enhancement of teacher training: key to improvement of English education in Japan. Procedia Soc. Behav. Sci. **217**, 16–25 (2016)
7. Kong, L., Wang, X., Yang, L.: The research of teaching quality appraisal model based on AHP. Int. J. Educ. Manage. Eng. (IJEME) **02**, 29–34 (2012)
8. Zhang, S.: Comprehensive evaluation of teaching quality based on cluster analysis and factor analysis. Int. J. Educ. Manage. Eng. (IJEME) **01**, 16–23 (2011)
9. Li, N., Zhang, Y.: Improvement and practice of secondary school geography teachers' informatization teaching ability based on the perspective of MOOCs. Int. J. Educ. Manage. Eng. (IJEME) **12**, 11–18 (2022)
10. Chen, X., Yang, X.: 14 basic problems in the development of new application-oriented undergraduate colleges. China Univ. Teach. **01**, 17–22 (2013). (in Chinese)
11. Pan, M., Che, R.: On the positioning of application-oriented universities. J. Higher Educ. **30**(05), 35–38 (2009). (in Chinese)
12. Chen, X.: Exploration of the construction of new applied technology university, pp. 378–388. Guangxi Normal University Press, Guilin (2019). (in Chinese)
13. Cai, H., Xiong, K.: Action learning: training path of the teaching ability of "double teacher" teachers in applied undergraduate colleges. Heilongjiang Res. High. Educ. **36**(06), 100–104 (2018). (in Chinese)
14. Wu, X.: The practice and reflection of the "Young teachers' teaching contest" to improve teaching ability of young teachers in university. Theory Pract Educ. **38**(09), 41–42 (2018). (in Chinese)
15. Zuo, J., Zhang, G., Huang, F.: Construction of ability and quality model of engineering talents based on analytic hierarchy process. In: Zhengbing, H., Zhang, Q., Petoukhov, S., He, M. (eds.) Advances in Artificial Systems for Logistics Engineering, pp. 356–366. Springer International Publishing, Cham (2022). https://doi.org/10.1007/978-3-031-04809-8_32
16. Hatice, Y., Murat, D.: Training program supporting language acquisition. Int. J. Mod. Educ. Comput. Sci. (IJMECS) **03**, 1–12 (2021)
17. Ma, L., Sheng, Q., ed.: Teacher Instructional Design Ability Development, vol. 36 Zhejiang University Press, Hangzhou (2016). (in Chinese)

# College Foreign Language Teacher Learning in the Context of Artificial Intelligence

Jie Ma[1,2(✉)], Pan Dong[3], and Haifeng Yang[4]

[1] School of Education, Huazhong University of Science and Technology, Wuhan 430074, China
361998096@qq.com
[2] School of Economics and Business Foreign Languages, Wuhan Technology and Business University, Wuhan 430065, China
[3] South-Central Minzu University, Wuhan 430074, China
[4] Faculty of Letters, Arts and Sciences, Waseda University, Shinjuku, Japan

**Abstract.** This study adopts a combination of quantitative and qualitative methods, uses the "teacher learning questionnaire" and interviews to explore and analyze the composition, characteristics and impact of AI on teacher learning of 148 college foreign language teachers in the context of AI language learning. The results are as follows: college foreign language teachers' learning includes three dimensions: intrinsic motivation, organizational support, and feedback and evaluation, and the participants' intrinsic motivation and feedback and evaluation reach the high-frequency experience level; the characteristics of college foreign language teachers' learning are ecological orientation, practical orientation and individual orientation; artificial intelligence technology guarantees the autonomy of teachers' learning through learner characteristics, supports the normalization of teachers' learning through effective supply of learning resources, and improves the collaboration of teachers' learning through virtual teaching and research rooms. Artificial intelligence integrates information technology and education, which helps to deepen and expand the research on teacher learning.

**Keywords:** Artificial intelligence · College foreign language teacher · Teacher learning · Ecological orientation · Lifelong learning

## 1 Introduction

"Teacher learning" is a concept embedded in the field of teacher education. With the increasingly prominent characteristics of teachers' initiative, daily learning and endogenous knowledge, teacher learning gradually replaces "teacher professional development" and becomes a new concept of teacher education [1]. In addition, under the influence of constructivism, situational learning theory and socio-cultural theory, with the occurrence of the "learning revolution", teacher learning research more clearly and consciously absorbs new learning theories to create a teacher learning environment and test its role in promoting teacher development, so as to make teacher learning become one of the frontier fields of teacher cognitive research and teacher education practice [2]. These provide a theoretical basis for further research on teacher learning.

© The Author(s), under exclusive license to Springer Nature Switzerland AG 2023
Z. Hu et al. (Eds.): ICAILE 2023, LNDECT 180, pp. 854–864, 2023.
https://doi.org/10.1007/978-3-031-36115-9_76

In 2018, the CPC Central Committee and the State Council issued the Opinions on Comprehensively Deepening the Reform of the Construction of Teachers in the New Era, which proposed: "By 2035, teachers' comprehensive quality, professional level and innovation ability will be greatly improved, and millions of backbone teachers, hundreds of thousands of outstanding teachers and tens of thousands of educators will be trained". Foreign language teachers are the subjects to "build new majors or directions, exploring new training modes, building new courses, and building new theories" [3]. Teacher learning is the only way to achieve these goals.

At present, a social ecosystem based on the flow of data and knowledge among social subjects is gradually taking shape, and big data driven education research, education management and decision-making have become the theme of the times [4]. With the further development of information and technology, AI participates in the breadth and depth of teachers' learning methods, quality assurance mechanisms, organizational support and operation modes, information and communication technology approaches and social support systems.

Based on these changes, this study aims to provide some inspiration and reference for the further promotion of teacher learning research driven by big data in the process of deepening the integration of information technology and education through sorting out foreign language teacher learning, especially the analysis of teacher learning turn in the context of artificial intelligence.

## 2 Literature Review

### 2.1 Teacher Learning

Teacher learning refers to "a holistic activity in which teachers actively seek to improve their overall quality under the support of external environment and continue to pursue the mutual unity of professional development and personal development" [5]; As an alternative concept of teachers' professional development, teacher learning emphasizes "teachers' initiative in their own professional development" [6], "the daily nature of the learning process" [7], and "the self-sufficiency of teachers' knowledge and ability" [8]. In the 1950s and 1960s, Taylor, an American scholar, prospectively put forward that "the future in-service training will not be regarded as 'training' teachers, but will help, support and encourage each teacher to develop the teaching ability he values and hopes to increase. The guiding and universally recognized spirit will be to put learning itself in the most important position" [9]. This study uses Fred Korthagen's *The Onion Reflection Model of Teacher Development,* which reveals the process of teacher learning: (1) In teacher learning, the cognitive, emotional and motivational dimensions of instructional behavior are intertwined, so teacher learning is multi-dimensional learning. (2) In the model, the learning process occurs at all levels, that is, teacher learning is multi-level learning. (3) At the core of the model is the core quality of teachers, that is, the role expertise in psychology [10].

Teacher learning has three characteristics: independent self-concept, rich individual experience, realistic demand and problem-solving drive. First of all, teachers as learners have independent self-concept. As adults and educators, teachers have gradually formed a mature self-concept from their previous growth and teaching experience, and have a clear understanding of their own abilities, personalities, attitudes, etc., so as to have an accurate grasp of the reasons and goals of their own learning. Secondly, teachers as learners have rich individual experience. The rich experience and corresponding practical knowledge that teachers have accumulated in the past learning process, social life and teaching career is an important basis for mobilizing teachers' thirst for knowledge. Finally, in addition to teachers as learners, teachers have multiple roles such as managers, organizers, communicators, etc., which endow teachers with multiple responsibilities such as caring for students, communicating with parents and leaders. In the process of assuming these responsibilities, teachers are faced with a variety of situational problems. The need to solve these problems, to a certain extent, urges teachers to actively participate in, consciously and automatically learn.

### 2.2 College Foreign Language Teacher Learning in the Context of AI

In the field of teacher learning research, AI brings a series of benefits by virtue of the advantages of big data computing. The typical applications of AI in the field of education include intelligent tutors to assist personalized teaching, educational robots, intelligent evaluation of real-time tracking and feedback, etc. In this research, foreign language teacher learning in the context of AI refers to foreign language teacher learning assisted by AI devices, which mainly include three categories: robots, professional software and online learning platforms. Robots, such as iFLYTEK translator, chat robot, etc.; Professional software, such as Google Translation, Kingsoft PowerWord, etc.; Network teaching platform, such as Chaoxing Fanya, moocs.unipus.cn, etc. In this study, foreign language teacher learning in the context of artificial intelligence is to investigate teacher learning based on the use of the above three types of devices with different degrees of intelligence.

## 3 Research Design

This study attempts to answer the following questions:

(1) What are the constituent factors of college foreign language teacher learning in the context of AI?
(2) What are the characteristics of foreign language teacher learning in the context of AI?
(3) What are the effects of AI on college foreign language teacher learning?

### 3.1 Subjects

The research includes two stages, questionnaire stage and interview stage.

Subjects in questionnaire stage are 148 college foreign language teachers, with an average teaching year of 11. The average time for participants to use the three types of artificial intelligence language learning devices: robots, professional software and online learning platforms is 2.33 years.

Based on the principle of maximum differentiation, 10 foreign language teachers from universities in different areas are interviewed (see Table 1).

**Table 1.** Basic information of subjects

| Subject | Gender | Age | Teaching Experience | Degree | Academic Title | University Type | Area |
|---|---|---|---|---|---|---|---|
| T1 | Female | 41 | 14 | Master | Lecturer | PGU-C | Central |
| T2 | Male | 50 | 24 | Doctor | Professor | PGU-C | Southwest |
| T3 | Male | 39 | 15 | Master | Associate Professor | PGU-C | East |
| T4 | Female | 25 | 1 | Master | Assistant | PGU-C | Central |
| T5 | Female | 34 | 8 | Master | Lecturer | PGU-F | South |
| T6 | Female | 40 | 15 | Master | Lecturer | DFU-C | Northeast |
| T7 | Male | 40 | 12 | Doctor | Lecturer | PMKU-C | East |
| T8 | Female | 39 | 14 | Doctor | Associate Professor | DKU-SE | Central |
| T9 | Female | 33 | 2 | Master | Assistant | PGU-C | Central |
| T10 | Female | 34 | 8 | Master | Lecturer | PGU-C | Central |

In the analysis of basic information about subjects, there are differences in teachers' gender, teaching experience, degrees, professional titles, school types and regions. The age distribution of participating teachers is uniform (1 teacher over 50 years old, 3 teachers between 40 and 49 years old, 5 teachers between 30 and 39 years old, and 1 teacher between 25 and 29 years old). The teaching age covers the early career (2 teachers between 0 and 3 years old), the middle career (7 teachers between 4 and 20 years old), and the late career (1 teacher above 20 years old). The proportion of gender (30% male and 70% female), degree (70% doctoral and 70% master) and professional title (20% junior, 50% intermediate, 20% associate senior and 10% senior) of teachers participating in the program is consistent with the basic situation of foreign language teachers in China's universities (Ministry of Education, 2017).

In order to facilitate the concise expression of the University type in the table, "Provincial General University" is abbreviated to PGU, "Provincial Key University" is PKU, "Provincial and Ministerial Key Universities" is PMKU, "Double First class" Universities "is DFU, and "Comprehensive" is represented by the letter C", "Normal Education" is NE, "Finance" is F, and "Science and Engineering" is SE.

### 3.2 Instruments

The questionnaire "Teacher Learning Questionnaire" is composed of two parts: the first part is background information, including teaching age, gender and usage time of AI language learning equipment; The second part is the "teacher learning questionnaire", which has 23 items in total. The questionnaire is compiled after referring to Sun Chuanyuan's "teacher learning questionnaire" and combining the specific situation of college teachers. The questionnaire adopts the Likert 5-point measurement method from "completely inconsistent" to "completely consistent". The data collected from the questionnaire was processed using SPSS 26.0.

In-depth interviews with 10 teachers were conducted through WeChat and QQ voice calls, involving the teacher learning of the interviewees, including the content and mode of learning, learning objectives, factors affecting learning, and the impact of AI on teacher learning. Interview questions are as follows:

(1) Please describe your learning content, learning duration and schedule.
(2) What goals do you want to achieve or achieve through learning?
(3) Please describe the people or things that affect you in the learning process.
(4) Do you use AI in your learning? If so, what is it? What effect does it have on your learning?

### 3.3 Data Collection and Analysis

The questionnaire was distributed online through the website https://www.wjx.cn in December 2022. A total of 150 questionnaires were collected. After eliminating the questionnaires with high repetition rate and abnormal response time, 148 valid questionnaires were retained with a recovery rate of 98.67%.

The interview was recorded and translated into words with the consent of the teachers interviewed in January 2023. The transcribed content shall be further confirmed by the teacher. In addition, the researcher also collected the personal profiles, published academic papers, weblogs and other materials of the interviewed teachers on the school's official website to supplement and verify the interview data. Data analysis refers to the practice of Wen Qiufang and Zhang Hong [11]. First, the bottom-up grounded analysis method is used to complete the first and second level coding, and then the theoretical constructs in the model are used for the third level coding. In the encoding process, the original text shall be used as the encoded byte to describe or summarize the data as much as possible (See Table 2).

**Table 2.** Example of text analysis

| Coding | Text Example |
|---|---|
| First Level Coding | You can also know what is happening in the world by learning the latest foreign magazine… Learning is a channel to keep pace with the times and is helpful for class teaching… On the one hand, I am very lazy and don't want to work hard. On the other hand, I still have a desire for new knowledge. I also want to make progress and learn some new knowledge. Once a week, I am not oppressed. If I can accept it, I can also implement my love of learning |

*(continued)*

**Table 2.** (*continued*)

| Coding | Text Example |
|---|---|
| Second Level Coding | Perception of environments, awareness of behavior change, awareness of ability improvement, supplement and update of beliefs, positioning of professional identity, sense of responsibility for undertaking a mission, and description of personal character |
| Third Level Coding | Environmental factors, changes in behavior, ability and belief, professional identity, teacher's mission and personal/Core quality |
| Theme | Learning objectives and content, learning process and influencing factors |

## 4 Results and Discussion

### 4.1 Factors of Foreign Language Teacher Learning

The first author of this article verified the structural validity and reliability of the Teacher Learning Questionnaire. The KMO value of the scale data is .811, and the associated probability value of the Bartlett sphere test is .000, indicating that the data is suitable for factor analysis. Exploratory factor analysis adopts principal component analysis method, and performs factor analysis on the data according to the maximum variance method. The load value of each item is above .40, and there is no item with a load above .40 on both factors, which is classified into three factors, explaining 64.309% of the total variance. The internal reliability analysis of the scale shows that the overall Cronbach's alpha coefficient of the scale is .872, and the Cronbach's alpha coefficients of the subscales are .906, .770 and .780 in order. The questionnaire structure is reliable. The three factors are named Intrinsic Motivation, Organizational Support, and Feedback and Evaluation according to variance contribution rate (see Table 3).

**Table 3.** Descriptive statistics of the teacher learning

| Scale | Factor | Mean | Std. Deviation | Cronbach's alpha |
|---|---|---|---|---|
| Teacher Learning | Intrinsic Motivation | 4.201 | 0.089 | .906 |
| | Organizational Support | 3.363 | 0.134 | .770 |
| | Feedback and Evaluation | 3.556 | 0.167 | .780 |

According to Oxford and Burry-Stock's classification standard of Likert's 5-point scale, the average value is equal to or higher than 3.5 for high frequency use, and the average value is introduced. The level between 2.5 and 3.4 is moderate, and the average value is equal to or lower than 2.4 is low-frequency use [12]. Table 3 shows that the participants' Intrinsic Motivation and Feedback and Evaluation reach the high-frequency experience level. First, teachers' the high-frequency experience on Intrinsic motivation corresponds to andragogy. Andragogy incorporates elements of cognitive and social constructivism, is problem-centered, goal oriented, practically-based, and although sometimes mandated (e.g., through an employer), usually self-initiated [13, 14]. Notably, it is a systematic and self-sustaining learning activity that is primarily based on self-motivation and commitment. The realism is that in all learning activities, intrinsic motivation, is an unavoidable paradigm that evokes and sustains effective learning, and as a result, positive job performance [15]. Second, the high level of teachers' experience of the Feedback and Evaluation dimension indicates that evaluation is an important factor to test the achievement of learning objectives and to maintain learning persistence. One of the important objectives of teachers' learning is to promote teaching, and the improvement of students' performance is an important and direct way to promote teachers' learning. Evaluating teacher's performance is essential to improve learning system as it is related to student's learning [16].

### 4.2 Characteristics of Foreign Language Teacher Learning

The three factors explored in the questionnaire research show that teacher learning presents three orientations: (1) ecological orientation. (2) practice orientation. (3) individual orientation, and this finding is also supported in the interview. Intrinsic Motivation, Organizational Support, and Feedback and Evaluation represents the process of teacher learning and reflects the interaction between external education environment and internal motivation. Specifically, ecological orientation is manifested as the purpose of adapting to the social and educational environment. For example, T1 describes two purposes of learning: one is to adapt to the needs of social development and growth, the other is because *"I am a little scared, I always feel that English is going backwards quickly, afraid of being eliminated"*; the second is to *"know what is happening in the world, feel disconnected from the world, and the ability of information retrieval and reception is declining. Learning is a channel to keep pace with the times"*. The practice orientation shows that the learning goal is to solve practical problems. For example, T3 believes that learning is *"helpful for class. After taking over extensive reading, the textbook has not made a breakthrough. This is just a supplement for students, who also need it. More and more foreign articles are selected for reading in the grade examination. Students are more interested in this than textbooks"*. This is consistent with the research on adult learning theory of teachers' learning motivation, that is, as adults, one of the main learning characteristics of teachers is the internal drive based on problems and practical needs [17, 18]. Individual orientation shows that teachers have a high degree of self cognition and independent self-concept. These are consistent with the theory of Korthagen's teacher reflection model. The purpose of teacher learning is to adapt to the surrounding environment, and then changes in behavior and belief. The dimensions of self-identity are also more abundant. The most important thing is to cause teachers to reflect on themselves and think about the quality of their own advantages. For example, T1 thinks that she *"loves learning, so she studies. To be honest, on the one hand, she is lazy and does not want to work hard. On the other hand, she is eager for new knowledge, wants to make progress, and wants to learn some new knowledge. She does not oppress me once a week. If she can accept it, she can also implement his love of learning"*. This is consistent with the discovery that teacher learning has "independent self-concept".

## 4.3 The Effects of AI on Foreign Language Teacher Learning

### 4.3.1 Learner Feature Mining and Intelligent Modeling Supported by AI

Through the collection, analysis, aggregation of multiple heterogeneous data, and the use of the information complementary mechanism between multimodal data, the external learning state and internal cognitive structure of learners can be characterized. Its logic is through multimodal data perception, learner state representation, and deep feature mining. AI focuses on teachers as learners' feature mining, provides personalized learning services for learners, and optimizes the models and strategies of personalized learning support services. AI uses teachers' learning evaluation data, psychological evaluation data, external behavior data, physiological information data and human-computer interaction data to realize intelligent evaluation and analysis of learners' behavior, knowledge, cognition, emotional state, and explore potential characteristics of learners' learning interest, learning preference, learning motivation, learning style, etc. For example, T1 describes that *"mobile phone and network search and information retrieval functions are used. Artificial intelligence should be used in Rubik's Cube dubbing. Each sentence you dub will give you a score. If the pronunciation is inaccurate, it will be marked as artificial intelligence. A little bit (similar to machine review), but at the beginning, it was a machine review, but only scores. Now it has been upgraded. The wrong words are marked red, the length of rhythm stress is marked yellow, and the correct green. Then add scores. "*

### 4.3.2 Normalization and Innovation of Teacher Learning Maintained by AI

Artificial intelligence technology supports the construction of a new mechanism for the representation, aggregation and supply of teachers' learning resources, and provides technical support for teachers' learning embedded work and continuous learning. First, AI has built a structured learning content system. The knowledge map is used to identify the upper and lower relationship, implication relationship, and fore and aft relationship between teachers' learning content, and AI is used to build a teacher's learning content system with clear structure and strict logic; Second, man-machine cooperation supports intelligent learning resource annotation. AI can solve the problems of diversified presentation methods of massive learning resources at the current stage, unclear resource labeling, and unclear logical relationship between resources, carry out human-computer collaborative classification, labeling and evaluation of educational resources, build a dynamic mapping relationship between learning resources and teaching practice, achieve quantitative evaluation of potential characteristics of learning resources, and provide support for systematic reorganization and personalized push of learning resources; Third, AI pushes intelligent learning resources based on learners' level and needs. AI builds an association analysis model between "learner knowledge resources", realizes intelligent matching of learning resources and learning needs, and provides accurate resource push according to specific practice situations, learners' needs, learning styles and other characteristics. For example, T9 believes that AI plays the role of *"resource guarantee service provider"* in her learning. She, a young teacher with two years of teaching experience, pays attention to the teacher's classroom language, teaches the course Introduction to Chinese Culture, and often uses the retrieval function. These websites and applications will occasionally push her information about China's excellent cultural traditions, cultures of various countries Professional knowledge, ideological and political elements and other related topics; T4 focuses on *"learning new teaching methods, including translation software, machine translation, etc."*. Her English learning software will push relevant articles according to her interests, and will also form a list of new words, develop regular review plans, etc.

### 4.3.3 Teachers' Collaborative Learning Promoted by AI

Artificial intelligence, through the virtual teaching and research room, has contributed to a network learning community, enabling foreign language (mainly English) teachers in colleges and universities at different career development stages to develop a platform for cross regional, interdisciplinary and normalized collaborative innovation and development with the help of QQ groups, WeChat groups and other media for the purpose of resource sharing and academic exchanges. The influence of online learning community on the ability development of foreign language teachers in colleges and universities focuses on three aspects: teachers' professional cognition and knowledge enrichment, professional skills enhancement, and professional quality improvement. Under the leadership of the person in charge of the task community, teachers orderly carry out and complete all teaching and research activities and tasks on time, and conduct formal and informal learning in the practice/knowledge community through self interaction, peer interaction and interaction with the community in terms of resources, cognition, emotion, value and behavior. No matter what form of interaction, reciprocity has become its prominent feature. T2 met teachers with the same interest through an online meeting, established and carried out cross regional, interdisciplinary and cross stage cooperation through QQ groups, WeChat groups and other media, and *"cooperated with workers who pay attention to English education in rural primary schools and teachers in charge of this work in primary education colleges, and planned to publish a book (about) English education in rural primary schools"*. T1 got to know *"English Rubik's Cube Show (dubbing APP) part-time operation, (her job is) selecting articles, typesetting, publishing videos after class, sending notices in groups, etc. through online classes"*.

## 5  Conclusion and Implications

Teacher learning has gradually replaced "teacher professional development" as a new concept of teacher education. Artificial intelligence integrates information technology and education, which helps to deepen and expand the research on teacher learning. College foreign language teachers' learning includes three dimensions: intrinsic motivation, organizational support, and feedback and evaluation, showing the characteristics of ecological orientation, practical orientation and individual orientation. Artificial intelligence technology guarantees the autonomy of teachers' learning through learner characteristics, supports the normalization of teachers' learning through effective supply of learning resources, and improves the collaboration of teachers' learning through virtual teaching and research rooms.

Teacher learning in the context of artificial intelligence emphasizes the construction of suitable soil for promoting teachers' professional development through the change of time, leadership, system, mechanism and teacher culture. The school-based research and practice community promoted in the current practice field all point to the paradigm of ecological transformation. AI promotes the research and analysis of individual teachers as learners and the construction of multidisciplinary teacher learning community. Facing the coming of knowledge economy and the development of information technology, teachers should be "learners" and lifelong learners. Teachers are not only adult learners, but also learners who lead students to learn, learn to teach and pursue their own initiative. The role of teachers as "learners" highlights the proposition that "teachers are the subject of learning", and highlights the life growth and development of "teachers as human beings".

**Acknowledgments.** This paper is supported by supported by Special Funding in 2021 of Hubei Educational Science Planning Project (2021ZA15), Humanities and Social Science Research Project of Hubei Provincial Department of Education (QSY17007) and Teaching Research Project of South-Central Minzu University (Jyx16033).

# References

1. Miao, P., Xiaoyan, L.: Interpretation of "teacher learning" from the perspective of adult learning theory: returning to the adult identity of teachers. Res. Teach. Educ. **26**(6), 16–21 (2014). (in Chinese)
2. Zhnetsky, V.: The social role of intellectuals. Translated by Bin Xiang, pp. 7–8. Nanjing: Yilin Press (2000). (in Chinese)
3. Fan Liming "New liberal arts": the needs of the times and the focus of construction. China Univ. Teach. (5), 4–8 (2020). (in Chinese)
4. Zhanjun, W., Gang, Q.: New paradigm of education research driven by big data. Peking Univ. Educ. Rev. **16**(1), 179–185 (2018). (in Chinese)
5. Guorui, F.: Educational Ecology, p. 27. People's Education Press, Beijing (2000). (in Chinese)
6. Chuanyuan, S.: Teacher learning: expectations and Realities-Take Shanghai primary and secondary school teachers as an example. Shanghai: Shanghai Normal University (2010)
7. Easton, L.B.: From professional development to professional learning. Phi Delta Kappan **89**(10), 755–761 (2008)
8. Fullan, M.: The new meaning of educational change (4th edition), pp. 283–291. Teachers College Press, New York (2007)
9. Lieberman, A., Mace, D.: Teacher learning: the key to educational reform. J. Teach. Educ. **59**(3), 226–234 (2008)
10. Gang, Z., Yuting, W., Jingliu, H., Yujuan, L.: How to understand the multiple essence and multiple levels of teachers' professional learning – a dialogue with Professor Fred Kosagen, a world-renowned educator. Mod. Distance Educ. Res. **3**, 32–43 (2021). (in Chinese)
11. Qiufang, W., Hong, Z.: Listening to the voice of young college English teachers: a qualitative study. Foreign Lang. Teach. **1**, 67–72 (2017). (in Chinese)
12. Oxford, R.L., Burry-Stock, J.A.: Assessing the use of language strategies worldwide with the ESL/EFL version of the strategy inventory for language learning. System **23**(2), 1–23 (1995)
13. Merriam, S., Brockett, R.: The Profession and Practice of Adult Education: An Introduction. Jossey-Bass, San Francisco (2007)
14. Spencer, B.: The Purposes of Adult Education: a Short Introduction, 2nd edn. Thompson Educational Publishing, Toronto (2006)

15. Sweden, S.D.M., Brendan, P.K.: A case study on adult and workplace learning. Int. J. Educ. Manage. Eng. (IJEME) **1**, 11–19 (2020)
16. Mahfida, A., Nusrat, J.L.: A web based automated tool for course teacher evaluation system (TTE). Int. J. Educ. Manage. Eng. (IJEME) (2), 11–19 (2020)
17. Yanju, J., Shaokang, X.: On the construction of adult learners' learning motivation. Rural Adult Educ. **8**, 7–8 (2000). (in Chinese)
18. Gilakjani, A.P., Leong, L.-M., Ismail, H.N.: Teachers' use of technology and constructivism. Int. J. Mod. Educ. Comput. Sci. (IJMECS) **5**(21), 49–63 (2013)

# The Innovation Integration Reform of the Course "Single Chip Microcomputer Principle and Application"

Chengquan Liang(✉)

School of Intelligent Manufacturing, Nanning University, Nanning 530200, China
759137827@qq.com

**Abstract.** The course of Single Chip Microcomputer Principle and Application is a required course for most colleges and universities in electrical engineering and automation, mechanical design and manufacturing and automation, robot engineering, artificial intelligence and other related majors. It is a comprehensive course that integrates the basics of electronic technology, computer language programming, electronic product design, etc. This paper is based on the teaching reform of the principle and application course of single-chip computer with the integration of specialty and innovation, and integrates the innovation and entrepreneurship project into the teaching process of the course. Students are no longer limited to memorizing theoretical knowledge, but combine the learning and practice of the course. The project training of innovation and entrepreneurship has stimulated students' enthusiasm for learning and confidence in innovation and entrepreneurship. They will be more interested in the course content and their comprehensive ability can be greatly improved. Such reform activities can cultivate more innovative and entrepreneurial talents and lay a solid foundation for their growth.

**Keywords:** One-chip computer · Professional and creative integration · Curriculum reform · Innovation and entrepreneurship training program

## 1 Introduction

To ensure that the course teaching achieves good results, the focus is on continuous construction and continuous optimization [1–4]. Only by clarifying the nature and characteristics of the curriculum and understanding the needs of talent training can we know the direction of curriculum reform and make the teaching process truly play its due role [5, 6].

Microcomputer integrating CPU, RAM, ROM, I/O, etc. with single chip microcomputer [7], its minimum system includes single chip microcomputer chip, power supply circuit, reset circuit of single chip microcomputer chip, clock signal generation circuit necessary for work, display module circuit, key circuit, etc. The minimum system of STC89C52 single chip microcomputer is shown in Fig. 1.

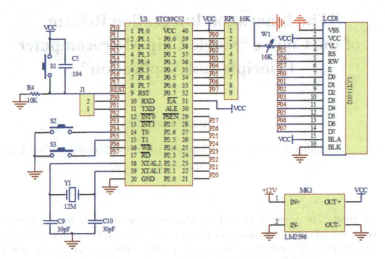

**Fig. 1.** Minimum System of Single Chip Microcomputer

The course of Single Chip Microcomputer Principle and Application is very practical. Microcontroller occupies a very important position in the field of industrial control, and its application is very extensive, almost everywhere, such as intelligent instruments, household appliances, intelligent robots, industrial control process equipment, aerospace equipment, medical equipment, etc. are widely used [8, 9]. Graduates of electrical, mechanical, robot engineering, artificial intelligence and other majors have many jobs related to SCM, such as electrical equipment development and installation, instrument detection, robot development, etc. Therefore, it is necessary for students to master the course of principle and application of single chip microcomputer [10].

The purpose of implementing the integration of specialty and innovation is to cultivate students' innovation and entrepreneurship ability, and to reconstruct and design the existing content, so that students can form innovation ability in future work and have a certain entrepreneurial ability [11]. How to take innovation and entrepreneurship as the guidance, take curriculum knowledge as the basis, reconstruct knowledge points in combination with practical projects, and achieve the organic integration of professional education and innovation and entrepreneurship education is the key to curriculum reform [12–14]. Therefore, it is imperative to explore an effective teaching method based on the integration of specialty and creativity in the principle and application of single-chip microcomputer.

## 2 Main Problems in the Course of Single Chip Microcomputer Principle and Application

The traditional teaching mode of the principle and application of single chip microcomputer is the theory teaching in the multimedia classroom and the experimental teaching based on the laboratory verification experiment. Generally, the theory and experimental teaching is carried out with the traditional 8051 chip as the core [15].

In the theoretical teaching, the teaching process of the principle and application of SCM in many schools is shown in Fig. 2. First, introduce the basic knowledge of SCM, including the basic concept, conversion of number system, and the development history of SCM [16]; Secondly, the internal structure and working principle of the hardware of 8051 single chip microcomputer (including the internal composition of the single chip microcomputer, clock circuit, reset circuit and internal memory) are introduced; Thirdly, the instruction system and C language programming of single chip microcomputer are introduced, and the data type, syntax and structure of C language programming are also introduced; Then, it introduces the application of SCM's internal resources, such as IO structure, interrupt system, timer, serial port, etc.; Finally, the memory expansion and parallel expansion technologies of single chip microcomputer are introduced [17, 18].

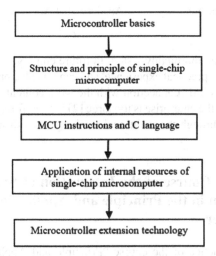

**Fig. 2.** Teaching Process of the Course of Single Chip Microcomputer Principle and Application

In the experimental teaching, we mainly completed several simple experiments as shown in Table 1: (1) The installation and simple application of keil and proteus software; (2) Use SCM to light LED water lamp; (3) Single chip microcomputer independent keyboard, 4 × 4 Matrix keyboard programming; (4) LCD1602 displays simple characters; (5) Single chip external interrupt independent key programming application; (6) The timer is used to realize the buzzer to play a piece of music; (7) Single chip microcomputer realizes serial communication with PC, and multi computer communication between single chip microcomputer and single chip microcomputer; (8) C language programming of analog to digital converter and digital to analog converter [19].

If only theoretical knowledge and operating procedures are taught in a straightforward manner, such teaching theory courses will be boring, the depth of practical training courses will be insufficient, and theory and practice will be disconnected. In the long run, the technology of single chip microcomputer can not be linked with the actual project and is separated from the actual application, so the technology of single chip microcomputer will stagnate, or even go backwards [20]. In the era of "mass entrepreneurship and innovation", the integration of innovation and entrepreneurship training is almost zero.

Table 1. Main Experimental Contents of Single Chip Microcomputer Experiment

| No. | Serial number | Contents of the experiment |
|---|---|---|
| 1 | Experiment 1 | Installation and use of software |
| 2 | Experiment 2 | LED stream lamps |
| 3 | Experiment 3 | Independent keys and matrix keyboard |
| 4 | Experiment 4 | LCD1602 display driven by Single chip |
| 5 | Experiment 5 | MCU external interrupt |
| 6 | Experiment 6 | Single chip timer |
| 7 | Experiment 7 | serial communication of Single chip |
| 8 | Experiment 8 | Application of ADC and DAC |

The theory has not been thoroughly tested. In future employment, there are often poor practical ability, lack of practical experience, and cannot solve practical problems. What you learn in school has little connection with the work content, and the deviation from the talents required by the enterprise is too large [21]. Therefore, teachers should reform their teaching, break through the traditional teaching methods and adopt new teaching methods.

## 3 Reform of the Course of the Integration of Specialization and Innovation in the Principle and Application of Single Chip Microcomputer

In the reform and practice of the course "Principle and Application of Single Chip Microcomputer", the cultivation of innovation consciousness is taken as the starting point [22]. The innovation and entrepreneurship competitions and professional competitions that students participate in mainly include the China International "Internet plus" Undergraduate Innovation and Entrepreneurship Competition (referred to as "Internet plus" competition), Undergraduate Innovation and Entrepreneurship Training Program, National Undergraduate Electronic Design Competition, National Undergraduate Smart Car Competition, China Undergraduate Engineering Practice and Innovation Ability Competition (referred to as "Work Training Competition"), Guangxi University Innovation Design and Production Competition, etc. The projects and topics involved in innovation and entrepreneurship competitions and professional competitions are often very new, and many innovative projects emerge every year. The curriculum reform is based on the innovation and entrepreneurship competition and professional competition projects to rebuild the projects suitable for curriculum teaching. The main reconstruction projects are: (1) design of multi-function LED water lamp; (2) The design of intelligent car race timer; (3) "Explorer" smart car design; (4) Design of waveform acquisition and display system based on PC display.

## 3.1 Design of Multi-function LED Stream Lamp

The reference circuit of the multi-function LED water lamp design project is shown in Fig. 3. In the figure, P1 port of single chip microcomputer is connected to matrix keyboard, P3.2 - P3.5 is connected to independent keyboard KEY1 - KEY4, and P0 is connected to LED1 - LED8.

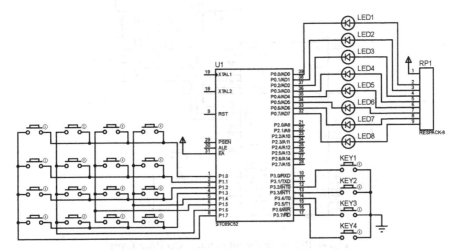

**Fig. 3.** Reference circuit of multi-function LED flow light

Project purpose: Learn the basic knowledge, internal structure and working principle of single chip microcomputer through the design of LED water light, understand the relationship between various decimal systems, complete the C language design of multiple modes display of LED water light driven by IO port, and complete the independent key press and matrix keyboard to change the LED water light mode.

Project requirements: (1) Press KEY1 on the independent keyboard to display binary numbers 10100101, 00001111, etc. on the LED 1-LED 8 water lamps; (2) Press KEY2 on the independent keyboard to turn on the LED 1 - LED 8 running lights in turn (using an array); (3) Press KEY3 on the independent keyboard to realize the display of LED 1 - LED 8 running lights with different requirements (each student is required to have different LED display functions); (4) Press KEY4 on the independent keyboard to enter the matrix keyboard to control the flow light speed mode (it is required to use the matrix keyboard to set at least 10 levels of different speeds).

## 3.2 Design of Intelligent Car Competition Timer

Since 2017, our school has organized students to participate in the National College Students' Smart Car Competition, in which a very important device - timer has been used. Therefore, LCD1602 display, external interrupt, timing counter knowledge, etc. are integrated into the SCM course with the timer of the competition as the starting point, which not only stimulates students' enthusiasm for the competition, but also improves their interest in learning. The reference circuit diagram of the intelligent car race timer is shown in Fig. 4.

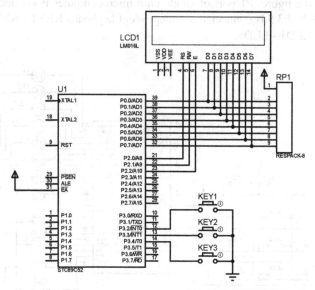

**Fig. 4.** Reference Circuit of Smart Car Race Timer

Project purpose: Through the design of timer, understand the basic knowledge of timer sensor, realize the independent key start, pause, and reset timer, display the timing time with LCD1602, use the internal timer counter of SCM to achieve timing, and use the SCM counter to record the number of turns.

Project requirements: (1) LCD1602 is used to display the time of the timer; (2) Press KEY1 on the independent keyboard to start, pause and reset the timer (using the external interrupt function); (3) Press KEY2 on the independent keyboard to increase the number of turns (using the external counting function of the SCM timer counter); (4) Press KEY3 to reset the number of cycles.

### 3.3 Design of "Explorer" Smart Car

It is an effective way to apply the learned knowledge to the practice link, so that students can really master the knowledge and remind themselves of their abilities and comprehensive qualities [23–25]. The history of education and countless facts can prove the important value of practice for knowledge learning [26].

The design of the "Explorer" intelligent vehicle originates from the project of "All terrain engineering vehicle design and production" of the 2020 engineering training competition. The design gives full play to the innovative thinking of students. The smart car passes through narrow bridges, pipes, stairs, grid carpets and other obstacles, and its performance is determined by the number and time of obstacles. The reference diagram of "Explorer" smart car race track is shown in Fig. 5.

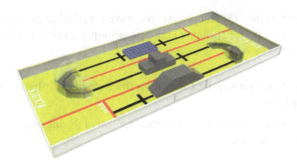

**Fig. 5.** Reference Diagram of the "Explorer" Smart Car Race Track

Project purpose: through the design of the "Explorer" smart car, master the working principle of the smart car tracking sensor and the tracking C language programming, the working principle of the smart car obstacle avoidance sensor and the obstacle avoidance C language programming, and the timer to achieve PWM control.

Project requirements: (1) 2–10 infrared sensors are used to track the black line; (2) 1–3 obstacle avoidance sensors are used to avoid obstacles; (3) PWM is used to control the intelligent vehicle to travel at different speeds in different sections.

### 3.4 Design of Waveform Acquisition and Display System Based on PC Display

The reference circuit diagram of waveform acquisition and display system design based on PC display is shown in Fig. 6. This circuit integrates functions such as ADC waveform acquisition, DAC waveform restoration, and serial port transmission of waveform data. It is a highly comprehensive project.

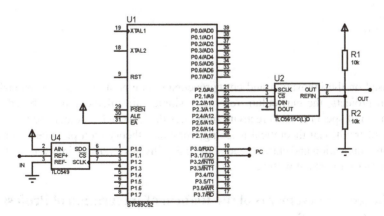

**Fig. 6.** Waveform acquisition and display system circuit based on PC display

Project purpose: Through the design of waveform acquisition and display system based on PC display, master the principle and programming of serial port transmission, master the waveform collected by ADC, and master the waveform output by DAC.

Project requirements: (1) Adopt 1Hz sine wave for ADC acquisition and transmit it to PC through serial port; (2) Single chip microcomputer controls DAC to restore the waveform collected by ADC.

### 3.5 Matrix of Teaching Content Based on the Reform of Integration of Specialty and Creativity

The teaching content based on the curriculum reform of the integration of specialty and creativity is shown in Table 2.

**Table 2.** Teaching content based on the curriculum reform of specialized and creative integration of single-chip microcomputer

|  | Basic knowledge | Structure and working principle | C programming | IO structure | Interrupt system | Timer | Serial port | ADC/DAC | SC expansion |
|---|---|---|---|---|---|---|---|---|---|
| Design of multi-function LED stream lamp | √ | √ | √ | √ | | | | | |
| Design of intelligent car race timer | √ | √ | √ | √ | √ | √ | | | |
| Design of "Explorer" Smart Car | √ | √ | √ | √ | √ | √ | √ | | |
| Design of waveform acquisition and display system based on PC display | √ | √ | √ | √ | √ | √ | √ | √ | √ |

Based on the innovation and entrepreneurship competition and professional competition projects, the curriculum reform of single-chip microcomputer based on the integration of specialty and creativity integrates the knowledge points and most experiments of traditional theoretical teaching, combines theory with practice, implements students' innovation and entrepreneurship, and cultivates students' innovation ability to achieve twice the result with half the effort.

## 4 Phased Achievements of the Reform of Integration of Professional and Creative Work

Through the reform of the integration of specialty and creativity in the course of the principle and application of single-chip microcomputer, we can expand and extend the teaching project based on the college students' electronic design competition, college students' smart car competition, college students' engineering training competition,

Guangxi college innovation design and production competition, and encourage students to actively participate outside the class. From 2019 to 2022, students from the Intelligent Manufacturing School of Nanning University participated in various college student competitions. Among them, our students have achieved good results in the National Undergraduate Electronic Design Competition, the National Undergraduate Smart Car Competition, and the Guangxi University Innovation Design and Production Competition, which are among the best in Guangxi. The results of students from the School of Intelligent Manufacturing of Nanning University in recent three years are shown in Table 3.

**Table 3.** Some competition results of students of Intelligent Manufacturing College of Nanning University in recent three years

| Items | Undergraduate Electronic Design Competition | Undergraduate Smart Car Competition | Guangxi University Innovation Design and Production Competition |
|---|---|---|---|
| Awards | 2 National second prizes; 4 District level first prizes; 5 District level second prizes; 8 District level third prizes | One national second prize; 2 district level first prizes; 2 district level second prizes; | 3 district level first prizes; 3 district level second prizes; 4 district level third prizes |

## 5 Conclusion

Traditional SCM teaching has many problems, such as boring theoretical courses, insufficient depth of practical training courses, and disconnection between theory and practice. Reform is imperative. Exploring the reform of integration of specialty and innovation in the course of principle and application of single chip microcomputer, based on the case of innovation and entrepreneurship competition and discipline competition, reconstruct several projects suitable for the teaching of single chip microcomputer, and effectively integrate the teaching content of single chip microcomputer into the project. This paper introduces the purpose and requirements of LED water light, competition timer, "Explorer" smart car, waveform acquisition and display system and other projects, and gives the project knowledge matrix (Table 2). Of course, these projects are not immutable, but constantly add new projects and new knowledge.

The innovation integration reform of the principle and application of single-chip microcomputer curriculum is based on the cultivation of students' project practice ability and innovation consciousness. It carries out the research and optimization of the teaching of single-chip microcomputer curriculum. In the teaching, it uses a variety of teaching methods and means to effectively combine the innovation ability with the curriculum system, effectively mobilizing the students' initiative and enthusiasm for learning, which not only improves the students' innovation ability, but also improves their theoretical knowledge and practical ability. This will provide a reference for the teaching of SCM courses in various colleges and universities. In the era of "mass entrepreneurship and innovation", we will continue to move forward and actively explore a more effective way to reform the principle and application courses of single chip.

**Acknowledgment.** This paper is the phased research result of the third batch of specialty innovation integration curriculum reform project of Nanning University, "Research on Teaching Reform of Single Chip Microcomputer Principle and Application Based on Specialty innovation Integration" (No. 2021XJZC03).

# References

1. Yuqing, H.: The network course construction of microcomputer principles and applications. Int. J. Educ. Manag. Eng. (IJEME) **1**(1): 51–57 (2011)
2. Ning, G.: Construction and implementation of innovation computer network practical teaching system. Int. J. Educ. Manag. Eng. (IJEME) **1**(2), 30–35 (2011)
3. Denggao, G., Jinghui, L., Jianping, L., Qiaoming, Y., Junfeng, L., Mei, Y.: Material physics & chemistry quality network curriculum construction and teaching practice. Int. J. Educ. Manag. Eng. (IJEME) **2**(5), 24–30 (2012)
4. Wang, X., Cao, Z.: Discussion reform of forestry panorama course teaching. Int. J. Eng. Manuf. (IJEM) **2**(12), 41–45 (2012)
5. Li, D., Zhu, W., Chen, Z., Wang, C.: Research of overall optimization based on a series of data structure courses teaching content. Int. J. Educ. Manag. Eng. (IJEME) **2**(11), 18–23 (2012)
6. Xu, X.: Training strategy of key innovation and entrepreneurship talents. Educ. Asia-Pac. Reg. **60**, 295–339 (2021)
7. Min, X.: Design and implementation of buzzer based on stc89c52 single chip microcomputer. Wireless Internet Technol. **19**(16), 95–98 (2022). (in Chinese)
8. Haiyan, H.: Practical research on teaching reform of single chip microcomputer technology course for vocational undergraduate students based on OBE concept. Mod. Vocat. Educ. **39**, 62–65 (2022). (in Chinese)
9. Huijun, L.: Reform and exploration of project-based SCM course. Educ. Teach. Forum **43**, 37–40 (2022). (in Chinese)
10. Yanhong, Y.: Research on teaching reform and practice of single chip microcomputer course integrating ideological and political education. Comput. Knowl. Technol. **18**(29), 153–155 (2022). (in Chinese)
11. Andrea, G.: Entrepreneurship and innovation design in education. an educational experience to train the new entrepreneurial designers. Des. J. **22**, 203–215 (2019)
12. Qichun, J., Yanli, Z.: Teaching reform of integrated curriculum design of single-chip microcomputer and embedded system under OBE-CDIO engineering education mode. J. Nantong Vocat. Univ. **35**(02), 57–61 (2021). (in Chinese)
13. Jianhua, Q., Zhicheng, Z., Zhang Xiong, W., Xiaojia.: Integrated teaching reform of "Microcomputer Principle" and "Single Chip Application" courses based on OBE concept. China Elect. Power Educ. **06**, 63–64 (2021). (in Chinese)
14. Dalin, Z.: Exploration of PAD teaching mode in teaching reform of SCM application technology course. J. Sichuan Vocat. Tech. Coll. **31**(02), 10–13+49 (2021) (in Chinese)
15. Fen, W.: Course construction and practice of "Microcontroller Application Technology" based on the integration of learning, learning and teaching. Jiangsu Educ. **21**, 69–74 (2021). (in Chinese)

16. Hongru, W., Wei, T., Houlian, W., et al.: Teaching reform and practice of "Microcontroller" course for mechanical major based on OBE concept. Sci. Technol. Innov. **05**, 66–68 (2021). (in Chinese)
17. Anyu, Z., Jun, W., Yanping, F.: Discussion on the teaching mode of SCM course based on the cultivation of innovation and entrepreneurship ability. J. Anhui Electr. Eng. Vocat. Tech. Coll. **25**(04), 108–111 (2020). (in Chinese)
18. Jie, X., Zhigao, L.: Research on the experimental reform of the principle and application of Single Chip Microcomputer based on the CDIO project background. Comput. Knowl. Technol. **16**(34), 141–143 (2020). (in Chinese)
19. Gaoqiang, L., Peng, L.: Teaching reform and practice of the course "Microcontroller Application Technology". Sci. Technol. Innov. **22**, 134–135 (2020). (in Chinese)
20. Yunyu, J.: Reform ideas for the course of single chip microcomputer application technology. Comput. Netw. **46**(20), 45 (2020) (in Chinese)
21. Fen, L., Huaizhuang, H., Guangqiang, L.: Exploration on the reform and innovation of the teaching mode of the course "Single Chip Application Technology". Popular Sci. Technol. **22**(09), 127–129 (2020). (in Chinese)
22. Xueming, L., Min. T.: Teaching reform and practice of single chip microcomputer principle and application course. China Educ. Technol. Equipment (17): 101–102+110 (2020) (in Chinese)
23. Yuxiang, Y., Wei, S., Ruipeng, G., et al.: Teaching reform and practice of single chip microcomputer course under the background of professional certification. Educ. Teach. Forum **35**, 197–198 (2020). (in Chinese)
24. Yinghua. Z.: Research on the course reform of Single Chip Microcomputer in the application oriented talent training mode. Comput. Prod. Circ. (09): 259+261 (2020) (in Chinese)
25. Shuting, G., Hua, S., Cuili, Y., Fengwei, Z.: Taking students as the center and starting from the course point - course reform and exploration of single chip microcomputer principle and application. Shanxi Youth **21**, 60–62 (2022). (in Chinese)
26. Wei, X.: Discovery and practice of EDA experimental teaching reform. Int. J. Educ. Manag. Eng. (IJEME) **1**(4), 11–19 (2011)

# Discussion and Practice on the Training Mode of Innovative Talents in Economics and Management in Women's Colleges

Zaitao Wang[1]($\boxtimes$), Ting Zhao[2], Xiujuan Wang[3], and Chuntao Bai[4]

[1] School of Business Administration, Shandong Women's University, Jinan 250300, China
wzt0813@163.com
[2] Library, Shandong Women's University, Jinan 250300, China
[3] Fundamental Science Section in Department of Basic Courses, Chinese People's Armed Police Force Logistics College, Tianjin 300309, China
[4] Management Co., Ltd., Tianjin 300405, China

**Abstract.** How to carry out the supply side reform of talent training and improve the social adaptability of applied innovative talent training is also one of the important issues faced by colleges and universities that set up economic management majors. Women's colleges and universities are an important part of China's higher education system. It is an inevitable choice for women's colleges to insist on running schools with characteristics and cultivate innovative talents. This paper explains the connotation of "economic and management applied innovative talents" through the method of "species plus genus difference", and on this basis, comprehensively reforms the objective elements, content elements and method elements in the talent training model. Under the guidance of the concept of talent training, we have formulated talent training objectives and specifications to adapt to the development of the industry, built a spiral integrated curriculum system and practical teaching system, explored research based teaching methods driven by teaching and research collaboration and school enterprise collaboration projects, and improved and built a diversified practical environment and a team of teachers with rich experience in enterprise management. Through the practice of some majors in Shandong Women's College, the reform of innovative talents training mode in economic management has achieved good results.

**Keywords:** Women's colleges · Economic and management applied innovative talents · Cultivation mode

## 1 Introduction

In China, the educational status of women has been greatly improved. According to <Outline for the Development of Chinese Women (2011–2020)>, which is showed as Table 1, it is not difficult to see from the table that women account for nearly half of all educational stages, and even more than men in the higher education stages of universities and masters. There is no doubt that the development level of women's higher education reflects the civilization and progress of society [1]. The mission of modern women's colleges and universities is to strive for women's equal access to higher education [2]. They have made great contributions to the training of women talents and the promotion of women's socio-economic status. Women's universities have become an indispensable part of the higher education system. The channels of cooperation and exchange between women's colleges and universities are not smooth enough and the fields are not wide enough; the social status and existing value of women's colleges and universities have not been fully recognized and widely recognized, and the training of innovative talents in women's colleges and universities is still facing many severe challenges [3–5].

Table 1. Proportion of women in different education stages

| Proportion of women (%) | Female postgraduate |  | General U and C | Adult U and C |  | Secondary education | Pri education |  | Pre education |  |
|---|---|---|---|---|---|---|---|---|---|---|
|  | D | M U | C | U | C | H | J | | Pri | Pre |
|  | 41.32 | 52.17 | 53.9 | 48.74 | 61.9 | 55.33 | 50.71 | 46.42 | 46.55 | 46.94 |

Note: Doctor abbreviated as D; Master abbreviated as M; Undergraduate abbreviated as U; College abbreviated as C; High school abbreviated as H; Junior high school abbreviated as J; Primary school abbreviated as Pri; Preschool education abbreviated as Pre.

Women's colleges and universities emphasize to cultivate women's comprehensive and applied talents suitable for social needs according to women's physiological and psychological characteristics. The implementation of applied innovative talent training in women's colleges and universities not only helps to promote gender equality in the field of higher education, promote the perfect development of higher education itself and cultivate high-level professionals with modern literacy, but also deconstructs gender bias, eliminates gender inequality, builds advanced gender culture, promotes women's development and promotes gender equality, promoting social civilization and progress will play an important role in leading and promoting [6, 7]. Different talent types should build different talent training models. Therefore, clarifying the connotation and essential characteristics of innovative talents in economic management of women's colleges and universities plays a key role in studying the innovative talent training model in economic management of women's colleges and improving the quality of talent training.

## 2 Definition of Economic and Management Applied Innovative Talents

This part mainly explains the connotation of "economic and management applied innovative talents" through the method of "species plus genus difference". First of all, we should put the defined object of "economic and management applied innovative talents" into a broader concept of "talents", which is called "species" [8]. Then, find out the unique nature of "economic and management applied innovative talents", which is different from other "talents" concepts. This nature is called "poor". "Economic and management applied and innovative talents" has three concepts: "innovation", "application" and "economic and management". The connotation of economic and management applied innovative talents is analyzed according to the "species" - "genus" relationship of talents - innovative talents, innovative talents - applied innovative talents, and applied innovative talents - economic and management applied innovative talents [9, 10].

The so-called talents refer to the workers who have accumulated more human capital in the form of knowledge and ability through learning and practice, and can create more social value in the same working time [11].

The word "innovation" comes from Latin, and its original meaning includes three meanings: update, create new things, and change. Innovative talents refer to those who constantly make breakthrough contributions to social progress or scientific and technological development by comprehensively using their own innovative thinking, innovative ability, and innovative quality and corresponding knowledge [12].

In the process of transforming objective laws into direct social interests, there are scientific "research" categories (requiring academic talents) that transform objective laws into scientific principles and scientific "application" categories (requiring applied talents) that apply scientific principles to social practice to transform them into products (material or non-material). Applied talents refer to the talents who directly apply the knowledge of applied science or newly discovered knowledge to the practice fields closely related to social production and life to seek direct benefits for mankind [13].

It can be seen that applied innovative talents are those who have diversified and overlapping knowledge structure, profound professional and technical ability, strong sense of social responsibility, critical spirit and innovative research consciousness, and can comprehensively use their own innovative thinking, innovative ability, innovative quality and corresponding knowledge to directly apply them to practical innovation activities closely related to social production and life, so as to make breakthrough contributions to the smooth transformation of the whole society.

At present, the training objectives and requirements for economic and management talents put forward by colleges and universities are mostly described as: innovative and application-oriented talents who have good humanistic quality and innovative spirit, have a solid theoretical basis of economics (management), skillfully use modern economic (management) analysis methods, and have strong practical ability to analyze and solve problems, and can engage in economic management in economic management departments, government departments, enterprises and institutions.

To sum up, economic and management applied innovative talents are innovative talents between academic talents and technical skills applied talents. They have broad basic theoretical knowledge, professional knowledge and technical reasoning ability, strong economic and management awareness, practical ability, independent learning ability and system thinking ability, team cooperation and communication ability, and economic and management design and innovation ability.

Secondly, what is "talent cultivation"? Cultivating talents is the primary task of higher education. Seven problems must be solved in talent cultivation: first, the proposition of educational concept; second, the determination of talent training objectives; third, the selection of talent training objects; fourth, the development of talent training subject; fifth, the use of talent training channels; sixth, the optimization of talent training process; seventh, the institutional guarantee of talent training. It can be seen that talent training is a systematic project, which includes seven elements of talent training: the concept, subject, object, goal, approach, mode and system [14]. Talent training mode is closely related to talent training, but talent training mode and talent training are two different concepts. As a systematic project, talent cultivation involves seven elements of talent cultivation, namely, the concept, subject, object, goal, approach, mode (process) and system. In order to cultivate innovative talents, the above seven elements must be reasonably developed and reorganized at the same time, including further updating the educational concept, further improving the work efficiency of the training subject, further mobilizing the learning initiative of the training object, further defining the training objectives of innovative talents, further innovating the talent training mode, further enriching the training channels, further reforming the training system and optimizing the training conditions. The talent training mode mainly refers to the design and construction of the training process, that is, the consideration of the education subject on the selection of majors and curriculum, the structure and procedures of teaching activities, and the determination of teaching organization and management forms. The so-called "talent training mode" refers to the theoretical model and operation mode of the talent training process designed by the training subject under the guidance of certain education concepts and the guarantee of certain training systems in order to achieve specific talent training goals, which is composed of several elements and has the characteristics of purposefulness, intermediary, openness, diversity and imitation.

## 3 Analysis of the Elements of the Training Mode of Economic and Management Applied Innovative Talents

The talent training mode is the most changeable and dynamic subsystem in the "talent training" system, and it is also the most complex subsystem of the constituent elements. The change of talent training mode is essentially the change of its constituent elements; The innovation of talent training mode is also mainly the innovation or reorganization of each constituent element. Therefore, in order to innovate the talent training model, we must carefully analyze the elements of the talent training model.

### 3.1 Talent Cultivation Concept

The concept of talent cultivation here refers to the educational concept at the macro (college) and micro (teacher) levels, that is, the rational understanding of the training subject on the essential characteristics, target values, functions, tasks and activity principles of talent cultivation, as well as the ideal pursuit of talent cultivation and various specific educational concepts formed by it. Such as quality, teachers and students, teaching, scientific research, activities and evaluation. The concept of talent training aims to answer the questions of "what should talents be in colleges and universities" and "how should talents be trained". From the perspective of philosophy, the function of talent training concept is to reveal the internal logic and ultimate value of talent training; from the operational level, it aims to guide the talent training process, including the design and conception of training procedures and links. The concept of talent training has an extremely important influence on the selection and determination of other elements of the talent training model [15].

### 3.2 Professional Setting Mode

Specialty setting mode is an important part of talent training mode. Major is mainly divided according to disciplines. Generally, major settings can be designed for shape change in terms of setting caliber, setting direction, setting time, setting space, so on and so forth [16]. The professional caliber refers to the coverage of the main disciplines or the basis of the main disciplines and the scope of business specified when the disciplines are divided. Setting direction refers to whether and how much specialized direction is differentiated within the professional caliber to stiffen or activate the specialty. The setting time refers to the time of specialty setting, whether the specialty is determined as soon as the students enter the school or the specialty training is determined after a certain stage of study. Setting space refers to whether there is still room for wandering and possibility of change after students' majors are determined, and whether students are allowed to change majors, departments, colleges or cross majors, departments or colleges.

### 3.3 Teaching Evaluation Methods

Teaching evaluation is an objective judgment and evaluation of the talent training process and its quality and benefits based on certain standards. Teaching evaluation is an important part of the talent training process, and it is also an effective form to test the effectiveness of talent training and an important means to stimulate teachers and students. The teaching evaluation involves two aspects: the middle evaluation of running a school and the micro evaluation of teaching and learning in teaching. Whether at the macro level or at the micro level, the current problems in teaching evaluation are shown in the following aspects: first, in the scope of evaluation, the evaluation of results is emphasized, while the evaluation of process is ignored; second, in terms of the purpose of evaluation, emphasis is placed on identification, selection and elimination rather than feedback, correction and regulation; third, on the basis of evaluation, pay more attention to the score of the examination than to creative thinking and practical ability; fourth, in terms of evaluation methods and means, we should pay more attention to examinations than other methods and means. This evaluation method restricts the teachers and students' independent choice of teaching and learning, constrains the free development of human personality, and cannot well adapt to the requirements of cultivating innovative talents. From single evaluation to multiple evaluation are inevitable, for the innovation of evaluation methods in training innovative talents.

## 4 Difficulties in the Implementation of Innovative Talent Training Mode in Economic Management

The School of Business Administration and the School of Innovation and Entrepreneurship are the two players in the game. Because of the administrative structure of their own policies and systems, inter-departmental competition is inevitable. At the same time, because participating in the issue of policies to solve major problems can enhance the power and authority of the department, the department will still choose to cooperate with other departments. The two colleges are both interconnected and relatively independent. Although they are smaller than the main body of the school, they can still serve as a unit for pursuing interests. Generally speaking, any institution pursues the maximization of its own interests, and colleges are no exception. The pursuit of departmental interests is the main driving force of its actions, and is also the core reason for departmental conflicts and inter-institutional games. In view of this, it is assumed that each participant is self-interested, and the purpose is to maximize its own interests, without caring about the interests of another participant. Both sides are in a cooperative relationship when cultivating talents, but the resources they invest in the process of cooperation can be controlled, and they are faced with the choice of more investment and less investment. The School of Business Administration is now marked as G and the School of Innovation and Entrepreneurship as C. There is a strategic relationship between the two. G invests more and C invests less, and C gains more income with less investment; on the contrary, if G invests less and C invests more, C will get less income with more investment. Or if both participants invest less resources, the training effect of innovative talents is poor, and both of them suffer losses. Both participants have invested more resources. At this time, the training effect of innovative talents is better, and both of them have obtained better benefits. The game problem between the two is expressed in Table 2 as follows with payoff table:

Table 2. Game between innovative talent training colleges

|  |  | Player C | |
|---|---|---|---|
|  |  | More investment | Less investment |
| Player G | More investment | 50, 50 | -100, 80 |
|  | Less investment | 80, -100 | -50, -50 |

Use mathematical expectation to analyze the income of the School of Business Administration:

$$E_{GM} = \frac{1}{2} \times 50 + \frac{1}{2} \times (-100) = -25 \tag{1}$$

$$E_{GL} = \frac{1}{2} \times 80 + \frac{1}{2} \times (-50) = 15 \tag{2}$$

Compare the expected payoffs in two cases:

$$E_{GM} < E_{GL} \tag{3}$$

It can be seen that the best strategy of the School of Business Administration is to invest less resources.

Similar to the expected income of the School of Innovation and Entrepreneurship:

$$E_{CM} = \frac{1}{2} \times 50 + \frac{1}{2} \times (-100) = -25 \qquad (4)$$

$$E_{CL} = \frac{1}{2} \times 80 + \frac{1}{2} \times (-50) = 15 \qquad (5)$$

Compare the expected payoffs in two cases:

$$E_{CM} < E_{CL} \qquad (6)$$

It can be seen that the best strategy of the School of Business Administration is to invest less resources.

From the above analysis, it can be seen that the best strategy for both the School of Business Administration and the School of Innovation and Entrepreneurship are to invest less resources. Finally, both parties choose to invest less resources as the Nash equilibrium point. At this point, the total income of the school in the cultivation of innovative talents is the lowest, at - 100, which is less than the total income of one party investing more resources, and the total income of one party investing less resources is - 20, which is far less than the total income of both parties investing more resources.

## 5 Taking the School of Business Administration of Shandong Women's University as an Example

The talent training mode refers to the theoretical model and operation mode of the talent training process, which is designed by the training subject in order to achieve specific talent training objectives, under the guidance of certain education concepts and under the guarantee of certain training systems, and is composed of several elements with the characteristics of system, purpose, intermediary, openness, diversity and imitation. The change of talent training mode is essentially the change of its constituent elements, and the innovation of talent training mode is mainly the innovation or reorganization of each constituent element.

### 5.1 Thoughts on the Reform of Talent Training Mode

Comprehensive reform should be carried out according to the purpose, content and method elements of the talent training model. Specifically, the objective elements include the training objectives and training specifications, the content elements include the curriculum system and its teaching content, and the method elements include the teaching and learning methods, teachers and practice environment. The idea of comprehensive reform of talent training mode is to determine the training objectives based on the training concept. The training specifications are designed according to the training objectives, and the curriculum system is designed according to the training specifications. Under the guarantee of a certain training system, certain teaching and learning methods are adopted according to the requirements of the curriculum system, as well as the construction of the teaching staff and practice environment. The training results reflected after the implementation of the training means will be transmitted to the society. The evaluation will be carried out according to the feedback of the society on the quality of talent training and the requirements of the school itself. The training objectives, training specifications, curriculum system and training means will be further optimized according to the evaluation results.

## 5.2 An Integrated Curriculum System Based on Spiral Ability Training

According to the talent training concept of "demand guidance, project driven, strengthening practice, teaching and research unity, highlighting innovation, and comprehensive development", according to the talent training specifications and characteristics, as well as the consistency, completeness, integration, progressiveness, and practicality of the curriculum system, and in combination with our own advantages and characteristics, we will build a talent training oriented, theory and practice oriented quality education is integrated into the training of professional knowledge and ability, and the knowledge, ability and quality are coordinated and unified. The curriculum system is oriented by ability cultivation, supported by centralized practice links, and based on core courses. A group of centralized practice links and core courses cultivate an application or design ability and related knowledge.

## 5.3 Establish an Effective Evaluation System for Training Innovative Talents in Engineering Application

The teaching quality assurance system is an important means to ensure the teaching level and the quality of talent training. It is an all-round and whole process quality management system project, which provides an effective way for talent training evaluation. Build a closed-loop teaching quality assurance system consisting of four subsystems: quality standard system, quality monitoring system, quality information and analysis system, and quality improvement system, promote the scientific, standardized, and information-based management of teaching quality, and thus improve the quality of teaching and personnel training. The principle in the evaluation process is the training quality standard. A quality standard system covering the main process of undergraduate teaching activities should be established, with the talent training goal as the highest program, the talent training program as the basic basis, and the daily teaching norms as the code of action. Form an all-round and whole process teaching quality monitoring system to evaluate and monitor teaching quality from multiple aspects. Evaluate the quality of talent training through the quality information and analysis system, and provide the basis for the quality improvement system.

# 6 Conclusion

How to cultivate high-quality innovative talents is a major issue facing higher education. Women's colleges and universities are mainly specialized in humanities and social sciences, with the goal of cultivating high-quality application-oriented professionals with gender awareness, sense of responsibility, innovative spirit, practical ability, and comprehensive development of morality, intelligence, physique and beauty, which are required by economic society and the development of women and children. The existing women's colleges and universities mainly focus on humanities and social sciences, which is consistent with the traditional female disciplines. In terms of major categories, they are mainly concentrated in literature, pedagogy, economics, law, management and other disciplines. Women's colleges and universities mainly think about what kind of education they should provide for women, such as "providing women with educational forms suitable for women's characteristics" to "meet the needs of women's growth and talent"; "cultivate women with the spirit of 'four self'"; employment oriented; based on the development of female students, compensation education should be carried out for female college students. In terms of specialty setting and course offering, we should think about how to teach courses with women's characteristics and cultivate talents with women's characteristics. In view of the problems encountered in the training of application-oriented talents, we should reform the training mode of application-oriented talents, and study and practice effective ways to improve the training quality of application-oriented innovative talents.

**Acknowledgment.** This research has been supported by the 2017 school level teaching reform research project of Shandong women's University: discussion and practice of innovative personnel training mode of economic management in women's University, and the Shandong Social Science Planning Project (Grant No. 18CGLJ48, 18CSJJ32).

# References

1. Offutt, S., McCluskey, J.: How women saved agricultural economics. Appl. Econ. Perspect. Policy **44**(1), 4–22 (2022)
2. Piva, E., Rovelli, P.: Mind the gender gap: the impact of university education on the entrepreneurial entry of female and male STEM graduates. Small Bus. Econ. **59**(1), 143–161 (2022)
3. Kyrpychenko, O., Pushchyna, I., Kichuk, Y., et al.: Communicative competence development in teaching professional discourse in educational establishment. Int. J. Educ. Manage. Eng. **4**, 16–27 (2021)
4. Hedija, V., Němec, D.: Gender diversity in leadership and firm performance: evidence from the Czech Republic. J. Bus. Econ. Manag. **22**(1), 156–180 (2021)
5. Bullough, A., Guelich, U., Manolova, T.S., et al.: Women's entrepreneurship and culture: gender role expectations and identities, societal culture, and the entrepreneurial environment. Small Bus. Econ. **58**(2), 985–996 (2022)
6. Sharm, D.P., Deng, D.K., Tigistu, G., et al.: A cloud based learning framework for eradicating the learning challenges of Ethiopian working professionals, disables and women. Int. J. Mod. Educ. Comput. Sci. **14**(2), 41–54 (2022)
7. Robledo, D.M.L.: Programming as an option for females in undergraduate studies. Int. J. Educ. Manage. Eng. **9**(1), 1–8 (2019)
8. Majumder, S., Chowdhury, S., Chakraborty, S.: Interactive web-interface for competency-based classroom assessment. Int. J. Educ. Manage. Eng. **1**, 18–28 (2023)
9. Ke, Z.: Research on innovation of the training mode of economic and management applied talents in local undergraduate colleges. Res. Higher Educ. Three Gorges **1**, 4 (2020). (in Chinese)

10. Ya, T., Yinghu, T., Jian, L.: Research and practice on the "cross-border integration" talent training system for application-oriented undergraduate economics and management majors in the context of new liberal arts -- taking Jinling University of Science and Technology as an example. Res. Manage. Univ. (1), 78–81 (2022). (in Chinese)
11. Adebayo, E.O., Ayorinde, I.T.: Efficacy of assistive technology for improved teaching and learning in computer science. Int. J. Educ. Manage. Eng. **5**, 9–17 (2022)
12. Hao, Y.: Research on the cultivation mechanism of high quality innovative talents in high level private universities. Educ. Mod. **5**(35), 28–30 (2018). (in Chinese)
13. Adigun, J.O., Irunokhai, E.A., Onihunwa, J.O., et al.: Development and evaluation of a web based system for students' appraisal on teaching performance of lecturers. Int. J. Inf. Eng. Electron. Bus. **14**(1), 25–36 (2022)
14. Nan, L., Yong, Z.: Improvement and practice of secondary school geography teachers' informatization teaching ability based on the perspective of MOOCs. Int. J. Educ. Manage. Eng. **1**, 11–18 (2022)
15. Ahmed, N., Nandi, D., Zaman, A.G.M.: Analyzing student evaluations of teaching in a completely online environment. Mod. Educ. Comput. Sci. **6**, 13–24 (2022)
16. Tendo, S.N.: Multimedia pedagogy among literature lecturers in two universities in Uganda post COVID-19. Int. J. Educ. Manage. Eng. **1**, 1–9 (2023)

# Cultivation and Implementation Path of Core Quality of Art and Design Talents Under the Background of Artificial Intelligence

Bin Feng[1,2] and Weinan Pang[1(✉)]

[1] School of Journalism and Cultural Communication,
Guangxi University of Finance and Economics, Nanning 530001, China
adrianbin@foxmail.com

[2] Cooperative Course of Performance, Film & Animation, Sejong University, Seoul 05006, South Korea

**Abstract.** AI has driven the process of industrial reform and economic development. China's huge network population has made Internet big data have a larger application market. A large amount of information data can improve the accuracy of AI operations. With the continuous adjustment of the industrial structure, the industrial division of labor and post responsibilities as well as the training mode of art and design talents are also facing new challenges. Starting from the current situation of art and design talents under the traditional training mode, the construction of the core quality of art and design talents and the implementation path of the core quality of art and design talents, this paper discusses the countermeasures of how to build the core quality of talent cultivation in the era of artificial intelligence. It aims to provide useful ideas and practical experience for China to cultivate high-quality creative talents in art design and strengthen professional construction.

**Keywords:** Artificial intelligence · Art and design talents · Core quality

## 1 Introduction

At the intelligence symposium of American universities in 1956, the professional name of "artificial intelligence" was first proposed by the famous professor John Change, which was also the first time that it was publicly used in academic discussions. In the following period of time, it did not attract the attention of the industry. It was not until more than 20 years later that "artificial intelligence" entered people's vision again and became the mainstream trend [1]. Although the development of traditional symbols is slow, scientific researchers have made great progress in speech recognition, human-computer interaction, neural network and other aspects by adopting the more advanced statistical probability model for the first time [2–4]. In today's increasingly mature computer technology, art and design education also gradually introduces and actively uses high-tech means such as artificial intelligence to cultivate "skilled, rich, diverse and comprehensive" talents on the basis of respecting the principles and laws of traditional art and design education, in order to pursue the best effect of art education [5, 6].

© The Author(s), under exclusive license to Springer Nature Switzerland AG 2023
Z. Hu et al. (Eds.): ICAILE 2023, LNDECT 180, pp. 886–898, 2023.
https://doi.org/10.1007/978-3-031-36115-9_79

In the era of rapid development of mobile network, with the increasing popularity of artificial intelligence technology, the field of big data network services has also been greatly expanded. With the change of people's information needs, media usage habits and communication behavior, image recognition technology, voice synthesis, human-computer interactive translation and virtual reality technologies have emerged [7–9].

On August 31, 2022, China Internet Network Information Center (CNNIC) released the 50th "Statistical Report on the Development of China's Internet Network", which showed that as of June 2022, the number of Internet users in China reached 1.051 billion, the Internet penetration rate reached 74.4%, and the per capita online time per week was 29.5 h; The proportion of Internet users using desktop computers, laptops, tablets and televisions to access the Internet was 99.6%, 33.3%, 32.6%, 27.6% and 26.7% respectively.

A large number of users and media usage habits have promoted the transformation and integration of the industry. The accelerated integration of various industries across fields has accelerated the development of mobile network technology. At the same time, the industrial division of labor, post functions and technical conditions have also changed accordingly [10].

The development of industry and society has put forward a greater and higher demand for the quantity and quality of talents, and also entrusted colleges and universities with a more sacred mission [11, 12]. The training mode of art and design talents based on meeting the needs of economic and social development has also been greatly challenged. While new technologies promote economic and social development, the education authorities should also comply with the trend of AI development and social needs, and carry out all-round changes in the university's professional setting system, focusing on the construction of emerging professional systems such as AI, art and technology. Data thinking, scientific and technological vision and core literacy have become important factors in today's scientific and technological innovation [13–15].

For the training of application talents related to art design, a correct analysis should be made from the aspects of knowledge structure, ability level, creativity, etc., which is the prerequisite and basis for cultivating high-level art design talents. The cultivation of core literacy of art and design courses in colleges and universities should grasp and construct its key factors, and the quality of core literacy should be continuously improved on the basis of adapting the application of new technologies to the development of the times [16–19].

To this end, the article discusses how to provide more art and design talents for the development of society and industry and provide a strong guarantee for high-quality teaching development from the perspective of the development of new technology, combining the background of the era of artificial intelligence and referring to the requirements of the cultivation of the core quality of art and design talents [20–22].

## 2 Challenges Faced by Art and Design Talents in the Era of Artificial Intelligence

The development of artificial intelligence technology is changing the standard of demand for art and design talents by employing enterprises. Art and design talents are facing the test brought by this new technological change [23].

## 2.1 AI technology Promotes the Innovation of Art and Design Talents

Artificial intelligence technology accelerates the elimination of traditional mechanical design work, and promotes artistic design talents to be more innovative.

With the continuous development of artificial intelligence, more intelligent design software is emerging. The emergence of software such as ShapeFactory, Withoomph, Designevo and Markmaker has greatly shortened the threshold of software operation. Not only that, the software itself can collect data according to users' preferences and provide users with a variety of color schemes. Users have replaced the traditional complex design and other processes by clicking and selecting, and the software operation of simple machines has been replaced by artificial intelligence, which requires contemporary art and design practitioners to have stronger comprehensive ability.

In the past, talents trained in art and design education in colleges and universities often used to copy other works directly and simply with the help of Internet resources, resulting in low quality and similar styles of works. Design is only based on mechanical reproduction, and related intellectual property rights are not protected.

In contrast, in the context of AI, today's art and design practitioners should not only have traditional painting and design skills, but also exchange roles between programmers and managers. At a certain level, they should not only understand the relevant technical knowledge, but also manage the team, constantly improve their skills, cultivate their ability to look at macro issues, and also learn computer-related code when necessary. Now more and more art and design practitioners complete the design task with the help of artificial intelligence. Traditional art and design practitioners urgently need to improve their creativity and uniqueness, improve design skills and increase the diversity of design methods. Only a solid design foundation and the ability to learn from others' strong points can we more calmly cope with the challenges brought by AI.

## 2.2 Intensified Contradictions in the Employment Market of Art and Design Talents in the Context of Artificial Intelligence

The replacement of human beings by artificial intelligence (at least partially) is an irreversible historical process. With the emergence of artificial intelligence, the relationship between human and artificial intelligence has undergone earth-shaking changes. AI not only improves the productivity of human beings, but also tries to replace them. It "robs" some of the work that should be done by human beings in traditional industries, thus replacing the traditional labor force. This is also the difference between the "revolution" of AI and the "machine replacing man" in the previous industrial revolution. With the continuous development of artificial intelligence, the scale of this alternative is bound to grow.

In the field of art and design, the number of employers has decreased, and the requirements for art and design talents have increased significantly. In the training of art and design talents in colleges and universities under the traditional training mode, due to the backward curriculum and teaching methods, it is difficult to match the requirements of employers in the context of artificial intelligence. After graduation, a considerable number of art and design talents have chosen other careers as their means of livelihood. In the questionnaire survey of "change of profession" and "not change of profession" of traditional art and design talents, the proportion of "change of profession" is as high as 46.9%; The proportion of "not changed" is 53.1% (Fig. 1). The high rate of "career change" of nearly half directly proves that there are problems in the cultivation of art and design talents in colleges and universities under the background of artificial intelligence. If this phenomenon does not change, the development prospects of the industry are worrying.

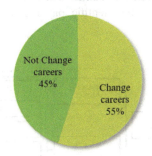

**Fig.1.** Proportion of career changes of art and design talents

Figure 2 is our monthly salary survey for graduates of art and design major after work.

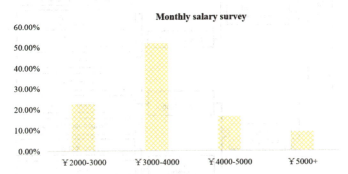

**Fig.2.** Monthly salary survey

The figure shows the monthly salary status of the people who have not changed careers, as shown in Fig. 2, the monthly salary "2000–3000 yuan" accounts for 22.5%, the monthly salary "3000–4000 yuan" accounts for 51.9%, the monthly salary "4000–5000 yuan" accounts for 16.4%, and the monthly salary "5000 yuan or more" accounts for 9.2%. Among them, the monthly income of "3000 - 4000 yuan" accounts for the most, reaching 51.9%, and the monthly income of "5000 yuan or more" accounts for the least, only 9.2%. This shows that most of the traditional art and design talents (68.3%) have a monthly salary of "3000–4000 yuan", and "2000–3000 yuan" (22.55%) is the normal. Compared with the five-figure salary of AI's new career in the context of AI era, it is verified that the advent of AI era has indeed lowered the value of traditional art and design talents and reduced their wages.

## 3 Core Quality of Art and Design Talents

The Ministry of Education of the People's Republic of China has repeatedly pointed out that the core quality of students' development mainly refers to the essential character and key ability that students should have to adapt to the needs of lifelong development and social development. It puts forward the core quality of cultivating "all-round development people", including three aspects, six qualities and 18 basic points (see Fig. 3). Its basic principles for a long time are to be scientific, pay attention to the times and strengthen the regionality.

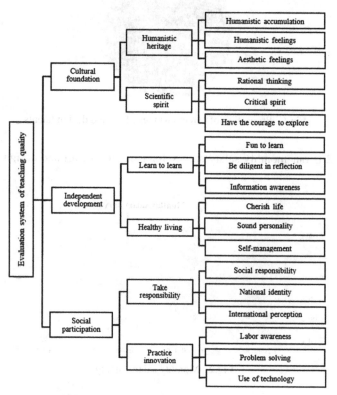

**Fig. 3.** The core quality structure of "all-round development" of students

The overall requirement for the construction of the core literacy of art and design talents is to meet the all-round development of human beings in the process of studying the cultivation of the core literacy of art and design talents under the background of the artificial intelligence era, and to conduct more professional and epochal research on art and design talents. Cultivate them to use the knowledge and concepts they have learned to solve the problems encountered in the era of artificial intelligence, turn the challenges into opportunities, and strengthen the construction of regional, contemporary and scientific core literacy learning system. The cultivation of core competence is the necessary quality and key ability of art and design talents.

## 3.1 Requirements of Art and Design Talents for Core Quality Training

Art design is an aesthetic change that people make artistic innovation on the objective things in life. The discipline of art and design integrates multiple disciplines, involving science and technology, humanities and economy. Its biggest feature is to serve people. Therefore, in the education of art and design, we should pay attention to the training of professional knowledge of art and design and the cultivation of practical operation ability.

Colleges and universities pay attention to cultivating students' practical ability, so the artistic design talents are application-oriented, that is, based on theory and aiming at practice. Therefore, while meeting the daily teaching requirements, the art design course teaching must pay attention to the development trend of the industry's demand for talents, not only grasp the applied teaching strategy of combining art design skills and theory, but also pay attention to the cultivation of students' core quality, knowledge, ability, attitude and other comprehensive qualities.

## 3.2 Elements of Core Quality of Art and Design Talents

The core quality of art and design specialty is reflected in the training of art and design talents, and the connotation of its core quality is combined with skill requirements. This paper discusses the core literacy of art and design professionals from four aspects: ideological character, social participation, personal growth, and high-level cognition, and puts forward two basic elements of the core literacy of art and design talents: comprehensive literacy and design practice ability.

### 3.2.1 Comprehensive Quality

Comprehensive quality is the main embodiment of the core quality of art and design talents, mainly including ethics, value orientation, social value and legal knowledge.

The ethics here refer to national consciousness, national art, patriotism, moral character and artistic morality. Art and design talents should respect the national customs and cultures of different countries, and pay attention to the integration of modern and traditional. Correct values and popular aesthetic orientation are reflected in the design works of designers. A designer without good moral quality and artistic ethics has no high design taste. Designers should have both creativity and innovation, both rights awareness and legal concepts, and rational use of design materials; Art design has the nature of cultural communication, so the practitioners must have a sense of social responsibility, active publicity, more participation in public affairs, and have more social responsibilities.

### 3.2.2 Design Practice

The practical ability of design is the concrete expression of social participation when cultivating core literacy. Design practice mainly includes three aspects (design expressiveness, unity and cooperation ability and communication and expression ability).

Design expressiveness is an important standard to measure the core quality of art and design talents, and also an important standard to measure art and design talents.

It includes aesthetic perception, practice and other aspects. Art and design is a process of combining art, technology, theory and practice. Team cooperation ability refers to the cooperation consciousness, organization and coordination ability of team members, while the art design industry is an industry with high demand for team cooperation and is also a team competition. Good design originates from the cooperation and division of labor of an excellent team. Therefore, art designers should not only have strong professional quality, but also have strong coordination and communication ability.

**Table 1.** Evaluation index system of core competence of art and design talents

| Goal | Level I indicators | Level II indicators | Level III indicators | Weight | Score |
|---|---|---|---|---|---|
| Evaluation of core competence of art and design talents | Comprehensive quality $c_1$ | Ethics $c_{11}$ | Homeland feelings $c_{111}$ | 5 | |
| | | | Art morality $c_{112}$ | 5 | |
| | | | National Art $c_{113}$ | 3 | |
| | | Values $c_{12}$ | Aesthetic orientation $c_{121}$ | 4 | |
| | | | Diligent in reflection $c_{122}$ | 4 | |
| | | Social value $c_{13}$ | Professional certificate $c_{131}$ | 8 | |
| | | | Competition winners $c_{132}$ | 8 | |
| | | Common sense of law $c_{14}$ | Copyright awareness $c_{141}$ | 4 | |
| | | | Intellectual property $c_{142}$ | 4 | |
| | Design practice $c_2$ | Design performance $c_{21}$ | Innovative design ability $c_{211}$ | 8 | |
| | | | Innovative capability $c_{212}$ | 8 | |
| | | | Aesthetic accomplishment $c_{213}$ | 4 | |

(*continued*)

**Table 1.** (*continued*)

| Goal | Level I indicators | Level II indicators | Level III indicators | Weight | Score |
|---|---|---|---|---|---|
| | | Unity and cooperation $c_{22}$ | Sense of responsibility $c_{221}$ | 5 | |
| | | | Service awareness $c_{222}$ | 5 | |
| | | | Risk awareness $c_{223}$ | 4 | |
| | | Communication expression $c_{23}$ | Technology application $c_{231}$ | 8 | |
| | | | Problem solving $c_{232}$ | 8 | |
| | | | English ability $c_{233}$ | 5 | |

On the basis of the research results of the "Research Results Conference on the Development of Core Literacy of Chinese Students", in view of the characteristics of the core literacy cultivation ability of art and design talents, according to the principles and purposes of the indicator system design, through a series of important questionnaires and interviews, the factors affecting the evaluation of the core literacy ability of art and design talents are summarized, Finally, it is determined that the indicator elements for the evaluation of the cultivation of the core competence of art and design talents should include the first-level indicators of "comprehensive literacy" and "design practical ability" (as shown in Table 1), which form an organic whole. In the process of practice, the two elements of comprehensive literacy and design expressive ability present an interactive situation, are interrelated and interact with each other.

By quantifying the three-level index points, the investigation is carried out according to the A-D index. Among them, if the score is above 95 points (excluding 95 points), it is rated as A, which means "excellent", if the score is 90–95 points, it is rated as B, it means "excellent", if the score is 80–89 points, it is rated as C, it means "pass", if the score is below 80 points (excluding 80 points), it is rated as D, and it means "unqualified". Through the evaluation of teachers, training objects and relevant enterprises, the interaction between the evaluation subjects is enhanced, the training object is emphasized to be a member of the evaluation subjects, and the evaluation system of mutual participation and interaction between teachers and students is established, so as to promote the development and development of the core competence training of art and design talents with multi-channel feedback information.

## 3.3 Construction of Core Quality Training System for Art and Design Talents

### 3.3.1 Research Significance of the Core Quality Training of Art and Design Talents

It is of great significance to study the cultivation of the core quality of art and design talents.

(1) Students' personal development needs

Core quality refers to the basic quality and essential character of cultivating students' lifelong development and adapting to social needs in the specific stage of education. It is a comprehensive reflection of students' theoretical knowledge learning, learning ability and learning attitude. Therefore, cultivating the lifelong development ability of students majoring in art design is the most basic quality and skill, and is the necessary condition to move towards and integrate into society.

(2) Teaching needs

Students majoring in art and design have a wide range of employment fields. Including large, medium and small enterprises and professional design companies in various industries. These enterprises and design companies have different needs. With the progress of the times. The needs of employers are also changing. How to keep up with the pulse of the times and cultivate talents meeting the needs of the industry? These all need to cultivate students' core quality in the course setting and actual teaching process. Make students have the necessary character and key ability for lifelong development.

### 3.3.2 Cultivation Concept of Core Quality of Art and Design Talents

The cultivation of the core quality of art and design talents needs to be carried out under the guidance of the cultivation concept of innovative talents combined with artificial intelligence thinking.

Artificial intelligence thinking logic is considered to be the principle that should be followed in the development of new products of the Internet and mobile Internet, and fully considers user design as its own use concept. With the passage of time and the progress of computer technology, new technologies and new development concepts blend and innovate constantly. In combination with AI technology and training objectives, colleges and universities should reform the training mode of art and design talents guided by social service thinking and platform logical thinking, and strive to improve the core quality of art and design talents, such as the design and production of adaptive learning environment. Colleges and universities can analyze the relationship between the needs of domestic enterprises and students' abilities based on the thinking logic of big data, scientifically develop curriculum and teaching plans, analyze students' social conditions in learning in combination with social service thinking, adopt team-based learning methods to cultivate team spirit and practical ability, and enable students to track actual projects through cooperation with design companies, carry out job division and realize the whole process experience of design, The platform logical thinking allows students to conduct online teaching and mixed teaching, creating a scene for students to deeply participate in learning.

### 3.3.3 Analyze and Refine the Core Elements of Design and Art Talent Cultivation

The improvement of knowledge, ability and comprehensive quality is the whole process of training art and design talents. In the process of training art and design talents, we should not only pay attention to the teaching of computer skills and theoretical knowledge, but also teach specific learning methods to students in school, so that students can form the habit of actively acquiring and mastering knowledge, and comprehensively improve cultural quality and personal taste.

In the era of artificial intelligence, more attention should be paid to the update of knowledge theory and computer technology, the design of new theoretical knowledge structure and the adjustment of teaching plan should be carried out in combination with the integration and development trend of artificial intelligence technology and art, so as to cultivate students' scientific literacy and the ability to use big data to learn. At the same time, colleges and universities should further improve the independent learning ability of art and design talents, enhance their opportunities for active learning, improve their practical ability and innovative spirit through an active learning attitude, and achieve the integration of theory and practice. In addition, they should also strengthen the humanistic quality of art and design talents, and explore the corresponding Chinese cultural elements of art and design majors through big data, Refine the core elements of art and design talent training.

## 4 Research on the Implementation Path of the Core Quality Training of Art and Design Talents

Through the research on the implementation path of the core literacy training of art and design talents in the context of artificial intelligence, the overall problem is divided into four sub-problems, and the corresponding research ideas are proposed. It mainly includes four aspects as shown in Fig. 4.

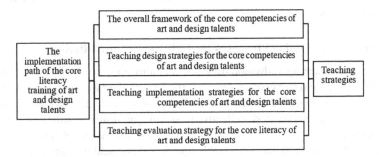

**Fig. 4.** Implementation framework of core literacy cultivation

### 4.1 Overall Framework and Specific Content of Core Literacy

Research the overall structure of the core quality training of art and design talents. Based on the core literacy of Chinese college students, the core literacy of art and design talents is defined in depth. In terms of the selection of curriculum content, according to the teachers' conditions and school positioning of each university, a comprehensive survey is conducted on relevant enterprises (large, medium and small enterprises) and graduates (new, previous employment and unemployed graduates) to determine the overall framework of individuality suitable for the cultivation and development of students in this major.

### 4.2 Teaching Design Strategy for Cultivating Core Literacy

Based on the basic quality system and specific content of art design major in colleges and universities, and combining with the characteristics and current situation of students in relevant colleges and universities, this paper analyzes it by using the methods of group discussion, interviews between teachers and students, and paper method. Adjust the professional curriculum. On this basis, we should further improve the curriculum of art design specialty in colleges and universities to make it more in line with the basic requirements of design education. On this basis, the teaching design of the course is discussed in depth. When carrying out the teaching design reform, the core quality and ability of the students of art design should be reflected.

### 4.3 Teaching Implementation Strategy of Core Literacy Cultivation

In college art design majors, how to effectively cultivate and improve their core quality is an important part of the teaching design of college art design courses, and also the key to the cultivation of core quality.

Therefore, the whole teaching implementation process should be reasonably designed to make the design of teaching implementation process reasonable and feasible. The teaching implementation strategy should keep pace with the development of the times, take the design industry consultation, social hot spots and students' concerns as the starting point, use the information teaching method, mobilize students' learning initiative, enhance the teaching quality, and cultivate the core quality and ability of art and design students.

### 4.4 Teaching Evaluation Strategies for Cultivating Core Literacy

The research on teaching evaluation strategies for cultivating the core quality of art design talents in colleges and universities aims to further verify the rationality and effectiveness of the research on teaching design and implementation strategies. In terms of research methods, case teaching method, collective lesson preparation, comparative research and other methods are adopted to develop a teaching evaluation model for cultivating the core competence of the major. The establishment of this teaching evaluation model pays attention to multi-angles (students, teachers, experts, enterprises) and enforceability while taking into account rationality.

## 5 Conclusion

With the popularization of artificial intelligence technology, how to cultivate a group of design and art talents that adapt to the times has become an important indicator to measure the quality of higher education in China. The development of new technology has created a new form of integration of art and science and technology. In the teaching application of art and design majors in colleges and universities, how to improve students' professional ability while also improving their core literacy is a very meaningful topic.

The talent cultivation of art design majors should not only enable students to master the knowledge structure that adapts to the times, but also have a positive creative spirit. Using new technologies such as artificial intelligence, we should cultivate students' data literacy, scientific and technological ability, accumulate humanistic literacy, condense the spirit of innovation and entrepreneurship, and improve the core quality of talents. Through the research on the implementation path of the core literacy of the construction of art and design courses, it can provide reference for similar domestic colleges and universities in teaching theory and practice research, evaluation system and mechanism exploration, so as to make beneficial exploration for the teaching reform, curriculum design, teaching practice and other aspects of colleges and universities.

## References

1. Arora, M., Bhardwaj, I.: Artificial intelligence in collaborative information system. Int. J. Mod. Educ. Comput. Sci. (IJMECS) **14**(1), 44–55 (2022)
2. Ramjeea, P., Choudhary, P.: State-of-the-art of artificial intelligence. J. Mobile Multimedia **17**(1), 427–454 (2021)
3. Bhaskara, M.: Artificial intelligence: state of the art. Intell. Syst. Ref. Library **172**, 389–425 (2019)
4. Ågerfalk, P.J., Conboy, K., Crowston, K., et al.: Artificial intelligence in information systems: state of the art and research roadmap. Commun. Assoc. Inf. Syst. **50**(1), 21 (2022)
5. Juana, Q.: Research on artificial intelligence technology of virtual reality teaching method in digital media art creation. J. Internet Technol. **23**(1), 125–132 (2022)
6. Chen, C.: Study on the innovative development of digital media art in the context of artificial intelligence. Comput. Intell. Neurosci. **1**, 1004204 (2022)
7. Rathi, J., Grewal, S.K.: Aesthetic QR: approaches for beautified, fast decoding, and secured QR codes. Int. J. Inf. Eng. Electron. Bus. (IJIEEB) **14**(3), 10–18 (2022)
8. Ebenezer, O.: Graphic design principles and theories application in rendering aesthetic and functional installations for improved environmental sustainability and development. Int. J. Eng. Manuf. (IJEM) **9**(1), 21–37 (2019)
9. Yunxuan, W.: Application of artificial intelligence within virtual reality for production of digital media art. Comput. Intell. Neurosci. **2022**, 3781750 (2022)
10. National Library Research Institute. China internet information center released the 50th statistical report on the development of internet in China. J. Natl. Library **31**(5), 12 (2022). (in Chinese)
11. Jian, L., Yuan, Z.: The exploration and practice in innovative personnel training of computer science and technology. Int. J. Educ. Manage. Eng. (IJEME) **2**(6), 47–51 (2012)
12. Jiang, Q., Jin, T., Chen, H., Song, W.: Research on cultivating undergraduates in the computer science based on students. Int. J. Eng. Manuf. (IJEM) **10**(6), 32–39 (2020)
13. Joseph, E.: Orn The Future of Education: Educational Reform in the Era of Artificial Intelligence. Translated by Li Haiyan and Wang Qinhui. Machinery Industry Press, Beijing (2019). (in Chinese)
14. Sun, Y.: Application of artificial intelligence in the cultivation of art design professionals. Int. J. Emerg. Technol. Learn. (IJET) **16**(8), 221–237 (2021)

15. Hu, J., Fu, L.: Innovation and development of environmental art design thinking based on artificial intelligence in culture, form and function. In: Jansen, B.J., Liang, H., Ye, J. (eds.) International Conference on Cognitive based Information Processing and Applications (CIPA 2021). LNDECT, vol. 84, pp. 635–642. Springer, Singapore (2022). https://doi.org/10.1007/978-981-16-5857-0_81
16. Yanzhang, X.: App interactive experience design thinking model based on "Internet+." Art and Design (Theory) **2**(08), 113–115 (2016). (in Chinese)
17. Jing, C.: Exploration and analysis of teaching strategies for cultivating core literacy of art and design talents in higher vocational colleges. Meihe Times (I) **8**, 126–128 (2017). (in Chinese)
18. Xiaoqun, C., Xiaomei, T., Yu, Z.: Research on the construction of the core quality connotation of art and design talents in adult colleges. J. Jiangxi Electric Power Vocational Tech. College, **34**(12), 126–128 (2021). (in Chinese)
19. Yanzhang, X.: Analysis of the core quality of design art talents in the context of artificial intelligence. Chinese Art **1**, 70–75 (2021). (in Chinese)
20. Jing, C., Chenxi, Z.: Construction of the core literacy system of art and design talents in the context of artificial intelligence. Beauty and Time (I) **9**, 123–125 (2021). (in Chinese)
21. Fufeng, C.: Research on art education in the context of AI era – Taking the course of cultural and creative product design as an example. Art Rev **24**, 361–362 (2019). (in Chinese)
22. Xiaoli, Y.: Reform and thinking of visual design mode in the context of artificial intelligence. Grand View Fine Arts **10**, 131–133 (2020). (in Chinese)
23. Amantha Kumar, J., Muniandy, B., Wan Yahaya, W.A.J.: Emotional design in multimedia learning: how emotional intelligence moderates learning outcomes. Int. J. Mod. Educ. Comput. Sci. (IJMECS), **8**(5), 54–63 (2016)

# Reform and Innovation of International Logistics Curriculum from the Perspective of Integration of Industry and Education

Xin Li[1], Meng Wang[1], Xiaofen Zhou[1], Jinshan Dai[2,3(✉)], Hong Jiang[1], Yani Li[1], Sida Xie[1], Sijie Dong[1], and Mengqiu Wang[1]

[1] School of Logistics, Wuhan Technology and Business University, Wuhan 430065, China
[2] Center for Port and Logistics, School of Transportation and Logistics Engineering, Wuhan University of Technology, Wuhan 430072, China
lixin@wtbu.edu.cn
[3] Department of Industrial Systems Engineering and Management, National University of Singapore, Singapore 117576, Singapore

**Abstract.** International Logistics is a professional course of logistics management and a school-level educational reform course of industry-education collaboration, which is an important support to cultivate international application-oriented talents. The course has the following pain points: the teaching content is too different from the enterprise practice, the cognitive load is too large due to the complicated combination of concepts and skills, and the knowledge points are too fragmented and fragmented to produce deep learning. With the general idea of school-enterprise collaborative education and deepening the concept of industry-education integration, we innovatively put forward the teaching strategy of "three-dimensional unity", one dimension: deepening the connection between concepts with theoretical knowledge; two dimensions: defining the cognitive steps with modular implementation; three dimensions: docking with enterprise practice with project system. Through the comprehensive innovation of objectives, contents, cases, resources, process and evaluation, the pain points are effectively solved. At the same time, it pays attention to the integration of course thinking and political elements in the process, including the shaping of values such as patriotism, responsibility, dedication and humanistic care, forming a teaching innovation model adapted to the application-oriented profession.

**Keywords:** International logistics · Industry-education integration · Three-dimensional integration · Teaching innovation mode

## 1 Introduction

International "Logistics" is a course of logistics management in Wuhan Technology and Business College (hereinafter referred to as WTBU), which is a school-level educational reform course based on the integration of industry and education, with a total of 32 h (2 credits), open for junior students. The course aims to cultivate international logistics talents with knowledge, ability, commitment, emotion and warmth for the society.

The course includes many abstract international logistics theoretical foundation, but also covers the bill of lading, import and export processes, and international trade practices and other practical areas, and also involves the domestic and foreign research drones technology and intelligent logistics and other frontier areas. If the practical field is the "backbone" of the course, the theory is the "support point"; the frontier field is academic and the practical application needs to be further developed, which is the "compass" of practice. How to carry out the integration of industry and education, the organic integration of theory, practice and frontier is a topic that the international logistics course team has been repeatedly thinking and practicing [1].

The characteristics of WTBU Logistics College are: professional docking with industry, talents serving industry, logistics management is located in the cultivation of "high-quality applied talents serving the needs of regional economy", through the analysis of students' learning situation research, it is found that students like the professional courses to a high degree, independent thinking ability, but think that the professional and However, they think that the connection between professional and practical work is not close, there are too many knowledge points in the course, they are afraid of difficulties, it is difficult to integrate and apply flexibly, and the depth of study of knowledge points is weak.

## 2 The Existing Problems of the Curriculum

In the past three years, through the evaluation of school students, interviews with graduate students, peer discussion and evaluation, research of similar colleges and universities, feedback from employers, etc., the "pain points" problems that restrict the development of this course have been summarized, mainly in the following three aspects:

### 2.1 The Teaching Contents Are Different from the Enterprise Practice

First of all, international logistics has not investigated the matching of courses and positions in terms of theoretical courses or practical course content, course hours allocation and practical effect evaluation, nor has it investigated the matching of courses and positions, so there may be greater subjectivity, which may be different from the practical requirements of enterprises or positions. Secondly, the design of practical syllabus and practical projects is generally organized, formulated and implemented by the instructors of the course, and lacks the participation of enterprises. In addition, the practical teaching process and assessment are described, guided and assessed by theoretical teachers, which may be different from the actual employment requirements of the enterprise or position. Finally, in various internships, students Although I have experienced it in enterprises (e.g., COSCO, JD.com, SF Express), the internship method mainly focuses on visiting the operation process, equipment, and explanation by enterprise personnel, and lacking practical experience and practical experience. As a result, the adaptation period of graduates is too long, and higher requirements are put forward for non-cognitive goals such as communication, coordination, and management, and students' career development is restricted [2].

## 2.2 Concepts and Skills Are Cumbersome and Complex that Cognitive Points Are Overloaded

Twelve chapters of the textbook, including hundreds of international logistics proper terms, and theories, basic knowledge and skills are scattered and difficult to understand systematically. In addition, students lack intuitive perception of all links of international logistics and "needle threads" of practical cognition, it is difficult for students to learn deeply [3].

The overall logical structure of most international logistics textbooks is "theory → case → exercises". Although the theory related to international logistics is the basis for the study of professional courses in logistics management, there are many difficult and difficult content such as "maritime bill of lading production" and "use of trade terms", and students are very prone to be afraid of difficulties. The course knowledge is complicated and the teaching time is limited, so it is impossible to achieve "all-inclusive". The current teaching method makes most students stay at the "introductory" level, which is not conducive to the systematic ability construction of students.

## 2.3 Fragmented Knowledge Points Are not Easy to Generate Deep Learning

At present, teaching mainly teaches knowledge by teachers, ignoring the cutting-edge development of disciplines and social production needs. Based on the development of the Belt and Road Initiative, it has led to higher standards and requirements for the international logistics industry in all walks of life. The knowledge system lacks systematic construction. Students often memorize by rote and forgetting rates are too high, which is not conducive to the cognitive structure network. The construction of the network and the cultivation of logical thinking system [4].

The logistics industry continues to integrate and develop, national policies are constantly updated, and the content of "static" courses cannot keep up with the times. As a result, students do not pay much attention to the cutting-edge development of the logistics industry, are not sensitive to the updating of national standards and policies, and the shaping of personal values is not enough, so that students face the truth after going to work. Difficulties and conflicts at work cannot be effectively resolved.

# 3 Innovation Strategy

In order to solve the pain points of the above courses, as presented in Fig. 1, this research takes school-enterprise cooperative education as the overall idea, deepens the concept of the integration of production and education, and innovatively puts forward the teaching strategy of "three-dimensional integration" as the main body. Two-dimensional: the theory system strengthens the correlation between concepts; Two-dimensional: cognitive steps are defined by modular implementation; Three-dimensional: to project system docking enterprise practice. Three-dimensional mutual promotion, explore the effective "three-dimensional integration" teaching strategy path [5].

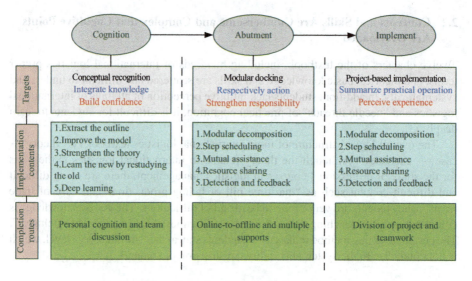

**Fig. 1.** "Three-dimensional integration" teaching strategy

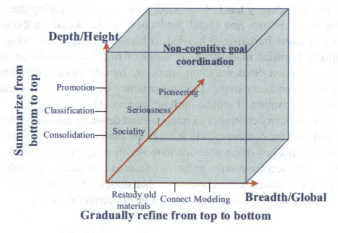

**Fig. 2.** "Conceptual knowledge" construction

## 3.1 Conceptual Understanding

Theoretical basis is the "support point" of professional courses, and also the basis of later modular and project-based learning. It needs to be compiled into the existing concept network of students, so as to promote the formation of long-term memory.

A top-down and bottom-up approach to conceptual architecture, Firstly, based on the existing learning content, the concept connection is strengthened from top to bottom to promote the construction of mental models. Then, the industry frontier, new policies and new regulations are integrated from bottom to reach a certain depth and width. Enable students to raise their theoretical knowledge to a certain height and comprehensiveness.

By strengthening, summarizing and improving the basic concepts, construct the collaboration of social, serious and pioneering non-identification goals, and form the conceptual understanding with the characteristics of renewal and iteration. Conceptual knowledge construction is shown in Fig. 2.

### 3.2 Modular Recombination

In order to undertake the "conceptual cognition" content, docking "project implementation" plan. The content of the textbook is integrated and reorganized into different modules, and each module is divided into several contents, each of which has different meanings and training objectives, so as to organize project-oriented teaching by modules. According to different tasks, different teaching methods are adopted according to the logical sequence suitable for the project, the work of all parties is broken down, and the process is coherent and integrated. Integrating learning Pass and other information teaching tools, forming a modular restructuring framework [6]. The modular recombination framework is shown in Fig. 3.

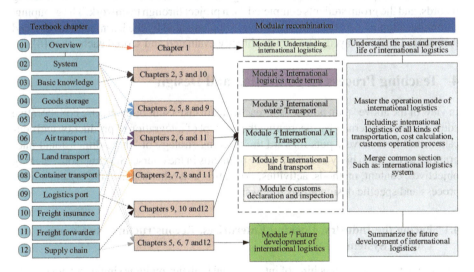

**Fig. 3.** "Modular recombination" frame diagram

### 3.3 Project-based Implementation

According to the content system of this course, the learning and growth process is planned, the comprehensive learning plan is designed and formed, and the learning path map is determined, shown in Fig. 4. Teachers ensure the implementation and promotion of project-based implementation chart through task supervision, step scheduling, process help, resource link and inspection feedback [7].

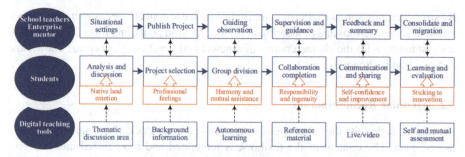

**Fig. 4.** The modular recombination framework

Assign the class to each group with a random sub-project to ensure the coverage of each group's project topics. The responsibility of each topic selection should be assigned to each student, and the team leader should guide each group to assign tasks. Members in the group should grasp the overall direction and content of the project. During the completion of the project, the group members made independent learning records, and the group students completed each project through teamwork. The grouping, communication and evaluation process was supplemented by the learning platform, and then shared and displayed. The groups directly evaluated and scored each other [8].

## 4 Teaching Process Organization and Design

In view of the above "pain points", the teacher team takes school-enterprise collaborative education as the overall idea, deepening the concept of integration of production and education, and innovatively puts forward the teaching strategy of "three-dimensional integration" as the main body to solve the key problems in the course. Starting from teaching objectives, content, methods, activities, evaluation and other aspects, the organizational process and specific design are as follows:

### 4.1 Facing the Industry, Docking Enterprises, Reconstructing the Course Content System

At present, the practical teaching of international logistics major is single and not closely connected with enterprises. Therefore, the formulation and design of the practical teaching curriculum system should be combined with enterprises, considering the actual needs of enterprises and relying on the power of enterprises, and the course content should be jointly designed and developed with enterprises for deep integration.

#### 4.1.1 Emphasis on Application and Enterprise Integration

According to the characteristics of international logistics curriculum, the curriculum system is modular and then project-based, and the curriculum ideological and political cases, international logistics frontier, industrial development trend and other contents are expanded into the integration. Curriculum modularization is not the simple recombination of traditional teaching courses, but the traditional knowledge input oriented teaching to knowledge, ability, quality output oriented teaching process, emphasizing the application of curriculum and enterprise integration [9]. This results in the strategy shown in Fig. 5.

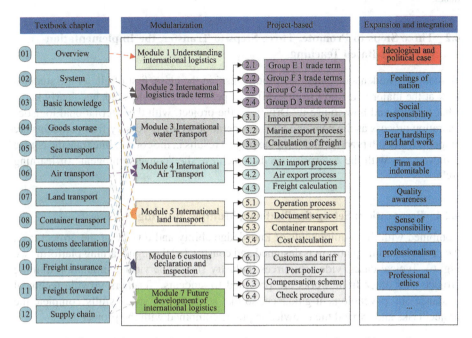

**Fig. 5.** Modular curriculum system, project system, expansion and integration

### 4.1.2 The Ideological and Political Cases of the Course Are Organically Integrated with the Teaching Content to Enrich the Course Connotation

Starting from the theoretical knowledge, value concept and spiritual pursuit of ideological and political education, such as dynamic attention to international logistics standards and industrial prospects, to promote students to build cultural confidence in the logistics industry and sense of responsibility, Based on the development trend of intelligent logistics, the paper analyzes the driving force of technological innovation on logistics industry and the significance of promoting industrial upgrading.

### 4.1.3 Industry Frontier, Policy Update into the Teaching Content, Expand the Course Extension

Under the impact of the global pandemic, the international logistics industry is already facing challenges such as business transformation and service improvement. The whole international logistics industry will present both "prosperous" and "difficult" situation. Industry standards have been constantly upgraded, and laws and regulations have been updated many times. Students are instructed to apply the mini program of "China Logistics Online Platform" and "Global Logistics Inquiry" to improve the efficiency of students in finding the latest policies, new laws and regulations and new guidelines in project tasks.

### 4.2 Three Steps to Promote, Three Steps to Transfer, the Implementation of Project-Based Teaching

To make the students become the "protagonist" in the class, first of all, let the students look at the major and study from a different perspective, let the students clearly know what to learn and how to use it? Driven by the project, with the international logistics operation process as the core and team cooperation as the necessary form, students are guided to in-depth practice and form the project results in the course. Integrate the course achievements with the enterprise needs to achieve the "conversion agent" of learning style.

#### 4.2.1 Preparation: Set up the Platform, Multiple Tutoring

Combined with the requirements of knowledge ability and quality of the project task, the project learning platform is designed and constructed through learning platform (see Fig. 6). Typical cases, video resources, industry standards, virtual simulation training projects and other frameworks are provided to students. Teachers provide technical guidance and psychological guidance, so that students can successfully complete the project tasks and expand the knowledge points combined with the project tasks.

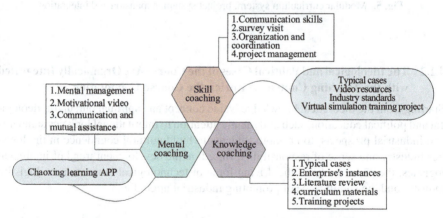

**Fig. 6.** Multi-tutoring mechanism and platform construction

#### 4.2.2 Implementation: Three Level Project, Step by Step

Taking Module 3: International logistics air transport as an example, docking items 3.1–3.2: Ocean import process, export process and liner billing [10].

Level 3 projects: In combination with online video learning, case analysis and discussion, multiple tutoring mechanisms and learning platform, the core content of shipping import and export process is completed, the liner billing method is calculated through group discussion and data collection, and finally the teacher feedback.

Level 2 Project: Each group of students completed the core content of the project design. Connect with the work flow of Marine import and export, division of labor, clear tasks, clear goals, strong operability, and share and display the progress of different modules after completion.

Main steps are listed:

1) Preliminary research stage: Through the investigation and analysis of the current situation of relevant enterprises (mainly school-enterprise cooperation units), the initial identification of logistics culture and professional value is formed. Through the background research of typical projects, students can feel the connection between theory and practice. From the research, understand the definition of sea transportation, its role, the formula of liner billing, factors to be considered, the causes of risks and preventive measures.

2) Medium-term promotion stage: By collecting and sorting out the preliminary research data, the team conducted internal discussion, communication and planning, and each group analyzed the current situation, market trend and future development mode in combination with the shipping industry, and promoted the writing of the project book in combination with shipping enterprises.

3) Project acceptance stage: Completion of the project, together with the results and expected value.

Level I project: Based on the team, as can be seen is Fig. 7, project report and achievement presentation were carried out around each group project. The specific process included: "Project report - teacher's questions - group mutual evaluation - project points". To enable students to form a holistic understanding of the international logistics industry, clear interdisciplinary integration of scientific literacy, innovative thinking and team consciousness.

Project Presentation    Teacher's Questioning    Peer-assessment    Project scores

**Fig. 7.** The team project presentation

### 4.2.3 Promotion: Self-construction, Deepening Cognition

Based on constructivism theory, it emphasizes student-centered, enabling students to actively explore and discover knowledge and actively construct the knowledge system they have learned. Promote in-depth teaching and expand the concept. For students, their time of contact with international logistics is still short, and their abilities are relatively poor. They often concentrate on learning a certain knowledge point, but lack the overall concept, forming a bad situation of "learning while forgetting". Deep learning is an important way to combat this adverse phenomenon. When explaining concepts, teachers should not only focus on the textbook itself, but also help students form a macro perspective, clarify cognition through the comparison of concepts before and after, construct knowledge network, and cultivate students' core qualities such as logical reasoning.

### 4.3 Process Motivation, Result Evaluation, to Ensure the Learning Effect

With the flexible application of tools such as Learning Pass, the trajectories of students' learning activities are recorded completely. Based on the response of students to teaching activities, such as the feedback rate of watching course videos, the average response time of learning, and the quality of replying topic discussion, students with weak learning drive are screened and given emphasis and encouragement [11]. Stimulating students' learning motivation has always been an important function of colleges and universities and teachers [12–14]. Only in this way can students truly become masters of learning and excellent talents [15].

The course evaluation system is shown in Table 1. The project task consists of teacher evaluation, group evaluation and self-evaluation. The evaluation includes whether the learning task has been completed as expected, whether the learning effect is good, the level of learning ability, the contribution of oneself in the group task, and the improvement suggestions for the next study, etc.

**Table 1.** The course evaluation system

| Evaluation process | Evaluation contents | | Proportion (%) |
|---|---|---|---|
| Usual performance (60%) | Course video | | 10 |
| | Homework or quizzes | | 20 |
| | Topic discussion and classroom interaction | | 10 |
| | Project tasks | Third-level project | 10 |
| | | Second-level project | 15 |
| | | First-level project | 15 |
| | | Project acceptance report | 20 |
| Final exam (40%) | | | |

## 5 Teaching Effect of Integration of Production and Education

The average score of students has been improved year by year, and students' satisfaction with the course is more than 93%. Students have won many excellent achievements in student competitions at all levels. Through the integration of industry and education and the docking of enterprises, many students are employed in the world's top 500 enterprises. The international logistics quality of students is also praised by employers. In the organic integration of ideological and political cases with teaching content, many students take the initiative to participate in volunteer and social practice activities such as going to the countryside and poverty alleviation, and the number of participant increases year by year. Specific materials are shown in the Fig. 8.

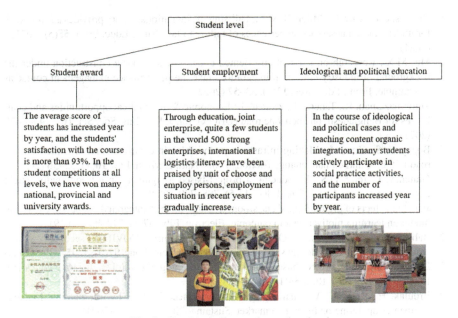

**Fig. 8.** Teaching effect at the student level

## 6 Conclusion

According to the thought of integration of production and teaching, a complete teaching content system related to professional courses has been constructed. I have compiled the supporting teaching syllabus, teaching cases, experimental course operation process, written research reports, teaching reform plans, and published relevant teaching reform papers, which have been promoted and applied in related majors of other colleges and universities.

It has been awarded the national first-class undergraduate course of virtual simulation experiment, which is shared and applied by universities all over the country. The course teaching model, teaching reform plan and teaching cases have been popularized and applied in other professional courses. At present, the teaching results of the "Design of Container iron-water Combined Transport Port Operation Plan" developed by this course based on the teaching reform of the integration of production and education of this course won the second prize of school-level teaching results.

**Acknowledgment.** This research was funded by School-level Teaching Reform Project (2022Z03) of Wuhan Technology and Business University.

# References

1. Hughes, L.J., Amy, N., Mitchell, M.L.: Utilising the invitational theory provides a framework for understanding assessors' experiences of failure to fail. Nurse Educ. Pract. **55**(5), 103135 (2021)
2. Mu, M.X.: Research on application-oriented undergraduate course construction under the background of industry-education integration- take the performance management course as an example. Front. Educ. Res. **12**(1), 53–57 (2022)
3. Tunviruzzaman, R., Tahera, T., Zannat, T.: Economic & geopolitical opportunities and challenges for Bangladesh: one belt-one road (OBOR). Int. J. Res. Bus. Soc. Sci. **10**, 2147–4478 (2021)
4. Bogojevi, S., Zou, M.: Making infrastructure "visible" in environmental law: the belt and road initiative and climate change friction. Trans. Natl. Environ. Law **10**(1), 35–56 (2021)
5. Gratchev, I., Jeng, D.S.: Introducing a project-based assignment in a traditionally taught engineering course. Eur. J. Eng. Educ. **43**(5), 788–799 (2018)
6. Jian, Q.: Effects of digital flipped classroom teaching method integrated cooperative learning model on learning motivation and outcome. Electron. Libr. **37**(5), 842–859 (2019)
7. Babincakova, M., Bernard, P.: Online experimentation during COVID-19 secondary school closures: teaching methods and student perceptions. J. Chem. Educ. **97**(9), 3295–3300 (2020)
8. Liao, D.: Enhancing project-based learning to the teaching of control engineering education. Boletin Tecnico/Tech. Bull. **55**(13), 124–128 (2017)
9. Brdulak, H., Brdulak, A.: Challenges and threats faced in 2020 by international logistics companies operating on the polish market. Sustainability **13**(1), 359 (2021)
10. Magnaye, R.B., Chaudhry, S.S., Sauser, B.J., et al.: Bridging the gap between practice and undergraduate teaching of operations management: the case of public liberal arts colleges. Int. J. Oper. Quant. Manage. **26**(1), 1–16 (2020)
11. Verbi, G., Keerthisinghe, C., Chapman, A.C.: A project-based cooperative approach to teaching sustainable energy systems. IEEE Trans. Educ. **60**(3), 221–228 (2017)
12. Fatima, S., Abdullah, S.: Improving teaching methodology in system analysis and design using problem based learning for ABET. Int. J. Mod. Educ. Comput. Sci. (IJECS) **5**(7), 60–68 (2013)
13. Kaviyarasi, R., Balasubramanian, T.: Exploring the high potential factors that affects students' academic performance. Int. J. Educ. Manage. Eng. (IJEME) **8**(6), 15–23 (2018)
14. Al-Hagery, M.A. Alzaid, M.A. Alharbi, T.S., Alhanaya, M.A.: Data mining methods for detecting the most significant factors affecting students' performance. Int. J. Inf. Technol. Comput. Sci. (IJITCS), **5**, 1–13 (2020)
15. Zhiqin, L., Jianguo, F., Fang, W., Xin, D.: Study on higher education service quality based on student perception. Int. J. Educ. Manage. Eng. (IJEME), **2**(4), 22–27 (2012)

# An Analysis of Talent Training in Women's Colleges Based on the Characteristics of Contemporary Female College Students

Ting Zhao[1], Zaitao Wang[2(✉)], Xiujuan Wang[3], and Chuntao Bai[4]

[1] Library, Shandong Women's University, Jinan 250300, China
[2] School of Business Administration, Shandong Women's University, Jinan 250300, China
wzt0813@163.com
[3] Fundamental Science Section in Department of Basic Courses, Chinese People's Armed Police Force Logistics College, Tianjin 300309, China
[4] Party and Government Office, Tianjin Beichen Science and Technology Park Management Co., Ltd, Tianjin 300405, China

**Abstract.** How to cultivate female college students and what kind of talents they will become is a question worth discussing and must be answered by women's colleges. By looking up the relevant literature at home and abroad, this paper clarifies the characteristics of women's colleges and universities, and points out their mission, value and development direction. In order to have a systematic understanding of female characteristics in colleges and universities, this paper briefly introduces the structural equation model method, and points out that their growth environment, values and consumption outlook are different from each other in the past, which is conducive to teaching women's colleges in accordance with their aptitude and completing the talent training goal better. This paper puts forward an integrated curriculum system that integrates the training specifications of knowledge, ability and quality, and takes the spiral ability training as the main line. Women's colleges and universities emphasize that, according to women's physiological and psychological characteristics, they should cultivate women's compound applied talents suitable for social needs, which will play an important role in leading and promoting the deconstruction of gender bias, eliminating gender inequality, building advanced gender culture, promoting women's development and gender equality, and promoting social civilization and progress.

**Keywords:** Women's colleges · Contemporary female college students · SEM · Personnel training

## 1 Introduction

The cause of women in the world has made great achievements, but still faces many challenges. More than half of the world's 800 million poor people are women; when wars and epidemics strike, women often bear the brunt. The main reason is that the historical stereotypes of gender inequality have not been completely eliminated, the social environment for women's development needs to be optimized, and women's participation in cultural education, economic participation, decision-making management, social security, legal protection and other aspects needs to be further improved [1]. According to < Outline for the Development of Chinese Women (2011–2020) >, which is showed as Fig. 1. It is not difficult to see from the table that from 2010 to 2020, the proportion of women in postgraduate education has continued to increase, and by 2020, its proportion has exceeded half. In the report of the 20th National Congress of the Communist Party of China in 2022, it was clearly pointed out that we should adhere to the basic national policy of gender equality and protect the legitimate rights and interests of women and children. Women's colleges and universities are educational institutions that provide higher education for women and are an important part of higher education in China. Building a strong country in higher education is an important task for women's colleges and universities. Training high-quality female talents with innovative ability should be an important task for women's colleges and universities.

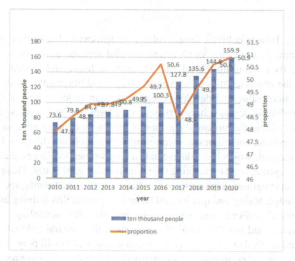

**Fig. 1.** Number and proportion of female graduate students in China from 2010 to 2020

## 2  Research on Characteristics of Women's Colleges and Universities

The characteristics of running a university is a systematic structure composed of multiple elements, and so is the characteristics of running a women's college. In this system structure composed of multiple elements, each element is interdependent and interacts with each other, which jointly affects the formation and development of school running characteristics. The following is a literature review of each element of school running characteristics.

## 2.1 Research on the Goal and Idea of Running a School

Kim et al. pointed out that female characteristics should be highlighted as the main body of school running, gender theory should be studied and gender awareness should be spread, which will further highlight its professionalism [2]. Dasgupta et al. believes that teaching should be carried out according to the needs of women as talents and the cognitive laws, and specialties and courses should be set up and taught according to women's characteristics [3]. The "female advantages" of specialties and courses will be paid more attention. He puts forward that we should surpass the tradition in school running goals, education models and school functions, and build a targeted environment for cultivating women's culture, so that there will be room for upward development and sustainable vitality.

## 2.2 Research on the Characteristics of the Main Body of Running a School

The research on the "characteristics of the main body of running a school" of women's colleges and universities is carried out from the two main bodies of teachers and students. (1) To train teachers in women's colleges and universities on gender awareness, so as to improve their consciousness and sensitivity of gender awareness; (2) This paper studies the scientific literacy, interpersonal communication and psychological anxiety of female college students. Sugianto pointed out that colleges and universities should take effective measures to strengthen the cultivation of college students' scientific literacy [4]. Mosleh et al. suggested that schools should start from reality and carry out mental health education to alleviate their psychological anxiety and help them improve their mental health [5]. Mozgalova et al. proposed improvement approaches from both the students themselves and the school [6].

## 2.3 Research on the Characteristics of Talent Training

The research focuses on the goal, mode and significance of female talent training. The training objectives formulated by women's colleges and universities mainly include "taking social and market needs as the basis, giving full play to women's advantages", "modern women who are excellent in quality, professional, physical, artistic, and aesthetic", and "opening up new employment channels for young women". Liu Jia and Yang Liu started from explaining the connotation and extension of the talent training model, analyzed the characteristics of the talent training model and the factors affecting the talent training model, it also puts forward some ideas on the talent cultivation of women's universities [7]. Tang Jianxiong discussed the characteristics and construction of the talent training model of modern women's colleges and universities, and pointed out that if women's colleges and universities want to really go hand in hand with other colleges and universities, they must build a talent training model that meets the needs of social development and has a distinctive female education color [8]. Choi et al. pointed out the defects of the traditional talent training model and the bottleneck in the reform of the talent model in women's colleges and universities, and discussed the employment oriented training model of applied talents for women [9]. Cai pointed out the significance of the "four self" talent training path of the school [10].

## 2.4 Research on the Characteristics of Disciplines and Specialties

There are two articles about "discipline and specialty characteristics". Warren et al. emphasized that the scientific setting of majors is an important way for universities to highlight the characteristics of running schools [11]. Women's universities can only ensure the training of outstanding women talents with characteristics by scientifically standardizing, fully demonstrating and building a scientific and characteristic professional system on the premise of in-depth discussion of teaching rules and the growth laws of women talents. Kyrpychenko et al. pointed out that the communicative competence development in teaching professional discourse is of special significance for women in colleges and universities [12].

## 2.5 Research on the Characteristics of Curriculum Construction

The research on the curriculum construction of women's colleges and universities focuses on the following two aspects: (1) The characteristic curriculum construction of women's colleges and universities. Sharm et al. aimed to apply exploratory applied research design, evaluate challenges through a unified cloud-based e-learning framework, and use hybrid research methods to propose a new cloud-based solution framework for women to be challenged by accidental injuries [13]. Li analyzed the problems of the development of the course of women's studies in China, the methods and objectives of the course construction of women's studies, and the ways of the course construction of women's studies to ensure the healthy development of the course construction of women's studies [14]. In addition, ADIGUN proposed that the integrated teaching of artistic dance and aerobics has the function of cultivating correct health outlook, improving artistic accomplishment, and meeting physiological and psychological needs of contemporary female college students [15]. (2) Cultural curriculum in women's colleges and universities. Koseoglu et al. based on the basic characteristics of the feminist curriculum, paid special attention to the gender inequality in the curriculum, advocated the construction of a curriculum serving the teaching purpose of women's colleges and opposed the gender bias in the curriculum [16].

Through the analysis of the above research, it is not difficult to find that the research mainly focuses on narrow aspects, with more repetitive research, less innovative research and less objective research attitude. There are few research results on the characteristics of foreign women's schools, which are not deep and thorough enough. The total number of research results on the characteristics of foreign women's colleges in the existing literature is relatively small, but only from two aspects. On the one hand, the characteristics of famous women's universities in the United States and South Korea are analyzed, and on the other hand, a comparative study of Chinese and foreign women's universities is conducted. There are few empirical research results, and most of the articles are the author's experience summary or theoretical overview. More empirical research is needed to more convincingly reflect the current situation of the running characteristics of women's colleges and universities in China, and accordingly put forward corresponding countermeasures and suggestions, in order to make our women's colleges and universities more robust, healthy development and growth.

## 3 Research Method

Based on the structural equation model (SEM), this paper will carry out the corresponding empirical research, build the structural equation model of college students' personality and behavior characteristics and the factors affecting the cultivation of innovative talents, and analyze and explore the main factors that affect the cultivation of innovative talents in economics and management.

Let $\xi$ be a vector of independent variables and $\eta$ be a vector of dependent variables, $\Lambda$ is a matrix of factor loadings, $\varepsilon$ a vector of residuals often known as unique variates. $B$, $L$ and $M$ are coefficient matrices, $I$ is the identity matrix of an appropriate order, Let us denote $B = \begin{pmatrix} L & 0 \\ 0 & 0 \end{pmatrix}$, $\Gamma = \begin{pmatrix} M \\ I \end{pmatrix}$, and $v = \begin{pmatrix} \eta \\ \zeta \end{pmatrix}$, the following equation can be obtained:

$$\eta = L\eta + M\xi \tag{1}$$

$$v = Bv + \Gamma\xi \tag{2}$$

To make sure the inverse of $I$–$B$ exists, we assume the $I$–$B$ is non-singular, then

$$v = (I - B)^{-1}\Gamma\xi \tag{3}$$

The components of the matrix $G$ are either 1 or 0 which connects $v$ to the observed variables $x$ such that $x = Gv$. Let $\mu = E(x)$, $\mu_\xi = E(\xi)$, $\Sigma = Cov(x)$, and $\Phi = Cov(\xi)$. Various covariances $\phi_{ij}$ are shown as two-way arrows in path diagrams. The full mean and covariance structure analysis model is given as:

Mean structure:

$$\mu = G(I - B)^{-1}\Gamma\mu_\xi \tag{4}$$

Covariance structure:

$$\Sigma = G(I - B)^{-1}\Gamma\Phi\Gamma'(I - B)^{-1'}G' \tag{5}$$

The measured variables x may be generated by the $\xi$

$$x = \mu + \Lambda\xi + \varepsilon \tag{6}$$

$$\xi = B\xi + \zeta \tag{7}$$

$$\xi = (I - B)^{-1}\zeta \tag{8}$$

$$x = \mu + \Lambda(I - B)^{-1}\zeta + \varepsilon \tag{9}$$

If the means are unstructured, $\mu = E(x)$. With a structure, we take $\mu = 0$ in (9) and let $\mu_\zeta = E(\zeta)$, and with the covariance matrix of the $\zeta$ and the $\varepsilon$ given as $\Phi_\zeta$ and $\Psi$, respectively, the mean and covariance structure of the model follow as:

$$\mu = \Lambda(I - B)^{-1}\mu_\zeta \tag{10}$$

$$\Sigma = \Lambda(I-B)^{-1}\Phi_\zeta\{(I-B)^{-1}\}'\Lambda' + \Psi \tag{11}$$

This representation shows the confirmatory factor analysis model

$$\Sigma = \Lambda\Phi_\zeta\Lambda' + \Psi \tag{12}$$

SEM is a multivariate statistical method, which combines factor analysis and path analysis, including leading variables, potential variables, interference or error variables, indirect effects or total effects for verification. Structural equation model can be divided into measurement model and structural model. It can not only study the structural relationship within variables, but also study the relationship between variables. Therefore, it is widely used in various studies by domestic and foreign scholars. In order to pursue preciseness, even if the established model can be fitted, the model must be modified according to the evaluation structure to obtain the most scientific, reasonable, most suitable and best explanation of the final model.

## 4 Analysis of Characteristics of Contemporary Female College Students

### 4.1 An Analysis of the Psychological Characteristics of Contemporary Female College Students

Influenced by self-gender role and external social and cultural atmosphere, the psychology of contemporary female college students mainly has the following characteristics.

(1) Strong self-awareness, incomplete self-awareness, relatively lack of team awareness, and urgent need to improve interpersonal skills. They tend to pursue and advocate individuals and personalities. In the survey, many female college students pay more attention to themselves and less attention to the feelings of others. They lack clear recognition of their roles and comprehensive understanding of their advantages and disadvantages. Many people think they lack interpersonal skills and need to improve their sense of teamwork and interpersonal skills.

(2) Confident, optimistic and frank, but sometimes lazy, lacking time concept. On almost all occasions, as long as they have the opportunity, they will publicize their personality. They dare to express their views without scruple. They are keen on various campus cultural activities and competitions and actively participate in various elections; but many students think they are lazy and have a weak sense of time.

(3) Strong learning ability, extensive hobbies, strong ability to accept new things, but affected by the physiological characteristics of late youth, emotional stability is not good, and lack of some self-control. Their average intelligence is higher than their previous peers, and they have strong learning ability.

### 4.2 An Analysis of the Characteristics of Contemporary Female College Students

Compared with the past, the family environment in which contemporary female college students grow up is relatively superior, and they have no worries about food and clothing since childhood, which clearly reflects the following characteristics.

(1) The vision is broad and the knowledge is rich, but sometimes the heart is empty and the goal is vague. Contemporary female college students have grown up with the Internet. Their familiarity and affinity with new media far exceed those of previous generations. They have a broader vision, more diversified channels to receive information, advanced mental development, and far more knowledge and precocity than previous generations. It also directly affects their lifestyle and values, which easily makes them feel empty. Moreover, the more familiar with such new media, the contrast between virtual and real will be felt. Some of them virtual communication ability has far exceeded their real communication ability. This virtual personality has a great influence on their socialization.

(2) Diversified values and more realistic values. The value orientation of contemporary female college students is more diversified. Some social realities have made them understand that they should focus on specific things rather than beliefs. Their aesthetic taste, lifestyle and moral boundaries have also gradually changed. The inner world has become "old" since childhood, and has a better understanding of the rules of the adult world.

(3) The consumption concept is ahead of time, and the consumption level is incompatible with the family economy. According to statistics, more than 99% of the nearly 15, 000 students in our college have bought mobile phones, and there are many high-end mobile phones; the average monthly living expenses given by parents to students reached 600–1, 500 yuan. And many students obviously showed that they could not bear hardships, and showed varying degrees of inadaptability in the face of setbacks. For example, in the centralized internship carried out by the school, some girls rent houses and even need help from others. Girls who practice in refrigerators of JD.com and other units generally say they are tired, and some even give up the internship in processing plants.

## 5 An Analysis of Talent Training in Women's Colleges Based on the Characteristics of Contemporary Female College Students

In combination with the teaching practice of women's universities and the campus culture of "advocating virtue, love, erudition, and beauty", women's colleges and universities have two main goals for talent cultivation: first, enrich professional education through liberal education, improve the quality of talent training, and establish a talent training model of "thick foundation, wide caliber", so as to improve the core competitiveness of women's colleges and universities; second, improve the liberal education curriculum system and model, and gradually form and reflect the unique and distinctive characteristics of women's colleges.

### 5.1 Training Specifications Integrating Knowledge, Ability and Quality

Training specification is the specific refinement and decomposition of training objectives. According to the talent training concept of "all-round development", the training objectives are refined based on knowledge, focused on ability and expanded by quality. According to the orientation of talent training objectives, determine the knowledge structure, ability structure and quality structure with horizontal correlation in spatial dimension. The same structure requires an integrated training specification with vertical correlation in time dimension:

(1) Knowledge structure requirements: a. master the knowledge of humanities and social sciences; b. master the basic knowledge of the discipline; c. master basic professional knowledge and theory; d. master professional technical knowledge. (2) Capability structure requirements: a. have the ability to understand the theoretical frontier, development trends and independently acquire knowledge in the professional field; b. have the ability to use engineering technology for system design; c. strong ability in scientific research, engineering practice and solving practical problems with the knowledge learned; d. have the ability to learn independently, innovate, work cooperatively and organize. (3) Quality structure requirements: a. have good ideological and moral cultivation, professional quality, physical and mental quality; b. have the spirit of dedication, interpersonal awareness and solidarity; c. have a certain literary and artistic accomplishment, scientific engineering practice methods d. have a certain international vision, realistic and innovative consciousness.

### 5.2 Integrated Curriculum System with Spiral Ability Training as the Main Line

According to the talent training concept of "demand guidance, project driven, strengthening practice, teaching and research unity, highlighting innovation, and comprehensive development", according to the talent training specifications and characteristics, as well as the consistency, completeness, integration, progressiveness, and practicality of the curriculum system, and in combination with our own advantages and characteristics, we will build a talent training oriented, theory and practice oriented Quality education is integrated into the training of professional knowledge and ability, and the knowledge, ability and quality are coordinated and unified. The curriculum system is oriented by ability cultivation, supported by centralized practice links, and based on core courses. A group of centralized practice links and core courses cultivate an application or design ability and related knowledge.

According to the talent cultivation concept of "demand guidance", the school and the enterprise jointly design the curriculum system, and the enterprise sends engineers to set up relevant engineering design courses in the school. The curriculum system has set up a series of centralized practice links, independent class experiments and in class experiments to implement the talent cultivation concept of "strengthening practice".

## 6 Conclusion

For thousands of years, women have been positioned as the role of procreation, housekeeping, helping husbands and taking care of children. They have been excluded from receiving education for a long time. "women without talent are virtuous" is a true portrayal of this phenomenon. It is only in recent decades that a large number of Chinese women have entered schools to receive formal education. It is precisely because the objects of education for thousands of years are all men (or basically men), so the current educational philosophy, teaching methods, curriculum structure and examination evaluation model are designed with men as the objects of education. The current education system mainly takes into account the psychological characteristics and cognitive laws of men, and rarely takes into account the characteristics of women.

This paper makes a deep analysis of the characteristics of contemporary female college students, and on this basis, puts forward some suggestions on talent training in women's colleges. What kind of education should be given to women, such as "providing women with educational forms suitable for women's characteristics" to "meet the needs of women's growth and talent"; "cultivate women with the spirit of 'four self'"; employment oriented; based on the development of female students, compensation education should be carried out for female college students. In terms of specialty setting and course offering, we should think about how to teach courses with women's characteristics and cultivate talents with women's characteristics. In view of the problems encountered in the training of application-oriented talents, we should reform the training mode of application-oriented talents, and study and practice effective ways to improve the training quality of application-oriented innovative talents.

**Acknowledgment.** This research has been supported by the 2017 school level teaching reform research project of Shandong women's University: discussion and practice of innovative personnel training mode of economic management in women's University, and the Shandong Social Science Planning Project (Grant No. 18CGLJ48, 18CSJJ32).

# References

1. Seymour, S.: In: Mukhopadhyay, C.C., Seymour, S. (eds.) Women, Education, and Family Structure in India, pp. 213–233. Routledge, New York (2021)
2. Kim, D., Yoon, M., Jo, I.H., et al.: Learning analytics to support self-regulated learning in asynchronous online courses: A case study at a women's university in South Korea. Comput. Educ. **127**, 233–251 (2018)
3. Dasgupta, U., Mani, S., Sharma, S., et al.: Effects of peers and rank on cognition, preferences, and personality. Rev. Econ. Stat. **104**(3), 587–601 (2022)
4. Sugianto, E.S.: The role of collaborative learning and project based learning to increase students' cognitive levels in science literacy. In: International Conference on Madrasah Reform 2021 (ICMR 2021), pp. 67–72, Atlantis Press (2022)
5. Mosleh, S.M., Shudifat, R.M., Dalky, H.F., et al.: Mental health, learning behaviour and perceived fatigue among university students during the COVID-19 outbreak: a cross-sectional multicentric study in the UAE. BMC psychol. **10**(1), 47 (2022)
6. Mozgalova, N.G., Baranovska, I.G., Hlazunova, I.K., et al.: Methodological foundations of soft skills of musical art teachers in pedagogical institutions of higher education. Linguist. Cult. Rev. **5**(S2), 317–327 (2021)
7. Jia, L., Liu, Y.: Research on guiding education of college students' interpersonal relations. Writer's World **22**, 110–111 (2021). (in Chinese)
8. Jianxiong, T.: Research and practice on the reform of new business talents training mode in higher vocational education -- Taking the International Economy and Trade Major of Guangdong Women's Vocational and Technical College as an example. Univ. Educ., 208–210+220 (2022) (in Chinese)

9. Choi, S.J., Jeong, J.C., Kim, S.N.: Impact of vocational education and training on adult skills and employment: an applied multilevel analysis. Int. J. Educ. Dev. **66**, 129–138 (2019)
10. Cai, X.: Research on the training mode of adaptive talents in higher vocational education under the background of educational informatization 2.0. In: 2022 6th International Seminar on Education, Management and Social Sciences (ISEMSS 2022), pp. 3542–3548, Atlantis Press (2022)
11. Warren, K., Mitten, D., D'amore, C., et al.: The gendered hidden curriculum of adventure education. J. Experiential Educ. **42**(2), 140–154 (2019)
12. Kyrpychenko, O., Pushchyna, I., Kichuk, Y., et al.: Communicative competence development in teaching professional discourse in educational establishment. Int. J. Educ. Manage. Eng. **4**, 16–27 (2021)
13. Sharm, D.P., Deng, D.K., Tigistu, G., et al.: A cloud based learning framework for eradicating the learning challenges of Ethiopian working professionals, disables and women. Int. J. Mod. Educ. Comput. Sci. **14**(2), 41–54 (2022)
14. Li, L.: Education supply chain in the era of Industry 4.0. Syst. Res. Behav. Sci. **37**(4), 579–592 (2020)
15. Adigun, J.O., Irunokhai, E.A., Onihunwa, J.O., et al.: Development and evaluation of a web based system for students' appraisal on teaching performance of lecturers. Int. J. Inf. Eng. Electron. Bus. **14**(1), 25–36 (2022)
16. Koseoglu, S., Ozturk, T., Ucar, H., Karahan, E., Bozkurt, A.: 30 years of gender inequality and implications on curriculum design in open and distance learning. J. Interact. Media Educ. **5**(1), 1–11 (2020)

# Solving Logistical Problems by Economics Students as an Important Component of the Educational Process

Nataliya Mutovkina(✉)

Tver State Technical University, Tver 170012, Russia
`letter-boxNM@yandex.ru`

**Abstract.** The article examines teaching students to solve logistical problems and also focuses on the importance of such skills in the professional activities of economists. Methodological teaching methods are considered in preparing students toward "Economics" at Tver State Technical University. Logistics asks in the educational process are presented to students as one type of optimization tasks and are solved as transport tasks, assignment tasks, backpacks, traveling sales agent. The author of the paper defines the role and practical significance of logistics tasks in the professional activity of economists. In training, students formalize logistics tasks, choose the most appropriate solution method, and analyze the results of the solution. The primary tools in this case are software tools available to students. The article defines the effects of a problem-oriented approach to training future economists on the example of logistics tasks. The novelty of the work is to identify the features of teaching individual disciplines, considering the need to include methods of solving logistics problems in their structure. In addition, the author's recommendations on teaching students the selected elements of logistics are new. In particular, students can choose tasks of different levels of complexity, as well as offer their tasks and methods of solving them.

**Keywords:** Logistics · Educational process · Logistics tasks · Optimization methods · Optimization tasks · Transport tasks

## 1 Introduction

Logistics tasks are an integral part of the economic tasks that economists-managers have to solve. First, this is because of the need for transport support for the business, delivery of raw materials, semi-finished products and finished products to both external customers and the movement of goods on the territory of the enterprise. Saving resources and time acts as a target function of most logistics tasks. Solving the tasks of delivering goods on time and at the lowest cost creates additional business opportunities for the company, attracting new customers, and improves the company's image. The main competitive advantages of the company are: cost reduction because of the optimization of logistics processes, because of reduction of transportation costs; guaranteed fulfillment of orders in the required volume and in the required time [1]. Cargo transportation is an important element of the economic system of any company. Transportation concerns not only the movement of goods outside the company but also the distribution of resources within it [2].

As practice shows, in their professional activities, graduates in Economics often have to solve problems related to the need to minimize the cost of delivering goods from suppliers to consumers, the tasks of transport logistics. To solve these problems, the following actions are necessary:

1) To collect and analyze information about the material flows available in the company.
2) To clarify the location of consumers and find out possible ways of delivering goods.
3) To determine the transport for delivery and choose the optimal transport, considering the cost of its operation and travel time.
4) To develop the optimal pathway to deliver goods to customers.

The work aims to prove the effectiveness of logistics tasks in the educational process for the professional training of economists.

The article is based on the results of a study designed to answer the fundamental question: "How does the solution of logistics tasks form students' knowledge, skills, skills and professional competencies, and contributes to effective professional activity?".

## 1.1 Literature Review

Logistics issues and solutions to logistics problems are considered in many educational and methodological works on optimization and transport logistics. So, in the work [2] a decision support system is proposed to assess the costs associated with logistics processes. This system allows you to calculate the economic, environmental, and social costs of the logistics process to ensure sustainable logistics. In the article [3]. The authors propose a solution to the problem of planning the movement of railway transport, considering the following features: besides planning service intervals for trains, the article additionally solves the problem of securing each train to the railway track. A mathematical model and a method for solving the scheduling problem are presented. The key feature of this mathematical model is that it does not use Boolean variables, but works with combinatorial objects (sets of permutations).

Sometimes delivering goods is complicated by the need to consider the specifics of the cargo being transported. For example, it can be extremely fragile or perishable goods. Here, a condition for the safety of goods is added to all other conditions. In the publication [4]. A system for monitoring the temperature regime of goods delivery is proposed. This system can be linked to the client's system so that all data on the temperature of processing, storage, and transportation can be sent to clients via the Extranet and the suppliers' Web server. The customer system can determine whether the temperature data is normal before the products are received. Similar technology is also considered in the works of [5, 6].

Many logistical problems are reduced to solving linear programming (LP) problems. Linear programming is a field of optimization theory that includes optimization problems in which the objective function and constraints have a linear form [7]. Linear programming is widely used to solve a variety of financial, marketing, production, and agricultural problems. In finance, LP is used for budgeting, asset allocation, financial planning, etc. [8, 9]. In marketing, LP is used to solve problems related to marketing research and media selection [10]. In addition, LP is used to compile an optimal production program, and optimize the product range. Similarly, LP is used in the agricultural sector to optimize the structure of crops, as well as the distribution of resources (water, land, fertilizers, etc.) [11].

LP is not a new concept, but the ease of use and accuracy of the results got to make it an indispensable tool in the hands of a decision-maker.

### 1.2 Software Tools for Solving Logistics Tasks

Linear programming problems can be solved using special software. Now the user may not even know the subtleties of methods for solving optimization problems to get the optimal solution in a short time. Only knowledge of the formalization of optimization problems is enough. Software having the specified functionality includes such programs as LINDO SS [12, 13], LINGO [14, 15], MathCAD, IOSO, Approx, Xpress Optimizer, Microsoft Excel, etc. In the article [16] a comparative analysis of several software products is presented. According to the results of the analysis, the most convenient programs for solving transport logistics problems are Microsoft Excel and MathCAD. The IOSO and Approx programs have only paid versions, and the convenience of working in these software tools for untrained users is difficult. Xpress Optimizer is a suitable programming environment, but because of the lack of a Russian interface and the lack of a free version, it cannot compete with Microsoft Excel and MathCAD.

The advantages of Microsoft Excel are also mentioned in [17, 18]. In this software, there is an optimization tool Solver, with a user-friendly and intuitive interface. Many researchers use Microsoft Excel to solve most optimization problems easily and quickly. For example, in [18] the solution of the problem of controlling uncertainty in income is presented from agriculture using a model of fuzzy multi-purpose linear programming.

In classroom classes with students at Tver State Technical University, Microsoft Excel is most often used to solve logistical problems. Here, this is due not only to the versatility and ease of use of this software, but also to the availability of a licensed version of Microsoft Office at the university.

## 2 Features of Solving Logistics Problems

### 2.1 Logistical Tasks in the Professional Activity of Economists

In the logistics activity of the enterprise and its subsequent optimization, the following mathematical problems are most often formed: forecasting demand for goods, determining the optimal stock of goods in stock, calculating the optimal order size, and forming a transportation plan from a group of suppliers to a group of buyers.

The definition of the type of task and its formulation isis carried out depending on the goals of the company. Therefore, students must first analyze the subject areas according to the principles of a systematic approach. Then it is necessary to understand what data is available to the analyst and what data is missing. If the numerical information is presented in full, then the logistic problem belongs to well-formalized problems and can be solved by standard mathematical methods. If there is an incompleteness of information, then two approaches are possible here. The first approach is to restore information using interpolation, extrapolation, and other methods. The second approach is used if information recovery is impossible and comprises expert evaluation. Both methods apply to problems that are difficult to formalize.

Logistics tasks are usually well formalized, but there are some uncertainties in their formulation and solution. For example, modern logistics concepts imply that determining the amount of demand for goods is the responsibility of the logistics services of the enterprise, therefore, demand forecasting can be attributed to logistics tasks [19]. However, forecasting the level of a time series is associated with difficulties in identifying trend, seasonal, and cyclical components. In addition, there is always the influence of a random deviation formed for various reasons that do not depend on the researcher.

In determining the optimal stock, the storage capacity is usually taken as a constant value, and the number of stocks stored on it is a variable value. However, this statement is correct with a certain conditionality. For example, it is assumed that the warehouse area cannot be physically increased, or that it is impractical from an economic point of view.

For the transport problem to be solvable, it is necessary and sufficient that the total cargo stocks at the points of departure are equal to the total needs at the destinations. The model of the transport problem satisfying this condition is called closed. If the specified condition is not met, then the model is open. If the stock exceeds the need, a fictitious destination is introduced, the cost of delivering the goods to which is zero. If the demand exceeds the stock, a fictitious departure point is introduced and delivery costs are also assumed to be zero [20].

Students must analyze the features of the problem and only after that make up its mathematical model.

## 2.2 Implementation of the Mathematical Model

A mathematical programming problem can be written:

$$\begin{cases} F = f(x_i) \to max(min), \\ \varphi_i(x_i) \le b_j, j = \overline{1,m}, i = \overline{1,n}, x_i \in D_i \end{cases} \quad (1)$$

The first expression is the objective function, in which $x_i$ there are the desired task variables. The second expression is the constraints of the task.

This model is used to formalize the transport task. However, when setting a transport task, it is necessary to consider its features:

1) The system of constraints is a system of equations, the transport problem is given in the canonical form.
2) The coefficients for the variables of the constraint system are equal to one or zero.

3) Each variable enters the restriction system twice: once for orders, the second time for needs.

Implementing the model is carried out in a software environment accessible to students and teachers. As an additional work that develops the student's thinking, implementing the model in other software environments can be considered, followed by a comparison of the results and features of the work.

Students develop a mathematical model of the problem first under the guidance of a teacher, then independently. When performing independent work, students can choose problem situations themselves for subsequent analysis, formulation, and solution of the optimization problem. Students can also choose methods for solving the optimization problem. Non-standard, creative approach to solving problem situations is welcome and has a positive effect on the evaluation of the student's work.

## 2.3 Analysis of the Results Obtained

In optimization theory, there is a concept of an acceptable, but not optimal, solution. An acceptable solution is a distribution of the desired quantities that satisfies all the conditions of the problem but does not provide an extremum of the objective function. Such situations occur often from the subjective preferences of participants in the decision-making process.

Objective factors can also influence the solution to the optimization problem. In particular, when solving a transport problem, the following circumstances may arise.

### 2.3.1 Blocking Delivery

If transportation from sender $i$ to consume $j$ is prohibited (for example, because of bad roads), then some large number A (A $\to \infty$) should be taken as the transport cost coefficient in the cell $(i, j)$. Then, in the optimal solution (on a minimum) this cell will be zero. If this cell still turns out to be the basic one, then this shows the insolvability of the problem because of prohibited transportation.

### 2.3.2 Bandwidth Limitation

Suppose, for example, it is required to consider the limited capacity of the road from sender $i$ to consume $j$.

Here, the column of the consumer $j$ in the distribution table should be divided into two parts. The left column shows the demand equal to the limited capacity, and the right column shows the remaining amount of demand. In the cell corresponding to the transportation from the sender $i$ to the consumer $j$ of the right column, the transportation should be blocked by adding the coefficient A. In this case, the limited capacity of the road will be automatically considered when solving the problem.

### 2.3.3 Accounting of Production Costs

There are tasks in which it is necessary to consider the costs not only for the transportation of products but also for the production of these products.

For example, there are $n$ points of production, from which the manufactured products must be delivered to $m$ points of consumption. Then the generalized costs from produce $i$ to consume $j$ are equal to the sum of the costs of producing products at point $i$ and the costs of transporting products from point $i$ to point $j$. It is required to find such an optimal product distribution plan so that the total costs are minimal.

Here, the coefficients of the objective function of a transport problem of this type are the costs equal to the sum of the costs of producing a unit of production at point $i$ ($c_i$) and the costs of transporting a unit of production from point $i$ to point $j$ ($p_{ij}$):

$$z_{ij} = c_i + p_{ij} \qquad (2)$$

Students are encouraged to consider various options for development and analyze how the solution to the problem will change.

## 3 An Example of a Lesson Dedicated to Solving Logistical Problems

The following is an example of a practical lesson conducted under the guidance of the author of the article in the course's framework "Marketing" on the topic "Sales policy of the company". The lesson is conducted with third-year students studying toward "Economics". At the beginning of the lesson, students are explained the theory, then a specific task is given, which is solved in the chosen software environment. Such a software environment is Microsoft Excel.

### 3.1 Initial Data and Problem Statement

Individual marketing strategies are usually developed for each type of market and its constituent types of goods. This is also typical for transport services, when the major objectives of marketing are the orientation of operational work to meet the needs of customers in the transportation of goods and passengers, increasing the coverage of the market with transport services. The solution to these tasks provides an increase in the company's income.

Based on the analysis of the problem area, students together with the teacher carry out the linguistic formulation of the problem.

There are three computer supply points: Warehouse No. 1, Warehouse No. 2 and Warehouse No. 3. There are also five stores: "Terabyte", "Leader", "Expert", "Digital Service", "Office Equipment", purchasing computers for retail sale. The warehouses have the following number of computers: Warehouse No. 1–200 pcs., Warehouse No. 2–250 pcs., Warehouse No. 3–200 pcs. It is required to deliver: to "Terabyte"–190 pcs., to "Leader"–100 pcs., to "Expert"–120 pcs., to "Digital Service"–110 pcs., and to the "Office Equipment" store 130 pcs. The cost of delivering one computer from each warehouse to each store is represented by the C matrix.

$$C = \begin{pmatrix} 28 & 27 & 18 & 27 & 24 \\ 18 & 26 & 27 & 32 & 21 \\ 27 & 33 & 23 & 31 & 34 \end{pmatrix}$$

It is necessary to find the best way to distribute computers.

## 3.2 Task Solving and Interpretation of Results

According to the method of teaching students to solve logistical problems presented in clause 2.2, the next stage is to form a block of source data on a Microsoft Excel sheet. Such a block is shown in Fig. 1. Attention should be paid to which type of transport task belongs: open or closed.

| Starting points | Destination points | | | | | Stocks |
|---|---|---|---|---|---|---|
| | "Terabyte" | "Leader" | "Expert" | "Digital Service" | "Office Equipment" | |
| Warehouse No. 1 | 28 | 27 | 18 | 27 | 24 | 200 |
| Warehouse No. 3 | 18 | 26 | 27 | 32 | 21 | 250 |
| Warehouse No. 3 | 27 | 33 | 23 | 31 | 34 | 200 |
| Needs | 190 | 100 | 120 | 110 | 130 | |

**Fig. 1.** Initial data

Students independently form a block of source data. This work takes 5–7 min. The location of some elements on the Excel sheet may differ for different students, but this only applies to the selected design style. The requirements of data integrity and ease of perception must be fully met. Then students prepare a block to solve the problem. The distribution table has exactly the same appearance as the table with the source data. The SUM() function is used in the "Needs" row and in the "Stocks" column. The cells in the distribution table remain unfilled (Fig. 2). Filling in them will occur automatically after using the "Solver" add-in. In some cell after the distribution table, the formula of the objective function is written: SUMPRODUCT(B3:F5;B10:F12). The first range contains computer delivery costs, and the second range contains computer distribution volumes. Students are also given 5–7 min to implement this procedure.

| Starting points | Destination points | | | | | Stocks |
|---|---|---|---|---|---|---|
| | "Terabyte" | "Leader" | "Expert" | "Digital Service" | "Office Equipment" | |
| Warehouse No. 1 | | | | | | 0 |
| Warehouse No. 3 | | | | | | 0 |
| Warehouse No. 3 | | | | | | 0 |
| Needs | 0 | 0 | 0 | 0 | 0 | |
| | | | | | | |
| **Target function** | 0,00 | | | | | |

**Fig. 2.** Distribution table of the transport task

After the students have implemented this stage, the teacher checks the correctness of the result got and the students search for a solution. In the "Search for a solution" window, students set all the parameters, restrictions and get the desired values for the distribution of computers in stores and the cost of their delivery (Fig. 3). When setting restrictions, it is necessary to consider that the quantity of goods transported cannot be fractional and negative.

Students are given 3 to 5 min to get the optimal solution in the software environment. After that, students analyze the resulting solution and possible alternative solutions.

|   | A | B | C | D | E | F | G |
|---|---|---|---|---|---|---|---|
| 1 | Starting points | | | Destination points | | | Stocks |
| 2 | | "Terabyte" | "Leader" | "Expert" | "Digital Service" | "Office Equipment" | |
| 3 | Warehouse No. 1 | 28 | 27 | 18 | 27 | 24 | 200 |
| 4 | Warehouse No. 3 | 18 | 26 | 27 | 32 | 21 | 250 |
| 5 | Warehouse No. 3 | 27 | 33 | 23 | 31 | 34 | 200 |
| 6 | Needs | 190 | 100 | 120 | 110 | 130 | |
| 7 | | | | | | | |
| 8 | Starting points | | | Destination points | | | Stocks |
| 9 | | "Terabyte" | "Leader" | "Expert" | "Digital Service" | "Office Equipment" | |
| 10 | Warehouse No. 1 | 0 | 100 | 30 | 0 | 70 | 200 |
| 11 | Warehouse No. 3 | 190 | 0 | 0 | 0 | 60 | 250 |
| 12 | Warehouse No. 3 | 0 | 0 | 90 | 110 | 0 | 200 |
| 13 | Needs | 190 | 100 | 120 | 110 | 130 | |
| 14 | | | | | | | |
| 15 | Target function | 15 080,00 | | | | | |

**Fig. 3.** Solving the transport task

### 3.3 Analysis of Alternative Solutions

The resulting solution shows that computers for retail sale should be delivered from the first warehouse to the stores "Leader" (100 units), "Office Equipment" (70 units) and "Expert" (30 units). The major consumer of the products stored in the second warehouse is the Terabyte store (190 pieces). From the third warehouse, computers are distributed to "Digital Service" and "Expert" stores.

To increase the assessment for the work done, students are recommended to consider the situations listed in the second section of the article. The results of the analysis are drawn up by students in writing, as a report on the work performed and submitted to the teacher for verification. Depending on the number of situations considered, 20 to 40 min can be provided for this work. According to the results of the check, the teacher announces the grades. The best works can be recommended for preparing reports in the framework of the following practical classes or scientific conferences.

## 4 Summary and Conclusion

The skills gained by students in solving logistical problems in the learning process allow them to subsequently:

1) To analyze the external and internal environment of the company, identify its key elements and assess their impact on the company.
2) To apply analytical and computational methods in practice for making managerial decisions.
3) To organize and ensure material and financial flows in a timely manner both in the company and abroad.
4) To use logistics tools in supply and distribution management.
5) To apply knowledge and solve problems in inventory management using various models of inventory control.

6) To evaluate the efficiency and develop the logistics process in the company's warehouses.

In the study's course, it was found that Microsoft Excel is a very effective tool for solving logistics problems. The built-in service "Solver" allows you to quickly and cost-effectively get solutions to logistical problems related to the distribution of shipments to customers.

## References

1. Anitha, P., Patil, M.M.: A review on data analytics for supply chain management: a case study. Int. J. Inf. Eng. Electron. Bus. **10**(5), 30–39 (2018). https://doi.org/10.5815/ijieeb.2018.05.05
2. Benotmane, Z., Belalem, G., Neki, A.: A cost measurement system of logistics process. Int. J. Inf. Eng. Electron. Bus. (IJIEEB) **10**(5), 23–29 (2018)
3. Grebennik, I., Dupas, R., Lytvynenko, A., Urniaieva, I.: Scheduling freight trains in rail-rail transshipment yards with train arrangements. Int. J. Intell. Syst. Appl. (IJISA) **9**(10), 12–19 (2017)
4. Ting, P.-H.: An efficient and guaranteed cold-chain logistics for temperature-sensitive foods: applications of RFID and sensor networks. IJIEEB **5**(6), 1–5 (2013)
5. Weihua, G., Tingting, Z., Yuwei, Z.: On RFID application in the information system of rail logistics center. IJEME **3**(2), 52–58 (2013)
6. Weihua, G., Yuwei, Z., Tingting, Z.: Research on RFID application in the pharmacy logistics system. IJEME **2**(8), 13–19 (2012)
7. Dantzig, G.B.: Linear Programming and Extensions. Princeton, NJ (1963)
8. Ajayi Dr, Ibrahim, D.: Application of linear programming in investment portfolio selection (using Microsoft Excel 13). Int. J. Acad. Res. Bus. Arts Sci. (IJARBAS.COM), **3**, 1–27 (2021)
9. Silva, P.M.S., Moreira, B.C.M., Francisco, G.A.: Linear programming applied to finance - building a great portfolio investment. Revista de Gestão, Finanças e Contabilidade **4**(3), 107–124 (2014). https://doi.org/10.18028/2238-5320/rgfc.v4n3p107-124
10. Uday, S.V., Hamritha Chaudhary, G.: Linear programming in market management using artificial intelligence. In: Vijayan, S., Subramanian, N., Sankaranarayanasamy, K., (eds) Trends in Manufacturing and Engineering Management. Lecture Notes in Mechanical Engineering. Springer, Singapore, pp. 845–851 (2021). https://doi.org/10.1007/978-981-15-4745-4_73
11. Alotaibi, A., Nadeem, F.: A review of applications of linear programming to optimize agricultural solutions. Int. J. Inf. Eng. Electronic Bus. (IJIEEB) **13**(2), 11–21 (2021)
12. Volchkov, V.M., Tarasova, I.A., Shvedov, E.G.: Solving integer programming problems in the LINGO computer package: textbook. Volgograd: VolgSTU (2020)
13. Kaur, J., Tomar, P.: Multi objective optimization model using preemptive goal programming for software component selection. Int. J. Inf. Technol. Comput. Sci. (IJITCS) **7**(9), 31–37 (2015)
14. Goldstein, A.L.: Modeling in LINGO. Bulletin of Perm National Research Polytechnic University. Electrical Engineering, Information Technology, Control Systems, **18**, 25–38 (2016)
15. Mikulich, E.M., Podina, K.V., Gudkov, V.A., Volchkov, V.M.: Application of the "LINGO" linear programming package for solving logistical problems of optimizing the cargo delivery process. Bull. Transp. **11**, 35–38 (2012)
16. Dubenetskaya, E.R.: Training of economic specialties students in the solving of optimization problems using specialized software. Sci. Bull. MSIIT **6**(38), 73–79 (2015)

17. Render, B., Stair, R.M., Hanna, M.E.: Quantitative analysis for management. 11TH EDITI. Pearson (2012)
18. Kumari, P.L., Reddy, G.K., Krishna, T.G.: Optimum allocation of agricultural land to the vegetable crops under uncertain profits using fuzzy multiobjective linear programming. J. Agric. Vet. Sci. **7**(12), 19–28 (2014)
19. Sergeev, V.I., Jeljashevich, I.P.: Logistics of Supply: Textbook for Bachelor's and Master's Degrees. Yurayt Publishing House, Moscow (2016)
20. Volkova, I.I., Prudnikova, O.M.: Methods of Optimal Solutions for the Bachelor of Economics: Textbook. Ukhta State Technical University Publishing House, Ukhta (2015)

# Exploration and Practice of Ideological and Political Construction in the Course of "Container Multimodal Transport Theory and Practice" for Application-Oriented Undergraduate Majors—Taking Nanning University as an Example

Shixiong Zhu, Liwei Li, and Zhong Zheng(✉)

Nanning University, Nanning 530200, Guangxi, China
zhengzhong2007@163.com

**Abstract.** As an important professional course for logistics majors, the course "Theory and Practice of Container Multimodal Transport" is a relatively new professional course to adapt to the rapid development of container multimodal transport. This paper defines the basic thinking based on the work process orientation. On this basis, it analyzes the necessity of strengthening ideological and political education in the field of container multimodal transport on how to implement the ideological and political requirements of the curriculum, excavates the ideological and political elements, determines the weight of different ideological and political elements, clarifies the thinking of ideological and political education of the curriculum, emphasizes the pertinence of ideological and political education of the curriculum, and discusses the specific problems. The research of this paper focuses on helping to improve the training quality of application-oriented logistics professionals. It ponders and expounds the necessity, construction ideas and construction paths of the ideological and political construction of this course, and carries out relevant practice in the teaching process, and has achieved good results.

**Keywords:** Container multimodal transport theory and practice · Curriculum ideological and political · Curriculum construction · Teaching practice

## 1 Introduction

In the new economic situation and social environment, the construction of new land and sea routes in the west is in full swing, and new requirements are put forward for the quantity and quality of qualified personnel for multimodal transport [1–3]. It is the mission of colleges and universities given by the times to continuously deliver more and better talents to the society [4–6]. In addition to professional knowledge, qualified talents must also be politically and ideologically competent. To improve the comprehensive quality of students in an all-round way, moral education contained in professional courses is duty-bound. Curriculum ideological and political education refers to a comprehensive educational practice that integrates ideological and political education elements, including theoretical knowledge, values and spiritual pursuit of ideological and political education, into the curriculum in professional courses and general courses, and exerts a subtle influence on students' ideological awareness and behavior, so as to achieve the goal of moral education and education [7, 8]. It is not a separate course, but a new education and teaching mode.

In recent years, the national education management department has repeatedly stressed that we should adhere to the central link of moral education and the cultivation of people, carry out ideological and political work throughout the whole process of teaching, achieve all-round education for all staff, and strive to create a new situation for the development of higher education in China [9]. We should take the comprehensive promotion of the ideological and political construction of the curriculum as a strategic measure to implement the fundamental task of building morality and cultivating people, integrate values into knowledge teaching and ability training, and help students shape correct world outlook, outlook on life and values [10–14]. It can be seen that the ideological and political construction of curriculum has been promoted to an important task of comprehensively improving the quality of talent training in colleges and universities. As people's teachers of professional courses in the new era, it is incumbent upon them to participate in and promote the ideological and political construction of courses [15]. Taking the course "Theory and Practice of Container Multimodal Transport" (hereinafter referred to as "Container Multimodal Transport", or CMT for short) as an example, this paper explores and practices the ideological and political construction of the course of logistics engineering in Nanning University (application-oriented undergraduate college).

## 2 The Necessity of Ideological and Political Construction of "CMT" Course for Application-Oriented Undergraduate

The training objectives of application-oriented undergraduate logistics engineering students are different from those of research-oriented undergraduate majors or higher vocational colleges [16]. The former focuses on training enough theoretical knowledge to guide production practice, strong hands-on ability to engage in post work, direct employment and "zero distance" employment. It emphasizes the direct impact of application-oriented undergraduates on production practice [17]. Therefore, the ideological and political construction of the CMT course for application-oriented undergraduates emphasizes not only knowing the actual performance of the field of the course, what problems exist, and what to do in action, but also knowing why to do so. Obviously, the ideological and political construction and implementation of curriculum are more urgent for application-oriented undergraduate students [18–20]. The necessity of ideological and political construction in the course of "container multimodal transport" for application-oriented undergraduate students is at least reflected in the following aspects.

(1) The ideological and political construction of the CMT course is conducive to enhancing the four self-confidence of application-oriented undergraduates [21, 22]. Transportation, including multimodal transport, plays a key role in the development

of the national economy. Compared with European and American countries, China's D multimodal transport has shown the world's first in container manufacturing capacity, technology research and development level and industrial supporting cluster level for 25 consecutive years. The contribution of "Made in China" to global multimodal transport cannot be underestimated. During the COVID-19, China's multimodal transport volume grew against the trend, and its export maintained a double-digit high-speed growth, resulting in a huge surplus. As an all-weather container transport product on the Eurasian Continental Bridge, which is less affected by climate and environment, has faster transport speed than sea transport, and is far cheaper than air transport, China Europe Express is increasingly welcomed by countries and regions along the "the Belt and Road".

When teaching the content of international container multimodal transport, the teacher showed the development achievements of China's multimodal transport to students, so that students could deeply understand the important role of China's multimodal transport in the world economic development and the economic development along the "the Belt and Road", and more aware of China's role as a major country in promoting the construction of a community with a shared future for mankind. In the process of knowledge learning, students should establish a global perspective, firmly adhere to the "four self-confidence", and are willing to demonstrate the strength and responsibility of China's road, theory, system and cultural self-confidence in future work and international exchanges.

(2) The ideological construction of the CMT course is beneficial to the application-oriented undergraduates to enhance their sense of cooperation. The operation process of container multimodal transport involves not only a variety of transportation modes such as road, railway, waterway and air, but also multi-way transport, multi-operator transport, multi-industry cooperation and the impact of multiple countries and their laws and regulations. It is necessary to establish a good cooperative relationship with all parties. It is necessary to encourage and teach students to firmly establish the principle of human kindness and improve their awareness and ability of cooperation through curriculum teaching.

(3) The ideological and political construction of CMT course is conducive to strengthening the spirit of inquiry of application-oriented undergraduates. At present, China's container multimodal transport still has many problems. For example, the application of information technology is backward, the information system supporting multimodal transport business and management has not yet been formed, and the intermodal transport business transactions still need complex paper documents; The efficiency and quality of multimodal transport services need to be improved; The loading and unloading facilities, documents and information of the freight station cannot be transferred in time. In addition, the coordination of multimodal transport system is not ideal. The combination of different modes of transport lacks research and planning, and has not formed a unified multimodal transport operation network. The lack of close cooperation between railway and maritime and inland water transport, and the poor connection between related enterprises are all problems facing multimodal transport. It is necessary for application-oriented undergraduates to participate in the research, development, upgrading and application of the system and strengthen the spirit of exploration.

(4) The ideological and political construction of the curriculum is conducive to enhancing the sense of hardship of application-oriented undergraduates. The container multimodal transport business, especially the international container multimodal transport business, often has many links, a long distance, is subject to many countries, regions and their laws and regulations, and has great risks. It is necessary to enhance the awareness of hardship and the ability to make emergency response for application-oriented undergraduates through the course of ideological and political education.

## 3 Basic Thinking of Ideological and Political Construction of the Course "CMT"

The essence of ideological and political courses is to establish morality and cultivate people. The goal is to cultivate college students who have both professional knowledge and correct ideals and beliefs, value orientation, political belief, social responsibility and patriotism, as well as talents who have the ability to reason and distinguish right from wrong. The basic idea of ideological and political construction of this course is to take the thought of socialism with Chinese characteristics in the new era as the guidance, focus on cultivating applied multimodal talents for the national development and national rejuvenation, clarify the responsibilities and requirements of the main body of ideological and political construction of the course, and build the carrier of ideological and political construction of the course in accordance with the principle of "full integration of professional knowledge teaching, ability training and value shaping", Control the key factors that affect the effect of ideological and political construction of the curriculum and strengthen the assessment of the implementation process of ideological and political construction of the curriculum to promote the effectiveness of ideological and political construction of the curriculum.

### 3.1 Mining Ideological and Political Elements of Curriculum Modules

The ideological and political elements of the curriculum should be thoroughly and comprehensively excavated to provide an important basis for improving the quality of ideological and political construction of the curriculum. The factors to be considered include: whether the textbooks, syllabus and courseware are applicable, whether the ideological and political elements of the curriculum module are mined in combination with the professional characteristics, whether the teaching team has sufficient awareness and attention, and whether the division of labor and cooperation is effective. The results are shown in Table 1.

In the table, "container transport", "multimodal transport" and "container multimodal transport" are abbreviated as "CT""MT" and" MTC" respectively. M1, M2, M3 and M4 represents different modules in turn. Task 1 to Task 8 are denoted by T1,…, T8 separately.

**Table 1.** Ideological and political elements and objectives of "container multimodal transport" course

| Module | Task | Mapping point | Objectives |
|---|---|---|---|
| M1:<br>Basic knowledge of CMT | T1:<br>Learn to know MT | 1) The current situation of CT in China;<br>2) China's MT increased during the epidemic;<br>3) Integrity of MT operators | 1) Set up correct three views and enhance professional confidence;<br>2) Cultivate the awareness of integrity, legal system and social responsibility, and be a qualified multimodal transport operator |
|  | T2:<br>Prepare equipment and tools for MT | 1) The necessity and scarcity of resource of MT<br>2) Domestic equipment on automatic CT;<br>3) Qinghai-Tibet Railway and other projects (video) | 1) Learn to cooperate;<br>2) The belief in building a country;<br>3) Establish craftsmanship spirit |
| M2:<br>CMT business process | T3:<br>Familiar with MT logistics nodes | 1) Global container port top 10 in 2021;<br>2) Automated CT domestic equipment;<br>3) The model team set a new world record | 1) Adhere to openness and innovation;<br>2) Loving and dedicated;<br>3) Craftsman spirit |
|  | T4:<br>Understand customs clearance management | 1) China's customs clearance speed;<br>(2) Customs information application;<br>(3) The impact of RCEP on CMT | 1) Keep learning;<br>2) Expand international vision |
|  | T5:<br>Familiar with MT billing, documents and operation process | 1) "Consignment, payment, document signing and bill" one-stop mode of MT;<br>2) Calculation of multimodal freight | 1) Keep improving, find and solve problems;<br>2) Have professional quality and meet the requirements of post ability |

(*continued*)

**Table 1.** (*continued*)

| Module | Task | Mapping point | Objectives |
|---|---|---|---|
| M3: Risk management of CMT | T6: Familiar with insurance and claims settlement | 1) Policies for import, export and customs declaration of goods; 2) Compliance of MT; 3) MT risk management | 1) Have legal awareness and bottom line awareness; 2) Serious and responsible attitude |
| M4: Design and implementation of CMT scheme | T7: Design MT scheme of container cargo | 1) Cost budget and control of container multimodal transport; 2) Market analysis; 3) Route selection, site selection and layout planning | 1) Pay attention to efficiency and fairness; 2) Good at objective and systematic analysis |
|  | T8: Organize and implement MT of container cargo | 1) The opportunity of the "the Belt and Road"; 2) Development history of international CMT; 3) The development trend of international trains | 1) Patriotic and dedicated; 2) Be bold in innovation |

## 3.2 Weight Analysis of Curriculum Ideological and Political Elements

Analytic Hierarchy Process (AHP) is a systematic method to solve complex multi-objective decision-making problems [23–26], which was put forward in the early 1970s for the US Department of Defense when studying the topic of "power distribution based on the contribution of various industrial sectors to the national welfare". The core principle of the method is to decompose the problem into different components according to the requirements of the overall goal, According to the interrelation, influence and subordination among factors, factors are aggregated and combined at different levels to form a multi-level analysis structure model, so as to rank the importance or advantages and disadvantages of different elements to the overall goal [27, 28].

In order to carry out the ideological and political construction of the curriculum more pertinently, according to the principle of AHP method, the importance of different indicators is determined by comparison between two groups, and the preliminary score is made, which provides basic data for the development of AHP. The course teaching team will discuss and grade the ideological and political elements of the course, absorb the opinions of similar course experts from other institutions, evaluate the importance of the course elements through research and telephone consultation, and compare and judge different indicators.

There are 8 main elements in the curriculum determined after the survey, as shown in Table 2.

The judgment matrix is constructed according to the results of comparison and judgment, as shown in Table 3.

**Table 2.** Main factors in course analysis and evaluation

| Code | Step | Content |
|---|---|---|
| $S_1$ | Step 1 | Understand multimodal transport |
| $S_2$ | Step 2 | Prepare multimodal transport tools and equipment |
| $S_3$ | Step 3 | Familiar with multimodal transport logistics nodes |
| $S_4$ | Step 4 | Understand multimodal customs clearance management |
| $S_5$ | Step 5 | Know multimodal transport billing, documents and operation process |
| $S_6$ | Step 6 | Be familiar with multimodal insurance and claim settlement |
| $S_7$ | Step 7 | Design multimodal transport scheme of container cargo |
| $S_8$ | Step 8 | Organize and implement multimodal transport of container goods |

According to the eight main factors identified in the table, the judgment matrix of the importance of ideological and political elements of the curriculum shown in Table 3 can be obtained. The analysis process is as follows.

After the weight score is obtained, hierarchical ranking analysis can be carried out. The core index is the determination of the eigenvector and the weight value. The eigenvector is obtained by the square root method, and the weight value is normalized so that the sum of all elements is 1. The AHP hierarchical analysis results are carried out to calculate the maximum characteristic root of the judgment matrix, and the consistency index CI is calculated. The analysis results are shown in Table 4. According to the principle of AHP method, CI equal to 0 indicates complete consistency. The closer CI is to

**Table 3.** Judgment matrix of the importance of ideological and political elements of the curriculum

| Index | $S_1$ | $S_2$ | $S_3$ | $S_4$ | $S_5$ | $S_6$ | $S_7$ | $S_8$ |
|---|---|---|---|---|---|---|---|---|
| $S_1$ | 1 | 1/2 | 1/3 | 1/4 | 1/8 | 1/6 | 1/9 | 1/9 |
| $S_2$ | 2 | 1 | 1/2 | 1/3 | 1/7 | 1/5 | 1/6 | 1/7 |
| $S_3$ | 3 | 2 | 1 | 1/5 | 1/6 | 1/4 | 1/6 | 1/6 |
| $S_4$ | 4 | 3 | 2 | 1 | 1/4 | 1/2 | 1/5 | 1/5 |
| $S_5$ | 8 | 7 | 6 | 4 | 1 | 2 | 1/2 | 1/2 |
| $S_6$ | 6 | 5 | 4 | 2 | 1/2 | 1 | 1/3 | 1/3 |
| $S_7$ | 9 | 7 | 6 | 5 | 2 | 3 | 1 | 1 |
| $S_8$ | 9 | 7 | 6 | 5 | 2 | 3 | 1 | 1 |

0, the more satisfactory consistency is displayed. The larger CI value, the more serious the inconsistency is.

**Table 4.** AHP hierarchy analysis results

| Items | Eigenvector | Weight (%) | Maximum characteristic root | CI |
|---|---|---|---|---|
| $S_1$ | 0.239 | 2.009 | 8.241 | 0.034 |
| $S_2$ | 0.344 | 2.88 | | |
| $S_3$ | 0.493 | 4.138 | | |
| $S_4$ | 0.767 | 6.444 | | |
| $S_5$ | 2.256 | 18.952 | | |
| $S_6$ | 1.382 | 11.61 | | |
| $S_7$ | 3.212 | 26.981 | | |
| $S_8$ | 3.212 | 26.981 | | |

For the purpose of further measurement of the size of CI, the random consistency index RI and consistency ratio CR (CR = CI/RI) are introduced. Generally, when CR < 0.1, the degree of inconsistency is considered to be within the allowable range. The consistency test of the analysis results is carried out according to the principle of AHP method, as shown in Table 5.

**Table 5.** One-time inspection results

| One-time inspection | | | | |
|---|---|---|---|---|
| Maximum characteristic root | CI | RI | CR | Results One-time inspection |
| 8.241 | 0.034 | 1.404 | 0.025 | Pass |

The calculation results of the analytic hierarchy process show that the maximum characteristic root is 8.241, and the corresponding RI value is 1.404 according to the RI table, so CR = CI/RI = 0.025 < 0.1. Through a one-time test, it shows that the analysis process is effective, and the weights of different ideological and political curriculum elements are determined according to the order of weight analysis results in Table 4.

## 3.3 Weight Analysis Results of Curriculum Ideological and Political Elements

According to Table 4 of the above analytic hierarchy process (AHP) analysis results, it is concluded that: (1) the weight of knowing multimodal transport is 2.009%, (2) the weight of preparing multimodal transport tools and equipment is 2.886%, (3) the weight of knowing multimodal transport logistics nodes is 4.138%, (4) the weight of knowing multimodal transport customs clearance management is 6.444%, (5) the weight of knowing multimodal transport billing, documents and operation process is 18.952%, (6) The weight of knowing multimodal transport insurance and claim settlement is 11.61%, (7) the weight of designing multimodal transport scheme of container goods is 26.981%, and (8) the weight of organizing and implementing multimodal transport of container goods is 26.981%. The weights of (5), (7) and (8) are the highest.

Therefore, in the ideological and political process of the course, it is necessary to strengthen the familiarity with multimodal transport billing, documents and operation processes; Design multimodal transport scheme of container cargo; Organizing and implementing the three aspects of container cargo multimodal transport to strengthen ideological and political construction and training is also the key link to improve the quality of ideological and political education in this course. Because these three aspects are also the core knowledge points of the course, it is necessary to combine the course elements and business links more closely, so that the ideological and political effect of the course can be more prominent.

## 4 Ideological and Political Construction Path and Practice of CMT Course

### 4.1 Select Teachers with Outstanding Ideological and Political Ability to Teach, and Create a Capable Subject of Ideological and Political Construction of Curriculum

The main body of curriculum ideological and political construction is the curriculum teachers, whose ability and level are directly related to the quality of curriculum ideological and political. Therefore, the first is that the teachers put forward the ideological and political requirements of the curriculum every time the curriculum is arranged. The teachers are required to take moral cultivation as the primary task of student education and training. The second is to give priority to professional teachers with outstanding ideological and political abilities and strong sense of responsibility. The third is to take the ideological and political ability requirements as the conditions for arranging further study and evaluating the best, and promote professional teachers to consciously improve the ideological and political ability of the curriculum. The curriculum ideological and political requirements of the selected teachers are clear, forming the ability requirement orientation. The teachers consciously improve the curriculum ideological and political ability, actively shoulder the responsibility of cultivating talents and educating people, let the ideological and political awareness take root in the heart, teach by example, and help the application-oriented undergraduates form correct values and ideals.

### 4.2 Compile Applicable and Visible Teaching Materials, Outlines and Courseware, and Build a Useful Course Construction Carrier Reflecting Ideological and Political Content

Textbooks, syllabus and courseware are the necessary carriers of curriculum teaching. They are not only required for the teaching of professional knowledge and vocational skills training, but also for the ideological and political construction and implementation of the curriculum. They need to be well constructed. In the process of teaching, we can use the task book, case analysis and other documents to arrange teaching activities, group training, watch with the help of network video, and further expand the teaching content, or transmit information, or assign tasks, enrich students' horizons, so that students can increase their interest in the content of ideological and political education, and improve the effect of ideological and political learning of the course.

### 4.3 Plan and Design a Number of Curriculum Implementation Rules, Including Teacher Evaluation and Collaborative System, to Form a Guarantee Mechanism for Establishing Morality and Cultivating People

The ideological and political construction of curriculum is a systematic project, which needs the support and help of all relevant aspects to achieve good results. At present, in addition to the practical needs and the requirements of the Party and the country, a number of curriculum construction and implementation rules need to be included, such as teacher evaluation, collaborative education system, interconnection mechanism, constraint mechanism, incentive mechanism, and so on. It needs to emphasize top-level planning and design, form a security mechanism based on moral education, promote the healthy and sustainable development of curriculum ideology and politics, and truly cultivate existing patriotism, broad international vision, innovation Craftsmen and patriots with good team spirit, professional quality and strong social responsibility, as well as socialist cause builders and new logistics comprehensive talents with rich knowledge and solid professional skills.

### 4.4 Add Value-Leading Indicators and Strengthen the Process Assessment and Evaluation of Ideological and Political Teaching

The application-oriented undergraduate students are the object of the ideological and political construction of the curriculum and the main body of the teaching class. All measures of the teaching reform are carried out around the adult talents of students, which is to make them move through the assessment of the main body of the teaching class. The assessment idea is to add an assessment index of "ideological and political performance", including the students' ideological and political performance (60 points) into the assessment scope, and assess them together with their personal performance (professional quality, 60 points), work performance (100 points), methods and skills (40 points) and others (40 points). Each task will be assessed once, with a full score of 300 points. At the beginning of the course, students should be clearly informed of the detailed rules of the process assessment to improve their learning initiative, enthusiasm and participation. The main observation points of the "ideological and political performance" assessment indicators are positive ideological performance, support for the Party, love for the motherland, and the absence of rebellious words and deeds of love for the Party and popular "good people in the workplace". There are multiple versions of the ten characteristics of "good people". After sorting out 15 characteristics, the 10 characteristics and full marks approved by the majority of applied undergraduates are selected as the evaluation indicators and criteria. The 10 indicators and full score criteria are: integrity: adhere to the truth and act with conscience; Honesty: seek truth from facts and do not resort to fraud; Kindness: always be grateful, be kind and do good deeds; Kindness: be willing to suffer losses and think for others; Humility: look down on yourself; Tolerance: can tolerate people with different opinions; Persistence: determination, patience and perseverance in doing things; Keep faith: words must be done, and faith must be rewarded; Capable: positive, optimistic, proactive, and able to handle things well; Innovation: strong sense of breakthrough, good at finding and solving problems.

## 5 Conclusion

The purpose of implementing curriculum ideological and political education is to train students to become excellent talents with morality, ideals and all-round development.

The ideological education of the "container multimodal transport" course explored the practical significance, basic thinking and path of the course construction, and implemented the practice of "creating the main body of ideological and political construction of lean courses, mining the ideological and political elements of the course modules, diversifying teaching forms, building a useful course carrier, forming a moral education guarantee mechanism, and strengthening the process assessment of ideological and political teaching of the course". Along the way, our teaching team has experienced the teaching of professional knowledge The integration of ability training and value shaping is indeed an effective way to build morality and cultivate people. After analysis, we need to be familiar with multimodal transport billing, documents and operation process; Design multimodal transport scheme of container cargo; Organize the implementation of three aspects of container cargo multimodal transport, strengthen ideological and political construction and training, strengthen the combination of courses and key business links, and improve the teaching effect of ideological and political education.

**Acknowledgment.** This research is supported by (1) The fourth batch of undergraduate core curriculum construction project of Nanning University "Exploration and Practice of Course Construction of 'Container Multimodal Transport Theory and Practice'" (2022BKHXK06); (2) Guangxi Higher Education Undergraduate Teaching Reform Project "Research and Practice on the Construction of Logistics Virtual Teaching and Research Office from the Perspective of Four-dimensional Linkage of 'Point, Line, Area and Body'" (2022JGB440); The subject of ideological and political construction of curriculum of Nanning University (2020SZSFK14).

## References

1. Hacène H.: Action research to enhance quality teaching. In: Arab World English Journal (AWEJ) May 2019 Chlef University International Conference Proceedings, pp. 4-12 (2019)
2. Vijayalakshmi, V., Venkatachalapathy, K.: Comparison of predicting student's performance using machine learning algorithms. Int. J. Intell. Syst. Appl. **11**(12), 34–45 (2019)
3. Rifat, M.R.I., Imran, A.A., Badrudduza, A.S.M.: Educational performance analytics of undergraduate business students. Int. J. Mod. Educ. Comput. Sci. **11**(7), 44–53 (2019)
4. Wang, X., Yingjie, W.: Exploration and practice for evaluation of teaching methods. Int. J. Educ. Manage. Eng. **2**(3), 39–45 (2012)
5. Zhiqin, L., Jianguo, F., Fang, W., Xin, D.: Study on higher education service quality based on student perception. Int. J. Educ. Manage. Eng. **2**(4), 22–27 (2012)

6. Skedsmo, G., Huber, S.G.: Measuring teaching quality, designing tests, and transforming feedback targeting various education actors. Educ. Assess. Eval. Accountability **32**(3), 271–273 (2020)
7. Xuejian, W., Yan, S.: The connotation, characteristics, difficulties and countermeasures of ideological and political education in the new era curriculum. J. Xinjiang Normal Univ. (Philos. Soc. Sci. Ed.) **2**, 50–58 (2020). (in Chinese)
8. Jiwei, Z.: Curriculum ideological and political: meaning, concept, problems and countermeasures. J. Hubei Univ. Econ. **17**(2), 114–119 (2019). (in Chinese)
9. Jinyao, R., Qing, W., Chao, G.: Exploration and practice of curriculum ideological and political systematic design in vocational colleges. China Vocat. Tech. Educ. **29**, 27–29 (2021). (in Chinese)
10. Yunbo, O.: Foster the sense of Chinese national community in the ideological and political curriculum. Ref. Polit. Teach. Middle Sch. **42**, 83 (2022). (in Chinese)
11. Liwei, P.: Shi Xiaorong exploration of the "four full coverage" model of curriculum ideological and political construction in the context of "new engineering." J. Nat. Insts Educ. Adm. **11**, 63–70 (2022). (in Chinese)
12. Shangzi, Z.: The organic combination of ideological and political courses and ideological and political courses: the three-dimensional proof of ideological and political principles. J. Henan Normal Univ. (Philos. Soc. Sci. Ed.) **49**(06), 124–130 (2022). (in Chinese)
13. Bailin, C., Yi, L., Jinqiu, Z.: Using data mining approach for student satisfaction with teaching quality in high vocation education. Front. Psychol. **12**, 1–8 (2022)
14. Suoming, H., Lijuan, L.: Problems and countermeasures in ideological and political teaching of new engineering courses. Educ. Theory Pract. **42**(36), 39–42 (2022). (in Chinese)
15. Chengwen, C., Yiming, W., Jibin, Z., et al.: Research and practice of ideological and political teaching of characteristic professional courses for engineering certification. Packag. Eng. **43**(S2), 54–60 (2022). (in Chinese)
16. Adam Lindgreen, C., Di Benedetto, A., Brodie, R.J., Zenker, S.: Teaching: how to ensure quality teaching, and how to recognize teaching qualifications. Ind. Mark. Manage. **100**, A1–A5 (2022)
17. Sal, C.: Practitioner research in a UK pre-sessional: the synergy between exploratory practice and student motivation. J. Engl. Acad. Purp. **57**, 1–6 (2022)
18. Jiao, L., Ji, F.: An analysis of the path of the ideological and political systematization of university curriculum. China Univ. Teach. **11**, 64–71 (2022). (in Chinese)
19. Lin, W.: A probe into the new theory and practical mechanism of curriculum ideological and political education. Ref. Middle Sch. Polit. Educ. **41**, 91 (2022). (in Chinese)
20. Fei, L., Dong, L.: Open a new perspective of "three complete education" in the ideological and political practice of college curriculum. Ref. Polit. Teach. Middle Sch. **40**, 110 (2022). (in Chinese)
21. Tao, J., Yujuan, S.: Problems and countermeasures in ideological and political construction of university curriculum. Sch. Party Constr. Ideological Educ. **20**, 44–46 (2022). (in Chinese)
22. Anna, S.: Perceptions of student motivation and amotivation. The Clearing House J. Educ. Strat. Issues and Ideas **94**(2), 76–82 (2021)
23. Xiaomin, X.: Application of analytic hierarchy process. Statistics and Decision (1), 156–158 (2008) (in Chinese)
24. Kong, L.A., Wang, X.M., Yang, L.: The research of teaching quality appraisal model based on AHP. Int. J. Educ. Manage. Eng. (IJEME) **9**(29), 29–34 (2012)

25. Haji, E., Azmani, A., Harzli, M.E.: Using AHP method for educational and vocational guidance. Int. J. Inf. Technol. Comput. Sci. (IJITCS) **9**(1), 9–17 (2017)
26. Cheng, J., Chen, S.: A fuzzy Delphi and fuzzy AHP application for evaluating online game selection. Int. J. Mod. Educ. Comput. Sci. (IJMECS) **5**, 7–13 (2012)
27. Listyaningsih, V., Utami, E.: Decision support system performance-based evaluation of village government using AHP and TOPSIS methods: secang sub-district of magelang regency as a case study. Int. J. Intell. Syst. Appl. (IJISA) **10**(48), 18–28 (2018)
28. Lotfi, F., Fatehi, K., Badie, N.: An analysis of key factors to mobile health adoption using fuzzy AHP. Int. J. Inf. Technol. Comput. Sci. (IJITCS) **12**(02), 1–17 (2020)

# A Study on Learning Intention of Digital Marketing Micro Specialty Learners Under the Background of New Liberal Arts—Based on Structural Equation Model

Yixuan Huang[1]([✉]), Mingfei Liu[1], Jiawei You[1], and Aiman Magde Abdalla Ahmed[2]

[1] School of Management, Wuhan University of Technology, Wuhan 430070, China
2635829647@qq.com

[2] Faisal Al Islam Bank, Khartoum 999129, Sudan

**Abstract.** In the context of the new liberal arts, the trend of interdisciplinary integration is becoming increasingly clear. Micro specialty model is a more in-depth exploration on the basis of interdisciplinary integration. Taking digital marketing micro specialty as an example, this paper constructs a study model of learning intention of digital marketing micro specialty learners based on stimuli-organism-response theory (S-O-R) and the technology acceptance model (TAM), and explores the relationship between variables using a structural equation model and a test of mediating effect. The findings show that: Self-efficacy, perceived ease of use and perceived usefulness directly affect learning intention, while perceived ease of use indirectly affects learning intention through self-efficacy and perceived usefulness. Educators significantly affect self-efficacy. Instructional media directly affect perceived ease of use and perceived usefulness, as well as self-efficacy and perceived usefulness through the mediating role of perceived ease of use. Instructional content significantly affects perceived usefulness.

**Keywords:** Digital marketing micro specialty · Learning intention · Structural equation model · S-O-R theory · TAM model

## 1 Introduction

Since the official launch of the "Six Excellence and One Top-notch" plan 2.0 of the Ministry of Education in October 2018, the construction of new liberal arts has attracted widespread attention from the society. The new liberal arts are characterized by strategy, innovation, integration and systematization. It is a real revolution in liberal arts education from discipline orientation to demand orientation, from adaptation service to support and guidance, and from specialty segmentation to cross-integration [1].

In this context, the construction of digital marketing micro specialty will undoubtedly be a feasible path for advancing liberal arts education. Facing the demand for innovative talents in national strategic emerging industries and enterprises' digital transformation, the digital marketing micro specialty closely combines cutting-edge theories and practices such as the Internet, big data and artificial intelligence, with the goal of cultivating senior compound innovative talents who adapt to the development of new technologies, new industries, new business forms and have the basic theoretical knowledge of digital marketing, market analysis and the ability to solve practical problems in marketing. The concept of micro specialty was first proposed in 2013 by an educational organization co-founded by Harvard and Massachusetts Institute of Technology. By determining 5–10 core courses, micro specialty provides a targeted curriculum system for the cultivation of talents in corresponding majors or positions [2]. Guided by career and competency development, micro specialty focuses on the integration of teaching content and the practicality of training effectiveness. It is not only refined and targeted, but also emphasizes the interdisciplinary [3–5].

Regarding the existing research achievements of micro specialty, Zhu Jie et al. took the construction of new engineering as the background and explored the construction path of micro specialty of computer [6]. Zhang Zhiping et al. studied the construction of big data micro specialty from the aspects of specialty positioning and planning, faculty team and curriculum construction [7]. In the context of MOOCs, Wang Xiaomin et al. took rail transit signal and control major of Southwest Jiaotong University as an example to build a micro specialty course system [8]. From the perspective of curriculum developers, researchers have mainly discussed the cultivation patterns of micro specialty and the construction of curriculum systems, while there has been little research from the perspective of curriculum learners. In addition, most researchers focus on the theoretical deduction and induction level, and little present research results through data analysis. Therefore, from the perspective of course learners, this paper analyzes the learning intention of digital marketing micro specialty learners based on survey data from course learners, in order to provide some reference for the specific construction of micro specialty.

## 2 Theoretical Basis and Research Hypothesis

### 2.1 Theoretical Basis

#### 2.1.1 Stimuli-Organism-Response Theory (S-O-R)

S-O-R theory holds that environmental factors, as external stimuli, will affect individuals' internal cognition and emotion, and then affect their behavioral responses [9]. The S-O-R theory has been applied mainly to the study of consumer decision-making behavior in the field of economic management, and in recent years it has also been applied to the study of learner learning behavior in the field of education [10–12]. The S-O-R theory provides a crucial theoretical foundation for this research.

#### 2.1.2 Educational Communication Theory

Educational communication is an activity in which educators transmit knowledge, skills, and ideas to specific educational objects by selecting appropriate information content and using effective communication channels in accordance with certain requirements [13].

Berol's communication model holds that the communication process should include receivers, information sources, channels and information, while the corresponding educational communication process should include learners, educators, instructional media and instructional content [14]. Therefore, the study of educational communication theory can help to identify the relevant factors or dimensions that affect the learning intention of digital marketing micro specialty.

### 2.1.3 Technology Acceptance Model (TAM)

The TAM model, or the technology acceptance model, was first proposed by Davis et al., drawing on the intention model and planned behavior theory of social psychology [15]. This model is used to predict the propensity of subjects to accept, use, or reject new information technology. Perceived ease of use and perceived usefulness are core variables in the TAM model, through which external environment variables in information system affect individual behavior and decision intention [16]. This model has been widely used by educational researchers to study learning intention and behavior. For example, Vidanagama suggested that university students tend to focus more on perceived ease of use on e-learning and the students' attitude has more influence on intention to use e-learning based on the modified TAM model [17].

### 2.1.4 Self-efficacy Theory

The self-efficacy theory was proposed by Bandura, who believed that self-efficacy refers to an individual's subjective judgment on whether he can complete a certain task or behavior [18]. Self-efficacy has an impact on learners' learning intention and learning outcomes. Nandang et al. found that computer self efficacy has a significant influence on computer anxiety, therefore it is expected to the university to be more often use the computer to provide lecture materials and assignments to students [19]. Liu Zhenyu et al. constructed a hypothetical model that self-efficacy influences learning outcomes based on the mediating effect of flow experience in the context of desktop virtual reality environment [20]. Lin Hongyi et al. found that self-efficacy can cultivate positive learning habits, enhance learning confidence and motivation, and thus reduce students' learning burnout [21].

## 2.2 Research Model and Hypothesis

In this research, the S-O-R theory is used as the basic framework to combine educational communication theory, TAM model, and self-efficacy theory, and to use educators, instructional media, and instructional content in educational communication links as external stimulus variables for the learning intention of digital marketing micro specialty. The self-efficacy, perceived usefulness and perceived ease of use of the learners in the link of education communication are taken as the variables of psychological experience, and the learning intention of digital marketing micro specialty is taken as the variable of behavioral response, so as to study the influencing factors and influencing mechanism of the learning intention of digital marketing micro specialty. The model constructed in this research is shown in Fig. 1.

# A Study on Learning Intention of Digital Marketing Micro Specialty Learners

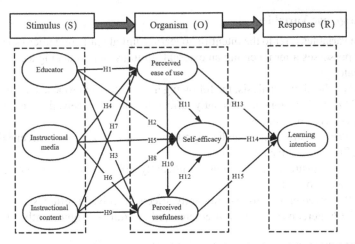

**Fig. 1.** The study model of learning intention of digital marketing micro specialty learners

### 2.2.1 Stimulus Variables

Stimulus variables are external environmental factors that influence the learning intention of digital marketing micro specialty, including educators, instructional media, and instructional content in the educational communication process. Due to the existence of educational tasks and training objectives, educators are in the position of leading, controlling and teaching in instructional activities and are indispensable subjects in micro specialty learning [22]. Instructional media refer to modern electronic media that can carry and transmit instructional information, including slide shows for offline learning, course platforms for online learning, etc. It facilitates the transfer of pedagogical information from the source to the learner and is an essential element in micro specialty learning. Instructional content is the primary message intentionally conveyed in the teaching process and is a critical prerequisite and conditional basis for the development of micro specialty learning. This research argues that external environmental stimuli have a direct or indirect effect on the learning experience and learning intention of digital marketing micro specialty.

### 2.2.2 Organism Variables

As learners' inner perception variables, organism variables include self-efficacy, perceived ease of use and perceived usefulness. In this research, perceived ease of use refers to learners' comprehensive judgment on the difficulty of knowledge learning and online platform operation in the learning process of digital marketing micro specialty. Perceived usefulness refers to the comprehensive value judgment of learners that educators, instructional media and instructional content can effectively improve learning outcomes. Self-efficacy is the level of confidence a learner has in their ability to successfully complete a digital marketing micro specialty with the skills they have. This research believes that the analysis of learners' internal experience is conducive to predicting their intention to learn digital marketing micro specialty.

### 2.2.3 Response Variables

Response variables refer to the intention to learn a digital marketing micro specialty that a learner possesses after receiving an external stimulus and producing a corresponding mental state.

Based on the above analysis, the following hypotheses are proposed in this research:

*H1/H2/H3*: Educators significantly affect learners' perceived ease of use/self-efficacy/perceived usefulness.

*H4/H5/H6*: Instructional media significantly affect learners' perceived ease of use/self-efficacy/perceived usefulness.

*H7/H8/H9*: Instructional content significantly affects learners' perceived ease of use/self-efficacy/perceived usefulness.

*H10*: Perceived ease of use significantly affects perceived usefulness.

*H11/H12*: Perceived ease of use/perceived usefulness significantly affects self-efficacy.

*H13/H14/H15*: Perceived ease of use/self-efficacy/perceived usefulness significantly affected the intention to learn.

## 3 Research Design

### 3.1 Research Object

This research was conducted for universities in Wuhan area, using a combination of stratified sampling and simple random sampling method to distribute online questionnaires to undergraduates. A total of 480 questionnaires were distributed, and 416 valid questionnaires were collected, with an effective recovery rate of 86.7%.

### 3.2 Research Tools

The questionnaire consists of seven main parts: educator, instructional media, instructional content, perceived ease of use, perceived usefulness, self-efficacy and learning intention. The items of the questionnaire are adapted from the maturity scale combined with the specific situation of digital marketing micro specialty. Likert's 5-point scale was used on all scales, with 1 to 5 indicating "strongly disagree" and "strongly agree" respectively. SPSS 26.0 was used to test the reliability of the data, and the Cronbach's α coefficient values of seven variables were respectively 0.702, 0.777, 0.776, 0.809, 0.830, 0.898, 0.873, and the overall Cronbach's α coefficient was 0.927, indicating good internal consistency of the questionnaire. The KMO value was $0.936 > 0.7$, and the significance level was $0.000 < 0.05$, indicating that all variables had good structural validity. AMOS 17.0 was used for the confirmation factor analysis, and the metrics for each fit scale were obtained to an acceptable level, with good overall fit results.

## 4 Data Analysis

### 4.1 Common Method Deviation Test and Correlation Analysis

Bias in common approaches is controlled programmatically by anonymous measurements and random permutations of latent variables. At the same time, Harman's single-factor test is used to detect common methodological biases in the collected data. The results of unrotated exploratory factor analysis showed that the maximum factor variance interpretation rate was 33%, lower than the judgment standard of 40% proposed by Harman [23], indicating that the common method deviation of the data in this study was within the acceptable range. Pearson correlation coefficients were used to find that there were significant correlations among all variables shown in Table 1, suggesting that external environmental stimuli may have a large impact on learners' internal learning experience and learning intention. Based on this conclusion, the structural equation model can be further tested.

Table 1. Structures and sources of the scale

|      | Educ   | IM     | IC     | PU     | PEU    | SE     | LI |
|------|--------|--------|--------|--------|--------|--------|----|
| Educ | 1      |        |        |        |        |        |    |
| IM   | 0.67** | 1      |        |        |        |        |    |
| IC   | 0.72** | 0.67** | 1      |        |        |        |    |
| PU   | 0.56** | 0.56** | 0.57** | 1      |        |        |    |
| PEU  | 0.34** | 0.34** | 0.42** | 0.62** | 1      |        |    |
| SE   | 0.46** | 0.42** | 0.50** | 0.59** | 0.72** | 1      |    |
| LI   | 0.69** | 0.60** | 0.68** | 0.62** | 0.54** | 0.65** | 1  |

Note: Educ: Educator; IM: Instructional media; IC: Instructional content; PU: Perceived usefulness; PEU: Perceived ease of use; SE: Self-efficacy; LI: Learning intention; ** indicates a significant correlation at 0.01 level;

### 4.2 Structural Equation Test

This research tested the hypotheses of each path of the theoretical model by AMOS 17.0. The parameter test results show that in the model, educators → perceived ease of use ($\beta = 0.083$, $P = 0.246$), educators → perceived usefulness ($\beta = 0.071$, $P = 0.228$), instructional media → self-efficacy ($\beta = 0.015$, $P = 0.339$) and other paths do not meet the significance criteria of the parametric test. That is, *H1, H3, H5, H7, H8,* and *H12* are not valid. After deleting the above paths and modifying the model, the final validated model is shown in Fig. 2, and the test results of main fit indexes are shown in Table 2.

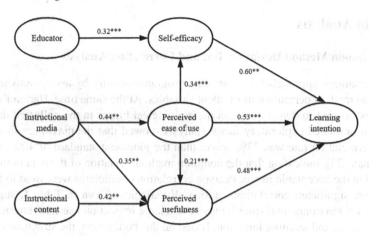

**Fig. 2.** Research model and path coefficients

**Table 2.** The modified model fit

| Fitting index | $\chi^2/df$ | RMR | SRMR | RMSEA | GFI | AGFI | TLI | CFI |
|---|---|---|---|---|---|---|---|---|
| Required value | 1–3 | < 0.05 | < 0.05 | < 0.08 | > 0.9 | > 0.9 | > 0.9 | > 0.9 |
| Actual value | 2.295 | 0.035 | 0.041 | 0.056 | 0.923 | 0.908 | 0.954 | 0.905 |

### 4.3 Mediating Effect Test

The mediating effect analysis of variables was shown in Table 3 and it was found that self-efficacy, perceived ease of use, and perceived usefulness have significant mediating effects between external learning environment stimuli and learning intention.

**Table 3.** Results of mediating effect test (Bootstrap = 5000)

| Mediating path | Coefficient | Bootstrap standard error | Bootstrap confidence interval | Significance |
|---|---|---|---|---|
| Educ → SE → LI | 0.192 | 0.026 | [0.142,0.244] | *** |
| IM → PEU → LI | 0.235 | 0.046 | [0.148,0.329] | *** |
| IM → PU → LI | 0.168 | 0.024 | [0.086,0.176] | *** |
| IC → PU → LI | 0.202 | 0.043 | [0.122,0.287] | *** |

(*continued*)

**Table 3.** (*continued*)

| Mediating path | Coefficient | Bootstrap standard error | Bootstrap confidence interval | Significance |
|---|---|---|---|---|
| IM → PEU → PU → LI | 0.044 | 0.015 | [0.019,0.076] | *** |
| IM → PEU → SE → LI | 0.090 | 0.020 | [0.053,0.131] | *** |

Note:Educ: Educator; IM: Instructional media; IC: Instructional content; PU: Perceived usefulness; PEU: Perceived ease of use; SE: Self-efficacy; LI: Learning intention; *** indicates a significant correlation at 0.001 level (bilateral)

## 5 Conclusions and Suggestions

Based on the above analysis, the following conclusions and suggestions are drawn:

Firstly, perceived ease of use, perceived usefulness and self-efficacy significantly affect the learning intention of digital marketing micro specialty. Meanwhile, perceived ease of use significantly affects perceived usefulness and self-efficacy. These results indicate that perceived ease of use plays a key role in enhancing learning intention to learn digital marketing micro specialty by promoting perceived usefulness and self-efficacy, and is an influential source of motivation for the generation and development of learning intention. Therefore, curriculum providers should take learner experience as the center, focus on the "zone of proximal development" of learners, and carry out the instructional content step by step, from the shallow to the deep. The selection or development of online learning platforms should also pay more attention to simplicity and convenience, so as to maximize the perceived ease of use for learners.

Secondly, the role of the educator significantly affects the learner's self-efficacy and ultimately their intention to learn. Educators positively influence learners' self-efficacy, that is, by providing instructional support, educators help to create a harmonious learning atmosphere, promote learners' confidence in completing learning activities and achieving learning goals, improve their internal motivation and sense of competence in learning, and ultimately generate positive intention to learn. Therefore, educators should give learners all-round and continuous attention in cognitive support, technical support and emotional support, so as to enhance learners' intention to learn digital marketing micro specialty.

Thirdly, the instructional content significantly affects the perceived usefulness, and ultimately affects the learners' intention to learn the digital marketing micro specialty. Perceived usefulness reflects the performance that learners believe learning digital marketing micro specialty will bring to their own development. The analysis results show that the learner's expectation of the instructional content lies in its practical value. Therefore, the design and implementation of digital marketing micro specialty for instructional content should constantly focus on the trends of learners' needs so that learners can feel its prominent role in improving learning outcomes.

Fourthly, instructional media have a positive impact on perceived ease of use and perceived usefulness, and further affect the intention to learn. In addition to directly affecting perceived ease of use and perceived usefulness, instructional media also indirectly affect self-efficacy and perceived usefulness through perceived ease of use. Instructional media are the enlargement and extension of the senses and the brain of the learner, and play an indispensable role in knowledge teaching, skill training and intellectual development. Therefore, course providers can enhance learners' perceived ease of use by continuously developing existing instructional media resources, simplifying and optimizing the operation process, so as to improve learners' learning confidence and perceived benefits, and finally realize the enhancement of learners' learning intention.

**Acknowledgment.** This project is supported by Research on the Construction of Teaching Team of Marketing Specialties in the Context of "Double First-Class" (w2019037) and Exploration of the Cultivation System of Excellence Talents in "Digital Marketing" Micro Specialty in the Background of New Liberal Arts (Teaching Research Project of Wuhan University of Technology).

# References

1. Junzong, Z.: The explaining of the new liberal arts from the four dimensions. J. Northwest Normal Univ. (Soc. Sci.) **56**(5), 13–17 (2019). (in Chinese)
2. Pasad, R., Traynor, C., Alabina, A.: Engaged IT experience course to enable the future workforce. In: Proceedings of the 18th Annual Conference on Information Technology Education, pp. 7–12 (2017)
3. Long, C., Nan, W., Lili, F.: A study and exploration on the talent training model of "Microspecialty" in local universities: taking D University as an example. J. Shijiazhuang Univ. **23**(02), 152–155 (2021). (in Chinese)
4. Yanyang, W., Yuan, Z., Yongming, W., Yangfeng, P.: Construction of micro specialty curriculum system in chemical engineering. High. Educ. Chem. Eng. **38**(05), 11–16+51 (2021). (in Chinese)
5. Xiaodong, T., Chengbo, Y., Linlin, D.: Research on interdisciplinary integration mode based on micro specialty. J. Hunan Post Telecomm. College **20**(04), 108–110 (2021). (in Chinese)
6. Jie, Z., Haiping, H.: Construction of micro-credential in Chinese universities under the background of new engineering. Softw. Guide **18**(11), 172–175+179 (2019). (in Chinese)
7. Zhiping, Z., Xiaoxiao, L.: A preliminary study on the construction of big data micro specialty. Comput. Era **08**, 68–70 (2019). (in Chinese)
8. Xiaomin, W., Wudong, Y., Qian, W., Yang, Y.: The construction of micro specialty curriculum system of rail traffic signal and control. Educ. Teach. Forum **05**, 35–37 (2019). (in Chinese)
9. Lazaeua, R.S.: Emotion and Adaption, pp. 212–215. Oxford University Press, New York (1991)
10. Wanxin, P., Qishen, L., Haiying, J.: Research on the influence of college students' willingness to learn under the online education model. Shanxi Youth **17**, 13–14 (2021). (in Chinese)
11. Xuesong, Z., Minjuan, W., Ghani U.: The SOR (stimulus -organism- response) paradigm in online learning: an empirical study of students' knowledge hiding perceptions. Interact. Learn. Environ. **28**(5), 586–601 (2020)
12. Huajun, W., Wenshuang, G., Juhou, H.: Research on the effect of teacher support on willingness to continue learning in MOOC courses. Mod. Distance Educ. **03**, 89–96 (2020). (in Chinese)
13. Guonong, N., Yunlin, L.: Educational Communication. Higher Education Press, Beijing (2005). (in Chinese)

14. Daqing, Z.: The construction of "Four-in-one" integrity education mode for college students based on Berlo's propagation mode. J. Jilin Agric. Sci. Technol. Univ. **28**(01), 18–21+116 (2019). (in Chinese)
15. Soepriyatna, Pangaribuan, C.H.: The direct and indirect influence of gamification on learning engagement: the importance of learning goal orientation (a preliminary study). Int. J. Inf. Eng. Electron. Bus. (IJEEB) **14**(04), 39–46 (2022)
16. Meihao, S.: A study on college students' willingness to use online learning behaviour in the post-epidemic era - based on technology acceptance model. Agric. Henan **30**, 32–34 (2021). (in Chinese)
17. Vidanagama D U. Acceptance of e-learning among undergraduates of computing degrees in Sri Lanka. Int. J. Mod. Educ. Comput. Sci. (IJMECS), **8**(04), 25–32 (2016)
18. Guanru, H., Xin, L.: Exploring the effects of self-efficacy on entrepreneurial intentions. Shanghai Manage. Sci. **44**(5), 74–79 (2022). (in Chinese)
19. Nandang, R., Budiman.: The influence of computer attitude, grade point average and computer self-efficacy against computer anxiety. Int. J. Educ. Manage. Eng. (IJEME) **9**(05), 10–17 (2019)
20. Zheyu, L., Yujing, L., Jihui, Z.: Research on the influence of self-efficacy on learning outcomes in desktop virtual reality environment: the mediating based on flow experience. J. Dist. Educ. **40**(04), 55–64 (2022). (in Chinese)
21. Hongyi, L., Fengyan, W.: The effect of discrimination perception on academic burnout among part-time graduate students: time management disposition and academic self-efficacy as chain mediator. China J. Health Psychol., 1–10 (2022). (in Chinese)
22. Le, Z.: On the status and function of educator and educate in vocational education. Hum. Resource Dev. **16**, 46–47 (2018). (in Chinese)
23. Podsakoff, P.M., MacKenzie, S.B., Lee, J.Y., et al.: Common method biases in behavioural research: a critical review of the literature and recommended remedies. J. Appl. Psychol. **88**(5), 879–903 (2003)

# Comparisons of Western and Chinese Textbooks for Advanced Electronic Packaging Materials

Li Liu[1], Guanchao Yin[1(✉)], Jin Wen[1], Qilai Zhou[1,2], Yao Ding[1], and Liqiang Mai[1(✉)]

[1] School of Materials Science and Engineering, Wuhan University of Technology, Wuhan 430070, China
{guanchao.yin,mlq518}@whut.edu.cn

[2] Faculty of Science, Shizuoka University, Shizuoka 422-8529, Japan

**Abstract.** Textbooks are the importance carrier of teachers' teaching and students' learning process. Recently, the traditional textbook forms cannot fully meet these needs, which requires more researches. This study chosen western and Chinese textbooks on the field of advanced electronic packaging materials to fully analyze and compare in the aspects of format structure, knowledge system design, knowledge point introduction and information technology application. From the general comparison results, this Chinese textbook only surpasses the western textbooks in the online course construction of information technology application. Unfortunately, there are several drawbacks that needs to be improved for the future textbook constructions, which are smaller numbers of references, pictures and tables, relatively old references, too much background knowledge and narrative content, less analysis of case studies, insufficient width and depth compared to the western textbook in this study. Through this comparison study between western and Chinese textbooks, the improvement directions of Chinese textbooks for materials and related majors were pointed out, which includes format structure, knowledge system design, knowledge point introduction and information technology application.

**Keywords:** Electronic Packaging Materials · Textbooks · Comparison

## 1 Introduction

Textbooks are the carrier of knowledge, the medium for students to learn knowledge and the foothold of teaching content and curriculum system [1, 2]. As an important medium for imparting subject knowledge, university textbooks need to not only highlight the basic knowledge of the discipline and reflect the cutting-edge knowledge in related fields, but also need to have a clear understanding of the teaching rules. However, the organization and compilation of textbooks require enormous knowledge including discipline professional theory, engineering application knowledge and modern education theory with Chinese characteristics and world-class advanced teaching concepts [3–5]. Thus, the researches on the teaching textbooks have always been a hot topic among the educators.

Especially for the training of novel engineering talents facing modernization in China, it is important to combine the development direction of China's new economic form with the requirements of international and future-oriented talents from an international perspective. Moreover, establishing the requirements for the training of outstanding talents in engineering that adapt to the industrial transformation and upgrading of China's traditional engineering industry and the development of strategic emerging industries is also of vital importance. However, the knowledge system of a single field is difficult to meet the needs of diversified knowledge structure of talents. For instance, the simple theoretical knowledge is difficult to fulfill the complex ability structure, while the traditional textbook forms cannot fully meet the needs of the new ecology of digital learning [6–8]. Therefore, the construction of teaching textbooks for engineering majors should closely follow the development of engineering sciences and related disciplines, especially for combining the new trends of the industry and the characteristics of China's engineering majors to build teaching textbooks that reflect the international and future-oriented for new engineering talent training.

One the one hand, the traditional textbook design mainly focuses on teacher teaching skills and knowledge system constructions. For cultivating high-level innovative talents, it is required to start from the student-centered and ability-oriented educational concept. Textbooks should provide space for students to take the initiative to build knowledge and the cultivation of students' comprehensive ability [9, 10]. On the other hand, with the application of modern education technology and information technology, the application of teaching methods such as heuristic teaching and research teaching has given birth to the construction of new forms of teaching materials. Therefore, it is necessary to absorb and learn from foreign advanced experience to enhance the construction and form of textbooks.

In this study, western (Materials for Advanced Packaging) and Chinese textbooks (Advanced Packaging Materials for Integrated Circuits) on advanced electronic packaging materials were fully compared in the aspects of format structure, knowledge system design, knowledge point introduction and information technology application to show the improvement directions for Chinese textbooks.

## 2 Descriptions of the Chosen Textbooks

In this study, the selected western textbook was <Materials for Advanced Packaging> with two edition versions that was written by Daniel Liu and C.P. Wong and published by Springer publisher in 2009 and 2017, respectively. This book is one of the few important works on electronic packaging materials. The authors of the original book are all authoritative experts in the field of electronic packaging. From the content point of view, this book not only includes the latest insights of internationally renowned scholars on packaging materials, including wire bonding materials, lead-free solder, substrate materials, flip chip underfills, epoxy molding compounds, conductive adhesives, thermal interface materials, nano-encapsulation materials, *etc*. It also covers the latest developments in electronic packaging technology, including three-dimensional integration, system packaging (silicon thinning, hole filling), nano-packaging and interconnect, wafer-scale packaging, MEMS packaging, LED packaging and other frontier fields. This book provides a large number of references, providing readers with comprehensive background information for reference to relevant research at home and abroad [11, 12].

In contrast, the Chinese textbook chosen in this study is <Advanced Packaging Materials for Integrated Circuits> written by Dr. Qian Wang and published by Publishing House of Electronics Industry in 2021. This book is one of the integrated circuit series of academic books. Recently, the overall scale of China's integrated circuit industry is experiencing unprecedented rapid development. However, due to the relatively weak foundation and some restrictions in foreign industries and technologies, China lags the international advanced level in most product subdivisions in all aspects of advanced electronic packaging materials, which still has very high foreign dependence.

Table 1 lists the general details of above western and Chinese textbooks for comparisons. In the following chapters, the comparisons of <Materials for Advanced Packaging Edition 2 (English)> and <Advanced Packaging Materials for Integrated Circuits (Chinese)> will be fully discussed from the aspects of format structure, knowledge system design, knowledge point introduction and information technology application.

**Table 1.** Comparisons and details of Western and Chinese Textbooks

| Name | Materials for Advanced Packaging Edition 1 | Materials for Advanced Packaging Edition 2 | Advanced Packaging Materials for Integrated Circuits |
|---|---|---|---|
| Language | English | English | Chinese |
| Publication year | 2009 | 2017 | 2021 |
| Publishers | Springer | Springer | Publishing House of Electronics Industry |
| Author Number | 44 | 57 | 5 |
| Page Number | 719 | 969 | 286 |
| Chapter Number | 19 | 22 | 13 |
| References | 1267 | 1789 | 289 |
| Ref Year | Within 10 years | Within 10 years | Within 15 years |
| Figure Number | 475 | 679 | 179 |
| Table Number | 97 | 113 | 44 |

## 3 Comparisons Between Western and Chinese Textbooks

### 3.1 Format Structure

By comparing the above two examples of Chinese and western electronic packaging material textbooks in the numbers of authors, chapters, references, pictures and tables listed in Table 1, the differences of format structure between Chinese and foreign textbooks can be summarized as follows.

It is noted that the author number of western textbooks is 57, which is far more than it of Chinese textbook (5 people). This is because that each chapters of this western textbook were written by experts in each specialized field, resulting in enormous people participating in the compilation. Moreover, the knowledge background and depth of this western textbook are wide.

Since electronic packaging materials are emerging cross-courses, the current knowledge system has not been fully built. Thus, there are no exercises in both Chinese and English textbooks as some theories were not been clarified totally.

Moreover, the shortcomings of domestic textbooks are also obvious. The numbers of references, pictures and tables are much smaller than it of western textbooks, which indicates that this Chinese textbook lag western textbook in terms of vividness. For example, there are chapters with fully plain text without figures and tables in this Chinese textbook. The narrative of the text with charts will make the teaching process of knowledge points more vivid to enable students learn professional knowledge more clearly. In addition, the gap in the reference number is also enormous. As reference can help readers expand their knowledge points during learning. Students would explore more about the knowledge they are interested in. Therefore, the new references can improve readers' enthusiasm for independent learning, which effectively completes the teaching goals.

### 3.2 Knowledge System Design

As shown in Fig. 1, the composition of different knowledge including background knowledge, general knowledge, application knowledge and frontier knowledge in western and Chinese textbooks is provided. The background knowledge includes introduction to industry background knowledge, pre-professional knowledge, *etc*. Basic knowledge means common knowledge and difficult points in professional knowledge. The definition of application knowledge is the application examples of basic knowledge, engineering application knowledge, industry application prospects and so on. Frontier knowledge includes cutting-edge knowledge of scientific research in emerging technology fields.

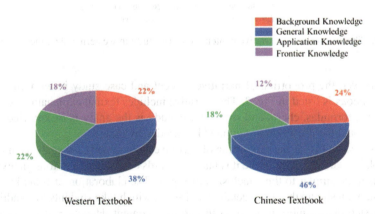

**Fig. 1.** The proportion of different types of knowledge in western and Chinese textbooks

In Fig. 1, the obvious difference between the western and Chinese textbooks for advanced electronic packaging materials lies in the aspects of basic knowledge, application and frontier knowledge. The length of background knowledge for Western and Chinese textbooks is quite similar. In terms of basic knowledge and extended knowledge, the proportion of application knowledge and frontier knowledge of western textbook is higher than its general knowledge. In contrast, domestic textbooks account for a higher proportion of basic knowledge than application and frontier knowledge.

Therefore, the focus of domestic and foreign textbooks is different. This Chinese textbook describes the basic knowledge more extensively and meticulously, so that students can learn and master the professional knowledge of related knowledge areas faster. However, for expanding knowledge, the content of domestic textbooks is relatively lacking due to the less and shallow descriptions of new technologies, frontiers, applications and corresponding extensions compared with the foreign electronic packaging material textbook. As a result, it can be concluded that this Chinese textbook can be significantly improved by introductions of expanding knowledge to further broaden students' horizons.

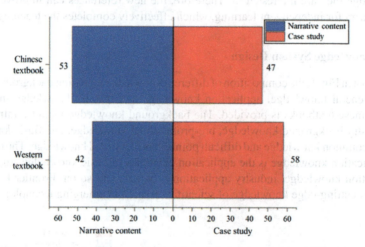

**Fig. 2.** The proportion of narrative content and case study in western and Chinese textbooks

Moreover, the proportion of narrative content and case study in western and Chinese textbooks is listed in Fig. 2. The narrative includes textual explanations of important terms, formulas, charts and so on. Case study is the analysis of knowledge point expansion and the illustration of applied knowledge by cases.

As shown in Fig. 2, the proportion of narrative content and case analysis in Chinese textbook is relatively even, while it is biased towards case analysis in foreign textbooks, which is quite similar to the conclusion of Fig. 1. The elaboration of basic knowledge points in domestic textbooks is detailed and solid, which take students to steadily learn step by step due to more proportion in narrative content. However, it is still slightly inferior to this foreign textbook for the expansion of knowledge points. In similar, there are more case studies in this foreign textbook, which also highlights the characteristics of foreign textbooks to enable readers to analyze and explore together. Therefore, Chinese textbook can add more case studies in the new edition to improve the interests of students.

### 3.3 Knowledge Point Introduction

Through carefully reading through < Materials for Advanced Packaging (English)> and <Advanced Packaging Materials for Integrated Circuits (Chinese)> textbooks, the comparative summary of Chinese and western textbooks in knowledge point introduction can be listed in Table 2.

Since the authors of each chapter in foreign textbook are independent, they all have their own introduction when introducing knowledge points. As a result, each chapter of foreign textbook is basically around a theme to describe from many aspects. Without the limitation to a fixed direction, a wider range content can be involved with more freedom. As for the Chinese textbook, it is basically introduced to students according to the three aspects of application, characteristics and development of electronic packaging materials, which are very clear and standardized at a glance.

**Table 2.** Comparison of Chinese and western textbooks in knowledge point introduction.

|  | Western textbook | Chinese textbook |
| --- | --- | --- |
| Structure | Chapters are written in different styles as they are written by varying authors | Similar structure and writing styles |
| Depth | More case studies to enhance students' independent learning skills | Insufficient depth |
| Breadth | High proportion of frontier field | Insufficient breadth |

### 3.4 Information Technology Application

Recently, due to the worldwide epidemic impacts, offline teaching across the world, especially for China, has become more and more difficult. To ensure that teaching work is not affected, growing offline classes are shifting to online teaching. After several years of online teaching exploration, the online teaching methods of various colleges and universities are also gradually improving and exploring growing suitable online teaching methods [13–15].

The combination of online and offline teaching can improve the implement teaching. However, it is quite important to require teachers and universities to constantly explore and improve how to combine online and offline teaching to achieve the best teaching effect [16–18]. The combination of paper textbooks and electronic materials can bring out the best teaching results. Therefore, in addition to paper textbooks, it is necessary to make a richer teaching resource library and more comprehensive teaching resources, such as electronic materials and teaching videos corresponding to paper textbooks. Students can gain more knowledge from the rich resource base and carry out deeper learning by themselves. As a result, teachers' teaching efficiency can be increased. For example, Beijing Institute of Technology designed a virtual simulation experiment of gold wire welding ball. Harbin Institute of Technology opened an English online course on microelectronic manufacturing technology. Moreover, the University of Maryland in the United States also offers online courses related to electronic packaging.

In terms of information technology applications in textbooks, universities from Chinese and foreign countries have not jointly established a digital resource library and quality evaluation system that can be shared. When college students want to find information, they not only have to face the problem that textbook resources cannot be shared between universities, but also must judge the quality of data by themselves, which greatly reduces students' learning efficiency.

At present, the digital educational resources owned by colleges and universities are mainly based on text and video. Most students just enter the video interface and fast forward or drag the progress bar to complete the viewing of the video, which makes the educational resources greatly wasted. If colleges and universities want to conduct online teaching, establishment of a rich digital education resource base can be a good option. The rich educational resource library can not only allow teachers to have more abundant teaching methods but also enable students to learn independently and efficiently. Colleges and universities should accelerate the construction of digital resources and develop various platforms to facilitate the use of teachers and students. In addition, some universities can also cooperate to jointly build a digital resource library, share teaching resources, and greatly enrich teaching content (Table 3).

Table 3. Comparison of Chinese and western textbooks in information technology application

|  | Western textbook | Chinese textbook |
| --- | --- | --- |
| Electronic textbook | Can be downloaded online | Basically accompanied by electronic textbooks but with fewer obtain ways |
| Website/QR code | Has specific links through publish house website | No |
| New forms | The construction is late than the Chinese textbook | Online course websites |

## 4 Effects of Textbook Improvement

By comparing the differences between Chinese and foreign materials textbooks, this study provides guiding suggestions for optimizing Chinese textbooks, which promotes the improvement of innovative talent training, high-level curriculum construction, and teacher training skills.

Based on the results, Wuhan University of Technology has carried out practice and reform in the following aspects. Firstly, for the main course of <Introduction to

Materials>, its textbook is updating to the new version, which integrates the latest application knowledge and case studies. Moreover, the modern science and technology are combined with latest theory can not only broadens students' knowledge, but also stimulates students' creativity.

Besides, all teaching courses in Wuhan University of Technology now require to complete the construction of online courses before their start. Therefore, with the advantages of online platforms, the construction of three-dimensional teaching materials can be promoted, improving students' self-learning efficiency and the teaching interaction between teachers and students.

## 5 Conclusion

The main purpose of this study was to compare the western and Chinese textbooks on topic of advance electronic packaging materials, which provides a direction for Chinese textbooks to improve afterwards. The conclusions can be drawn as follows.

1) In the aspect of format structure, although Chinese textbooks is concise, clear and logical due to small authors, the English textbook has rich information, knowledge and distinct level.
2) For the design of knowledge system, this Chinese textbook can be significantly improved by bringing more expanding knowledge to further broaden students' horizons.
3) As for knowledge point introduction, the Chinese textbook shows its insufficient in depth and breadth of textbooks, which can be enhanced by adding more case studies in various aspects.
4) The Chinese textbook surpasses the western textbooks in the online course construction of information technology application.

**Acknowledgment.** This work is supported by the Teaching Reform Project of Wuhan University of Technology (Grant No. w2021063).

## References

1. Vojíř, K., Rusek, M.: Science education textbook research trends: a systematic literature review. Int. J. Sci. Educ. **41**(11), 161354 (2019)
2. Sikorová, Z., Bagoly-Simó, P.: Textbook as a medium: impulses from media studies for research on teaching materials and textbooks in educational sciences. In: International Conference on Textbooks and Educational Media: Perspectives from Subject Education, 1–22 (2015)
3. Rottensteiner, S.: Structure, function and readability of new textbooks in relation to comprehension. Procedia Soc. Beh. Sci. **2**, 3892–3898 (2010)
4. Marsono, W.M.: Design a digital multimedia interactive book for industrial metrology measurement learning. Int. J. Mod. Educ. Manag. Eng. **8**(5), 39–46 (2016)
5. Fan, L.: Textbook research as scientific research: towards a common ground on issues and methods of research on mathematics textbooks. ZDM **45**, 765–777 (2013)

6. Scott, K., Morris, A., Marais, B.: Medical student use of digital learning resources. Clin. Teach. **15**(1), 1–86 (2018)
7. Shafiee, N.S.M., Mutalib, S.: Prediction of mental health problems among higher education student using machine learning. Int. J. Educ. Manage. Eng. **10**(6), 1–9 (2020)
8. Adebayo, E.O., Ayorinde, I.T.: Efficacy of assistive technology for improved teaching and learning in computer science. Int. J. Educ. Manage. Eng. **12**(5), 9–17 (2022)
9. Sievert, H., Ham, A.K.V.D., Heinze, A.: The role of textbook quality in first graders' ability to solve quantitative comparisons: a multilevel analysis. ZDM Math. Educ. (53): 1417–1431 (2021)
10. Hadar, L.L.: Opportunities to learn: mathematics textbooks and student' achievements. Stud. Educ. Eval. **55**, 153–166 (2017)
11. Lu, D., Wong, C.P.: Materials for Advanced Packaging. Springer, New York (2009). https://doi.org/10.1007/978-0-387-78219-5
12. Lu, D., Wong, C.P.: Materials for Advanced Packaging, 2nd edn. Springer, Cham (2017). https://doi.org/10.1007/978-3-319-45098-8
13. Hilton, J.: Open educational resources and college textbook choice: a review of research on efficacy and perceptions. Educ. Tech. Res. Dev. **64**, 573–590 (2016)
14. Berry, T., Cook, L., Hill, N., Stevens, K.: An exploratory analysis of textbook usage and study habits: misperceptions and barriers to success. Coll. Teach. **59**(1), 31–39 (2010)
15. Bowen, W.G., Chingos, M.M., Lack, K.A., Nygren, T.I.: Interactive learning online at public universities: evidence from a six-campus randomized trial. J. Policy Anal. Manage. **33**(1), 94–111 (2014)
16. Krämer, B.J., Neugebauer, J., Magenheim, J., Huppertz, H.: New ways of learning: comparing the effectiveness of interactive online media in distance education with the European textbook tradition. Br. J. Edu. Technol. **46**(5), 965–971 (2015)
17. Almazova, N., Andreeva, S., Khalyapina, L.: The integration of online and offline education in the system of students' preparation for global academic mobility. In: International Conference on Digital Transformation and Global Society, pp. 162–174 (2018)
18. Cheung, S.K.S, Lee, K.K.W., Chan, K.K.L.: A review on the development of an online platform for open textbooks. In: International Conference on Hybrid Learning and Continuing Education, pp. 196–207 (2014)

# Innovation and Entrepreneurship Teaching Design in Application-Oriented Undergraduate Professional Courses – Taking the Transportation Enterprise Management Course as an Example

Yan Chen[1], Liping Chen[2(✉)], Hongbao Chen[3], Jinming Chen[4], and Haifeng Yang[5]

[1] College of Traffic and Transportation, Nanning University, Nanning 530200, China
[2] Pubei Middle School, Qinzhou 535300, China
2676501161@qq.com
[3] School of Architecture and Engineering, Guangxi City Vocational University, Chongzuo 532100, China
[4] School of Electronic Information and Automation, Civil Aviation University of China, Tianjin 300300, China
[5] Faculty of Letters, Arts and Sciences, Waseda University, Shinjuku, Japan

**Abstract.** Integrating innovation and entrepreneurship teaching into application-oriented undergraduate professional courses can help college students connect with career development and cultivate application-oriented professional and technical talents for the development of local economy. By analyzing the current development situation of innovation and entrepreneurship education for college students, analyzing the students' learning situation, and analyzing the role and significance of transportation enterprise management as the core course of transportation specialty in cultivating application-oriented talents with integrating innovation and entrepreneurship education, this paper proposes to build a link between the teaching content of the course and the work content of the vocational post with innovation and entrepreneurship as a bridge, and excavate the innovation and entrepreneurship elements in the transportation enterprise management course. According to the work content of the management post of the transportation enterprise and the teaching methods such as team-based learning and task-driven approach, the innovation and entrepreneurship teaching design is carried out.

**Keywords:** Applied undergraduate · Professional courses · Innovation and entrepreneurship teaching · Transportation enterprise management

## 1 Introduction

Promoting innovation and entrepreneurship education in university education and cultivating college students' innovation and entrepreneurship ability can help college students better connect with career development and promote employment.

## 1.1 The Development Status of Innovation and Entrepreneurship Education for College Students

The proposal and promotion of innovation and entrepreneurship education for college students has developed for more than 20 years. In 1998, the United Nations Educational, Scientific and Cultural Organization (UNESCO) clearly proposed that college students should be trained to be excellent college students with entrepreneurial thinking, entrepreneurship and entrepreneurial skills. After graduation, they can not only find jobs, but also create jobs themselves. Driven by this educational concept, innovation and entrepreneurship education for college students at home and abroad has shown different characteristics through experiments and reforms. American colleges and universities take the concept of innovation and entrepreneurship as an indispensable ideological concept for infrastructure construction, set innovation and entrepreneurship education as a professional research direction, and formulated a complete set of class system and teaching plan for characteristic innovation and entrepreneurship education. According to 2016 statistics, the entrepreneurship rate of American college graduates has reached nearly 25%. The innovation and entrepreneurship education in Japanese universities has gone through the stage from utilitarian education to mass education. It is led by the government, assisted by universities and society, and trained from high school students, with strong coherence and regional characteristics. Innovation and entrepreneurship education in Britain is more about creating a cultural atmosphere. In China, there are three main ways for colleges and universities to carry out innovation and entrepreneurship education, namely, college students' innovation and entrepreneurship courses, college students' innovation and entrepreneurship training programs and college students' innovation and entrepreneurship competitions, which can guide students to carry out career planning in combination with their majors and clarify their career development direction, which can play a role in enhancing students' innovative thinking, practical ability, communication ability, etc., but only 1.5% of college students actually start businesses after graduation, There are few successful cases of entrepreneurship [1].

## 1.2 The Relationship Between Application-Oriented Talent Training and Innovation and Entrepreneurship Teaching

The cultivation of application-oriented talents is a comprehensive reform of the teaching model based on capacity output. The specific reform path is based on career orientation, industry-teaching integration, workshops, professional spirit, industry background, etc. The ultimate goal of innovation and entrepreneurship education is to build the professional ability of college students. The goals of the two are the same. However, in the college education system, at present, these two parts of teaching work are still two teaching systems, not well integrated, and there are bottlenecks such as lack of entrepreneurial experience of professional teachers, parallel innovation and entrepreneurship curriculum system and professional curriculum system, and lack of platform support [2–4]. In order to solve this problem, experts and scholars put forward relevant countermeasures. In addition to reforming the talent training mode, in the aspect of curriculum construction, they propose to build dual-qualified curriculum, specialized and innovative integrated curriculum with integrating innovation and entrepreneurship education into professional construction based on the countermeasures of systematic and applied courses in the work process [5, 6].

### 1.3 Problems to be Solved in this Paper

The application-oriented curriculum is the carrier of cultivating application-oriented talents. Transportation enterprise management (hereinafter referred to as "the course") is a basic course of transportation specialty, and also a practical course of profession and innovation & entrepreneurship integration reform supported by Nanning University. It is one of the key courses for the cultivation of application-oriented talents in transportation specialty. The course covers all aspects of enterprise operation and management, such as strategy, corporate culture, brand, market research, marketing, production operation, team management, customer management, human resource management, equipment management, quality management, financial management, et al., corresponding to relevant positions in production & business & administration & other departments of the enterprise, and is closely related to career development. The reform of the teaching content and teaching method of this course based on the idea of innovation and entrepreneurship can explore how to carry out the integration of specialty and entrepreneurship, cultivate students' innovation and entrepreneurship ability in professional development based on the course, connect professional development and job requirements, lay the foundation for employment and entrepreneurship, and also provide reference for how to integrate professional courses into innovation and entrepreneurship education and teaching.

## 2 The Role and Significance of Integrating Innovation and Entrepreneurship Teaching into the Course of Transportation Enterprise Management

Integrating innovation and entrepreneurship education into the course of transportation enterprise management not only enriches the breadth and depth of the course itself, promotes the development of the content of the course, but also guides students to pay attention to the industry and enterprises, and develops students' professional abilities. It also cultivates application-oriented professionals for the society, and promotes local economic development, which has an important role and significance.

### 2.1 Analysis of Students

The course is set up in the second semester of the third year of the college. Students have completed part of the professional courses, and also completed the school's innovation and entrepreneurship courses. They have certain professional knowledge and entrepreneurship foundation. However, how to transform these knowledge into ability should be gradually transformed after docking with the industry.

Before the beginning of the class, a survey was conducted on students in 2019. After analysis, it was found that students' understanding of enterprise operation and management is superficial, and they have not yet understood the rules of enterprise economic operation. In addition, some students have experience in innovation and entrepreneurship activities, but most of them have taken relevant courses such as entrepreneurship foundation, but still lack strong innovation and entrepreneurship awareness and clear future career and job development intentions. The survey data are shown in Table 1.

Table 1. Statistics of Survey Data

| S/N | Items | Attitude | |
|---|---|---|---|
| | | Affirm | Negate |
| 1 | Have Taken Entrepreneurship Foundation and Other Related Courses | 100% | 0% |
| 2 | Participated in Innovation and Entrepreneurship Projects | 30% | 70% |
| 3 | Participated in Practical Activities inside and outside the School | 70% | 30% |
| 4 | Have a Strong Sense of Innovation and Entrepreneurship | 30% | 70% |
| 5 | Clear Future Career and Post Development Intention | 10% | 90% |
| 6 | Be Willing to Expand after Class | 60% | 40% |

Therefore, it is necessary to integrate innovation and entrepreneurship teaching in the course of transportation enterprise management. First of all, it can promote students' cognition and understanding of the industry and enterprises, and then it can guide students to clarify their career intentions and connect with career development.

### 2.2 The Promotion of Innovation and Entrepreneurship Teaching to Professional Courses Teaching

The integration of innovation and entrepreneurship teaching can better clarify the ability goal orientation of professional curriculum teaching, and clearly cultivate the skills and qualities of students through career development. The cultivation of application-oriented professionals is mainly around employment. The ability requirements of application-oriented talents are different from those of research-oriented and academic talents. Taking the transportation enterprise management course as an example, first of all, the course introduces the main line of "double entrepreneurship". Whether it is employment or entrepreneurship, we should treat our work with the attitude of innovation and entrepreneurship, and the attitude determines our actions [7]; Secondly, strong innovation and entrepreneurship ability can make the work more successful [8]. Therefore, the ability training of the course will be carried out around the aspects of program design, publicity, communication and analysis, so as to clarify the ability and quality objectives of the course. Therefore, the integration of innovation and entrepreneurship education is a good promotion for the teaching of application-oriented professional courses.

### 2.3 The Course Meets the Needs of Enterprise Operation and Management

The course of transportation enterprise management is a course that closely combines theory and practice. It connects with the posts of production, business, administration and other departments of transportation enterprises. The teaching content of each unit is closely related to the work content of the post. In order to stimulate the students' interest in learning, it is necessary to establish a link channel between them. Innovation and entrepreneurship teaching can just take on this channel task, taking the task/project as the carrier, Encourage students to combine theory with practice, pay attention to and analyze actively, deepen knowledge and transform ability by completing tasks/projects.

## 3 Innovation and Entrepreneurship Elements Contained in the Course

To integrate innovation and entrepreneurship teaching into professional courses, it is necessary to excavate the innovation and entrepreneurship elements contained in the courses and integrate them into the teaching content and process of the courses, so as to achieve the purpose of integration of specialty and entrepreneurship.

### 3.1 Teaching Requirements for the Course

The teaching objective of this course is to enable students to understand the enterprise management methods, processes and tools of passenger and freight transportation enterprises, transportation science and technology service enterprises, etc. in the field of domestic transportation comprehensively and systematically through the study of all the teaching contents specified in the syllabus, and master the strategic management, business analysis, customer management, brand marketing management, production and operation, business process The knowledge and methods of human resources, quality control, team management, material management, equipment management, information system and financial management will inspire innovative thinking and further strengthen the professional foundation and comprehensive ability of students majoring in transportation.

### 3.2 Main Teaching Contents of the Course

The teaching content of the course mainly includes enterprise management theory, strategic management of transportation enterprises, business management of transportation enterprises, resource management of transportation enterprises and development management of transportation enterprises.

1) Enterprise management theory

Combining the general enterprise management theory with the characteristics of transportation enterprises, and taking the five management functions of "planning, organization, command, coordination and control" of the process management theory of the process management school as the core, this paper analyzes the contents and methods of transportation enterprise management.

2) Strategic management of transportation enterprises

Carry out enterprise strategic analysis of passenger and freight enterprises and transportation operation enterprises, list possible strategic options, make strategic choices and carry out strategic implementation. Learn EFE matrix, IFE matrix, SWOT matrix and other strategic analysis tools, and learn the application of Porter's competitive strategy selection model and Boston matrix's diversification strategy selection model. Analyze the possible obstacles and possible measures for the implementation of enterprise strategy. Master the brand building process, brand equity model and brand management strategy of transportation enterprises. Use the "four levels" structure model of enterprise culture to construct the content level of transportation enterprise culture, master the implementation steps of enterprise culture, and refine the characteristics of transportation enterprise culture industry. Analyze the front and back service systems of passenger and freight transportation business.

3) Business management of transportation enterprises

Use scientific methods to conduct transportation market survey and research, and find ways to solve the core business problems such as freight pricing, service satisfaction, transportation demand analysis, etc.; Understand the characteristics of production and operation of transportation enterprises, set production indicators, and conduct production and operation process management; Straighten out the passenger and freight transport business process, analyze the key points in the business process, optimize the process in combination with customer satisfaction, service efficiency, service quality, and other aspects, and use the evaluation index system to make a comparative analysis before and after optimization; Apply the "4P" theory model of enterprise marketing to establish the transportation marketing system, compile the marketing planning plan, control the marketing process, and evaluate the marketing work; Learn customer relationship hierarchy theory and three-level customer relationship marketing theory, establish appropriate customer relationships, understand what behaviors will affect customer loyalty, learn to use customer concentration index to analyze customer distribution, conduct targeted customer management, understand the key points of forming positive word-of-mouth communication and how to pay attention to customer word-of-mouth in the service process.

4) Resource management of transportation enterprises

We need to use scientific management methods to effectively support business activities with human, financial, material, information and other resources in the enterprise. Specific learning contents include: talent selection and motivation, enterprise staffing method, Maslow's demand hierarchy model, and two-factor motivation theory; The division of teams, site management, safety management and work responsibilities of team leaders, and how to select team leaders; Material management and production control JIT method, ABC inventory management method, ordering method, BOM list and other methods related to material management, MRP system cognition; Methods and management contents for selection, maintenance and renewal of transportation equipment; The functional structure, deployment, application and design of the transportation enterprise management information system, as well as the enterprise information construction;

Investment and financing management, current asset management and financial analysis of transportation enterprises, draw financial analysis index system, and recognize DuPont analysis method [9, 10].

5) Transportation enterprise development management

Learn service quality statistical analysis, service quality gap analysis, SERVQUAL service quality model, service quality management cost analysis, etc., and be able to scientifically formulate service quality evaluation index system; How to correctly formulate project objectives, decompose project tasks and manage progress; How to carry out innovative management in the operation process of transportation enterprises.

### 3.3 Analysis of Elements of Innovation and Entrepreneurship

From the teaching objectives and contents of the course, it can be concluded that the innovation and entrepreneurship elements contained in the course mainly include the following categories.

1) Method class

Including the management tools and methods used in the management process of transportation enterprises, which is the industry knowledge reserve for innovation and entrepreneurship.

2) Skills

Including operational management skills such as scheme design, coordination and organization, leadership and command, process control and target management, as well as working skills for innovation and entrepreneurship.

3) Quality

Including team communication, roadshow, analysis and judgment in the process of completing the project/task, which is a necessary quality for innovation and entrepreneurship.

4) Occupation

Including corporate ethics, service awareness, concern for the environment and sustainable development, courage to endure hardship, self-discipline and other industrial corporate culture appeals, which are the industrial attributes of innovation and entrepreneurship.

## 4 Innovation and Entrepreneurship Teaching Design

### 4.1 Project Design

After the students are divided into groups, each study group first selects a business direction, and after completing the study of management theory knowledge of each unit, take the innovation and entrepreneurship of "transportation enterprise management practice" as the main line, and use the knowledge learned to complete a task (project), which is jointly completed by the team. Each large class (80 min) is about 20 min for task practice. The project design is shown in Fig. 1.

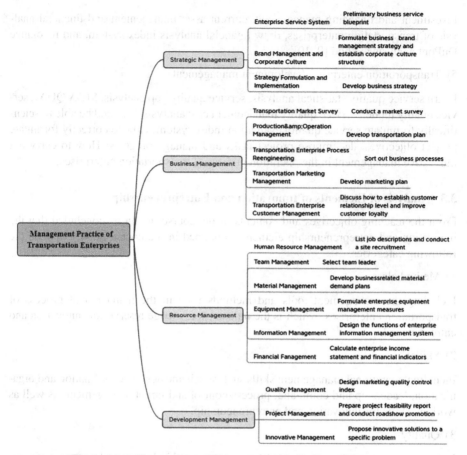

**Fig. 1.** List of Innovation and Entrepreneurship Projects of the course

### 4.2 Comprehensive Practical Design

By selecting a certain transportation business, from market research, strategic analysis and selection, to the formulation of marketing strategies, service process optimization and service quality control, an operation and management plan with distinct themes and detailed contents will be formed.

In the process of completing the comprehensive practical task, the students have carried out an innovative thinking on enterprise operation and management and carried out a creative work [11].

### 4.3 Teaching Design

The teaching design is carried out by the combination of team-based learning and task-driven method. The team-based learning requires to guide students to participate in the teaching process and play the interactive role between students, so as to help students acquire knowledge and skills [12]. Task-driven By setting up project development tasks for comprehensive application of knowledge, effectively organizing teaching materials, promoting students' autonomous learning in classroom teaching, and improving the application ability of knowledge learned [13]. Therefore, the combination of team-based learning and task-driven method to design the teaching organization process can well stimulate the creativity of students and cultivate the awareness of innovation and entrepreneurship. The relevant teaching process design is shown in Fig. 2.

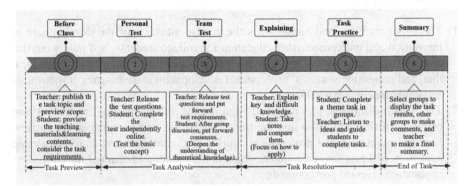

**Fig. 2.** Teaching Process Organization

## 4.4 Assessment Design

The course assessment is also integrated into the innovation and entrepreneurship teaching design. In terms of course assessment, three assessment units A, B and C are designed. Unit A is mainly for personal testing to assess students' mastery of theoretical knowledge of enterprise management, integrate industry and enterprise knowledge into the assessment content, and increase students' awareness and understanding of industry and enterprise; Unit B, mainly for team task assessment, takes innovative thinking and entrepreneurial practice as the main assessment basis, and evaluates students according to the completion of each class task practice, and integrates innovation and entrepreneurship teaching into professional teaching; Unit C is mainly for comprehensive assessment. Students are required to answer questions and complete analysis and design related to the course content within a limited time to evaluate the teaching effect of the integration of specialty and creativity [14]. The evaluation formula is as follows.

$$S = \frac{\alpha}{2} \times (A + \frac{t}{\overline{T}} \times B) + (1 - \alpha) \times C \tag{1}$$

In formula (1), S means the comprehensive evaluation score, A, B, C respectively refer to the assessment score of each unit in A, B, C, $\alpha$ refers to weight coefficient, usually 40% to 60%, t refers to individual learning engagement performance score, and $\overline{T}$ refers to the average score of learning performance of all members of the individual's group.

For example, the weight coefficient set for the course is 50%, the assessment scores of each unit of A, B and C of a student are 90, 88 and 85 respectively, the individual learning performance score is 95 points, and the average group learning performance score is 90 points. According to formula (1), the student's comprehensive assessment score is 88.72 points.

## 5 Conclusion

1) The paper analyzes and summarizes the current situation of the development of innovation and entrepreneurship education for college students, and points out that entrepreneurship is not the only purpose of innovation and entrepreneurship education, and promoting students' professional development is the essence of innovation and entrepreneurship teaching.
2) The paper discusses the role and significance of integrating innovation and entrepreneurship education into the transportation enterprise management curriculum, and points out that professional courses and innovation and entrepreneurship education will promote each other, which is the carrier and channel for cultivating application-oriented professionals.
3) By analyzing the teaching content of the course of transportation enterprise management, the paper extracts the elements of innovation and entrepreneurship, and lays the foundation for the teaching design of innovation and entrepreneurship.
4) The thesis carries out a specific innovation and entrepreneurship teaching design in combination with the course of transportation enterprise management.

**Acknowledgment.** This project is supported by The second batch of school-level teaching team in 2019 "Applied Effective Teaching Design Teaching Team" (2019XJJXTD10) and Nanning University's second batch of teaching reform project of specialized innovation integration curriculum (2020XJZC04).

## References

1. Qi, C.: Talking about the current situation of innovation and entrepreneurship education for college students at home and abroad. Modern Econ. Inf. **18**, 394 (2016). (in Chinese)
2. Wang, C., Gong, C., Liang, Z.: The status quo and improvement measures of innovation and entrepreneurship education for college students in the new era. New West **04**, 136–138 (2022). (in Chinese)
3. Wang, Y.: Analysis of the current situation of college students' innovation and entrepreneurship education and research on countermeasures. Res. Pract. Innov. Entrepreneurship Theory **4**(07), 182–184 (2021). (in Chinese)
4. Li, R., Chang, P.: Analysis and suggestions on the development of innovation and entrepreneurship education for college students. Inner Mongolia Sci. Technol. Econ. **01**, 38–39 (2021). (in Chinese)
5. Cao, L., Yang, H.: Cultivation of artisan design professionals by industry masters as the main body of teaching guidance – exploration of the cultivation of application-oriented talents in design majors of Nanning University. Educ. Teach. Forum **33**, 99–102 (2022). (in Chinese)

6. Ping, X.: Research on the status quo and countermeasures of innovation and entrepreneurship education for undergraduate students in application-oriented universities. Res. Pract. Innov. Entrepreneurship Theory **3**(06), 74–75 (2020). (in Chinese)
7. Zheng, Y.: Research on innovation and entrepreneurship teaching system of engineering graphics course group. In: Proceedings of 4th International Conference on Social Science and Higher Education (ICSSHE 2018) (Advances in Social Science, Education and Humanities Research), (181), 492–494 (2018)
8. Yu, J.: Research on the construction of innovation and entrepreneurship teaching system based on computer science and technology. In: Proceedings of the 2018 8th International Conference on Management, Education and Information (MEICI 2018), pp. 927–931 (2018)
9. Liu, X.: Research on the application of humanistic management thought in modern enterprise management. Account. Corpor. Manage. **4**(3), 87–94 (2022)
10. Aggarwal, A., Verma, R., Singh, A.: An efficient approach for resource allocations using hybrid scheduling and optimization in distributed system. Int. J. Educ. Manage. Eng. (IJEME) **8**(3), 33–42 (2018)
11. Harijanto, B., Apriyani, M.E., Hamdana, E.N.: Design online learning system for kampus merdeka: a case study web programming course. Int. J. Educ. Manage. Eng. (IJEME) **11**(6), 1–9
12. Zhao, H., Shen, Y.: Innovative exploration of peer learning to integrate international talent training in colleges and universities. Beijing Educ. (Higher Educ.) **10**, 50–52 (2022). (in Chinese)
13. Ding, X.: Practice and exploration of task-driven flipped classroom in the course of "Web programming." Sci. Technol. Wind **496**(20), 121–124 (2022). (in Chinese)
14. Berrabah, F.Z., Belkacemi, C., Ghomari, L.Z.: Essential and new maintenance KPIs explained. Int. J. Educ. Manage. Eng. (IJEME), **12**(6), 11–20 (2022)

# The Construction of University Teachers' Performance Management System Under the Background of Big Data Technology

Fengcai Qin[1] and Chun Jiang[2(✉)]

[1] Nanning University, Nanning 530299, Guangxi, China
[2] College of Digital Economics, Nanning University, Nanning 530299, Guangxi, China
6113186@qq.com

**Abstract.** Big data has been applied to student learning, teacher teaching, resource construction and university management. Teacher performance management is an important part of university management and an important means to improve the quality of education. However, there are many problems and defects in the performance management of university teachers. The performance evaluation of university teachers is mainly carried out by hand-filling and statistical summary. In the process of statistics, the audit of indexes is time-consuming and laborious, which is easy to produce data errors and low efficiency. Based on big data technology, through big data analysis to evaluate and optimize talent training, it is divided into three different important modules: the information-based performance evaluation of Talent Innovation Training, the information-based application performance evaluation of university scientific research capability and the information-based capability evaluation of university comprehensive services, and the establishment of data warehouse, to build an open network management system, to build an open network sharing platform, to strengthen the dynamic study of teaching management, to formulate various objectives and indicators of performance evaluation, to realize the flat, dynamic and integrity of performance evaluation management.

**Keywords:** Data mining · Performance management · Big data

## 1 Introduction

Globalization, competition, and reform calls offer several serious problems for higher-education institutions in terms of funding sources and allocations, teaching quality, and operational administration. It is vital to effectively analyze performances in order to maximize operational and management advantages [1]. The performance appraisal of university teachers is an important work of the school. Mastering the current situation of teachers, measuring teachers' scientific research achievements and teaching workload, and supervising teachers' teaching process are important means to reflect teachers' teaching level and teaching quality [2]. Through the comprehensive evaluation of teachers' education and teaching performance, scientific research performance, social service performance and teachers' professional development, teachers can effectively guide and stimulate the effect, and promote the improvement of teachers' comprehensive quality and professional skills, which makes the ability of schools and teachers develop together [3]. Big data can collect a large number of different types of data, and seek the hidden relationship and value behind the data in deep mining and scientific analysis [4]. With the continuous development of information construction in universities, many universities have established and run various kinds of database systems, including all kinds of basic data needed for performance evaluation [5, 6]. How to effectively integrate the basic data and conduct in-depth analysis with the index system has become an urgent problem to be solved in the further deepening of university performance evaluation [7]. The design of evaluation index system, the collection of basic data and the construction of evaluation methods are the key factors of university performance evaluation [8]. The massive data in the information age provides a reference and basis for the development of education. Under the impact of the information age, the performance evaluation and information construction of education have become the driving force for the development of current higher education [9].

This study starts with the current situation and problems of University Teachers' performance management, and uses the concept of big data for reference, analyzes the Enlightenment of big data thinking on University Teachers' performance management, and puts forward the construction method of university performance management system based on data mining technology.

## 2 Problems of Performance Management in Colleges and Universities Under the Background of Big Data

The fragmentation of data in the digital era is full of university teaching environments, and it also lays out the path for university teaching decision-making. Yet, how to mine the critical data required for university personnel training and management innovation using high-tech information data has become the emphasis and challenge of information technology usage in a big data environment. Teaching indicators are primarily used to assess instructors' instructional burden. Computer technology is permeating university instructional administration. The use and practice of digital management significantly improves teaching quality. On the one hand, it is entirely reliant on information technology; on the other hand, it is continually supported by big data. A cloud computing system is a data information system that consists of virtualized network data storage, dispersed network data terminals, and integrated information technology. Its main advantage is that it can perform large-scale information data management as well as comprehensive evaluation and analysis on large talent training data models and university parameter indicators, allowing it to effectively build a virtual, high-scalability, high-speed, and high-speed cloud computing system. Reliable, low-cost information-integrated network-based performance evaluation system for teaching management.

At present, the performance evaluation of university teachers is mainly carried out by manual filling and statistical summary. The statistics of basic data is cumbersome and complex, and the audit of indicators in the statistical process is time-consuming and laborious, which is easy to produce data errors and low efficiency. From the content of indicators, teaching indicators are mainly the workload of teachers' teaching, and scientific research indicators are mainly the number of papers published, monographs, topics, etc. the determination of such indicators is too much emphasis on the results of indicators, lack of evaluation of specific work process. From the evaluation criteria of indicators, the quantity of some indicators cannot represent the quality of indicators, especially in academic indicators. Overemphasizing the quantity of annual achievements deviates from the essential characteristics of academic research. At the same time, to a certain extent, it will lead to increasingly strong internal friction and utilitarianism among teachers, seriously affecting the physical and mental health of teachers and the construction of university teachers Design. The university teaching management is complex, so whether it is performance evaluation or information transmission, it is always inseparable from the support of big data, especially the university work information transmission needs timeliness and dynamic, so as to carry out performance evaluation for the cultivation of educational technology talents. Figure 1 reflects the main structure of school public governance.

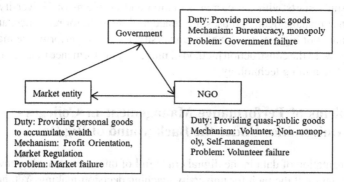

**Fig. 1.** The main structure of public governance

Although some colleges and universities have established information-based performance management platforms, due to the lack of teachers' information literacy, the function setting of the management platform is not comprehensive enough, and it is difficult to fully play the role of the performance management platform, so the work efficiency has not improved greatly [10]. In practice, through the sharing of big data resources, data mining, analysis and statistics, the teaching management information in different links such as talent innovation, management innovation and performance evaluation is analyzed and quantified, and through the information docking between universities, the openness of the market and the effectiveness of information transmission are increased. University resource information and related teaching management information can be uploaded to the cloud platform through virtual servers to share data resources on campus and off campus, thus providing intellectual support for innovative training of talents.

## 3 The Influence of Big Data on University Teachers' Performance Evaluation Platform

The quality of university teachers plays a vital role in the quality of education. Education level not only determines the level of training talents, but also relates to the improvement of national quality [11]. At present, the problems in the construction of university teachers. Schools need to formulate corresponding measures to maintain high-quality talents, stimulate teachers' work enthusiasm, and improve teaching quality, scientific research level, social service level and their own professional development level [12]. From the content point of view, the index teaching indicators mainly include teachers' teaching work, and the research measures mainly include the number of published papers, the number of books, and the problem of quantity. As a result, the university lacks the specific working process of evaluation, which cannot be reflected by the quality index.

As time goes by, the accumulation of data increases the amount of basic data, and even establishes a big data system library for each college department, so as to more effectively quantitatively manage the performance evaluation of teachers and college departments. The purpose of performance management is to promote the development of teachers themselves. However, the most important thing in the university performance management system is to provide reference or play a role in the development of teachers, which leads to the form of performance evaluation. The main purpose of performance management is to promote the development of teachers, improve the efficiency of university management and promote the rapid development of universities. However, performance management in many universities does not help teachers' development, and performance evaluation has become a superficial form.

## 4 Implementation of University Performance Evaluation Management Platform Based on Big Data

### 4.1 Architecture of University Performance Evaluation Management Platform

The framework of a university educational technology talent training performance evaluation system should include information infrastructure construction, information application analysis, information resource utilization and sharing, and information organization structure management, all based on big data technology. Nevertheless, the informatization construction must always review and optimize the associated talent training mode through big data analysis in order to deploy resources wisely. The system's business function has flaws of its own. When it comes to acquiring vendors, each system's institutions are uneven. Third-party services are often offered by mature products rather than the school's enhanced unified data standard, and some do not even provide data interfaces and table domains [13]. Of course, the development performance of the University's educational technology talent training should always be consistent with the construction of informatization, focusing on the intelligent training of talents and the innovation of University's scientific research strength, as well as the comprehensive strength research in the process of University's talent training [14]. The performance assessment management platform based on big data analysis must link with the school's more relevant business system in order to facilitate data sharing and sharing. The important connection is the successful integration of information and data resources. The relatively independent data statistics and analysis of business systems is a major barrier to big data, so we should integrate the information resource management system, particularly for existing data integration, to further accelerate quantitative evaluation based on big data analysis in university performance management.

The university performance evaluation system is divided into three different important modules, the informatization performance evaluation of innovative cultivation of talents, the informatization application performance evaluation of university scientific research ability and the informatization ability evaluation of university comprehensive service. The specific process is shown in Fig. 2.

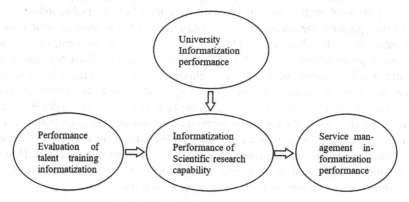

**Fig. 2.** Performance evaluation system

Because big data is open in exploring the internal laws of things, that is, it is not necessary to preset the conclusions in advance, and more unknown laws and multiple systems can be obtained by studying and analyzing the collected fine data. The performance evaluation analysis layer evaluates the performance of universities by applying big data mining algorithm, deeply analyzes the internal relationship of evaluation data. Performance evaluation and analysis layer can use data cloud storage and distributed management to integrate system architecture. In the age of big data, universities must integrate the information background development trend, promote the effectiveness and modernization of university teaching work through various technical means and methods, change the traditional teaching management practice mode through computers and data management systems, and realize the effective disclosure and sharing of data resources [15].

## 4.2 Construction of University Performance Data Warehouse

Therefore, the integration of information resources is a key step. The current university network system is uneven, and each subsystem is relatively independent, just an entrance to the employee account, which is lack of data sharing. In the long run, the implementation of quantitative assessment of teachers' performance needs the accumulation and storage of basic data. With the further development of education informatization, big data has also been applied to university teaching management, but many universities are difficult to design big data independently, so the government needs to do a good job in the top-level design first, and then gradually implement the big data design of each university. The structural relationship between the independent perceived familiarity and the score of local samples and non-local samples is shown in Fig. 3.

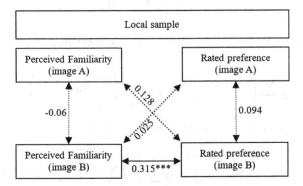

**Fig. 3.** The relationship between local samples and non-local samples

University data integration layer is the basic data source of university performance evaluation, including heterogeneous data sources distributed in various systems. After establishing the starting mechanism of heterogeneous data sources, the data needed for

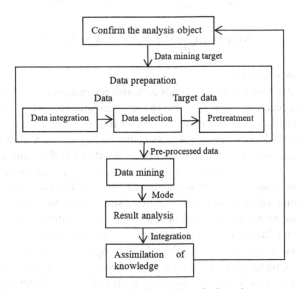

**Fig. 4.** Data mining process in financial analysis and management

university performance evaluation can be incrementally integrated into the data warehouse to ensure the accuracy of basic data. The data mining process of performance management is shown in Fig. 4.

Suppose the expected output is, and the global error between expected and actual output is defined as L:

$$L = \frac{1}{2}\sum_{k=1}^{m}\left(z_k - z'_k\right)^2 \tag{1}$$

Through the back propagation process, the error is expanded to the hidden layer as:

$$L = \frac{1}{2}\sum_{k=1}^{m}\left[f(\lambda_k) - z'_k\right]^2 = \frac{1}{2}\sum_{k=1}^{m}\left[f\left(\sum_{k=1}^{t}w_{jk}y_j + b_k\right) - z'_k\right]^2 \tag{2}$$

Finally, the reverse transmission to the input layer is:

$$L = \frac{1}{2}\sum_{k=1}^{m}\left[f\left(\sum_{k=1}^{t}w_{jk}y_j + b_k\right) - z'_k\right]^2 = \frac{1}{2}\sum_{k=1}^{m}\left[f\left(\sum_{k=1}^{t}w_{jk}\left(\sum_{j=1}^{n}w_{ij}a_i + b_j\right) + b_k\right) - z'_k\right]^2 \tag{3}$$

The network error is a function of the weights wij and wjk. Therefore, the error E can be changed by changing the weight of the neuron, thus:

$$\Delta w_{ij} = -\varepsilon\frac{\partial L}{\partial w_{ij}}(i = 1\ldots m, j = 1\ldots n) \tag{4}$$

$$\Delta w_{jk} = -\varepsilon\frac{\partial L}{\partial w_{jk}}(j = 1\ldots n, k = 1\ldots t) \tag{5}$$

where ε represents the rate.

Teachers are the subject and object of performance management in universities. The main role is that teachers can participate in the formulation and implementation of performance management system, while the object role is that teachers are the managed objects. This dual identity makes teachers occupy a key position in the application of new technologies, so performance management based on big data puts forward requirements for teachers, and teachers must have informatization awareness, so as to stably carry out informatization performance management [16]. The cultivation of educational technology talents within universities requires a strong emphasis on scientific research. Therefore, it necessitates the integration of high-quality scientific research resources and the establishment of an open network management system to allow scientific researchers and experimental instruments to be shared on an open network platform, facilitating dynamic research on teaching management and developing various objectives and evaluation indicators for performance assessment, to realize the flattening, dynamic and integrity of performance evaluation management [17].

## 5 Conclusion

The performance evaluation and analysis system of university teachers needs to be continuously improved according to the practice results [18]. This paper introduces the importance of university performance management evaluation to university management, analyzes the impact of big data on university teachers' performance management under the background of big data, and puts forward the step plan for building university teachers' performance data warehouse, providing reference for future performance management of universities:

1) Collect basic data, including the performance evaluation of talent training informatization, the performance evaluation of university scientific research capacity informatization application and the evaluation of university comprehensive service informatization;
2) Integrate information resources and information of various business systems;
3) Gradually implement the top-level design and coordinate the informatization construction at the school level;
4) Enhance teachers' information literacy, promote teachers to accept the performance management system and apply it in teaching and management.

**Acknowledgment.** This project is supported by the 2022 private higher experts project of the 14th five-year plan of Guangxi educational science (2022ZJY3220) and the 2021 Professor Training project of Nanning University (2021JSGC15, 2021JSGC08).

## References

1. Tsen, Y.-J.: An exploratory analysis in the construcion of college performance indices. Int. J. Organizat. Innov. **5**(3), 98–132 (2013)
2. Yan, X.: University performance evaluation management platform based on big data. Microcomput. Appl. **033**(012), 3–6 (2017)
3. PeiJiayin, X., Hanwen, L.Y.: Business curriculum reform and teaching practice based on OBE concept and data mining. Int. Public Relat. **106**(10), 124–125 (2020)
4. Jian, S.: The design and implementation of hospital personnel salary management sysem. Electron. Technol. Software Eng. **175**(05), 203–205 (2020)
5. Youxing, X., Huilin, H.: Evaluation of the business performance of listed companies in my country's big data concept sector. Data Min. **007**(001), 16–25 (2017)
6. Donghui, C.: Reform and innovation to promote performance management, and actual results are shown. Finan. Super. **484**(22), 16–18 (2020)
7. Ma, L., Zhou, X.: Research on the evaluation mechanism of innovation and entrepreneurship team management based on data mining classification. Modern Electron. Technol. **042**(011), 178–180+186 (2019)
8. Yi, Z.: Enterprise human resource performance management innovation in the era of big data. Shangxun 146(13): 100+102 (2018)
9. Dongying, C.: Research on the application of decision tree data mining technology in the performance evaluation of public utilities management. Digit. Des. **6**(09), 265–266 (2017)
10. Amjad, M., Linda, N.J.: A web based automated tool for course teacher evaluation system (TTE). Int. J. Educ. Manage. Eng. (IJEME) **10**(02), 11–19 (2020)

11. Zhu, X., Zhu, Z.: An application of the wiles test in the formulation of education strategy for the newly-upgraded colleges of China. Int. J. Mod. Educ. Comput. Sci. (IJMECS) **3**(02), 15–21 (2011)
12. Dafeng, C., Haiyong, C.: Analysis of information system trend auditing under big data environment. Finan. Account. Monthly **861**(17), 118–125 (2019)
13. Hongrun, G.: Research on human resource assessment management system based on data mining. Enterp. Reform Manage. **385**(20), 57–59 (2020)
14. Zhao, L.: A preliminary study on enterprise human resource performance management in the big data era. China Manage. Inform. Technol. **22**(08), 72–73 (2019)
15. Jingru, Z.: Research on enterprise human resource performance management innovation in the big data era. China Manage. Inf. Technol. **022**(011), 70–72 (2019)
16. Xinxin, W.: Discussion on the construction of enterprise human resource performance management system in the era of big data. Times Finan. **786**(32), 101–103 (2020)
17. Patil, M.M., Hiremath, B.N.: A systematic study of data wrangling. Int. J. Inf. Technol. Comput. Sci. (IJITCS) **10**(01), 32–39 (2018)
18. Sharma, K., Marjit, U., Biswas, U.: PTSLGA: a provenance tracking system for linked data generating application. Int. J. Inf. Technol. Comput. Sci. (IJITCS) **7**(04), 87–93 (2015)

# Curriculum Evaluation Based on HEW Method Under the Guidance of OBE Concept

Chen Chen[1(✉)] and Simeng Fan[2,3]

[1] School of Business Administration, Wuhan Business University, Wuhan 430056, China
15720381@qq.com
[2] Department of Education, Anyang University, Anyang 14028, South Korea
[3] Department of Environmental Art and Design, Hubei Institute of Fine Arts,
Wuhan 430204, China

**Abstract.** OBE (Outcomes-Based Education) concept emphasizes learning achievement-oriented, and combining OBE concept to construct curriculum quality evaluation system is the basic activity of reverse design and forward implementation in OBE. The construction and evaluation of curriculum index system is a complex and systematic engineering. The HEW (Hierarchy Entropy Weight) Method evaluation method is designed by combining the hierarchical structure of AHP (Analytic Hierarchy Process) and EWM (Entropy Weight Method). A curriculum quality evaluation model suitable for the concept of OBE is constructed. Based on the three basic features of OBE concept, namely, reverse design, forward implementation and learning outcomes, an evaluation system with 27 indicators of 3 layers is constructed. EWM was used to convert experts' opinions into evaluation weights of index. In the evaluation system, learning outcomes are critical index of curriculum evaluation, which includes goal achievement, achievement quality, and feedback. In the reverse design process, the level of teaching team, the rationality of teaching objectives and teaching organization are the most important concerns. In the forward implementation process, students' attitude to learning is the most important factor. Under the student-centered concept, teacher needs to use various methods and means to stimulate students' inner learning drive.

**Keywords:** Curriculum evaluation · OBE · Entropy Weight Method

## 1 Introduction

Curriculum evaluation is an important content to guarantee the quality of teaching implementation, which has been widely paid attention by teaching managers and teachers. Under the concept of OBE, the curriculum is based on learning results to complete reverse design and forward implementation, so it is more important to establish a scientific curriculum evaluation system. Redesigning the index system of curriculum quality evaluation according to the concept of OBE, and setting the scientific weight for evaluation index, thus carrying out objective evaluation of curriculum, is not only the basis of the reverse design of OBE, but also the key of the teaching design in the forward implementation process.

In the process of course evaluation, scientific and reasonable evaluation system is the most important thing. The evaluation system is like a commanding baton, which will affect the whole process of curriculum design and implementation under the concept of OBE. In the evaluation system, it is necessary to make subjective selection and judgment depending on the experts, but also to guarantee the objectivity of the system construction, so it is necessary to choose scientific methods to establish the curriculum evaluation system under the concept of OBE, which plays an important role in enriching the teaching theory and promoting the teaching practice.

## 1.1 OBE

### 1.1.1 Correlation Coefficient Derivation

OBE was first put forward by Spady scholars in 1981 and explained as "the design of every activity in the education system should be based on the final learning result" [1]. OBE includes three core concepts: student-centered, reverse design and continuous improvement. In the modern information society with rapid change of talent demand, the role of curriculum is to create a formal knowledge about ideation and process of start-up, university curriculum invest more time to guide students, but the effect is not ideal [2]. Higher education needs to be more connected with social needs and training goals. Under such background, more and more educational scholars pay attention to the educational concept of OBE.

The OBE philosophy emphasizes the focus on learners' actual learning needs, clearly focus on expected learning outcomes, and focus on learning outcomes. Following the principle of reverse design and forward implementation, the course structure is constructed, the teaching is designed and implemented, and the dynamic adjustment is realized by evaluation, which ensures the achievement of expected learning results and the cultivation of students' ability. His implementation consists of four steps: Defining - Realizing - Assessing - Using [3]. Assessing is a connecting step from the top to the bottom, from the reverse to the forward link, scholars have done a lot of research on curriculum evaluation. Osman discussed the perceptions of the undergraduate students from Civil and Structural Engineering Department who have undergone their industrial training program, in order to assurance to the effectiveness of OBE [4]. Alghamdi determine a learning strategy using an Outcomes-based Education approach for E-Business Strategies and redesign the curriculum using an Active Learning Approach [5]. Premalatha describes the framework of OBE and detailed survey on CO (Course Outcomes)-PO (Program Outcomes) mapping and its attainment models [6].

Under the background of OBE teaching concept, course assessment should pay attention to the core characteristics of OBE, combined with the teaching process of forward design and reverse implementation, design index of response evaluation system, and analyze the weight of the index to guide the process of continuous improvement.

### 1.1.2 EMW

EMW is a theoretical mathematical method, which originates from thermodynamicsand is introduced by Shennong. It is a measure to describe the uncertainty of system states. From the point of view of information theory, information is a measure of the degree of order and entropy is a measure of the degree of disorder. According to this property, the smaller the information entropy, the smaller the degree of information disorder and the larger the value of useful information. The larger the index weight, the larger the entropy weight, and the opposite, the smaller the entropy weight. According to the viewpoint of information theory, the weight of the index is calculated according to the information amount carried by each index. The varying degree of each factor in the index depends on the information amount. The larger the variation degree of each factor, the larger the information amount carried by the index. That is, the greater the difference, the greater the weight of the counter. Conversely, the smaller the weight, the weight will contain the information itself characteristics, has a certain objectivity, such a calculation method applied in subjective evaluation method can make the evaluation have a certain objectivity.

In the process of project evaluation and target decision-making, appraisers often consider the relative importance of evaluation indicators and reflect the weight of indicators. According to entropy, the quality of information obtained in decision-making process will directly affect the precision and reliability of decision-making. Entropy weight method is an objective value assignment method, based on information quality, the entropy weight of each index is calculated by information entropy, and then the weight of each index is calculated by entropy weight. Among all kinds of subjective evaluation methods for analyzing project scheme, entropy weight method has the characteristics of high precision, strong objectivity and good evaluation. It is widely used in many fields such as scheme selection [7], algorithm optimization [8], safety production [9].

### 1.1.3 Curriculum Evaluation

Hierarchical structure can help the decision-maker to make more reasonable evaluation under the framework of the hierarchy of specific problems, and make the evaluation problems of multiple indexes and multiple factors more systematic and scientific. The hierarchical index system is the choice of many scholars in the evaluation process [10–13].

At the same time, in the course quality evaluation, Liu chooses an innovative perspective that considers course evaluation as a multiple criteria decision-making problem, they proposed a model with AHP and FIS (fuzzy inference system) to measure the course effectiveness regarding various indicators [12]. Nielsen studied the relationship between teachers' and students' styles and course evaluation, and found that their styles, especially for complex content, are not compatible, which may cause teachers' and students' different evaluations of curriculum [14]. Irina Integrated Application of Multiple Criteria Decision Making (MCDM) and Analytical Hierarchy Process Fuzzy (AHPF) Method, design a new methodology for evaluating the quality of distance learning courses [15]. Maarten analyzed over 3000 courses taught in a large European university and found that the traditional use of student evaluations to evaluate courses may not be a good choice, they found that course evaluations is upward biased, and that correcting for selection bias has non-negligible effects on the average evaluation score and on the evaluation-based ranking of courses [16]. From the research of scholars, it is a complicated problem how to evaluate the curriculum from an effective angle, and different quality indicators will analyze the curriculum quality from different perspectives such as the objective and personnel.

## 2 Hierarchy Entropy Weight Method Model

### 2.1 Hierarchy Entropy Weight Method Model Process

The HEW model mainly consists of four steps. First of all, establish the hierarchical assessment model, and then introduce the entropy as the weight of the evaluation system when determining the weight of assessment criteria. The operation processes are shown in Fig. 1.

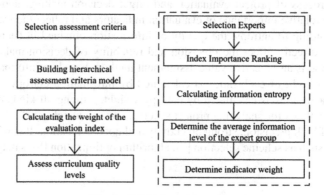

**Fig. 1.** HEWM model process

1) Considering the feasibility and systematicness of the evaluation index, the course assessment criteria are selected according to the essential characteristics of OBE.
2) Combined with the hierarchical model of AHP method, the indexes of assessment are hierarchically constructed.
3) Experts in the field are selected, the importance evaluation of the index weights is collected, and the index weights are determined by the information entropy of the evaluation, and the evaluation system is constructed.
4) The evaluation index system is used to evaluate the course.

### 2.2 Assessment Criteria for Curriculum Under OBE Concept

The first step of constructing evaluation system is to select evaluation indicators. The existing mature evaluation structures such as CIPP (context, input, process, product) education evaluation indicators and hybrid learning evaluation cannot fully reflect the core characteristics of OBE, so we need to design a more targeted index system for OBE. The curriculum evaluation index for OBE should include the following three parts, reverse design, forward implementation evaluation and outcome and output evaluation.

### 2.2.1 Reverse Design Index

The core of the concept is to carry out the reverse design of the curriculum based on the objectives and expected learning outcomes. The whole environment and concept starting from teaching design must reflect the core of the reverse design based on learning outcomes. For example, objective of teaching design, teaching organization and arrangement, and learning resource design should support the curriculum design and implementation around OBE.

### 2.2.2 Forward Implementation Evaluation

The forward implementation of hybrid curriculum should take teachers and students as the center, and carry out online and offline teaching activities according to the teaching scheme formulated in the reverse design stage, so as to realize the learning output. Therefore, we should pay attention not only to the process evaluation of students, but also to the performance of teaching process, such as teaching attitude, teaching activities and teaching means application. Teachers and students, as two main subjects in teaching activities, are bound to have interactive behavior. The essence of classroom teaching is the process of dialogue between teachers and students. Therefore, the frequency, time and depth of interaction between teachers and students in teaching should also be the important content of course design and evaluation.

### 2.2.3 Outcome and Output Evaluation

The evaluation of learning output is based on the comprehensive evaluation of the expected and non-anticipated achievements of the course, and further analysis of the achievement of the curriculum objectives, so as to understand the learning effect of the curriculum comprehensively. It is used to reflect the development of knowledge, ability and comprehensive quality of students and the application of learning results in the curriculum of learning. Teachers reflect on the problems in the course design and implementation, and improve the curriculum continuously.

## 2.3 Hierarchical Assessment Criteria

According to the evaluation index of OBE curriculum, combining systematicness and measurement, the index system is constructed according to the hierarchical structure, which includes three levels of index system, as shown in Table 1.

**Table 1.** Assessment criteria for OBE curriculum

| Level 1 | Level 2 | Level 3 |
|---|---|---|
| $F_1$: Reverse design | $S_{11}$: Teaching Objectives | $T_{111}$: Feasibility |
| | | $T_{112}$: Developmental |
| | | $T_{113}$: High-order |
| | $S_{12}$: Teaching programme | $T_{121}$: Design Concept |
| | | $T_{122}$: Teaching Content |
| | | $T_{213}$: Application of teaching methods |
| | | $T_{124}$: Teaching Method |
| | $S_{13}$: Learning Resources | $T_{131}$: Teaching Team |
| | | $T_{132}$: Hardware Condition |
| | | $T_{133}$: Software Resources |
| $F_2$: Forward implementation | $S_{21}$: Teaching Organization | $T_{211}$: Teaching Plan Execution |
| | | $T_{212}$: Teaching Activity Organization |
| | | $T_{213}$: Application of Teaching Methods |
| | | $T_{214}$: Teaching Attitudes |
| | $S_{22}$: Teaching atmosphere | $T_{221}$: Teacher-student Interaction |
| | | $T_{222}$: Student–student Interaction |
| | $S_{23}$: Learning Performance | $T_{231}$: Classroom Learning Attitude |
| | | $T_{232}$: After-Class Time |
| $F_3$: Learning outcomes | $S_{31}$: Target achievement | $T_{311}$: Knowledge Acquisition |
| | | $T_{312}$: Capability Development |
| | | $T_{313}$: Value Improvement |
| | $S_{32}$: Curriculum Satisfaction | $T_{321}$: Satisfaction with the Curriculum |
| | | $T_{322}$: Satisfaction with Teachers |
| | | $T_{323}$: Satisfaction with Self-improvement |
| | $S_{33}$: Quality of results | $T_{331}$: Scientific |
| | | $T_{332}$: Innovative |
| | | $T_{333}$: completeness |

## 2.4 EWM Process

EWM combines the subjective evaluation of experts and the objective entropy method to determine the index weight. The main steps are as follows.

Firstly, the ranking opinions of the expert group are collected, and the ranking data set $(a_{i1}, a_{i2}, ..., a_{in})$ of the expert group on the importance degree of a certain indicator set is $C = (c_1, c_2, ..., c_n)$ assumed to be a ranking matrix $A = (a_{ij})_{k \times n}$ of the expert group, where $a_{ij}$ is an evaluation of the expert $i$ on the indicator $j$.

In order to eliminate the uncertainty of experts in the ranking process, the blindness analysis is carried out and the total cognition degree is calculated. The transformation entropy function is used as formula (1), where $I$ is the importance ranking value given by the expert, $m = j + 2$ is the number of transformation parameters, and $j$ is the actual maximum order number.

$$\mu(I) = \frac{\ln(m - I)}{\ln(m - 1)} \tag{1}$$

The values in the sorting matrix $A = (a_{ij})_{k \times n}$ are brought into the formula (1) to perform corresponding transformation, where the membership matrix $B = (b_{ij})_{k \times n}$ of the sorting matrix $A$ is obtained.

Assuming that each expert has the same discourse power, the average degree of cognition of the index $b_j = \frac{\sum_{i=1}^{k} b_{ij}}{k}$ is assumed.

Define the blindness of experts' perceptions of indicators $c_j$ arising from uncertainty as formula (2)

$$Q_j = \frac{[\max(b_{1j}, b_{2j}, \ldots, b_{kj}) - b_j] + [b_j - \min(b_{1j}, b_{2j}, \ldots, b_{kj})]}{2} \tag{2}$$

The evaluation vectors $X = (x_1, x_2, \ldots, x_n)$ of the experts on the indicators $c_j$ are obtained, where $x_j = b_j(1 - Q_j)$.

Normalize the indicators to obtain the weight matrix $W = (w_1, w_2, \ldots, w_n)$, where $w_j = \frac{x_j}{\sum_{j=1}^{n} x_j}$.

## 3 Curriculum Evaluation Under OBE Concept

### 3.1 Expert Selection

Six teaching experts were invited to evaluate the importance of curriculum indicators. All of them had senior titles or positions and had been engaged in teaching or management for more than 15 years, including both teaching managers and first-line teachers, which ensured the comprehensiveness and objectivity of evaluation.

If the number of indicators is too large, experts' relative importance may be affected. According to the indicator system, the indicators at different levels are grouped into five groups: $F_1$–$F_3$ (three indicators), $S_{11}$–$S_{33}$ (nine indicators), and $T_{111}$–$T_{133}$ (ten indicators). There are eight indexes for $T_{211}$ to $T_{232}$ and nine indexes for $T_{311}$ to $T_{333}$. Experts are invited to rank the relative importance of the indicators in the group. If the indicators are not different, they can rank the indicators in the same order.

### 3.2 Weight of Assessment Criteria Base on EWM

Collect the original evaluation and evaluation of experts and sort it out to form an importance ranking matrix. The importance of indicators is shown in the Table 2.

**Table 2.** Assessment criteria weight and ranking table

| Level 1 | Level 2 | Level 3 | relative weight | weighting | Ranking |
|---|---|---|---|---|---|
| $F_1$: Reverse design (Relative weight: 0.330) | $S_{11}$: Teaching Objectives (0.282) | $T_{111}$: Feasibility | 0.460 | 0.0427 | 7 |
| | | $T_{112}$: Developmental | 0.307 | 0.0285 | 19 |
| | | $T_{113}$: High-order | 0.233 | 0.0217 | 26 |
| | $S_{12}$: Teaching programme (0.401) | $T_{121}$: Design Concept | 0.253 | 0.0336 | 15 |
| | | $T_{122}$: Teaching Content | 0.258 | 0.0342 | 14 |
| | | $T_{213}$: Application of teaching methods | 0.306 | 0.0406 | 11 |
| | | $T_{124}$: Teaching Method | 0.182 | 0.0241 | 22 |
| | $S_{13}$: Learning Resources (0.318) | $T_{131}$: Teaching Team | 0.465 | 0.0488 | 3 |
| | | $T_{132}$: Hardware Condition | 0.304 | 0.0319 | 16 |
| | | $T_{133}$: Software Resources | 0.231 | 0.0242 | 21 |
| $F_2$: Forward implementation (0.236) | $S_{21}$: Teaching Organization (0.456) (0.456) | $T_{211}$: Teaching Plan Execution | 0.276 | 0.0297 | 18 |
| | | $T_{212}$: Teaching Activity Organization | 0.362 | 0.0390 | 12 |
| | | $T_{213}$: Application of Teaching Methods | 0.360 | 0.0388 | 13 |
| | | $T_{214}$: Teaching Attitudes | 0.344 | 0.0220 | 25 |
| | $S_{22}$: Teaching atmosphere (0.270) | $T_{221}$: Teacher-student Interaction | 0.433 | 0.0277 | 20 |
| | | $T_{222}$: Student–student Interaction | 0.222 | 0.0142 | 27 |

(*continued*)

**Table 2.** (*continued*)

| Level 1 | Level 2 | Level 3 | relative weight | weighting | Ranking |
|---|---|---|---|---|---|
| | $S_{23}$: Learning Performance (0.274) | $T_{231}$: Classroom Learning Attitude | 0.632 | 0.0408 | 10 |
| | | $T_{232}$: After-Class Time | 0.368 | 0.0238 | 23 |
| $F_3$: Learning outcomes (0.433) | $S_{31}$: Target achievement (0.478) | $T_{311}$: Knowledge Acquisition | 0.374 | 0.0774 | 2 |
| | | $T_{312}$: Capability Development | 0.407 | 0.0845 | 1 |
| | | $T_{313}$: Value Improvement | 0.219 | 0.0454 | 5 |
| | $S_{32}$: Curriculum Satisfaction (0.271) | $T_{321}$: Satisfaction with the Curriculum | 0.393 | 0.0463 | 4 |
| | | $T_{322}$: Satisfaction with Teachers | 0.254 | 0.0300 | 17 |
| | | $T_{323}$: Satisfaction with Self-improvement | 0.352 | 0.0415 | 8 |
| | $S_{33}$: Quality of results (0.250) | $T_{331}$: Scientific | 0.411 | 0.0447 | 6 |
| | | $T_{332}$: Innovative | 0.379 | 0.0412 | 9 |
| | | $T_{333}$: completeness | 0.210 | 0.0229 | 24 |

## 3.3. Characteristics of Curriculum Evaluation Indexes Under OBE Concept

## 4 Examples of Curriculum Evaluation

### 4.1 Curriculum Evaluation

The evaluation of the course is carried out in logistics management specialty of W university. W University was founded in 1985, is a local, applied undergraduate university in the middle of China. The W university has more than 10 thousand students, and the logistics management specialty, which is a first-class major in Hubei Province, with more than 500 students.

The course of "Logistics System Analysis" was selected for evaluation. This course is the basic course of logistics management major. It is opened in the 5th semester, with 48 h and 3 credit points. In assessment, 2 teaching management personnel, 2 professional teachers and 5 students who have selected the course are invited. Each item in evaluation system, form $T_{111}$ to $T_{333}$, is scored from 1 to 5 points, with a full score of 5. The arithmetic mean value is taken for the evaluation staff of the same type. The total score of the three categories is 40% for teaching management staff, 40% for professional teachers and 20% for students respectively.

The final evaluation score of Logistics System Analysis is 4.296, among which the average evaluation score of teaching managers is 4.317, the average evaluation score of professional teachers is 4.425 and the average evaluation score of students is 3.996. The course has been reformed by OBE, and the curriculum syllabus based on OBE concept has been developed, and the teaching is carried out according to the curriculum syllabus, and the quality evaluation of OBE concept has obtained a high score. At the same time, the evaluation result of the curriculum reform is excellent, which also proves the effectiveness of the evaluation result of the curriculum reform.

### 4.2 Characteristics of Curriculum Evaluation Under OBE Concept

Learning results are the most important factor in evaluation, which means that the results are the core goal of OBE. The second important reverse design process and the lowest positive implementation process indicate that more energy should be invested in reverse design of courses under OBE, especially the design and presentation of learning results, which is the most important index to evaluate the quality of courses.

It should be noted that in the learning outcomes part with the highest importance, the weight of each factor is increased, and the target achievement, including the knowledge attainment and ability development of students, are the most important factors affecting the curriculum quality, which fully reflects the core essence of OBE, which also indicates that teachers' teaching is not only the completion of the whole teaching process, but also the significance and final achievement of the teaching activities.

On the other hand, in the second important reverse design part, teaching team, teaching goal, teaching organization and teaching content are the key indicators that need to pay attention to, which indicates that the concept of OBE teaching design actually puts forward better requirements for the teaching team, and teachers need to fully understand its connotation from the teaching concept, and complete the deep-level transformation of teaching objective and teaching organization on this basis.

At last, students' attitude to learn, teacher's activity organization and means application are the two most important indicators in the process of curriculum implementation. Student's input is the most important part of the indicators, and teachers are more like the directors of teaching activities, so they should promote and stimulate independent learning ability of student, and use various means and methods to guide and promote their positive attitude to learn.

It is worth noting that in the course of evaluation of Logistics System Analysis, students' scores of $T_{311}$-$T_{313}$ are obviously lower than those of teaching managers and professional teachers. Through the interviews with students, we know that students are actually quite satisfied with the overall course, but not satisfied with their own input, so they are not satisfied with the value enhancement and ability development they have gained. This dissatisfaction is due to their unsatisfactory to themselves. It can be seen that students are eager for self-improvement, which also promotes teachers to stimulate student's self-dissatisfaction in the teaching process and turn them into positive learning attitude and behavior.

## 5 Conclusion

In order to better adapt to the scientific evaluation of courses under the concept of OBE education, a quality evaluation model of curriculum based on HEWM was constructed. Firstly, the evaluation Indicators were organized more clearly by using the hierarchical structure of AHP, and then experts were invited to perform the weight analysis on the indexes. The weights of the indexes are determined by EWM method, and the scientific evaluation of the curriculum is carried on.

1) An evaluation system based on the concept of OBE, which includes 27 specific indexes of 3 levels, is established. The index system reflects the core characteristics of OBE forward design, reverse implementation and emphasis on learning outcomes.
2) The HEW Method firstly establishes the hierarchical index content, and then converts the expert's subjective opinion into objective entropy value by EWM in order to establish the scientific weight of indicators.
3) In the evaluation system, learning outcomes, as core of OBE concept, become critical content of evaluation, from the side of the side, it shows that evaluation indicators embody the core of OBE concept.
4) In the reverse design process of curriculum construction, teachers as the main designers, the level of teaching team, the rationality of teaching objectives and the organization of teaching is the most important index, while in the forward implementation process, the cooperation between students and teachers is emphasized, among which students' attitude to learning is the most important factor, and teachers need to use various methods to stimulate intrinsic learning drive of student.

**Acknowledgments.** This project is supported by the Ministry of Education Collaborative Education Project "Cultivation Effect of Logistics Management Professionals Based on OBE Concept" (220905522223325).

## References

1. Spady William, G.: Outcome-based education: critical issues and answers. Arlington: American Association of School Administrators, p. 212 (1994)
2. Wiradinata, T., Antonio, T.: The role of curriculum and incubator towards new venture creation in information technology. Int. J. Educ. Manage. Eng. (IJEME) **9**(5), 39–49 (2019)
3. Acharya, C.: Outcome-based education (OBE): a new paradigm for learning. Centre Dev. Teach. Learn. **7**(3), 7–9 (2003)
4. Osman, S.A., Zaidi, M., Mat, K., et al.: Outcome based education (OBE) curriculum assessment for industrial training program: based on students' perception. New mark. Res. J. **12**(6), 454–463 (2009)
5. Alghamdi, T., Jamjoom, A.: Developing E-business strategies curriculum case study in the information systems department. Int. J, Mod. Educ. Comput. Sci. (IJMECS) **4**(2), 1–7 (2012)
6. Premalatha, K.: Course and Program Outcomes assessment methods in outcome-based education: a review. J. Educ. **199**(3), 111–127 (2019)
7. Bruno, M.S., Leoni, P.G., Lucila, M.S.C.: Performance evaluation of green suppliers using entropy-TOPSIS-F. J. Clean. Prod. **207**(10), 498–507 (2019)

8. Susan, S., Ranjan, R., Taluja, U., Rai, S., Agarwal, P.: Neural net optimization by weight-entropy monitoring. In: Verma, N., Ghosh, A. (eds.) ICCOML: International Conference on Computational Intelligence, pp. 201–213. Springer, Singapore (2019). https://doi.org/10.1007/978-981-13-1135-2_16
9. Xiangxin, L., Kongsen, W., Liwen, L., Jing, X., Hongrui, Y., Chengyao, G.: Application of the entropy weight and TOPSIS method in safety evaluation of coal mines. Procedia Eng. **26**, 2085–2096 (2011)
10. Asli, L., Gokturk, S.: Quality evaluation of graduates. Soc. Behav. Sci. **70**(25), 1009–1015 (2013)
11. Jianting, S., Chunyou, D., Jiuyang, H.: Research and design of teaching evaluation system based on fuzzy model. Int. J. Educ. Manage. Eng. **2**(10), 45–51 (2012)
12. Yan, L., Xin, Z.: Evaluating the undergraduate course based on a fuzzy AHP-FIS model. Int. J. Modern Educ. Comput. Sci. (IJMECS) **6**(12), 55–66 (2020)
13. Suartini, N.K.Y., Divayana, D.G.H., Dewi, L.J.E.: Comparison analysis of AHP-SAW, AHP-WP, AHP-TOPSIS methods in private tutor selection. Int. J. Modern Educ. Comput. Sci. (IJMECS) **1**(15), 28–45 (2023)
14. Nielsen, T., Kreiner, S.: Course evaluation for the purpose of development: What can learning styles contribute? Stud. Educ. Eval. **54**(9), 58–70 (2017)
15. Irina, V., Romualdas, K.: Methodology for evaluating the quality of distance learning courses in consecutive stages. Soc. Beh. Sci. **191**(7), 1583–1589 (2015)
16. Maarten, G., Anna, S.: Measuring teaching quality in higher education: assessing selection bias in course evaluations. Res. Higher Educ. **58**(7), 341–364 (2017)

# The Relevance of a Systematic Approach to the Use of Information Technologies in the Educational Process

Nataliya Mutovkina[1]([✉]) and Olga Smirnova[2]

[1] Department of Management and Social Communications, Tver State Technical University, Tver 170012, Russia
`letter-boxNM@yandex.ru`
[2] Institute of Economics and Management, Tver State University, Tver 170021, Russia

**Abstract.** The object of research in this work is modern information technologies and software, which is advisable to use for solving different tasks in the educational process. The authors focus on the positive aspects of the use of information technologies and consider them an integral part of the educational process. Digital competencies gained by students in the process of computer-based learning are necessary for their professional activities. The better students' information technology skills are developed, the more competitive they will be in the labor market. The article presents an overview of information technologies and software that are successfully used in teaching students. The paper provides examples of the use of modern information technologies to solve the problems of information retrieval, data analysis, and visualization. It is established that information technologies and software should be applied, considering the continuity of the educational material, as well as the interdisciplinary nature of the educational process. The novelty of the research lies in the proposal of a systematic approach to the use of information technologies in the educational process. Under the systematic approach, information technologies should ensure not only the continuity of educational material but also the continuity of each other.

**Keywords:** Educational process · Information technology · Software · System approach · Analysis tasks · Visualization · Digitalization

## 1 Introduction

Using information technologies and computer software in the educational process is an integral part of the digitalization of the education system. Some 15 – 20 years ago, the educational process in most Russian educational organizations was implemented only through classroom communication between teachers and students. The major tools of the teacher were a wooden board and chalk, and students made notes of lecture material and solved problems in notebooks. Now traditional teaching methods are gradually being replaced by interactive, remote methods, and on-line technologies. This does not mean that the blackboards and notebooks have sunk into oblivion. But now students are increasingly replacing ordinary paper notebooks with laptops or tablets. Some students even take notes on smartphones. Interactive whiteboards, computers, projectors, and other technical devices came to the teachers' aid. Each teacher can broadcast educational material using presentations displayed via a laptop and a projector on a special screen. Over the past five years, distance learning has gained undeniable importance.

Information technologies implemented in various software environments are actively used by students to solve educational tasks in various directions. Information technologies are simply irreplaceable for processing large amounts of data, their analysis and visualization, in solving optimization and forecasting problems. The results of solving educational tasks are also drawn up by students as reports and presentations, and then they are accepted for defense.

Thus, modern information technologies in the educational process are used not only as teaching tools, and organization of the educational process but also as tools used by students to solve educational problems.

However, many questions arise in this area. For example, choosing software products for solving educational tasks issues of continuity of information technologies in the educational process; the need to consider the interdisciplinary nature of educational material when using information technologies, and others. The primary task that the authors of this study solved can be planned: what should the complex information technologies used in the educational process be in order to ensure a high level of learning efficiency? The existing limitations should be considered: the cost of software; the complexity of mastering individual software products; the time required for training to work in separate software environments.

The purpose of the work is to study the possibilities of a systematic approach to the use of modern information technologies and software in the educational process.

## 1.1 Literature Review

The relevance of the use of information technologies in the educational process is because of the social need to improve the quality of education, as well as the practical need to use modern computer programs. The main purpose of using information technologies in the educational process is, first, to strengthen the intellectual capabilities of students in the information society, as well as individualization and improvement of the quality of education. This is stated in the works of many scientists, for example, in [1–5] and others.

In the article [6], the authors point out the main positive effects of the use of information technologies in the educational process. They include high-speed processing of large amounts of information, extensive visualization capabilities, multiple reproducibilities of educational material, time savings, and increased learning efficiency.

Some researchers are considering the possibilities of information technology for teaching a specific course, for example, mathematics [7].

The paper [8] considers the role of information technologies in creating business incubators and innovative enterprises based on educational organizations. In Indonesia, as in many other countries, information technology is used as a tool to activate and promote entrepreneurship and contribute to a smooth transition from training to real business.

In the economic sphere, the creation of digital companies is of increasing interest [9]. But the creation and operation of a digital company is impossible without a workforce with digital competencies.

In some works, the authors talk about the experience of using certain information technologies and software in the educational process. So, the article [4] tells about the InetRetriever tool, which students can use to get any necessary information from the Internet in real time. Students can also use InetRetriever to implement, execute and test their projects using real data. The article [1] demonstrates the technology of using the DirectumRX software product in the educational process of studying the disciplines "Content Management in an organization" and "Document management in an organization".

Software products such as Tableau [10], Microsoft Excel, STATISTICA [11], programs for building a decision tree: Deductor, See5, WizWhy [12], GIS technologies [13] and other software can be used for analysis and forecasting. There are choosing software tools. However, acceptable alternatives may be significantly limited.

### 1.2 Directions of Digitalization of the Educational Process

The main directions for digitalization of the educational process are the creation of an interactive learning environment in each educational organization with free access to scientific libraries and databases; introducing modern information technologies and software into the educational process to solve educational tasks that turn into professional tasks; the creation of digital educational organizations in which students will receive an education without leaving them at home.

Information technologies make it possible to implement such teaching methods as modeling and simulation work [6]. These methods are becoming increasingly common in the formulation and solution of educational tasks. For example, future economists solve tasks ranging from optimization and management of the company's activities to analysis of the country's foreign economic activity. Information technologies are used already in the first year of training. And here, a systematic approach is important. With this approach, the teacher ceases to be a translator of information, but becomes an intermediary between the student and the means of information technology, instilling skills in independent work.

The analysis of thematic publications and the results of the experience of using information technologies in educational organizations has shown that the greatest efficiency of the educational process can be got through the systematic use of information technologies in an optimally selected combination. When forming an information technology system in the educational process, it is necessary to consider its specifics, direction and planned learning outcomes, the level of the material and technical base of the educational organization and the level of training of the teaching staff.

## 2 A Systematic Approach to the Use of Information Technologies in the Educational Process

### 2.1 The Essence of the System Approach

Using a systematic approach in teaching involves the relationship between the components of the educational process, each of which can function with maximum efficiency, relying on internal connections in the educational system. The content of the studied material is one of the structural components of the educational process, the development of which is associated with the selected methods, forms, and means of teaching. The teacher manages this system, also being its component. The effectiveness of the educational system depends on which technologies the teacher will use [14].

A systematic approach to the use of information technologies in the learning process should not be replaced by inefficient "digitized" pedagogical practice, when a teacher withdraws from the educational process, for example, by transferring an electronic version of his lectures to students for independent study. Here, the student closes himself in his computerized environment, to the detriment of interactive and group forms of work, which reduces the effectiveness and the strategy of educational activity does not change.

A systematic approach that meets the needs of the information society assumes that university teachers and students become active "subjects of higher education" at the same time. The teacher acts as the subject of the organization of the educational process at the university, and the student is the subject of the educational process. In forming universal and professional competencies, a student in modern conditions becomes not only an object of training and education but also an initiative subject of educational activity. It is assumed that the manifestation of purposeful activity in personal formation and self-realization through a set of actions, and implementation ensures the ascent to the next stage of professional development [2].

### 2.2 Modern Information Technologies in the Educational Process

According to the system approach, information technologies in the educational process can be defined as a set of methods, hardware, and software for collecting, organizing, storing, processing, transmitting, and presenting information that expands students' knowledge and develops their capabilities to manage socio-economic, technical and other processes in the future.

Each information technology can be considered as a system comprising the elements shown in Fig. 1.

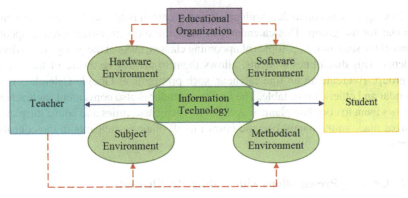

**Fig. 1.** Components of information educational technology

The hardware environment is a set of technical means used (computer, projector, peripherals, and so on). The software environment includes basic and application software. The subject environment is the content of a specific subject area, science, technology, and knowledge. The methodological environment is formed by the instructions for the use of the elements listed above.

Currently, the following information technologies can be successfully applied in the educational process.

### 2.2.1 Working with Web Resources

Students are invited to analyze Web resources on the themes offered by the teacher and those resources that the students themselves will find. At the beginning of the lesson, the teacher briefly talks about the problem and the possibilities of its solution, and makes a statement about the problem. Then students, using specific Web resources, carry out a thematic search, analysis and processing of information under the task. At the end of the lesson, students summarize the results got from several Web sources, and evaluate the effectiveness of Web resources. Such classes contribute to the development of students' critical thinking.

A variety of Web resources that can be used in the educational process are virtual excursions. Students can visit thematic exhibitions, museums, and research institutes without leaving the classroom, and get acquainted with the objects of their study at no costs.

On-line boards can be used for students to work in groups, for brainstorming, and for students to develop joint solutions. The digital approach to brainstorming allows you to better organize a discussion, discuss problems, and record and visualize intermediate and final results.

Web resources for organizing surveys allow the teacher to find out the opinions of students regarding the organization of the educational process, the availability of presentation of educational material. To organize on-line surveys, you can use Socrative, Google Forms, SurveyMonkey, Poll Everywhere and several other services. Based on the results he got, the teacher can react quickly and improve the quality of classes.

To keep students up to date with upcoming events, it is advisable to create an on-line calendar for the group. This calendar will mark the most important events, upcoming events of the student group, topics of upcoming classes, tasks of the group, and individual students. This disciplines students, allows them to always be aware of the events of the group. To create an on-line calendar, such programs as Google Calendar, Yandex. Calendar and others are suitable. The on-line calendar is also convenient for teachers, as it allows them to correlate planned events with their capabilities and adjust the plan. The changes made immediately become known to all other participants in the educational process.

### 2.2.2 Creating Presentations with Multimedia Elements

Lectures are conducted with the help of presentations of educational material. Presentations can be developed in software environments, for example, Microsoft PowerPoint, Photo Show PRO, Apple Keynote, Kingsoft Presentation, Impress, Hippani Animator, Google Slides, Piktochart, Zoho Show and others. The primary requirement for the software used is its accessibility for all participants in the educational process. The presentation file should be opened both on the teacher's computer and on the students' computers.

To effectively hold the attention of students and activate their interest in the topic, ordinary presentations comprising text and some diagrams no longer help. Presentations containing various media elements that can be turned on at the right moment are much more useful. These include images, podcast clips, pictograms, interactive graphics, sound effects, websites and others. Adding high-quality media objects makes lectures more attractive. The same applies to presentations of student reports prepared as part of practical classes or scientific conferences.

### 2.2.3 Educational Video

This information technology refers more to passive learning tools. However, it is very effective for mastering educational material if it is discussed at the end of watching the video.

There are extensive collections of educational videos on the Internet. Many of them are well structured, and have an internal search, which allows you to quickly find the video.

Well-chosen educational videos can positively affect the development of a student's competencies, memory, critical thinking, and ability to solve problems.

### 2.2.4 Conference Communication Technologies

These information technologies have become relevant to the COVID-19 pandemic. With the weakening of the pandemic, interest in them has not disappeared but even intensified. The educational process has gone beyond the classroom. Now teachers and students often use services such as Microsoft Teams, Skype, Zoom.

Besides conducting classes within the curriculum, videoconferences can be organized to communicate with experienced teachers, and experts from other countries, to conduct joint work with students from other educational organizations. Such communication allows students to get acquainted with new ideas, the opinion of eyewitnesses.

It is advisable to invite students to prepare questions on the videoconference in advance. This way they will feel like full-fledged participants of the video meeting and not passive viewers.

### 2.2.5 Virtual Simulators

Virtual simulators are indispensable for training specialists in medical, engineering, and natural science fields. However, more and more opportunities are opening up for their use in the training of economists [15], sociologists, psychologists, and teachers.

For example, the TeachLive virtual simulator, developed at the University of Central Florida, allows you to virtually immerse yourself in the reality of teaching a lesson, form practical skills related to lesson organization, and managerial decision-making in various pedagogical situations [16].

### 2.2.6 Use of Students' Mobile Devices

Smartphones and tablets are now available to every student. It's no secret that many teachers see a serious problem because students are often distracted by these devices during classes. Indeed, this interferes with the conduct of classes, the development of educational material. However, students' addiction to mobile devices can be turned to good by delivering educational content via mobile devices.

It is necessary to get used to this reality and use mobile devices for educational purposes. There are several educational platforms using students' mobile devices. These include Edap, Skill Pill, Bridge, Kitaboo Insight, SoloLearn, Quizlet and other information resources.

### 2.2.7 Information Technologies of Knowledge Assessment

Testing systems are mainly used to assess students' knowledge, which are embedded in the electronic information and educational environment of the university, for example, based on Moodle. Their major advantage is fast, convenient, impartial and automated processing of students' answers. But there is also a drawback, comprising the fact that the answer system does not allow the student to show his creative abilities.

Interactive knowledge monitoring systems, for example, PROClass, Mentimeter, act as an alternative to such test technologies [17].

## 3 Application of Information Technologies in the Course "Economic Theory"

The systematic application of these technologies, software and hardware makes it possible to form an integral educational complex that considers the cognitive characteristics of the student, because of the systematic accumulation and processing of data on the cognitive activity of students. The following are some examples of the use of information technologies listed in the second section of the article in the course "Economic Theory".

### 3.1 Use of Internet Resources

The following Web resources can be used in teaching economic theory:

1) Search engines in economics (for search, collection, extraction, processing, verification, analysis and storage of information and digital content to solve tasks).
2) Reference and legal databases (Guarantor, Consultant plus) for the search, analysis of regulatory documents in economics.
3) Cloud services (Google Drive, Yandex. Disk, Dropbox, etc.) for storing and sharing information.
4) Communication services (MS Team, Zoom, Skype, etc.).
5) Collaboration platforms (Trello, Miro).

An important addition to the educational process toward "Economics" is the analysis of information from the official websites of state authorities, the Central Bank of Russia and the application of this information in solving specific tasks.

### 3.2 Use of Data Analysis and Visualization Technologies

When conducting practical exercises, for example, when performing analytical tasks, it is advisable to supplement them with various visualization tools by preparing a dashboard [18].

Figure 2 shows a screenshot of the money supply analysis task in Google Data Studio. This dashboard is interactive.

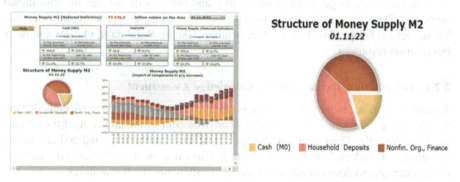

**Fig. 2.** The structure of the money supply in the Russian Federation

Using dashboards and visualization tools is relevant in all areas of professional activity, including personnel management, but in each of them, they will take a unique form adapted to the tasks of a particular business or project.

As part of the independent preparation for the business games, it is proposed to use the Miro platform. The Miro platform has a wide functionality with which students can perform their part of the team task both jointly and individually. In addition, the business game involves preparing the performance of teams and here it is possible to use the MS Teams platform, which has already become familiar during the distance learning period,

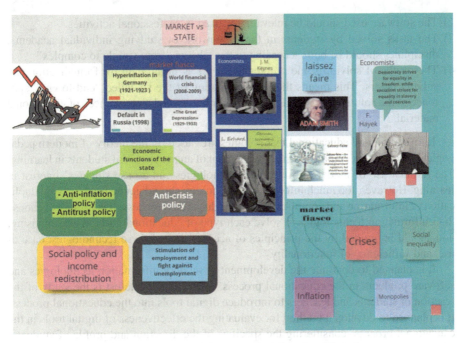

**Fig. 3.** "Market VS State" on the Miro virtual board

on which each team prepares a speech with the defense of its own position. Figure 3 shows a version of the task for the business game "Market VS State".

These tools are aimed at developing visualization and teamwork skills (soft-skills) and increasing its effectiveness.

### 3.3 Use of Knowledge Assessment Technologies

When conducting the test asks, it is advisable to use Google forms as a tool. This makes it possible to keep the test execution time under control, does not require printing out assignments on paper, and also reduces the complexity of checking assignments by the teacher and generates a summary statement based on the results of passing the test.

In order to conduct a survey, vote or receive feedback from the audience in real time, both at lectures and in practical classes, it is possible to use the service Mentimeter. This allows you to determine the mood of the audience, get a slice of residual knowledge, involve students in an active discussion and conduct a live dialogue with the audience, which makes classes more communicative and digital at the same time.

## 4 Summary and Conclusion

The educational process should be built in such a way that the following principles are fulfilled:

1) All educational tasks should, if possible, be solved with the help of modern software in order to save time and develop skills.

2) The software used should be relevant and used in professional activities.
3) The tasks solved by students in the framework of studying individual academic disciplines should be built according to the scheme "from simple to complex".
4) The results of solving some problems can act as initial data for formalization and solving others. This means the implementation of the so-called "end-to-end" approach to the solution, as well as ensuring continuity and "inheritance" of educational material.

Information and educational technologies are based on the theory of modern pedagogical theories that emphasize the development of an active role of students in learning. The rational use of modern information technologies can arouse students' interest in learning. However, too much information technology should not be used to avoid duplication of information and complication in the educational process. It is enough to choose 2−3 information technologies to solve each type of educational task. When choosing, you should be guided by the principles of accessibility of these technologies, ease of use, and sufficiency of functionality.

It is planned to continue the development of modern information technologies and software products in the educational process. Of particular interest is the study of the reaction of students and teachers to introduce digital tools into the educational process. It is planned to develop a method for evaluating the effectiveness of digital tools in the educational process, considering the specifics of the direction and profile of students' education.

# References

1. Alekseeva, T.V., Gubina, L.V.: Application of modern software products in the field of education. Baltic Hum. J. **3**(32), 24–28 (2020)
2. Shibaeva, N.A., Voronkova, L.V.: Digital technologies application in higher education as a social innovation of the modern information society. Drucker's Bulletin **2**, 70–80 (2020)
3. Dabas, N.: Role of computer and information technology in education system. Int. J. Eng. Techniques **4**(1), 570–574 (2018)
4. Mohamed, N., Al-Jaroodi, J., Jawhar, I.: Enhancing the project-based learning experience through the use of live WEB data. IJMECS **4**(11), 33–43 (2012)
5. Wang, J., Liang, H.: Discussion on domestic universities construction of digital teaching platform. IJEME **2**(6), 1–6 (2012)
6. Li, W., Zhou, R., Deng, P., Fang, Q., Zhang, P.: Construction of case teaching model for Management specialty supported by information technology. IJEME **2**(9), 44–48 (2012)
7. Chang-Xing, L.: Research and reflections on college mathematics teaching based on information educational technology. IJEME **1**(2), 43–47 (2011)
8. Wiradinata, T., Antonio, T.: The role of curriculum and incubator towards new venture creation in information technology. Int. J. of Educ. Manage. Eng. (IJEME), **9**(5), 39–49 (2019)
9. Al-Samawi, Y.: Digital firm: requirements, recommendations, and evaluation the success in digitization. Int. J. of Inf. Tech. Comput. Sci. (IJITCS) **11**(1), 39–49 (2019S)
10. Bibhudutta, J.: An approach for forecast prediction in data analytics field by Tableau software. Int. J. of Inf. Eng. Electr. Bus. (IJIEEB) **11**(1), 19–26 (2019)
11. Yakovlev, V.B.: Econometrics in Excel and Statistica: A Textbook. KnoRus, Moscow (2020)

12. Gabdulin, R.R., Lyaskovskaya, E.A., Korovin, A.M., Rets, E.A.: Forecasting demand in the market of road construction equipment using data mining. Bulletin of the South Ural State University. Ser. Computer Technologies, Automatic Control, Radio Electronics **22**(3), 117–131 (2022)
13. Shumakova, A.S., Vdovin, S.A., Ubozhenko, E.V.: Application of ArcGIS and ArcView software products to solve applied economic problems. Regul. Land Property Relations in Russia: Legal and Geospatial Support, Real Estate Valuat. Ecol., Technol. Solut. **3**, 213–218 (2021)
14. Fedulova, M.A., Karpov, A.A.: System approach in the design of educational process with the application of information technologies. In: The Collection: Science. Informatization. Technologies. Education. Materials of the XIII International Scientific and Practical Conference, PP. 416–419 (2020)
15. Reutov, V.Ye., Reutova, V.V., Kravchenko, L.A., Troyan, I.A.: Business simulation as an interactive method for training economists. Finance, Banks, Investments **1**, 162–171 (2021)
16. UCF Center for Research in Education Simulation Technology. University of Central Florida. https://sites.google.com/view/teachlive/home?pli=1. Accessed: 08 Jan 2023
17. Kravchenko, L.: Online resources that will decorate any activity. Mentimeter. https://novator.team/post/869. Accessed 09 Jan 2023
18. Smirnova, O.V.: To the question of updating the content of economic disciplines using digital technologies. Bulletin Tver State University. Series: Econ. Manage. **2**(54), 249–257 (2021)

# Construction and Practice of "CAD/CAM Foundation" Course Based on Learning Outcome

Ming Chang[1(✉)], Wei Feng[1], Zhenhua Yao[1], and Qilai Zhou[1,2]

[1] School of Materials Science and Engineering, Wuhan University of Technology, Wuhan 430070, China
chm@whut.edu.cn
[2] Faculty of Science, Shizuoka University, Shizuoka 422-8529, Japan

**Abstract.** This paper introduces the course design and reform of "CAD/CAM Foundation" based on learning outcome. This paper analyzes the challenges brought by the change of learning environment, industry development, students' learning intentions and habits, and the change of teaching philosophy under the current engineering education background. The design of the course starts from the course objectives. By refining the course objectives, ILOs (Intended Learning Outcomes) were set for the teaching content of each chapter and corresponding practical tasks, assignments, and interactive sessions were designed to examine ILO achievement. Through the quantitative evaluation of ILO, a detailed learning outcome evaluation mechanism was established, the process assessment was enhanced, and the comprehensive evaluation of knowledge objectives and capability objectives was realized. At the same time, it also introduces the teaching strategies and practical experience combining with students' learning characteristics. Through the outcome-oriented curriculum design and reform, the learning outcome has been improved, and the instructional design and product evaluation of the curriculum can be continuously improved in this way.

**Keywords:** CAD/CAM · Curricula construction · Learning outcome

## 1 Introduction

The ability to apply computer technology is an essential part of university education. With the rapid development and popularization of computer technology, the requirement of students' computer ability in application field, the character changes of pupils and the development of teaching concept, innovation in CAD/CAM course are required urgently, including the teaching objectives, course content, teaching organization, teaching strategies, etc.

Nowadays, few industries can operate without the aid of computer. More and more software were developed to meet the requirement of specific application or used as common platform software. Obviously, it is impossible to embrace all the software operation or theories of algorithm in CAD/CAM course. The aim of learning CAD/CAM technology is no longer to handle software but to have a clear understanding of CAD/CAM technology, and utilize CAD/CAM technology to analyze and solve problems [1]. CAD/CAM software should be defined as tools to realize the ideas of user. Students should have the ability to choose proper tools and grasp the usage of tools rapidly. Knowledge competence is not equivalent to competitive competence [2]. Curriculum objectives have become not only knowledge objectives, but also competence objectives [3]. And in some way, the latter will be increasingly important.

As to the pupils in nowadays, they are named as the original inhabit of internet. They grew up with the company of electronic products and the internet. The habit of get information via electronic products from internet was deeply ingrained [4, 5]. The rich of resources on internet provide the facility to quick study for students; meanwhile, inappropriate searching keyword and Lack of a complete knowledge framework may leads to misunderstanding. Furthermore, students become less patient to listen to a long speech; especially under the condition they can get the answer quickly via searching. They prefer to logical think rather than boring memory and deduction. These changes are conducive to the realization of curriculum objectives, but put forward higher requirements for curriculum organization. In addition, along with the spread of computer application in education, the uses of course website become normal state. Affected by the COVID epidemic in recent years, appropriate course construction and adaptive teaching organization become more and more important [6, 7].

Furthermore, the ideas of teaching also changed dramatically in recent years [8, 9]. With the gradual implementation of engineering education certification in china, the OBE (Outcomes-based Education) concept has been familiar to educators. And as the core of OBE concept, production-oriented, learning-centered and continuous improvement has been widely adopted in university education [10–12]. According to the curriculum objectives and the characteristics of the curriculum, production-oriented education can be realized in different path, such as project-based teaching, online learning and teaching, and even game-based teaching methods [13–16]. In the practice of OBE concept, the evaluation of learning production is the key point, and it is the foundation for continuous curriculum improvement. Some quantitative calculation methods of curriculum objectives were proposed [17, 18]. Obviously, the rationality of evaluation approach and criterion are the key to realize effective evaluation. Major of the authors attended engineering education certification in two years. The course "CAD/CAM basis" had been reconstructed and designed based on production-oriented approach. In this paper, the practice in the reform of "CAD/CAM basic" course was introduced.

## 2 Background and Course Analysis

### 2.1 Course Objective

As a course to establish students' basic knowledge of CAD/CAM, "CAD/CAM basic" is offered to sophomore students. After revisions in recent years, the curriculum objectives have been set as: a) Able to understand and deal with the relationship between human and computer; has clearly understanding of the purpose and significance of CAD/CAM application in professional field. b) Master the basic principles of CAD/CAM in the course and have the ability to apply these basic principles. c) Able to use CAD/CAM technique in the analysis of engineering application; has a certain sense of innovation and innovation ability. It can be seen that multiple objectives such as engineering ethics, knowledge ability and innovation ability were included.

## 2.2 Course Content

The main content of the course are listed in Table 1. It can be seen from the table that the course mainly contains basic theories of CAD/CAM, the application methods of basic principles and the practical application of CAD/CAM software.

**Table 1.** Main content of the course

| Chapter | Main knowledge points |
|---|---|
| Basic understanding of CAD/CAM system | Basic concepts, Man-machine relationship, System function, The relationship between operating system, supporting software and application software, Development of application software |
| Engineering data processing | Data storage, Data classification and processing method, |
| Graphic image processing | Basis of graphic transformation, Image processing |
| Geometric modeling | Geometric modeling method, Curves and surfaces, |
| Introduction of modeling software | Basic modeling function, Secondary development |
| CAM and its applications | CAM basis, Example of CAM application |

## 2.3 Learning Basis and Learning Effect of Previous Sessions

Before the course, a questionnaire survey about learning basis and learning willingness was conducted. The result was summarized in Table 2. It can be seen that most students have more or less access to CAD/CAM software. But few students can handle one or two CAD/CAM software deftly although most of them had tried to teach themselves. They are eager to master CAD/CAM technology, but they confuse mastering CAD/CAM technology with mastering software usage. Meanwhile, quick access to information via the internet did not help them much in their self-study. Boring commands, complex operations and learning without specific goals lead to poor learning results.

**Table 2.** Learning basis and learning willingness (survey results)

| Question | Majority answer |
|---|---|
| Do you ever have access to CAD/CAM software more or less? | Yes |
| Can you handle one or more CAD/CAM software expertly? | Cannot |
| Have you ever tried to learn CAD/CAM software through self-study? | Yes |
| What approaches have you used to learn CAD/CAM techniques? | Textbook or video tutorials |
| What do you want to learn from course? | Usage of one or two software |

Another survey was conducted on the performance of previous students in subsequent studies. Most of them have a good grasp of CAD/CAM basic knowledge and can use certain CAD/CAM software to complete certain tasks. But they didn't do as well when confronted with software they hadn't encountered before. This indicates that their understanding of CAD/CAM technology was not sufficient which results in the insufficient of learning ability.

## 3  Practice of Curriculum Reform

In production oriented learning, students are the main body of learning, so the design and implementation of learning should be carried out on this basis. The focus of this reform is to combine curriculum objectives with students' learning characteristics, fully mobilize their initiative, provide appropriate learning materials, conduct appropriate teaching organization, and carry out assessment in a reasonable way.

### 3.1  Curriculum Planning Based on Outcome Objectives

The course was reorganized in accordance with curriculum objectives. To ensure the objective evaluation of learning production, the curriculum planning was conducted logic is shown in Fig. 1.

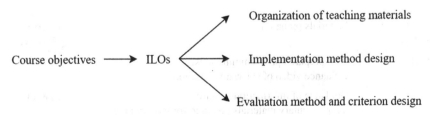

**Fig. 1.** The logic of curriculum planning

The relationship between the main content of the course and the course objectives were established via setting the intended learning outcomes (ILOs) of main content. The ILOs were written to describe the content to be learned, the level to be achieved and descriptions enacted to outcome [19]. The main content was divided into four categories. The ILOs setting and the course objective supported are listed in Table 3. The course objective can be subdivided into several ILOs according to the classification of content. In this way, the course objectives were broken down into smaller objectives, and the instructional design became more targeted.

**Table 3.** Main content, ILOs and Course objective

| Main content | ILOs | Course objective |
| --- | --- | --- |
| Cognitive information | Understanding of basic concept, consistence, importance or meaning | a) |
| Foundations (mathematical, computer skills and engineering knowledge) | Have a ground foundation to support the application of CAD/CAM | b) |
| Methodologies related to CAD/CAM | Understand how CAD/CAM works; Grasp certain Methodologies in CAD/CAM | b) & c) |
| Experiment section | To complete specific tasks; Extended learning and design | b) & c) |

According to ILOs, teaching materials were reasonably organized. In addition to the teaching materials of knowledge content, the extended reading materials, data reference guidance, basic knowledge-related exercise materials, and experimental guidance materials were organized according to the needs of ILOs. For example, experiments are responsible to the realization of course objects b) and c). Step by step approach has been adopted in the design of experiment [20], to gradually realize the purpose from the practice of theoretical content to independent learning, practice ability and creative thinking. Every experiment task contains three parts and materials prepared are listed in Table 4.

**Table 4.** Teaching material organization of experiment section

| Experiment content | Materials prepared | Course objective |
| --- | --- | --- |
| Part1 | Well prepared programming files, Guidance video of software operation | b) |
| Part2 | Guidance of information acquire, Supplementary materials prepared for inquiry purposes | b) & c) |

(*continued*)

Table 4. (*continued*)

| Experiment content | Materials prepared | Course objective |
|---|---|---|
| Part3 | Examples<br>Recommendation of design target | b) & c) |

Part 1 is the verification part, which mainly realizes the basic methods taught in the course through software. The part was designed to be easy to complete, which helps to increase students' interest and serves as basis for students to make their own design. So, well prepared programming files and guidance video of software operation were provided. The part 2 is advanced part, which can be realized by modify the given program to meet certain requirement. In this part, learners need to find and use methods which were not given in the example. Supplementary materials and guidance of information acquire were provided. The part 3 is a free design part and some recommended topics and examples were offered.

The implementation methods were divided as class lecture and interactive sessions, group discussion, after class reading, software practice, and so on. The adoption of implementation methods depend on concrete knowledge type and the learning characteristics of students. Along with the implementations, the evaluations of ILOs achievement were also designed to quantify the degree of achievement. For example, the results of the experiment were quantified from four aspects: design idea, implementation degree, discussion of results, and information acquisition and application. For other approach, the quantitative evaluations were also defined in the same way. After such setting, the evaluations of learning outcome can be more detailed and accurate.

## 3.2 Mobilize the Initiative of Students

Making students clearly recognize the outcome goal of learning is the basis for realizing learning-centered. The course objectives and ILOs of each chapter should be posted to students in class and course website. So does to the evaluation method. In this way, clear learning goals were established for students, as well as how to get a better score. Besides, a clear framework of knowledge should also be presented to students. This helps students to establish a complete knowledge system, master the correlation of knowledge, and then be able to analyze problems from the perspective of systems engineering. In practice, the knowledge frame was presented in the way of mind map.

In the learning process, the students' interest relate to challenge and motivation. Proper balance of challenge and motivation will keep students interested in learning. Too difficult challenges will greatly increase the probability of students to give up, while too easy will become boring and lose interest. In classroom teaching, teaching time should be allocated according to the type of content. Contents that need logical understanding should be focused, and specific implementation steps or procedures should be minimized. For example, when introducing the data processing method, the focus should be on the basic principle and characteristics of the method, while the program implementation of the method can be minimized or omitted. This is in line with the characteristics of students' learning interests. Too much detailed teaching will only make students feel bored and lose interest. Appropriate interaction such as questions and quizzes were adopted. With the convenience of the course website, these interactions can take place in real time to all or to individuals. In the experiment teaching, well prepared experimental guidance materials can help students to get the goal in few minutes and it is helpful to arouse the interest of students. At the same time, it also reduces the obstacles in program operation, command use, especially when using unfamiliar software. In the second and third parts of the experiments, it is necessary to give priority to several students who are progressing faster. It turns out that students are willing to mentor and share with each other when there are gaps in their progress. Especially in the third part, discussing and sharing experiences will further stimulate students' creative desire. In the study of Wang [21], this kind of learning environment would adjust students' learning approaches and study behaviors, regardless of individual differences. Besides, it is also helpful to review students' works. In the course, comments were made by teachers and other students. For students, this process can not only give them a sense of achievement but also gain new harvest [22].

### 3.3 Outcome Evaluation

In order to reasonably evaluate the learning outcome, a variety of evaluation approaches was adopted. As mentioned above, this reform sets ILO for every teaching links in the curriculum, and each ILO can be quantified based on the assessment of corresponding tasks. The data sources of evaluation included the learning performance in classroom and on course website, the completion of course assignments, the completion of experiment tasks and score of final exams, etc. The evaluation of learning process was strengthened and the criteria for evaluating were refined. The changes of evaluation are listed in Table 5.

**Table 5.** Changes of evaluation

| Items | Portion (previous) | Portion (present) | Details (previous) | Details (present) |
|---|---|---|---|---|
| Final exam | 70% | 50% | Knowledge mastery | Knowledge mastery, Ideas for application |
| Experiment | 20% | 30% | Process and Result | Design, Process, Discuss, Information acquire and usage |
| Learning process | 10% | 20% | Assignment | Assignment, quizzes, Learning performance |

In this way, the achievement of course objectives is quantitatively calculated and documented, which serves as the basis for course effectiveness analysis and continuous improvement.

## 4  Results and Discussion

After two years' curriculum construction and practice, the mode of teaching and evaluation of the curriculum has been fixed. The results of the achievement evaluation were used in continues improvement. The achievement of each course objectives and the achievement of each student were calculated and graphed. The results of last term are shown in Fig. 2. The achievement degree of each course goal exceeds 0.8, reaching the preset goal. It can also be clearly seen from the figure that the achievement of course objectives of some individual students is lower than 0.8. By analyzing these scores, why and what need to be improved can be found out.

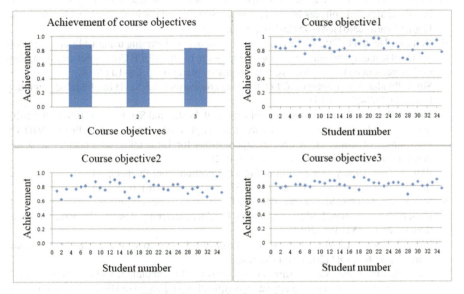

**Fig. 2.** Course objectives achievement

From the perspective of students' performance in the learning process, time spending on the course website increased about 20%, classroom teaching is more active than before, and students show more creative performance in the course, especially in the experimental section.

## 5  Conclusion

To realize the transformation of the curriculum to be outcome-oriented, learning-centered and continuous improvement, "CAD/CAM basis" course was redesigned. After two-year practice, following conclusions can be drawn:

1) The refinement of curriculum outcome objectives into specific ILOs is conducive to the effective quantification of curriculum objectives.

2) Organizing teaching resources around ILOs is conducive to the achievement of outcome objectives.
3) Curriculum planning aiming at learning production is conducive to continuous improvement of curriculum.

## References

1. Ye, X., Wei, P., Chen, Z., Cai, Y.: Today's students, tomorrow's engineers: an industrial perspective on CAD education. Comput. Aided Des. **36**(14), 1451–1460 (2004)
2. Hyland, T.: Competence, knowledge and education. J. Philos. Educ. **27**(1), 57–68 (1993)
3. Chen, J., Liu, D., Li, X., Zeng, D.: Curriculum system of construction management specialty in universities and colleges based on core knowledge and competence. J. Arch. Educ. Inst. Higher Learn. **23**(01), 41–45 (2014). (in Chinese)
4. Bere, A., Deng, H., Tay, R.: Investigating the impact of eLearning using LMS on the performance of teaching and learning in higher education. In: 2018IEEE Conference on e-Learning, e-Management and e-Services (IC3e), Langkawi, Malaysia, pp. 6–10 (2018)
5. Shivanibindal, S.V.: Impact of ICT on teaching and learning. ZENITH Int. J. Multidisciplinary Res. **3**(1), 262–273 (2013)
6. Jandrić, P., et al.: Teaching in the age of Covid-19. Postdigital Sci. Educ. **2**, 106–1230 (2020)
7. Sobaih, A.E.E., Hasanein, A., Elshaer, I.A.: Higher education in and after COVID-19: The impact of using social network applications for E-Learning on students' academic performance. Sustainability. **14**(9), 5195 (2022)
8. Zhu, Y.: The Supply-side reform of higher education from the perspective of government function transformation. J. Higher Educ. **37**(08), 16–21 (2016). (in Chinese)
9. Mihaela, A., George, M.T., Ligia, V.V.: The Exposure of Chinese higher education to the development of international education system. Ovidius University Annals, Economic Sciences Series, **xiii**(1), 353–358 (2013)
10. Wang, J., Zhu, Z., Li, M.: Result-oriented: From certification concept to teaching mode. China Univ. Teach. **6**, 77–82 (2017). (in Chinese)
11. Xiaoying Lin, Sheng Xu, Zhipeng Liu. An analysis of curriculum teaching reform based on the background of engineering education certification[J]. Journal of Higher Education, 2022, 8(6): 136–138,142. (in Chinese)
12. Wang, J., Zhang, Q., Li, C.: Research and discussion on teaching reform of modern control theory based on OBE concept. Univ. Educ. **8**, 81–83 (2022)
13. Sola Guirado, R.R., Vacas, G.G., Alabanda, Ó.R.: Teaching CAD/CAM/CAE tools with project-based learning in virtual distance education. Educ. Inf. Technol. **27**: 5051–5073 (2022)
14. Berselli, G., Bilancia, P., Luzi, L.: Project-based learning of advanced CAD/CAE tools in engineering education. Int. J. Int. Des. Manuf. (IJIDM) **14**, 1071–1083 (2020)
15. Akhtar, S., Warburton, S., Xu, W.: The use of an online learning and teaching system for monitoring computer aided design student participation and predicting student success. Int. J. Technol. Des. Educ. **27**, 251–270 (2017)
16. Goli, A., et al.: Architectural design game: a serious game approach to promote teaching and learning using multimodal interfaces. Educ. Inf. Technol. **27**, 11467–11498 (2022)
17. Wang, Z., Ma, Y., Gao, Y.: Calculation method of evaluation target value for the achievement degree of curriculum objectives: for engineering education professional certification. Educ. Teach. Forum **21**, 41–44 (2021)
18. Liu, Z., Feng, G., Zhao, H.: Achievement evaluation of practical courses for engineering education certification: practical courses of computer science and technology. Educ. Teach. Forum **37**, 108–109 (2020)

19. Biggs J. Teaching for Quality Learning at University [M]. SRHE and Open University press, 2003, 2$^{nd}$ edition
20. Yixian, D., Tian, Q., Xuan, D., He, K.: CAD/CAM courses integration of theoretical teaching and practical training. Procedia. Soc. Behav. Sci. **116**(1), 4297–4300 (2014)
21. Wang, X., Yelin, S., Cheung, S., Wong, E., Kwong, T.: An exploration of Biggs' constructive alignment in course design and its impact on students' learning approaches. Assess. Eval. High. Educ. **38**(4), 477–491 (2013)
22. Gelmez, K., Arkan, S.: Aligning a CAD course constructively: telling-to-peer and writing-to-peer activities for efficient use of CAD in design curricula. Int. J. Technol. Des. Educ. **32**, 1813–1835 (2022)

# Research and Practice of Ideological and Political Education in the Context of Moral Education and Cultivating People

Geng E. Zhang[1,3] and Liuqing Lu[2(✉)]

[1] Faculty of Intelligent Manufacturing, Nanning University, Nanning 530000, China
[2] Personnel Office of Nanning University, Nanning 530000, China
78704108@qq.com
[3] Faculty of Engineering, Universiti Malaysia Sabah, 88400 Kota Kinabalu, Sabah, Malaysia

**Abstract.** Curriculum ideological and political construction is an important guarantee for the implementation of the fundamental task of strengthening moral education and cultivating people, "for whom to cultivate people, what people to cultivate, how to cultivate people" is always the fundamental problem of education. How to integrate professional education with moral character education to achieve ideological and political education of all members and the whole process. By consulting literature and under the guidance of curriculum moral character construction, this paper studies and practices the reform of curriculum ideological and political teaching on the teaching system, teaching content, teaching methods and teaching evaluation. After a semester of teaching practice, the students and peer teachers' evaluation forms are collected and analyzed, and the average score is 95 points, indicating that the moral character reform of the curriculum has achieved a certain effect, which can play a very good supporting role in cultivating students in moral education.

**Keywords:** Curriculum ideological and political education · Strengthen moral education and cultivate people · Teaching reform

## 1 Introduction

Since the curriculum moral character education was proposed in 2016, it has gone through six years. The state attaches great importance to it, all levels and departments take joint actions, colleges and universities focus on the implementation, constantly promote the teaching reform of curriculum ideological and political education. On August 14, 2019, the General Office of the CPC Central Committee and The General Office of the State Council printed and distributed *Several Opinions on Deepening the Reform and Innovation of Ideological and Political Theory Courses in Schools in the New Era*. Its guiding ideology is to accelerate the modernization of education, build an educational powerhouse, run an education that the people are satisfied with, and strive to cultivate new generation to take on the great task of national rejuvenation, train socialist builders and successors with all-round development of moral, intellectual, physical, aesthetical and labor education. On April 25, 2022, the CPC Secretary General Xi Jinping made an inspection of Renmin University of China and delivered an important speech, stressing that "for whom, who and how to cultivate people" is always the fundamental issue of education. We will carry out the fundamental task of strengthening moral education and cultivating people, and cultivate new generation to take on the great task of national rejuvenation for the party and the country. Textbook [2020] 6 the notice of Propaganda Department of the CPC Central Committee and Ministry of Education on Printing and Distributing *the Implementation Plan for the Reform and Innovation of Ideological and Political Theory Courses in Schools in the New Era.* The university stage focuses on enhancing students' mission, further enhancing the pertinence and effectiveness of teaching according to the requirements of different types of schools and different levels of talent training, consciously practicing socialist core values. The notice of *Guidance Outline for the Construction of Ideological and Political Courses in Colleges and Universities* No. 3 of Higher Education Office [2020] emphasizes that professional courses are the basic carrier of the construction of moral character courses. It is necessary to combine the education of Marxist standpoint, viewpoint and method with the cultivation of scientific spirit in the course teaching, so as to improve students' ability of correctly understanding, analyzing and solving problems. For engineering courses, emphasis should be placed on strengthening engineering ethics education, cultivating students' craftsman spirit of striving for excellence, and inspiring students' feelings and mission of serving the country through science and technology.

Under the guidance of various documents, the related research on curriculum ideology and politics has been deepened continuously, more and more colleges and universities have participated in the research boom of curriculum ideology and politics, and the research results have increased exponentially.

**Table 1.** The distribution statistics of curriculum ideological and political research topics from 2017 to 2022

| Year of publication | Theoretical research | Practical research ||||||| Total |
|---|---|---|---|---|---|---|---|---|
| | | School | Majors and courses | Teaching | Text-book | Teacher | Resources | |
| 2017 | 10 | 2 | | | | | | 12 |
| 2018 | 12 | 15 | 3 | 1 | | 4 | | 35 |
| 2019 | 26 | 15 | 25 | 18 | | 6 | 3 | 93 |
| 2020 | 69 | 35 | 91 | 51 | 1 | 15 | 7 | 266 |
| 2021 | 119 | 61 | 156 | 92 | 4 | 21 | 4 | 457 |

Table 1 shows the distribution statistics of subjects of moral character studies in the curriculum from 2017 to 2022 [1]. It can be seen from Table 1 that shows on morality studies in the curriculum are mainly focused on majors and courses. The majority of colleges and universities continue to implement the concept of strengthening moral education and cultivating people, and actively carry out morality reforms and studies in professional courses, with 156 papers increased by 156 times from none in 2017 to 156 in 2021, effectively solve the organic integration of profession and ideological and political areas.

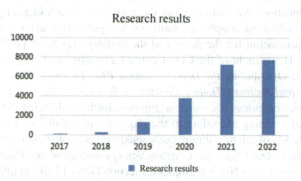

**Fig. 1.** Statistics of academic ideological and political papers from 2017 to 2022(Unit: papers)

Figure 1 shows the published length of periodical papers on moral character themes from 2017 to 2022 according to the statistics of China National Knowledge Network. In 2019, under the guidance of the Ministry of Education's notice on the *Guidelines for the Construction of Ideological and Political Construction in Higher Education Curriculum,* 1314 journal papers on the subject of curriculum moral character education were published, and 7728 journal papers were published in 2022, with an increase of 488%. There are 34 references in the search terms of course ideology and politics + automotive electrical appliances on CNKI. There is only one paper on the moral character research of *Automotive Electrical Appliances*. In the paper *Exploration on the Course Ideological and Political Thinking of Automotive Electrical Appliances in Higher Vocational Colleges from the Perspective of Three Education* [2]. Li Jing discussed the combination of techniques and methods in the course of *Automotive Electrical Appliances* in higher vocational colleges, but did not mention the course morality research on the course of *Automotive Electrical Appliances* in ordinary undergraduate colleges. This paper studies and practices how to carry out moral character teaching in the course of *Automobile Electrical Appliances* in application-oriented colleges and universities.

## 2  Problems in Curriculum Ideology and Politics

Yang Mei et al. put forward the following problems of moral character education in college curriculum [3]. First, teachers of professional courses in colleges and universities do not have strong ideological and political awareness of the curriculum, they lack a comprehensive knowledge and understanding of moral character thinking in the curriculum. Professional teachers think that morality education in the curriculum is the responsibility of ideological and political teachers, which is a burden to themselves. They do not pay enough attention to curriculum moral character education and lack the initiative and enthusiasm to carry out curriculum ideological and political reform. Second, the overall consciousness of moral character construction in college curriculum is not strong, the boundaries of moral character construction in different professional courses are blurred, lack of personality, cannot reflect the professional characteristics, cannot carry out curriculum ideological and political research in specific courses under the awareness of professional integrity, most of the research reflect as generic research. Curriculum moral character education can be divided into "explicit ideological and political education" and "implicit ideological and political education". Wang Yiqing et al. pointed out that "implicit ideological and political education" has some problems, such as unclear principles and concepts, lack of close integration with majors, lack of abundant educational resources, and lack of ideological and political literacy of education subjects, and the paper proposed corresponding solutions [4]. Dong Hui et al. elaborated the difficulties of moral character teaching from the perspective of teachers' ability, and pointed out that there were problems to be solved in the aspects of teachers' consciousness, knowledge level, teaching ability, teaching wisdom and morality experience in curriculum ideological and political construction [5]. Liang Jin et al. analyzed the practical predicament of integrating curriculum ideology and politics into teaching in colleges and universities in the new era from the three aspects of colleges, secondary colleges and professional teachers [6]. This paper also pointed out that the main problems hindering the integration of curriculum moral character education into teaching are the insufficient supply of supporting policies in colleges, the insufficient play of the main front role of secondary colleges, and the lack of curriculum ideology and politics ideas, abilities and feelings of college teachers. This paper also put forward the corresponding solution strategy. Sun Gang et al. pointed out that teachers and students generally lack a deep understanding of the curriculum moral character training mode, the relevant mechanism is not perfect, the overall design and implementation ability is not mature, the endogenous motivation of professional teachers is insufficient, evaluation standards are not clear and other factors affect the construction of professional curriculum ideological and political teaching, and put forward corresponding solutions [7].

## 3 Research on the Path of Ideological and Political Teaching Reform of *Automobile Electric Appliances* Course

Major courses should reflect the characteristics of their major, as a core course of transportation related majors [2], the moral character education reform of the curriculum should organically integrate the national education objectives, college education objectives, professional education objectives, curriculum teaching objectives and teaching unit objectives. In addition, the national, college and professional objectives should be deepened into the curriculum and detailed to each teaching unit. Reflect the curriculum ideological and political education in every teaching unit, so as to enrich the students' quality education invisibly. Diversified reform should be carried out from personnel training program, curriculum teaching system, teaching content, teaching method and teaching evaluation, so that explicit moral character education can be carried out simultaneously with implicit moral character education, constantly improve morality education, and fully implement the idea of strengthening moral education and cultivating people.

## 3.1 Reform of ideological and Political Teaching System of *Automobile Electric Appliances* Course

Fang Caihong et al. proposed that the construction of curriculum moral character system in colleges and universities should follow three principles and three dimensions, from the three principles of combining theory and practice, explicit education and implicit education, overall promotion and key breakthrough, and from the three dimensions of educational operation mechanism, educational guarantee mechanism and educational evaluation mechanism to build a perfect system of curriculum morality education [8]. Constantly excavate the history, culture and industry spirit of different majors and refine them into the core value system of majors, so as to integrate them into the curriculum teaching system. According to the characteristics and advantages of transportation majors, we should scientifically and reasonably expand the breadth, depth and temperature of professional morality education, and build a curriculum moral character system with comprehensive coverage, rich types, progressive levels and mutual support, as shown in Fig. 2. In view of the course learning objectives into task design, the task-driven main line online and offline mixed teaching mode is used to implement the teaching process, the curriculum moral character education is integrated into it, it can be OBE-oriented and combined with the curriculum objectives to carry out diversified assessment.

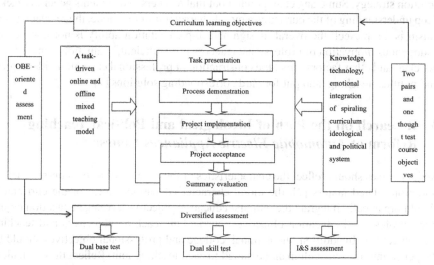

**Fig. 2.** Ideological and political teaching system of knowledge, technology, emotional integration

## 3.2 Ideological and Political Teaching Content Reform of Automobile Electric Appliances Course

Automobile electrical equipment is one of the standards to measure the advanced level of automobiles. The electronic control technology of automobiles is gradually developed with the development of electronic technology and the improvement of automobile related regulations (fuel consumption regulations, emission regulations and safety regulations). This course mainly introduces the structure, composition, working principle, maintenance experiment and fault diagnosis method of automotive power system, starting system, ignition system, instrument system, lighting and signal system and auxiliary electrical system.

The specific teaching content mainly includes the structure, type, working principle, working characteristics, capacity and influencing factors, charging and fault of battery; structure, type, working principle, characteristics of silicon rectifier generator; principle and classification of regulator; protection circuit of automotive power system; structure, working principle, characteristics and models of DC series motor; starter transmission mechanism; starter drive protection circuit; requirements of automotive ignition system; composition and working principle of traditional ignition system; analysis of the working process of ignition system, factors affecting secondary voltage; structure, composition, working principle of instrument system, lighting and signal system, auxiliary electrical system; composition and characteristics of automotive electrical circuit; distribution devices of automotive electrical circuit; expression and analysis methods of automotive circuit diagram; typical car circuit analysis and other contents. After summarizing the content of the course, it mainly includes three parts: the structure, working principle and maintenance of automotive electrical components. How to integrate moral education elements into each teaching content is shown in Table 2, the morality elements of the curriculum are reflected in the selection and arrangement of teaching content [9].

### 3.3 Reform of Ideological and Political Teaching Methods of Automobile Electric Appliances Course

Teaching methods directly affect the degree of realization of teaching objectives and excellent grade of teaching effect. In the process of teaching implementation, it is necessary to pay attention to the diversification of teaching methods and effective interaction between teaching and learning, adhere to the student-centered, teacher-led. The organic integration of various teaching methods reflects the diversity and flexibility of teaching methods, as well as two-way interaction between teachers and students [10]. The reform of ideological and political teaching method of *Automobile Electric Appliances* course mainly includes two aspects:

1) BOPPPS effective teaching model is adopted for teaching design. The BOPPPS teaching model mainly includes six elements, among which the most critical one is the design of learning objectives, which can be divided into cognitive objectives, skill objectives and emotional objectives. Integrating moral education elements into emotional objectives can reflect curriculum moral character education.

2) In the P2 participatory teaching link of BOPPPS teaching model, a variety of effective teaching methods are used to change students' passive learning into active learning. Through a variety of effective teaching methods such as scene teaching, group discussion, smart site, fish tank teaching method, carousel, peer teaching method, students' participation can be improved, teacher-student interaction and student-student interaction can be strengthened, students' interest in learning can be improved, so that can

**Table 2.** Ideological and Political Teaching Contents

| Serial number | Course content | Ideological and political elements |
|---|---|---|
| 1 | Introduction | From the development of automobile electronic technology leads to the development and application of our country electronic technology, cultivate students' spirit of scientific and technological power |
| 2 | Maintenance, use and overhaul of storage battery, generator, starter and electrical system | Moral education elements are added to the contents related to the use, maintenance and overhaul of various electrical systems to train the students of automobile service engineering, cultivate the spirit of great craftsman, the consciousness of safe operation based on professional ethics, and the spirit of team cooperation |
| 3 | Structure and principle of storage battery, generator, starter and electrical system | Cultivate students' rigorous logical analysis ability and scientific research spirit, and cultivate students' awareness of environmental protection and ecological civilization based on automobile electrical appliances |
| 4 | Complete vehicle circuit repair | Combined with various models, the vehicle line overhaul, interspersed comparison of various countries of automobile enterprises in the development of automotive electrical system, put forward the advantages of the development of automotive electrical in our country, as well as the need to improve the place, encourage students to become strong in science and technology |

achieve effective teaching. Cultivate students' sense of teamwork, dialectical thinking, spirit of craftsman, spirit of science and technology, and awareness of safety and environmental protection.

## 3.4 Ideological and Political Teaching Evaluation Reform of Automobile Electric Appliances Course

The subject of curriculum moral character teaching evaluation is both the teacher and the student, and the evaluation of the effect of curriculum moral character teaching should be multi-dimensional, multi-level and comprehensive [11]. Teaching evaluation should be carried out in the whole process of learning and teaching, closely combined with cognitive goals, using more authentic evaluation methods in the evaluation, taking students as the center, experiencing participatory evaluation based on problems, and giving timely feedback to the evaluation results. The quality evaluation system of curriculum ideological and political "cloud teaching" based on 4E theory is constructed from the four dimensions of economy, efficiency, participation and effectiveness, which improves the depth and breadth of theoretical research on curriculum moral character quality evaluation [12]. The "collaborative evaluation" of professional course teachers and ideological and political counselors has certain effectiveness in solving the problem of morality evaluation of professional courses [13].

The moral character teaching evaluation of *Automobile Electric Appliances* course should reflect the elements of safe operation, the elements of team cooperation and craftsman spirit. In the group discussion and report, safe operation accounted for 10%, team cooperation accounted for 20% and craftsman spirit of great power accounted for 20%. Morality elements should be deeply integrated into the teaching evaluation.

## 4 Effect of Ideological and Political Teaching Reform of *Automobile Electric Appliances* Course

There are various ways to evaluate the effect of teaching reform, including questionnaires, quality evaluation forms, multiple choice questions and learning experiences. In order to fully understand the ideological and political reform of *Automobile Electric Appliances* course, this paper adopts the form of quality evaluation table for data collection and analysis. After the end of the course, students and fellow teachers will evaluate according to the lecture content. Table 3 shows the statistical table of teaching quality evaluation in the first semester from 2022–2023. There are 20 items related to ideology and politics in the quality evaluation form, the maximum score is 5 points for very good, the minimum score is 1 point for not very good, and the total score is 100 points.

**Table 3.** Evaluation table of ideological and political teaching quality

| Serial number | Evaluation content | 5 points | 4 points | 3 points | 2 points | 1 point |
|---|---|---|---|---|---|---|
| 1 | This course takes moral cultivation as the central link, educating people in the whole process and in all aspects | 5 | | | | |
| 2 | It can improve students' professional skills and professionalism | 5 | | | | |
| 3 | The moral character content integrated into the course keeps up with the trend of The Times and can be attractive to students | 5 | | | | |
| 4 | The course content design conforms to the requirements and rules of science and engineering education | 5 | | | | |

(*continued*)

**Table 3.** (*continued*)

| Serial number | Evaluation content | 5 points | 4 points | 3 points | 2 points | 1 point |
|---|---|---|---|---|---|---|
| 5 | Teachers in the course of moral character construction has a certain research result | | 4 | | | |
| 6 | The content of professional knowledge has synergistic effect with the content of moral character theory | 5 | | | | |
| 7 | Moral character material content is novel, teaching courseware, cases, exercises and other teaching materials are rich | 5 | | | | |
| 8 | It can integrate knowledge imparting, ability cultivation and quality education | 5 | | | | |
| 9 | It can integrate moral character elements into the teaching content invisibly | 5 | | | | |
| 10 | It can enhance the affinity and pertinence of moral character education | 5 | | | | |
| 11 | When it comes to the teaching of moral character content, students can keep a good mental state | | 4 | | | |
| 12 | Students take the initiative to ask questions to teachers or discuss moral character content | | 4 | | | |

(*continued*)

**Table 3.** (*continued*)

| Serial number | Evaluation content | 5 points | 4 points | 3 points | 2 points | 1 point |
|---|---|---|---|---|---|---|
| 13 | Curriculum assessment forms are diverse and not simple | 5 | | | | |
| 14 | Establish a teaching feedback mechanism and have regular face-to-face communication with students or student representatives | 5 | | | | |
| 15 | Homework or examination questions have both professional assessment and moral character assessment | 5 | | | | |
| 16 | Can use a variety of modern teaching methods to carry out curriculum teaching | 5 | | | | |
| 17 | Be able to master the knowledge points required by this course and apply them skillfully | | 4 | | | |
| 18 | It has a high recognition of the necessity of moral character courses in science and engineering | | 4 | | | |
| 19 | The course cultivates students' practical ability and innovation ability | 5 | | | | |
| 20 | Students can put the moral character requirements in the curriculum into practice | 5 | | | | |

As can be seen from Table 3, the average score is 95, indicating that the moral character reform of the curriculum has achieved certain effects and can play a good supporting role in cultivating students in moral education.

## 5 Conclusion

Ideological and political education plays a vital role in the talent training of contemporary colleges and universities. Education and politics is not the same as ideology and politics. It refers to students who receive moral education in the study of professional courses and learn how to be a man before learning. There is a Chinese saying in China that "ten years to grow trees, and a hundred years to cultivate people". Moral education is not only the content of learning in the political course, but also should be permeated into each course, so that students can subtly learn knowledge, understand the truth, set up the correct socialist core values, and become the successors of the new era. This paper analyzes and research the documents related to morality education. Found that few studies were conducted on the Automotive Appliances course. For this purpose, this paper carries out the research and practice of curriculum ideological and political reform from the teaching reform path, teaching system, teaching content, teaching method and teaching evaluation, Put forward the curriculum system of the deep integration of knowledge, technology and situation. The teaching content was designed using the BOPPPS teaching model. The morality content throughout, and designed a complete set of course evaluation forms. Some results have been achieved. It is a reference for the ideological and political reform of other courses.

**Acknowledgment.** This paper is supported by: (1) The second batch of Nanning University "Curriculum Ideological and Political Education" demonstration course construction Project, No. 2020SZSFK19; (2) Diversified "Collaborative Ideological and Political Education" teaching team, No. 2022SZJXTD03.

## References

1. Lu, D.: The research progress, difficulty focus and future trend of curriculum ideology and politics in the new era. J. Xinjiang Normal Univ. (Philos. Soc. Sci. Ed.) **03**, 1–16 (2022)
2. Li, J., Xie, D.: The ideological and political exploration of automobile electrical appliance course in higher vocational colleges from the perspective of San Quan education. Time Motor **08**, 62–63 (2022)
3. Yang, M., Zhzo, S., Li, J.: Current situation, problems and countermeasures of ideological and political development in college curriculum – based on visualization map and Cite Space. J. Shaoyang Univ. (Soc. Sci. Ed.) **21**(02), 95–99 (2022)
4. Wang, Y., Li, L., Hao, L.: Problems and improvement strategies of "recessive education" in ideological and political teaching in colleges and universities. Educ. Theory Pract. **40**(3), 37–39 (2020)
5. Dong, H., Du, J.: The difficulties of ideological and political promotion of curriculum and its solutions. Ideol. Theoret. Educ. **5**, 70–74 (2021)

6. Ling, J.: The realistic dilemma and mechanism innovation of college curriculum ideological and political integration into professional teaching in the new era. Occupation (04), 33–36 (2022)
7. Sun, G., Li, H., Fang, Y., Zhao, C.: Restrictive factors and effective ways of ideological and political thinking in applied university curriculum. J. Changchun Normal Univ. **41**(02), 93–96 (2022)
8. Fang, C.: Research on the construction of ideological and political system of college curriculum under the background of strengthening moral education and cultivating people. J. Soc. Sci. Jiamusi Univ. **40**(01), 233–235+239 (2022)
9. Zhang, J., Mao, X.: Exploration and practice of implementing curriculum ideology and Politics in the teaching of Automobile Electrical and Electronic Control Technology. Time Motor **01**, 99–100 (2022)
10. Meng, Z., Li, L.: Some problems and improvement paths in ideological and political teaching practice of curriculum. Chin. Univ. Teach. **03**, 51–57 (2022)
11. Jun, C.: Thinking on Constructing Ideological and political evaluation system of college curriculum. J. Chongqing Univ. Sci. Technol. (Soc. Sci. Ed.) **03**, 106–112 (2022)
12. Hongmei, M.: Construction of curriculum ideological and political "cloud teaching" quality evaluation system based on 4E theory. J. Soc. Sci. Jiamusi Univ. **40**(02), 224–227 (2022)
13. Tianqi, Z.: Research on the effectiveness of the "Collaborative Evaluation" mechanism of ideological and political effects of professional courses. J. Higher Educ. **8**(04), 168–171 (2022)

# A Quantitative Study on the Categorized Management of Teachers' Staffing in Colleges and Universities

Zhiyu Cui[✉]

Personnel Department, Wuhan University of Technology, Wuhan 430072, China
cuizy@whut.edu.cn

**Abstract.** The reasonable allocation of university faculty has always been one of the important contents of university management. Based on the literature review of university faculty staffing management, this paper clarifies the principles and basis that should be followed in university staffing. By using the mathematical statistical methods such as exponential smoothing, which changes with time, the student equivalent of a university is predicted, and the number of teachers and the corresponding number of teaching assistants are allocated accordingly. Through dynamic management of university staffing, adjustment of the position proportion of various types of personnel, and analysis of the reasonable degree of the allocation of the number of personnel in each post, in order to achieve the purpose of making full use of staffing benefits. The paper expounds the need of social and economic development and the importance of building a high-level university faculty, and puts forward corresponding suggestions for the future staffing management, aiming at creating conditions for promoting the characteristic development of university management.

**Keywords:** Universities · Stuffing management · Cubic exponential smoothing method · Quantitative study

## 1 Introduction

Higher education is related to the development of the national economy. The quality of education directly determines the future of the country [1, 2]. With the further deepening of the reform of the higher education system, every university regards continuous optimization of the structure, stable improvement of quality and overall improvement of efficiency as the goals of reform and pursuit [3]. To achieve the reform objectives and promote the reform of university management system, one of the contents is to do a good job in university staffing and post management [4]. A research team has conducted a questionnaire survey and evaluation on the progress of university reform based on expert consultation. In one survey, a total of 230 questionnaires were distributed in colleges and universities across the country, and 203 valid questionnaires were recovered. Only 3.4% of the respondents believed that the reform effect was very significant, and 17.2% of the respondents commented that the result was very insignificant, while the rest of the respondents commented that the reform effect was not high. Among the eight evaluation indicators listed, the reform of institutions and staffing system of colleges and universities is considered as "not significant" and "not significant". Such investigation and evaluation results have caused many scholars' deep reflection [5–7]. It can be seen that it is imperative to implement the staffing and post setting of institutions of higher education. A new mechanism should be established to not only dilute and simplify the establishment of staffing, but also strengthen the self-restraint and self-control of staffing management.

The construction of teachers is the most important guarantee for the quality of higher education [8–10]. Headcount management is an effective way to reasonably allocate human resources in colleges and universities. It can effectively control the total number of personnel, scientifically plan the proportion of personnel structure, dynamically adjust the structure of staff and workers, and has a strong guiding role in the overall development of colleges and universities. Scientific and reasonable staffing verification and post setting management is the key to building a reasonably structured, capable, efficient and dynamic talent team, and also the key to promoting the discipline construction of colleges and universities, which has a guiding role for the scientific development of the university [11]. The establishment of posts is not simply an increase or decrease in the number, but a reorganization and optimization of resources. From the perspective of economics, the staffing and posts of colleges and universities have economic characteristics [12, 13]. By strengthening the accounting of personnel costs, we can achieve the best school-running benefits with the minimum employment costs. Scientific staffing management and strict on-demand post setting can control and adjust the proportion and quantity of all kinds of personnel, can eliminate the disorder in management, and is also the key to improve the scientificity of university recruitment and achieve the organizational purpose with minimum consumption [14].

The establishment of institutions of colleges and universities involves all aspects of university work, mainly including the establishment of internal institutions and the determination of functions, the establishment and staffing of disciplines and professional posts, as well as the efficiency, effect and benefit in actual activities. In accordance with the spirit of the Guiding Opinions of the CPC Central Committee and the State Council on Promoting the Reform of Public Institutions by Classification (2011) and the Several Opinions of the Ministry of Education and other five departments on deepening the reform of streamlining administration, delegating powers, delegating powers and optimizing services in the field of higher education (2017), in order to promote the reform of the employment system of colleges and universities and further promote the scientific and comprehensive development of colleges and universities, in-depth discussion is made on the methods of staffing and posts suitable for the situation of colleges and universities and highlighting the characteristics of each college, Undoubtedly, it has important practical significance.

## 2 Research Methods

The research methods used in this paper mainly include empirical research, comparative analysis and mathematical statistics.

## 2.1 Methods of Empirical Research

Headcount management is a highly practical administrative practice. For the history and current situation of headcount management in colleges and universities, including the situation of institutions and personnel, the documents of headcount management, the headcount needs of departments and secondary units, etc., it is necessary to obtain detailed information and relatively accurate data through interviews, symposiums, questionnaires and other empirical research methods.

## 2.2 Methods of Comparative Analysis

Human resource allocation and benefit control in colleges and universities are issues of universal concern to global higher education, and also the focus of practice and exploration in domestic colleges and universities in recent years [15–17]. Through document retrieval and data analysis, we can understand the historical evolution, current situation, management systems and methods of university headcount management in various countries, which can broaden our horizons and provide reference for the research of university headcount management [18–20]. At the same time, by comparing and analyzing their practices in organization setting, post division, staffing verification standards, and fund management, we explore the innovative thinking of staffing management.

## 2.3 Methods of Mathematical Statistics

The logical starting point of staffing management is the objective evaluation and reasonable setting of workload, and how to measure the workload of teaching, scientific research and school affairs is a very complex problem. As the tested objects, such as teachers, their work will change randomly due to various internal or external conditions, and their work efficiency will also be randomly distributed. Therefore, mathematical statistics methods such as regression analysis, statistical mean, simple proportion, etc. are widely used in the analysis of research variables [21, 22]. The widely collected data can be used to study the fractional function and numerical characteristics of random variables as well as the relationship between various types of compilation, and then infer regular conclusions. This has been proved to be effective in the management of education around the world [23–25].

# 3 Prediction Results

The student equivalent data of each college in recent years are shown in Table 1. It can be seen from the table that the change of student equivalent of each college in consecutive years presents a quadratic curve trend, so the exponential smoothing method can be used to predict. Among them, the cubic exponential smoothing method uses the weighted average of the observation values of each period in time order as the prediction value, showing the characteristics of the impact of the future value of historical data changing with time. This method is more suitable for the research content of this paper.

The prediction model of the cubic exponential smoothing method is:

$$\hat{y}_{t+T} = a_t + b_t T + c_t T^2, \ T = 1, 2 \ldots \quad (1)$$

where,

$$\begin{cases} a_t = 3S_t^{(1)} - 3S_t^{(2)} + S_t^{(3)} \\ b_t = \frac{\alpha}{2(1-\alpha)^2}[(6-5\alpha)S_t^{(1)} - 2(5-4\alpha)S_t^{(2)} + (4-3\alpha)S_t^{(3)}] \\ c_t = \frac{\alpha^2}{2(1-\alpha)^2}[S_t^{(1)} - 2S_t^{(2)} + S_t^{(3)}] \end{cases} \quad (2)$$

While, $S_t^{(1)}$, $S_t^{(2)}$, $S_t^{(3)}$ is the first, second and third exponential smoothing value separately, and its calculation formula is:

$$\begin{cases} S_t^{(1)} = \alpha y_t + (1-\alpha)S_{t-1}^{(1)} \\ S_t^{(2)} = \alpha S_t^{(1)} + (1-\alpha)S_{t-1}^{(2)} \\ S_t^{(3)} = \alpha S_t^{(2)} + (1-\alpha)S_{t-1}^{(2)} \end{cases} \quad (3)$$

**Table 1.** Prediction of student equivalent trend in 2023 based on cubic exponential smoothing method

| College | Historical data of student equivalent |  |  |  |  |  |  | Exponential smoothing | Forecast trend |
|---|---|---|---|---|---|---|---|---|---|
|  | 2016 | 2017 | 2018 | 2019 | 2020 | 2021 | 2022 | Optimal value of α | 2023 |
| A | 4420 | 4397 | 3869 | 3846 | 3814 | 3838 | 4062 | 0.990000 | 3837 |
| B | 2974 | 3028 | 3025 | 3051 | 3036 | 3058 | 4362 | 0.876947 | 3058 |
| C | 2196 | 2179 | 2111 | 3811 | 3828 | 3853 | 3899 | 0.971271 | 3853 |
| D | 3953 | 3854 | 3801 | 2923 | 3133 | 3283 | 3397 | 0.85211 | 3281 |
| E | 1408 | 1463 | 1440 | 2201 | 2172 | 2174 | 2511 | 0.939704 | 2175 |
| F | 2040 | 2087 | 2126 | 1755 | 1843 | 2027 | 3185 | 0.014609 | 2025 |
| G | 1732 | 1739 | 1736 | 3447 | 3058 | 3066 | 2051 | 0.77394 | 3085 |
| H | 1227 | 1221 | 1236 | 333 | 769 | 874 | 1902 | 0.45272 | 738 |
| K | 776 | 749 | 692 | 1288 | 1656 | 889 | 1137 | 0.346483 | 1249 |

Considering the known student data, the smoothing coefficient α Search and optimize within its value range. In the process of optimization, the mean square error of the predicted student equivalent is used as the optimization objective function. The expression of the objective function is as follows:

$$\sigma = \sqrt{\frac{1}{n}\sum_{i=1}^{n}(X_i - Y_i)^2} \quad (4)$$

In formula (4), $X_i$ is the original student equivalent, $Y_i$ is the predicted student equivalent, and $n$ is the number of original data.

When the value of the smoothing coefficient $\alpha$ changes, there is a continuous functional relationship between the predicted student equivalent result error $\sigma$ and the smoothing coefficient $\alpha$, which can be expressed as

$$\sigma = f(\alpha) \tag{5}$$

The optimal smoothing coefficient $\alpha$ can be obtained by solving the following equation:

$$f'(\alpha) = 0 \tag{6}$$

In accordance with the formulas (1)–(6), the cubic exponential smoothing prediction equation is established, and the optimal values of the smoothing coefficients of each college are listed in Table 1. The student equivalent prediction results based on the optimal smoothing coefficients are displayed in Fig. 1.

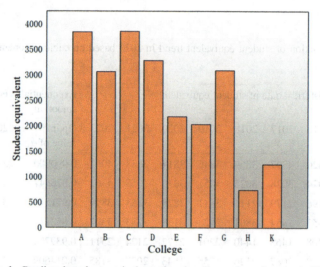

**Fig. 1.** Predicted student equivalent of each college of a university in 2022

It can be clearly presented from the above figure that the predicted student equivalent of College A and College C in 2023 is relatively large. Combining the actual student equivalent data of these two colleges from 2016 to 2022, it can be found that when the smoothing coefficient $\alpha$ When it is close to 1, the weight of student equivalent in recent years in the predicted value is larger; The best smoothing coefficient of F College $\alpha$ When it is close to 0, the weight of student equivalent in the distant years in the predicted value is larger; When the best smoothing coefficient $\alpha$ When it is close to 0.5 (e.g., H College), the student equivalent in the middle year of the predicted value has a larger weight.

Based on the above forecast data, in line with the "Regulations on the Management of Headcount of General Institutions of Higher Education (Draft)" jointly issued by the

Office of the Central Organization Headcount Committee, the Ministry of Education and the Ministry of Finance, the staffing of institutions of colleges and universities is divided into the staffing of teachers, the position of teaching and auxiliary staff, the personnel of full-time scientific research, the member of staff performing various management responsibilities, and the staffing of affiliated units including logistics and undertaking public welfare community services. It is also emphasized that all kinds of personnel in colleges and universities should form a reasonable proportion of the structure, among which teaching and research personnel and teaching and auxiliary personnel should account for more than 80% of the total number of college personnel. At the same time, it is stipulated that the total number of faculty and teachers should be determined considering the ratio of students to teachers. On the basis of the relevant standards of the Ministry of Education, combined with the actual situation of school management, based on the student equivalent data, the teacher preparation plan of each teaching unit can be calculated, as shown in Table 2.

**Table 2.** Number of teachers according to the requirements of different student-to-teacher ratios

| College | Student equivalent | 14: 1 Number of teachers | 16: 1 Number of teachers | 18: 1 Number of teachers | 20: 1 Number of teachers |
| --- | --- | --- | --- | --- | --- |
| A | 3837 | 274 | 240 | 213 | 192 |
| B | 3058 | 218 | 191 | 170 | 153 |
| C | 3853 | 275 | 241 | 214 | 193 |
| D | 3281 | 234 | 205 | 182 | 164 |
| E | 2175 | 155 | 136 | 121 | 109 |
| F | 2025 | 145 | 127 | 113 | 101 |
| G | 3085 | 220 | 193 | 171 | 154 |
| H | 738 | 53 | 46 | 41 | 37 |
| K | 1249 | 89 | 78 | 69 | 62 |
| Total | 23301 | 1664 | 1456 | 1295 | 1165 |

Taking the student equivalent predicted in Fig. 1 as the base, the teacher staffing plan of each college calculated according to the ratio of students to teachers is indicated in Fig. 2. Therefore, it can be seen that the number of teachers in each college in Fig. 2 is consistent with the trend of student equivalent in Fig. 1. Since the Ministry of Education has made clear quantitative requirements for human resources investment in the process of running colleges and universities in the undergraduate teaching evaluation program, defined the connotation and scope of the student-teacher ratio, and new innovations have also emerged in the staffing management of colleges and universities, making the human resources investment of colleges and universities meet the minimum standards to ensure the quality of education and teaching.

The data predicted by the exponential smoothing method lacks the ability to identify the turning point, i.e., if the student equivalent of a certain college has an obvious turning trend, the three-time exponential smoothing method cannot accurately predict the success, but this can be compensated by the survey prediction method or expert prediction method.

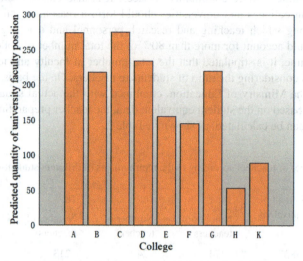

**Fig. 2.** The number of teachers predicted according to the standard student-teacher ratio of 14:1 required by the Ministry of Education

Moreover, the algorithm has strong adaptability and can automatically identify data patterns and adjust the prediction model. For the current situation that the ratio of students to teachers is too high in most colleges and universities, and the shortage of staff in some colleges and universities is serious, we should expand the team of full-time teachers, full-time counselors, and full-time scientific researchers through the introduction and appointment system, so that the ratio of students to teachers can be reduced from the current 20:1 to about 18:1 as soon as possible, laying the groundwork for the realization of the standard of 14:1 issued by the Ministry of Education in the next stage. The number of teachers in each college predicted in the light of the standard student-teacher ratio of 14:1 of the Ministry of Education is shown in Fig. 2.

## 4  Analysis and Suggestions

With the continuous development of the reform of higher education, the reform of the staffing management in colleges and universities has also accelerated. The purpose of the establishment management of university teachers is to improve educational efficiency and meet the requirements of scientific management of education and teaching. At present, the common pursuit of all major universities in the reform of teacher staffing management is to improve the overall level of teachers and promote the healthy development of education. Among them, the main measures include the following three aspects.

## 4.1 Entry and Exit Mechanism of Teachers

Based on the cubic exponential smoothing algorithm, this paper can predict the number of student equivalents of each college in the future, and then predict the number of teachers. In fact, in order to alleviate the shortage of teachers caused by the expansion of undergraduate and graduate enrollment, major universities have increased the introduction of teachers, and the number of new teachers has increased year by year. At the same time, the entry of non-teachers should be strictly controlled. The vacancy of non-teachers should be solved by internal adjustment as much as possible. For other urgently needed professional technicians and managers, they should be recruited to the public through personnel agency. There will be no new headcount for work posts, and try to gradually reduce such headcount through labor outsourcing and retirement.

Through the implementation of the above measures, gradually adjust the proportion of students and teachers to make it moderate and reasonable. At the same time, the employment system of combining fixed and mobile headcount is implemented. The fixed headcount is steadily rising. The number of mobile headcount is increased, a relaxed employment atmosphere is created, and the employment environment is improved. Attention is paid to "career retention", "emotional retention", and "treatment retention". Use scientific methods to manage teachers, fully mobilize and explore the enthusiasm and potential of teachers, form a good academic atmosphere and an academic environment of fair competition, explore and establish a relatively stable backbone layer and an orderly flow layer management mechanism.

## 4.2 Enrichment of Staffing Methods

At present, the way of staffing in colleges and universities is relatively single, lacking scientific and reasonable staffing methods. Although the cubic exponential smoothing method in this paper can be used to predict the student equivalent, the relatively rough method of determining the ratio of students to teachers has gradually become inapplicable to the current level of university management and development. Various specific and quantifiable factors, such as teaching duration and quality, teaching content and effect, scientific research level and achievements, should be appropriately included in the process of making a comprehensive assessment.

In addition, on the basis of the approval scheme of other types of post staffing approved by the executive unit, we began to make preparations for the major adjustment of the next stage of staffing, such as the fine-tuning of organizational structure, the sorting and optimization of post responsibilities, publicity and learning, skill training, and internal rotation, to improve the structure of university faculty. At this stage, the number of overstaffed positions is mainly frozen without adding new employees, so that the number of management, work attendance and other positions can be controlled, and retirement and job transfer can be added to maintain the state of natural reduction.

## 4.3 Establish and Improve Relevant Supporting Measures

Some colleges and universities have made some innovative attempts to explore the management of teacher staffing, such as gradually abolishing the career staffing, implementing the employment system of labor contracts, or the personnel agency system. These measures are opportunities and great challenges for current colleges and universities. The channels of talent flow are smooth to a higher degree, and excellent talents, especially high-level talents, are easily attracted to these colleges and universities. At the same time, the pre-employment and long-term employment system of "go up or down" makes the faculty lose its due stability, and reduces the sense of belonging and happiness of college teachers who take the mission of moral education as the core, In the long run, it will affect the realization and healthy development of the school's talent training objectives.

In fact, the management of teachers' staffing should be reflected in the qualification, employment, assessment, evaluation and other links. By clarifying the professionalism of teachers, teachers can comprehensively promote their confidence in their own work and enhance their sense of responsibility and enterprising spirit. At the same time, colleges and universities should establish an open and transparent evaluation mechanism in line with their own development strategies to reduce the structural contradictions between the school staff, which can not only achieve the reasonable allocation of school resources, but also make talents in line with the sustainable development of the school stand out.

## 5 Conclusion

Scientific and reasonable staffing verification and post setting management in colleges and universities is the key to building a reasonably structured, capable and energetic talent team in colleges and universities, and also the key to promoting the discipline construction in colleges and universities, which is of great significance to the overall development of colleges and universities. This paper uses the cubic exponential smoothing algorithm to predict the equivalent of future students, and then determines the number of teachers according to the ratio of students to teachers, which better realizes the combination of qualitative and quantitative, and improves the prediction accuracy of the number of posts in colleges and universities. On this basis, several suggestions and measures for the management of university teachers' staffing are put forward. On the basis of in-depth analysis of the current situation of university teachers' development and talent training objectives, the scientific allocation of teachers is carried out to provide assistance for the reform of university teachers' staffing management system and the development of their own quality.

Post staffing is a very complex system engineering, which is closely related to the future human resource allocation of the school and the vital interests of the existing staff. Scientific and reasonable post setting and effective post staffing management are of great significance to innovation of personnel management system in colleges and universities, optimization of human resource allocation, and mobilization of enthusiasm, initiative and creativity of faculty and staff. Therefore, in line with the idea of taking into account the current situation and long-term development of the team, the new round of post staffing optimization should be practical and realistic, and at the same time, the transition should be stable and gradually in place. Adopt a combination of qualitative and quantitative methods to implement staffing according to posts, optimize and adjust the structure of human resources based on the principle of "adjusting the internal structure and reaching consensus", so that the structure of human resources tends to be more optimized and reasonable, which not only promotes the development of key disciplines in colleges and universities, ensures the "double strength" of teaching and scientific research, but also achieves the goal of streamlining and improving management level.

# References

1. Zhiqin, L., Jianguo, F., Fang, W., Xin, D.: Study on higher education service quality based on student perception. Int. J. Educ. Manage. Eng. **2**(4), 22–27 (2012)
2. Singh, V., Dwivedi, S.K.: Two way question classification in higher education domain. Int. J. Mod. Educ. Comput. Sci. **7**(9), 59–65 (2015)
3. Beltadze, G.N.: Game theory - basis of higher education and teaching organization. Int. J. Mod. Educ. Comput. Sci. **8**(6), 41–49 (2016)
4. Hemei, L.X.: Research on the issue of post and staffing of management posts in colleges and universities. Yunmeng Acad. J. **36**(2), 88–91 (2015). (in Chinese)
5. Lan, L.: Problems and countermeasures of post management of university teachers. Educ. Guide **11**, 39–41 (2010). (in Chinese)
6. Wang, B., Zheng, Y., Jing, Y., et al.: On the reform and management of the establishment of institutions in colleges and universities. J. Handan Vocational Tech. College **34**(4), 59–61 (2021). (in Chinese)
7. Cunhong, Y.: Promote the reform of teacher staffing policy and release the vitality of education human resources. Jiangxi Educ. **31**, 14–15 (2019). (in Chinese)
8. Shi, L., Yao, Y., Xia, X., Xie, X., Lanlan, Wu.: Practical teaching staff construction under new situation. Int. J. Educ. Manage. Eng. **1**(2), 57–61 (2011)
9. Meng, X.: Survey of research quality and ability of vocational school teachers. Int. J. Educ. Manage. Eng. **2**(10), 9–16 (2012)
10. Kumar, S.: A fuzzy based comprehensive study of factors affecting teacher's performance in higher technical education. Int. J. Mod. Educ. Comput. Sci. (IJMECS) **5**(3), 26–32 (2013)
11. Luis, R.: Understanding tenure reform: an examination of sense-making among school administrators and teachers. Teach. Coll. Rec. **122**(11), 1–42 (2020)
12. Jiagan, D.: The characteristics of teacher staffing management from the perspective of human resources allocation in Japan. Comp. Educ. Res. **1**, 56–58 (1999). (in Chinese)
13. Danan, W.: Discussion on the performance management of university teachers under the post appointment system. Liaoning Educ. Res. **2**, 89–91 (2006). (in Chinese)
14. Maddin, B.W., Mahlerwein, R.L.: Empowering educators through team-based staffing models. Phi Delta Kappan **104**(1), 33–37 (2022)
15. Wang, J., Pu, S., Wang, H., et al.: A tentative discussion on the classified management of university teachers' posts – The experience of foreign first-class universities and the practice of Chinese universities. Journal of Sichuan Univ. (Philosophy and Social Sciences Edition) **2**, 127–136 (2014). (in Chinese)
16. Paradise, L., Zhu, Y., Zhang, Y., et al.: Value identification of classified management of university teachers. Higher Eng. Educ. Res. **5**, 59–64 (2015)
17. Wang, J., Pu, S.: Analysis on the classified management of university teachers: based on the perspective of modern university personnel management system reform. China Univ. Teacher Res. **1**, 1–6 (2014). (in Chinese)
18. Jianmin, H.: An analysis of the space for improving the efficiency of human resources management in colleges and universities – The development of university staffing function from the perspective of the evaluation scheme's student-teacher ratio index. China Higher Educ. Res. **1**, 56–59 (2009). (in Chinese)

19. Lianzhang, Z.: Construction of human resource management system for teachers of Xi'an Foreign Affairs University. Lanzhou Jiaotong University, Lanzhou (2014). (in Chinese)
20. Gong, Y., Liu, D., Liu, W., et al.: Farewell to the old personnel management system - the record of Tsinghua University's implementation of post appointment and post allowance system. China Higher Educ. **2**, 4–7 (2000)
21. Lanping, L.: Quantitative evaluation of university teachers' post and staffing. J. Gansu Union Univ. (Nat. Sci.) **26**(2), 30–32 (2012). (in Chinese)
22. Zhu, X., Jiang, B., Chang, W., et al.: Discussion on staffing method of a large military hospital. J. Hosp. Manage. People's Liberation Army **4**, 373–375 (2005). (in Chinese)
23. Kotok, S., Knight, D.S.: Revolving doors: cross-country comparisons of the relationship between math and science teacher staffing and student achievement. Leadersh. Policy Sch. **21**(2), 345–360 (2022)
24. Christopher, R., Tuan, N.: Recent trends in the characteristics of new teachers, the schools in which they teach, and their turnover rates. Teach. Coll. Rec. **122**(7), 1–36 (2020)
25. Rasheed-Karim, W.: The Influence of policy on emotional labour and burnout among further and adult education teachers in the U.K. Int. J. Emerg. Technol. Lear. (iJET) **15**(24), 232 (2020)

# Course Outcomes and Program Outcomes Evaluation with the Recommendation System for the Students

Khandaker Mohammad Mohi Uddin[1(✉)], Elias Ur Rahman[1], Prantho kumar Das[1], Md. Mamun Ar Rashid[1], and Samrat Kumar Dey[2]

[1] Department of Computer Science and Engineering, Dhaka International University, Dhaka 1205, Bangladesh
`jilanicsejnu@gmail.com`
[2] School of Science and Technology, Bangladesh Open University, Gazipur 1705, Bangladesh

**Abstract.** Different kinds of educational systems have been applied in learning and teaching to maximize the slope of the learning curve of the students. One of those educational systems is the OBE system or Outcome-based education. Outcome-based methods have been adopted in education systems around the world, at multiple levels. In this education system, students and teachers have a clear view of what needs to be accomplished by the end of the course. However, in this education system, as many of the students are trying to get the same outcome, weak students are getting behind to achieve their goals. Hence creating inequity, the proposed work uses the prior student data to apply the Course Outcomes and Program Outcomes (CO-PO) attainment model and recommends the students to focus on their individual weaknesses. So as the students are studying through their courses, their results are recorded. Then the data of those students are used to train the application in our proposed work. Consequently, the application can identify pupils who need to learn their needed topics. As a result of this proposed method, students with different abilities can not only rely on the outcome, but also have to develop their required subjects to thrive in their learning process.

**Keywords:** OBE · CO-PO · attainment model · Recommend system · Outcome

## 1 Introduction

Goals are the focal point of each educational system element according to the educational paradigm known as outcomes-based education (OBE). By the end of the learning session, every student should have been able to complete the task [1]. There is no single specified technique of instruction or evaluation used in OBE; rather, all of the courses, opportunities, and exams should help students achieve the predetermined outcomes. The role of the faculty person may shift to include that of an instructor, trainer, facilitator, or mentor depending on the desired outcomes [2, 3]. Outcome-based methods have been adopted in education systems around the world, at multiple levels. Australia and South Africa adopted OBE policies from the 1990s to the mid-2000s but were abandoned in the face of substantial community opposition [4, 5]. The United States has had an OBE program in place since 1994 that has been adapted over the years [6]. In 2005, Hong Kong adopted an outcome-based approach for its universities [7]. Malaysia implemented OBE in all of its public school systems in 2008 [8]. The European Union has proposed an education shift to focus on outcomes, across the EU. In an international effort to accept OBE, The Washington Accord was created in 1989; it is an agreement to accept undergraduate engineering degrees that were obtained using OBE methods [9]. As of 2017, the full signatories are Australia, Canada, Taiwan, Hong Kong, India, Ireland, Japan, Korea, Malaysia, New Zealand, Russia, Singapore, South Africa, Sri Lanka, Turkey, the United Kingdom, Pakistan, China, and the United States.

The outcome-based education system embodies the commonsense thinking and practice of effective instructional design and delivery that can be recognized in high-performing learning systems across our society. In this approach, the outcomes of the students' labor are more precisely targeted. Due to the fact that all students must be prepared for the demands of continuous learning and improvement in the Information Age job market, outcome-based models address a clear need in our society for learning systems that promote rather than limit the learning potential of all students. Although in OBE, the individuals who implemented the system in place decide the results. Even while the same results were claimed to have been attained, across programs or even amongst instructors, they could be perceived differently, resulting in a disparity in education. Assessments may become overly mechanical in assessing if an outcome has been met by just checking to determine if the student has learned the material. The assessment may not be primarily concerned with how well a student can use and apply knowledge in various contexts [10]. Students may never be taught how to apply the knowledge they have learned as a result of the emphasis on determining whether the outcome has been reached.

Our research focuses on this concern with the outcome-based education system. The information from a class from a department's course at a university is considered to be evaluated in this research. Our application will then use this data to determine which students need to increase their knowledge in their particular classes. As a consequence of this assignment, every student will be able to pinpoint their areas of weakness. All underachieving students will be helped to improve their courses by being made aware of their limitations.

## 2 Related Works

In a study, Md Nujid et al. [11] evaluated the results of the Geotechnic course in the Bachelor of Engineering (Hons.) Civil program at the school of the department in a university. This study evaluates the result of the students and their survey. It shows that not everyone is correctly evaluated in OBE. The average COPOs percentages are 67 and 81. Lavanya et al. [12] in a study defined the true use of CO-POs in an Outcome-based education system. In this study, they also did the survey from the students and then they used that survey answers to find the attainment of the CO. To attain the PO they took the same approach. This study finds the CO and PO scores of each student and then they attain them. If the scores are below the threshold then the required step must be taken.

This approach shows how the attainment of the CO and PO can help weak students to become better. Mawandiya et al. [13], in their study, went for the same approach. But in this study, the scoring of the CO and PO is very comprehensive. This study found that the CO attainment level for each student and for the entire class can be estimated giving ample scope for multiple possibilities of corrective actions.

In a study, Sahar et al. [14] created a project named Quality Assurance System (QAS). This project is a web-based project where students, the dean of the faculty, and QAS Coordinators will have different views on the website. From this project, the students of the university can see their educational reports. Devasis Pradhan [15], in his study, showed how the Outcome-based education system can change the learning curve of the students. He showed that outcomes-based education can create students more skilled when susceptible to the real world. Shaikh et al. [16] created an application using Python programming language. This application is designed with staff and admin credentials where the admin can upload, edit and delete student lists for the current academic year and staff has the option of inserting data such as CO-PO matrix, target values, and direct and indirect assessment values. After accepting all values, the system generates CO and PO attainment sheet which shows whether the attainment has been achieved or not. This in turn can also help in designing the curriculum. Khwaja [17] built a web-based tool for assessing the CO-POs of the students. ASP.Net templates provided by Visual Studio are used for this project. With this application, the assessment of the results of the students can be easily done and the students of the institution can see their results.

The UGC (University Grants Commission) has now mandated OBE-based education in Bangladeshi institutions. The main problem is that, if we try to accomplish this manually, it is a very challenging procedure to align the program outcomes with the course outcomes. The method by which this manual procedure can be automated using a special application is the theme of our research. We develop our online application, which utilizes the proposed optimum algorithms, to address this research problem.

## 3 Methodology

Admin, Teacher, Respective Authority (Department Head or Dean), and Student are the four users of this system. Admins can edit courses, teachers, and CO/PO in addition to assigning CO/PO, COs, and teachers. The instructor will issue grades based on CO/PO, and the suggested method will then calculate and report the outcome. The outcome, which is known as a progressive report, will be produced for both the entire batch and each individual student. Our system's main goal is to give each student and the entire batch a progress report so that the appropriate authorities can care for each student and the entire batch. The outcome appears as a pie chart. Students can only view their scores; whereas department heads and faculty deans can access batch-wise results individually. The created findings will all be available for download at the end, and each user will be able to obtain their individual progress report as a document file. The work flow of the proposed system is shown in Fig. 1.

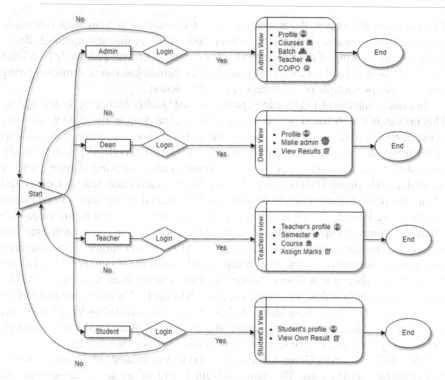

**Fig. 1.** Workflow diagram of the proposed system

The following task is carried out by users in the suggested system.

- Admins can access the system by logging in and control all of its features.
- Except for the outcome, the admin can add, modify, and delete all of the information and users.
- The teacher will issue grades based on CO/PO.
- The system will compute grades and produce a pie chart of the overall results.
- Both batch-wise and individual outcomes can be observed by the appropriate authorities.
- Students are only able to see their results and the system notifies them to improve their POs which are very important in the next semester.

Every sort of user that uses our system must first log in. If they enter the proper username and password in the login section, they can access the system; if not, they are returned to the login page. The flowchart shows that the teacher can access their profile, see the courses they have been assigned, and can only give grades. Assigning courses, course teachers, batches, and CO/PO, on the other hand, is up to the admin. Algorithm 1 depicts an admin's action in the suggested system. If an administrator tries to log in three times without success, their account will be deactivated.

***Algorithm 01:*** *Algorithm for an admin*
1. Begin
2. While i is equal to 1 to 3 do
    a. Input username of email_address and password
    b. If username of email_address and password match with the database
        i. Input choice
        ii. If choice is batch
            1. assign_batch()
            2. edit_batch_info()
        iii. Else if choice is course
            1. assign_course()
            2. edit_course()
        iv. Else if choice is teacher
            1. add_teacher()
            2. assign_teacher()
        v. Else if choice is CO-PO
            1. add_CO_PO()
            2. edit_CO_PO()
            3. assessing_CO_PO()
    c. Else
        i. Enter username or email_address and password
3. End while
4. If the value of i is 4
    a. Block_the_user()
5. End

## 3.1 Course Outcome (CO)

The course outcomes effectively outline how each student would benefit from the course uniquely upon completion and how the same knowledge may be applied in the workplace and in academics after graduation. According to the curriculum, the university selects the many courses that students must complete within four years. Each course's syllabus provides references to specific books, and before deciding on the course's outcomes, the instructors study the prefaces of those books. The program outcomes are then matched to the course outcomes. That mapping establishes the connection between the program result and the course outcome. The four levels of the mapping are High (H), Medium (M), Low (L), and blank. Each CO is assigned to each PO independently if a course has three to twelve program outcomes.

The CO determines what students should be able to do after they achieve the learning goal of a particular course. For each course, 3(three) to 4(four) CO are being used in our system. Every course has a unique CO. For instance, students who complete the CSE-407 (Artificial Intelligence & Neural Networks) course will be able to:

- tackle significant challenges in the real world, integrate and apply networking analysis and artificial intelligence principles.
- determine and examine various problems, then seek answers.

- develop artificial intelligence-based solutions to a variety of issues.
- prepare for local, regional, and global hackathons and project-based competitions.

### 3.2 Program Outcomes (PO)

The POs in our system incorporate the knowledge, effectiveness, and abilities that graduates of the Computer Science and Engineering Program must have. The BSc in CSE program's alumni will be qualified to:

- Build on the understanding of problem-solving, teamwork, and communication skills obtained via the curriculum to advance as a computer science and engineering professional.
- By continuing education, including graduate studies, practical qualifications, and licensure, engage in lifelong learning activities that advance their growth both professionally and personally.
- Show your commitment to social justice, ethics, and leadership via your personal and professional contributions to the society.

For our system, 12 (twelve) POs are predetermined [18–20]. The list of POs for our suggested system can be seen in Table 1.

Table 1. List of POs in the proposed system

| | |
|---|---|
| PO 1 | Engineering Knowledge |
| PO 2 | Problem Analysis |
| PO 3 | Design/development of solutions |
| PO 4 | Investigation |
| PO 5 | Modern tool usage |
| PO 6 | The engineer and society |
| PO 7 | Environment and Sustainability |
| PO 8 | Ethics |
| PO 9 | Individual work and teamwork |
| PO 10 | Communication |
| PO 11 | Project Management and Finance |
| PO 12 | Life-long learning |

## 3.3 CO/PO Evaluation Process

The system measures the percentage, pie chart, and bar graphs for the students in a course after receiving the total marks for the relevant CO. Course teacher will give the marks to the students for a specific course's COs. Then the system calculates the percentage of the obtained marks of corresponding COs using an optimal algorithm. Though these CO's are aligned with predefined POs, the system will evaluate the POs for the students after getting the marks of COs.

# 4 Design and Implementation

We used a variety of tools to create our web application, choosing the ones we were most accustomed to working with. The development tools used to build this application are PHP (LARAVEL), HTML, TAILWIND CSS, VUE.JS, JAVASCRIPT, and MYSQL.

## 4.1 Modules of the System

This app has four different types of users. Admins will be supreme users, teachers will only be capable of completing duties that have been allocated to them, and the other users will only be able to observe them. Admin, Teacher, Head of the Department (HoD)/Dean and student are the four different sorts of users of the proposed system.

## 4.2 Module Description

**Admin:** The administrator will be able to perform superuser functions. In addition to assigning users and departments, he can also allocate courses to teachers and batch, semester, and co-op courses. Besides adding, updating, and deleting them, an admin can make whatever adjustments he wants.

**Teacher:** The teacher is one of the few who makes use of this system. The teacher will be crucial in carrying out numerous system functions. They must first log onto this system using their name and an authorized password. After successfully signing in, he will be directed to the instructor panel page, as shown in Fig. 2. There, he will only see the courses he took and have the ability to input his grades for those courses (course, midterm, final, lab).

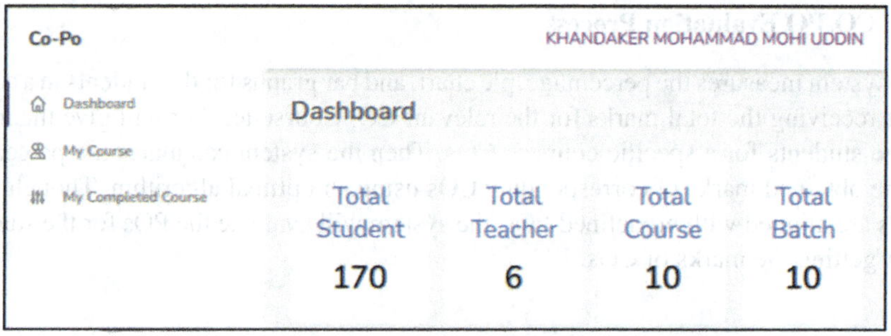

| Co-Po | | | | KHANDAKER MOHAMMAD MOHI UDDIN |
|---|---|---|---|---|
| 🏠 Dashboard | **Dashboard** | | | |
| 👤 My Course | | | | |
| 📋 My Completed Course | Total Student | Total Teacher | Total Course | Total Batch |
| | 170 | 6 | 10 | 10 |

**Fig. 2.** Teacher's panel of the proposed application

When a teacher visits the instructor panel page, he can edit a student's grades by selecting the "my course" option, then going to their batch course. The grades for every student in that batch will then be shown to him, along with an assign update option next to each student's number that may be changed by clicking on it. One CO will satisfy one PO (program outcome), which is a set requirement, and each course will have four COs (course outcomes). The teacher can choose which CO in that course receives certain assignments or exam questions.

**Dean:** Both the individual batch results and the total results will only be visible to DEAN. So that he may quickly assess each student's performance on the CO-PO for each course and identify the poorer individuals. After successfully logging in, he will be taken to the dean panel shown in Fig. 3. He can view the course results for each batch both individually and collectively using the bar and pie charts in his dashboard.

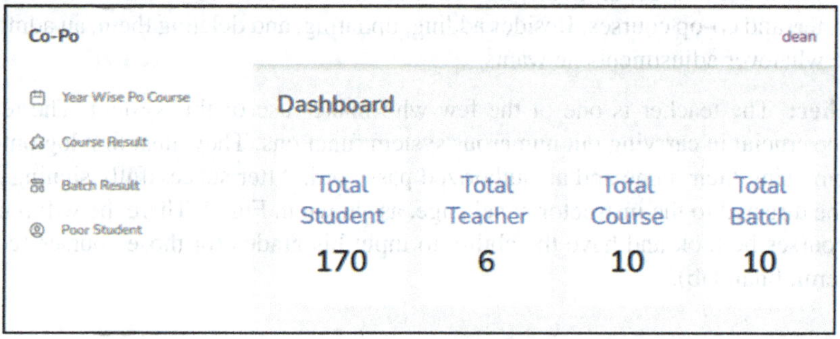

| Co-Po | | | | dean |
|---|---|---|---|---|
| 📅 Year Wise Po Course | **Dashboard** | | | |
| 📊 Course Result | | | | |
| 📋 Batch Result | Total Student | Total Teacher | Total Course | Total Batch |
| 👤 Poor Student | 170 | 6 | 10 | 10 |

**Fig. 3.** DEAN/Head of the department panel

**Student:** The CO-PO scores for the courses that make up their batch must be viewed by students. After successfully logging in, he will be directed to the student panel. He can see a bar chart that displays the course results for every student in his batch. He can also see a bar chart of the course results for the entire members of his batch.

## 5  Result and Discussion

The dean panel offers four choices. He has access to both the overall batch results and the weak student outcomes for each course in each batch. When he chooses the individual option for a course, the CO-PO result for that course will be displayed if he clicks on the result option next to them, as seen in Fig. 4.

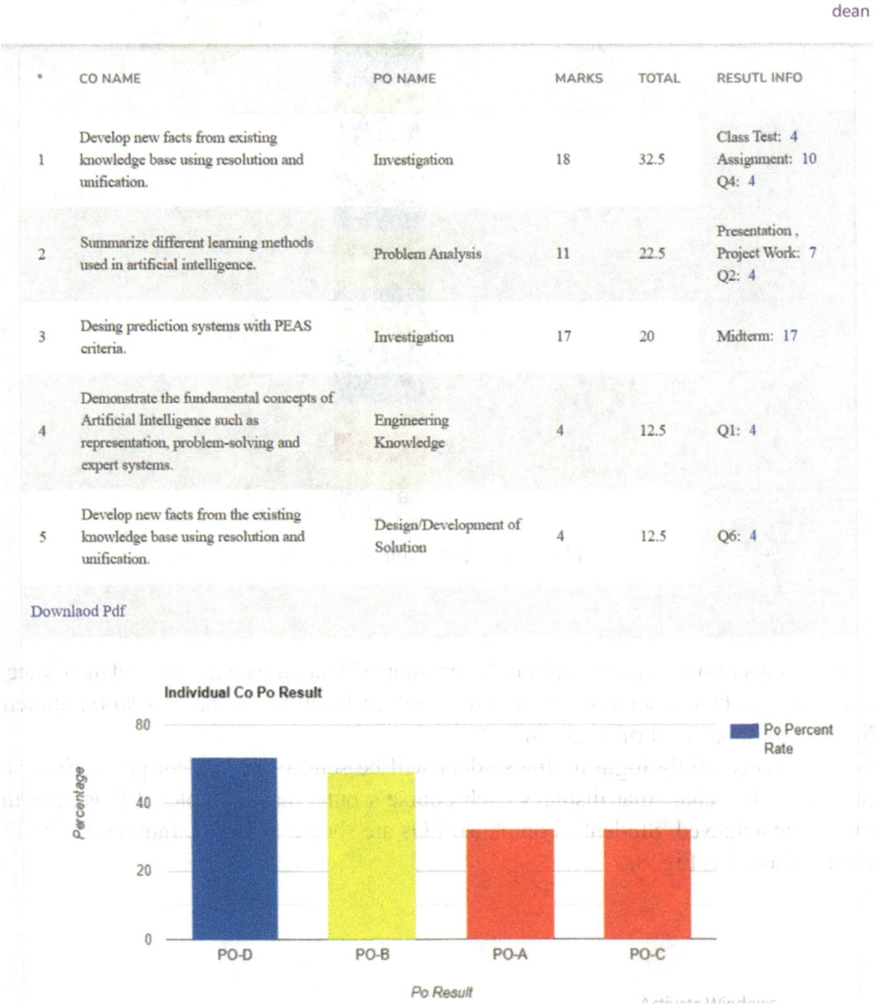

| * | CO NAME | PO NAME | MARKS | TOTAL | RESUTL INFO |
|---|---|---|---|---|---|
| 1 | Develop new facts from existing knowledge base using resolution and unification. | Investigation | 18 | 32.5 | Class Test: 4<br>Assignment: 10<br>Q4: 4 |
| 2 | Summarize different learning methods used in artificial intelligence. | Problem Analysis | 11 | 22.5 | Presentation,<br>Project Work: 7<br>Q2: 4 |
| 3 | Desing prediction systems with PEAS criteria. | Investigation | 17 | 20 | Midterm: 17 |
| 4 | Demonstrate the fundamental concepts of Artificial Intelligence such as representation, problem-solving and expert systems. | Engineering Knowledge | 4 | 12.5 | Q1: 4 |
| 5 | Develop new facts from the existing knowledge base using resolution and unification. | Design/Development of Solution | 4 | 12.5 | Q6: 4 |

Downlaod Pdf

**Fig. 4.** Individual CO-PO result

As shown in Fig. 5, the CO-PO results for a course are presented even before the dean chooses the entire batch results option for that course.

**Fig. 5.** CO-PO result of a batch

Dean logs into the system, selects "poor student" from the menu bar, and then selects a department, batch, semester, and course before locating the batch's worst student. Figure 6 contains a list of weak students.

After successfully login in, the student will be sent to the student panel. Students can view a bar chart that displays each course's outcomes, completed POs, and the percentage achieved. Student's completed POs are shown as Fig. 7. Individual CO-PO result is shown as Fig. 8.

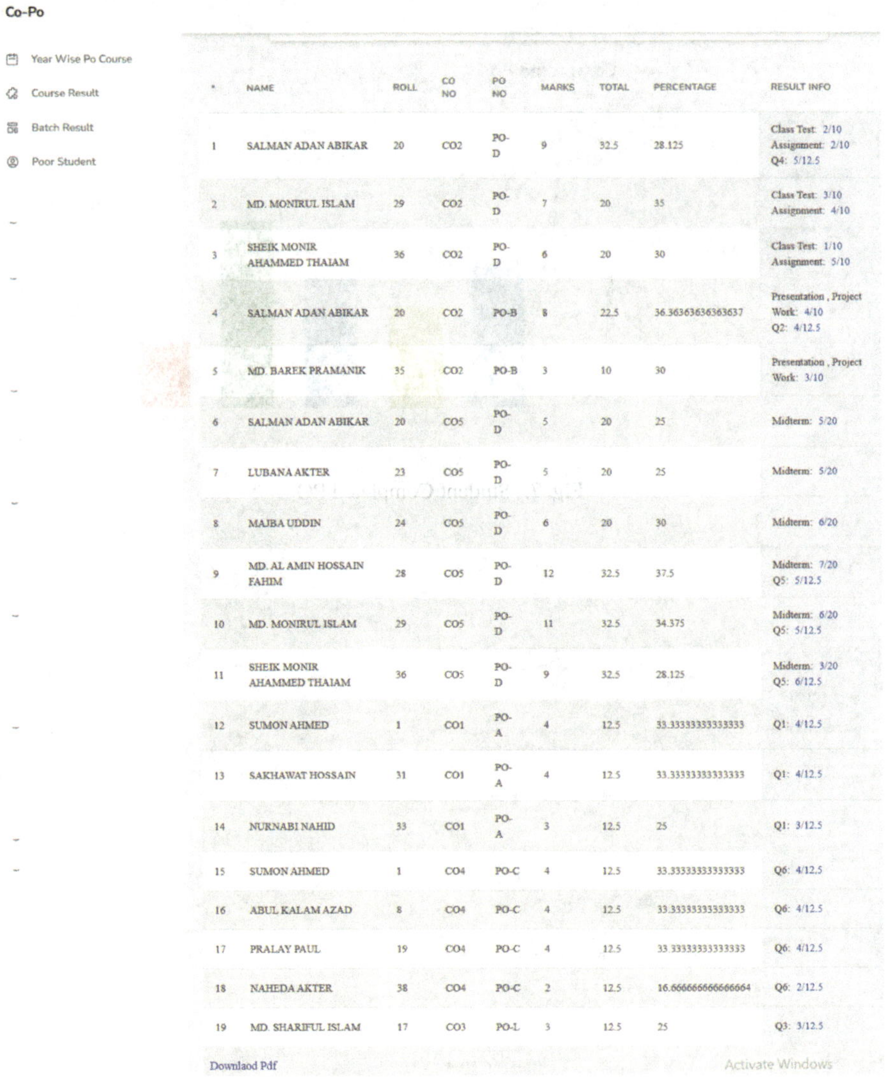

Course Outcomes and Program Outcomes        1049

**Co-Po**

- Year Wise Po Course
- Course Result
- Batch Result
- Poor Student

| * | NAME | ROLL | CO NO | PO NO | MARKS | TOTAL | PERCENTAGE | RESULT INFO |
|---|------|------|-------|-------|-------|-------|------------|-------------|
| 1 | SALMAN ADAN ABIKAR | 20 | CO2 | PO-D | 9 | 32.5 | 28.125 | Class Test: 2/10 Assignment: 2/10 Q4: 5/12.5 |
| 2 | MD. MONIRUL ISLAM | 29 | CO2 | PO-D | 7 | 20 | 35 | Class Test: 3/10 Assignment: 4/10 |
| 3 | SHEIK MONIR AHAMMED THAIAM | 36 | CO2 | PO-D | 6 | 20 | 30 | Class Test: 1/10 Assignment: 5/10 |
| 4 | SALMAN ADAN ABIKAR | 20 | CO2 | PO-B | 8 | 22.5 | 36.363636363637 | Presentation, Project Work: 4/10 Q2: 4/12.5 |
| 5 | MD. BAREK PRAMANIK | 35 | CO2 | PO-B | 3 | 10 | 30 | Presentation, Project Work: 3/10 |
| 6 | SALMAN ADAN ABIKAR | 20 | CO5 | PO-D | 5 | 20 | 25 | Midterm: 5/20 |
| 7 | LUBANA AKTER | 23 | CO5 | PO-D | 5 | 20 | 25 | Midterm: 5/20 |
| 8 | MAJBA UDDIN | 24 | CO5 | PO-D | 6 | 20 | 30 | Midterm: 6/20 |
| 9 | MD. AL AMIN HOSSAIN FAHIM | 28 | CO5 | PO-D | 12 | 32.5 | 37.5 | Midterm: 7/20 Q5: 5/12.5 |
| 10 | MD. MONIRUL ISLAM | 29 | CO5 | PO-D | 11 | 32.5 | 34.375 | Midterm: 6/20 Q5: 5/12.5 |
| 11 | SHEIK MONIR AHAMMED THAIAM | 36 | CO5 | PO-D | 9 | 32.5 | 28.125 | Midterm: 3/20 Q5: 6/12.5 |
| 12 | SUMON AHMED | 1 | CO1 | PO-A | 4 | 12.5 | 33.33333333333 | Q1: 4/12.5 |
| 13 | SAKHAWAT HOSSAIN | 31 | CO1 | PO-A | 4 | 12.5 | 33.33333333333 | Q1: 4/12.5 |
| 14 | NURNABI NAHID | 33 | CO1 | PO-A | 3 | 12.5 | 25 | Q1: 3/12.5 |
| 15 | SUMON AHMED | 1 | CO4 | PO-C | 4 | 12.5 | 33.33333333333 | Q6: 4/12.5 |
| 16 | ABUL KALAM AZAD | 8 | CO4 | PO-C | 4 | 12.5 | 33.33333333333 | Q6: 4/12.5 |
| 17 | PRALAY PAUL | 19 | CO4 | PO-C | 4 | 12.5 | 33.33333333333 | Q6: 4/12.5 |
| 18 | NAHEDA AKTER | 38 | CO4 | PO-C | 2 | 12.5 | 16.666666666664 | Q6: 2/12.5 |
| 19 | MD. SHARIFUL ISLAM | 17 | CO3 | PO-L | 3 | 12.5 | 25 | Q3: 3/12.5 |

Downlaod Pdf

**Fig. 6.** Weak students' list

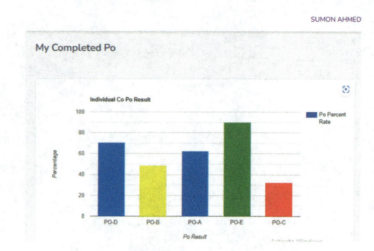

**Fig. 7.** Student Completed PO

**Fig. 8.** Individual CO-PO Result

## 6 Recommendation for the Students

If a student receives a poor CO-PO score, he will be made aware of this CO PO, and moving forward, he must perform better in this course. Algorithm 2 demonstrates the logic behind recommending the courses the student should take to get excellent grades the upcoming semester and create standard POs.

*Algorithm 02:* Recommend the students based on POs percentages
1. Start
2.    From session get stu_id, roll, semester_id
3.    for i: 1 →12   // Total number of PO is 12
4.       PO_wise_obtain_percentage( i, stu_id, roll)
5.       for i: 1 →12
6.          if PO_wise_obtain_percentage( i, stu_id, roll) is less than 40%
7.             Notification_forPO (i)
8.          end if
9.       end for
10. end for
11. End

Figure 9 Shows how notifications will be sent to students' notification panels advising them to improve their CO-PO. They must raise their score after receiving the notifications.

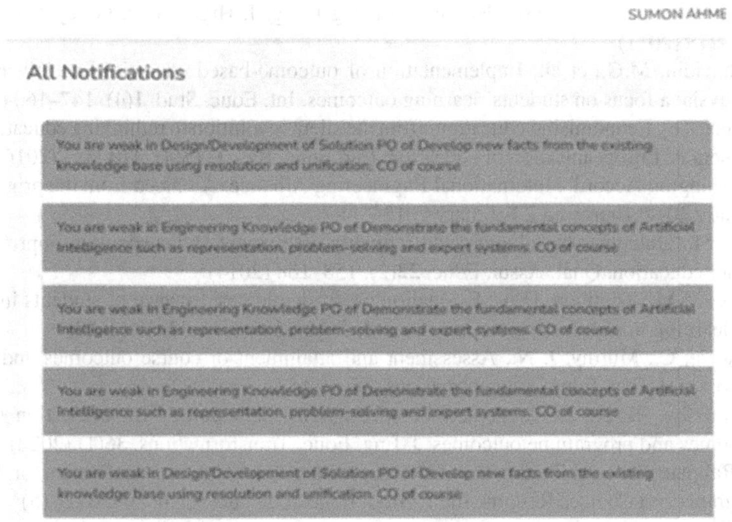

**Fig. 9.** Notifications to a weak student to make good POs scores in the upcoming semester

## 7 Conclusion

The effective implementation of outcome-based education depends on the assessment of program outcomes (POs), and one of the most important instruments for achieving this is the analysis of course outcomes (COs). The proposed study applies the CO-PO attainment paradigm to previous student data and advises students to concentrate on their unique areas of weakness. The analysis helps the student identify their areas of weakness before they follow through with assignments, complete extra coursework, or enroll in other classes to strengthen their understanding and grade. This recommendation is put into practice for the particular course outcome that the student struggles with. Generally, this activity will aid in assessing the institution's teaching and learning process, which will then aid in estimating the successes of several batches of a certain program. A mobile application will be developed in the future to improve the system's usability.

## References

1. Spady, W. G.: Outcome-Based Education: Critical Issues and Answers. American Association of School Administrators, ERIC, (1994)
2. Donnelly, K.: Australia's adoption of outcomes based education: a critique. Issues Educ. Res. **17**(2), 183–206 (2007)
3. Allais, S.M.: Education service delivery: the disastrous case of outcomes-based qualifications frameworks. Prog. Dev. Stud. **7**(1), 65–78 (2007)
4. Austin, T.: Goals 2000: the Clinton administration education program. Retrieved. April. **4**, 2005 (2005)
5. Borkar, H. G., Lakhandur, D. B.: OUTCOME-BASED EDUCATION
6. Kennedy, K.J.: Conceptualising quality improvement in higher education: policy, theory and practice for outcomes based learning in Hong Kong. J. High. Educ. Policy Manag. **33**(3), 205–218 (2011)
7. Mohayidin, M.G., et al.: Implementation of outcome-based education in Universiti Putra Malaysia: a focus on students' learning outcomes. Int. Educ. Stud. **1**(4), 147–160 (2008)
8. Antunes, F.: Economising education: from the silent revolution to rethinking education. a new moment of Europeanisation of education. Eur. Educ. Res. J. **15**(4), 410–427 (2016)
9. "Washington Accord". International Engineering Alliance. Archived from the original on 26 January 2012. Retrieved 2 February 2012
10. Tam, M.: Outcomes-based approach to quality assessment and curriculum improvement in higher education. Qual. Assur. Educ. **22**(2), 158–168 (2014)
11. Nujid, M.M., Tholibon, D.A.: Evaluation on academic performance of students in teaching and learning in engineering course. Asean J. Eng. Educ. **6**(1), 33–39 (2022)
12. Lavanya, C., Murthy, J. N.: Assessment and attainment of course outcomes and program outcomes. J. Eng. Educ. Transformations, **35**(4), 158–168 (2022)
13. Mawandiya, B. K., et al.: A new comprehensive methodology for evaluation of course outcomes and programme outcomes. J. Eng. Educ. Transformations, **36**(1) (2022)
14. El_Rahman, S. A., Al-Twaim, B. A.: Development of Quality Assurance System for Academic Programs and Courses Reports. Int. J. Mod. Educ. Comput. Sci. **6**, 30–36 (2015)
15. Pradhan, D.: Effectiveness of outcome based education (OBE) toward empowering the students performance in an engineering course. J. Adv. Educ. Philos. **5**(2), 58–65 (2021)
16. Shaikh, H.M., Kumar, A.: Implementing an automated application for attainment calculations of program outcomes in outcome based education. J. Positive Sch. Psychol. **6**(2), 6006–6016 (2022)

17. Khwaja, A.A.: A web-based program outcome assessment tool. In: 2018 21st Saudi Computer Society National Computer Conference (NCC), pp. 1–6. IEEE, April (2018)
18. Chandna, V. K.: Innovative methodology for the assessment of Programme Outcomes. In: 2014 IEEE International Conference on MOOC, Innovation and Technology in Education (MITE), pp. 27–31. IEEE, December (2014)
19. Rajak, A., Shrivastava, A.K., Bhardwaj, S., Tripathi, A.K.: Assessment and attainment of program educational objectives for post graduate courses. Int. J. Mod. Educa. Comput. Sci. **11**(2), 26–32 (2019)
20. Liu, Y., Zhang, X.: Evaluating the undergraduate course based on a Fuzzy AHP-FIS model. Int. J. Mod. Educ. Comput. Sci **12**(6), 55–66 (2020)

# Methodology of Teaching Educational Disciplines to Second (Master's) Level Graduates of the "Computer Science" Educational Program

Ihor Kozubtsov[1]✉, Lesia Kozubtsova[1], Olha Myronenko[1], and Olha Nezhyva[2]

[1] Kruty Heroes Military Institute of Telecommunications and Information Technology, Kyiv 01011, Ukraine
kozubtsov@gmail.com

[2] Taras Shevchenko National University of Kyiv, Kyiv 04053, Ukraine

**Abstract.** Purpose and objectives of the article. To substantiate the unified methodology of teaching the discipline of educational components to applicants of the second (master's) level of higher education of the educational program "Computer Science" of the field of knowledge 12 Information Technologies of the specialty 122 Computer Science of full-time and part-time forms of study. To achieve this goal, the following tasks are set: 1. Analyze the current state of research and publications; 2. To substantiate the unified methodology of teaching the discipline of educational components to applicants of the second (master's) level of higher education of the educational program "Computer Science" of the field of knowledge 12 Information Technologies of the specialty 122 Computer Science of full-time and part-time forms of education on the basis of a computer game. Research result. The unified methodology of teaching the discipline of educational components to applicants of the second (master's) level of higher education of the educational program "Computer Science" of the field of knowledge 12 Information Technologies of the specialty 122 Computer Science of full-time and part-time forms of education on the basis of the teacher's game is justified. Through the application of the reverse transformation of a student into a teacher, the educational and developmental goal of students acquiring primary educational quasi-professional teaching experience has been achieved. The scientific novelty lies in the fact that the first developed a unified methodology for teaching the discipline of educational components to applicants of the second (master's) level of higher education of the educational program "computer science" on the basis of a game (gamified) approach interested in acquiring educational quasi-professional experience in teaching. Thus, the method proposed by the authors for teaching educational disciplines of educational components to students of the second (master's) level of higher education of the educational program "Computer Science" of the field of knowledge 12 Information Technologies of the specialty 122 Computer Science provides a practical implementation of the theory of "anticipatory learning". As a result of its application, a large reserve of time budget allocated for independent work will allow students to deeply study the educational material, prepare for a lecture class and, as a final result, acquire educational quasi-professional experience for future activities. In the case of applying

© The Author(s), under exclusive license to Springer Nature Switzerland AG 2023
Z. Hu et al. (Eds.): ICAILE 2023, LNDECT 180, pp. 1054–1067, 2023.
https://doi.org/10.1007/978-3-031-36115-9_94

the methodology based on gamification, it allows to create such an informational and educational environment that contributes to the independent, active pursuit of higher education students to acquire knowledge, professional skills and abilities, such as critical thinking, the ability to make managerial (managerial) decisions, work in a team, be ready to cooperate; helps reveal abilities and motivates self-education. The methodology reveals excellent properties in the case of a mixed application of distance education and advance training, which will not only ensure the implementation of the curriculum for mastering the educational component, but also provide maximum opportunities for students to preserve life and health in war conditions.

**Keywords:** teaching methods · academic discipline · methodology of scientific research · applicant · student · computer science

# 1 Introduction

## 1.1 Problem Statement

According to the conditions of intensive growth in the amount of scientific and technical information, rapid change and updating of the system of scientific knowledge, there is a need for a qualitatively new theoretical training of future highly qualified specialists (master's) level of higher education of the educational program "Computer Science" in the field of knowledge 12 Information technologies, specialty 122 Computer sciences. These qualified specialists will be capable of independent creative work, implementation of knowledge-intensive technologies in production and adaptation to the conditions of market relations.

Knowledge of the methodology of scientific research, methods and organization of scientific research activity will help masters of the specialty 122 Computer sciences to easily get involved in professional activities, to translate scientific knowledge into a practical plane and will contribute to the development of rational and creative thinking.

Scientific activity in higher educational institutions is an integral part of the educational process and is carried out with the aim of integrating scientific, educational and industrial activities in the higher education system. The Law of Ukraine "On Higher Education" (2014) [1] defines the main tasks of scientific activity in higher educational institutions, including:

- organic unity of the content of education and programs of scientific activity;
- direct participation of the subjects of the educational process in scientific and research works conducted in the higher educational institution;
- organization of scientific, scientific-practical, scientific-methodical seminars, conferences, competitions of scientific research, coursework, diploma and other works of participants in the educational process.

The COVID 19 pandemic created an unprecedented challenge to the education system, intensively prompted the use of distance education in institutions of higher education (HEIs) across the country, and became more active and the subject of widespread use [2]. The existing traditional methods of teaching do not allow providing high-quality training. Rote teaching-learning methods are not as effective as far as the curriculum design in the computing field is concerned [3].

In this regard, there was a need to substantiate a unified author's method of teaching the educational discipline of educational components to students of the second (master's) level of higher education of the educational program "Computer Science" for the implementation of effective distance learning [4].

### 1.2 Literature Review. Analysis of Recent Research and Publications

Students' and cadets' motivation for traditional teaching methods began to fade paradoxically with the advent of computer technology. Game methods of teaching adults and gamification technologies in education came to the fore. These ideas were reflected in the publications of both foreign and domestic researchers. We will conduct a historical outline of the main works.

In order to maximize learning in a constrained amount of time, the paper [5] investigates the application of active learning pedagogy. The potential for boosting student engagement and learning with active learning is quite promising. It is not a novel idea and has been encouraged and promoted since the 1980s. Active learning is used in classrooms by many professors due to its numerous advantages. Faculty are urged to examine their own pedagogical approaches and work to enhance them in order to better engage students and pique and hold their interest. Active learning has been used as a tool to pique students' interest and ultimately boost learning, but it has rarely been used to directly affect learning in terms of time. The active learning approach used in this research is based on traditional pedagogies that were created using a variety of psychological theories of learning, motivation, and engagement. A survey of the students conducted following the use of several instances of this active learning technique showed an improvement in student learning.

As the teaching and learning system is not a collection of knowledge that is packed in the mind, the development of the world today forces scholars in the area of education to examine the teaching techniques and tactics. The usage of educational teaching games is one of the most current developments. Games boost student motivation and assure engagement with instructional information, which provides a pleasant and engaging way to accomplish the intended goals. The writers of the article [6] make an effort to highlight the need of using games to enhance learning in practical ways and raise the bar of the educational process in a collaborative setting for both instructors and students. The suggested approach in this research was evaluated using a survey technique, and the findings are quite positive for academic professionals.

The prototype of the development of the methodology of teaching certain disciplines in a game form was the justification of the methodology of teaching electrical engineering disciplines by the method of a virtual computer game [7], which was further developed in the concept of independent training of cadets of the Ground Forces on educational and training tools by the method of a game on a virtual computer [8].

Further practical steps in the gamification of teaching based on game methods and strategies of teaching and education can be found in the work of K. Kapp [9].

V. Bugaeva rightly notes that gamification is an educational technology which is rapidly developing, having a huge potential to positively influence the effectiveness of the educational process [10, p.135].

From the point of view of our research, it is relevant for the preparation of future highly qualified specialists (master's) level of higher education of the educational program "Computer Science" of the field of knowledge 12 Information Technologies of the specialty 122 Computer Science to consider gamification: as a way of forming the active professional behavior of future specialists in the IT industry [10]; as formal and informal space [11]; as an innovative pedagogical educational technology [12, 13] and learning technology in educational activity [14].

This confirms the opinion of N. Volkov that gamification is one of the trends of modern higher education [15].

### 1.3 Highlighting Aspects that are not Sufficiently Studied

The analysis of previous studies revealed that in modern pedagogy in terms of the development of this topic, there is no unified method of teaching educational disciplines of educational components to students of the second (master's) level of higher education of the educational program "Computer Science" of the discipline. Based on this, the authors have chosen this current direction of research.

### 1.4 Purpose of the Article

The purpose of the article is to substantiate the methodology of teaching the discipline of educational components to students of the second (master's) level of higher education of the educational program "Computer Science" of the field of knowledge 12 Information Technologies of the specialty 122 Computer Science of full-time and part-time forms of education.

### 1.5 Research Objectives (Goals)

To achieve the goal, the following tasks are set:

1. To analyze the current state of research and publications.
2. To justify the methodology of teaching educational disciplines of educational components to students of the second (master's) level of higher education of the "Computer Science" educational program of the field of knowledge 12 Information technologies of the specialty 122 Computer sciences of full-time and part-time forms of education based on the teacher's educational game.

## 2 Research Methods

### 2.1 Research Tools

Basic research tools include methods of theoretical analysis and generalization of scientific literature on the topic of the research; generalization to formulate conclusions and recommendations for effectiveness.

## 2.2 Reliability and Accuracy of Results

Reliability of the results of the study is ensured by the correctness of the use of mathematical apparatus and research methods.

## 2.3 Methodological Basis of the Study

The object of scientific and theoretical research is not just a single phenomenon, a specific situation, but a whole class of similar phenomena and situations, their totality.

The methodological basis of the research is the ideas of L. Vyhotskyi, P. Halperin, Y. Babanskyi, S. Rubinshtein, (pedagogical psychology), V. Bespalko, Y. Mashbits, (cybernetic approach in pedagogy, programmed learning and automated training systems). Thus, the basic theory for effective learning is proposed by the theory of "advanced learning" (M. Nechkina, 1984 [16]; S. Lisenkova, 1988 [17]) or the "inverted class" theory (J. Bergmann, A. Sams, 2012 [18]). If blended learning is backed by quality material, it may be successfully implemented. The topic, performance tasks, discussion forums, and quiz questions all make up good content if they are presented in an engaging and organized way. The ability of pupils in the cognitive, emotional, and psychomotor domains must also be evaluated using high-quality material [19].

# 3 Research Results

## 3.1 Theoretical Foundations of the Construction of Teaching Methods

Scientific and research activities in higher education institutions of Ukraine are carried out on the basis of the current laws of Ukraine "On higher education" [1], "On scientific and scientific and technical activity" [20], statutes of higher education institutions and provide for the wide involvement of students to research work, enrichment of their knowledge with new scientific data, development of abilities for creative thinking, scientific analysis of phenomena, processes, which is a fundamentally important task not only for the department of "Computer Sciences".

The methodology of teaching the educational discipline of educational components to students of the second (master's) level of higher education of the educational program "computer science" is based on the understanding of the students of higher education of the concept of "methodology" as a study of the organization of activities. The innovators of this point of view are leading scientists such as A. Novikov [21], L. Marakhovskyi, B. Sus, S. Zabara, I. Kozubtsov, I. Gevko, A. Stepaniuk, H. Tereshchuk, Yu. Khlaponin, etc.

The combination of forms of conducting classes is a mandatory element of the successful assimilation of the educational material of the educational components of the students' discipline of the second (master's) level of higher education of the educational program "Computer Science" (the field of knowledge 12 Information Technologies of the specialty 122 Computer Science of full-time and part-time forms of education). It should be noted that currently the choice of pedagogical learning technologies is a key problem for the teacher – the subject of the educational process [22]. At the same time, the lack of recommendations regarding formalization in their selection expands the lecturer's degree of freedom in creative search and experimentation.

As can be clearly seen from Table 1 the following learning methods are the most accepted for the formation of knowledge: didactic games; practical training; teaching others (mutual learning) and independent work.

**Table 1.** Comparative characteristics of different teaching methods

| Teaching methods | | Solved tasks | | | | | |
|---|---|---|---|---|---|---|---|
| | | form | | develop | | | gaining experience |
| | | knowledge | skills | thinking | memory | language | |
| Verbal | (lecture) | 5% | ++ | – | – | – | ++ |
| | reading | 10% | | | | | |
| | listening | 20% | | | | | |
| Visually | Work with multimedia (audiovisual) | 20% | + | ++ | + | + | – |
| | Viewing the drawing | 30% | | | | | |
| | Demonstration | 30% | + | + | + | ++ | – |
| | Video viewing | 50% | | | | | |
| Practical | Working with a book (reading) | 10% | + | + | + | + | + |
| | Educational discussions | 50% | ++ | – | ++ | + | ++ | ++ |
| | Didactic games | 70% | ++ | – | ++ | + | ++ | ++ |
| | Practical training | 75% | + | ++ | ++ | + | – | ++ |
| | Teaching others is the application of what has been learned | 90% | ++ | – | ++ | + | ++ | ++ |
| Independent work | | 80% | ++ | ++ | ++ | + | + | |
| Oral and written control | | | ++ | – | + | + | ++ | |

Note: ++ - solve very well; + - solve partially; – – solve poorly

The author's method of teaching the educational disciplines of the educational components provides for the creation of positive pedagogical conditions for students of the second (master's) level of higher education of the educational program "Computer Science" (the field of knowledge 12 Information Technologies of the specialty 122 Computer

Sciences of full-time and part-time forms of education) to encourage them to acquire the primary practical experience in the process of educational quasi-professional activity.

Educational quasi-professional activity enriches students with the ability to direct them to self-development in the direction of future professional activity, which includes: management (managerial) activity; scientific (research) activity; pedagogical (teaching) activity.

Educational quasi-professional management (managerial) activity is formed in students through a responsible attitude to the self-organization of independent work.

Independent work is the main means of mastering educational material in the time free from standardized educational classes, that is, lectures and practical classes (auditory work). During independent work, students should pay attention to: work on processing and studying the recommended literature; preparation for discussions and other tasks proposed by the teacher; work on an abstract (educational article, theses, report); work on an individual research project, etc.

The methods of conducting classes in the educational disciplines of the educational components with indications of approximate time are given in Table 2.

**Table 2.** Methods of conducting classes

| Type of lesson | The structure of the lesson ||||||| Methodical support |
|---|---|---|---|---|---|---|---|---|
| | Introductory part || Main part ||| Final part || |
| | t' | method | t' | | | t' | method | |
| | | | | main | additional | | | |
| Lecture | 10 | C | | EL | S, Cl, D | 5 | | The text of the lecture |
| Practical class | 20 | C | | E | IW | 10 | C | Methodical development. Software tasks |
| Independent work | – | – | – | IW | – | – | – | Methodical development. Tasks in the IW |

## 3.2 Peculiarities of the Methodology, Application of Teaching Methods and Tools

Features and methods of conducting classes with indications of estimated time are given in Table 2. The table uses the following abbreviations: C – conversation; EL – educational lecture; S – story; Cl – clarification; D – demonstration; DS – display; E – exercise; IW – independent work.

The introductory lecture is conducted by the leading lecturer, one of the professors. The goal, task structure of the educational discipline is explained to the students at this lecture. Besides, it is justified in the professional need for its study by future teachers and scientists in the field of Information technologies.

The lecturer must practically show the method of conducting the lecture, demonstrate at the highest level his/her own pedagogical skill, methodological culture, which should be reflected in the students as a certain standard to which one must go and try to exceed over time by forming his/her own pedagogical skill and methodological culture.

The professional activity of students as future scientific and pedagogical workers begins with the scientific understanding of the object, subject and purpose of the work and the final result. Besides, the very course of work is a project. For students of the specialty 122 Computer science, in most cases, the presentation is interesting in a game form with an illustrative example in the form of an algorithm of the reporting stages of the academic discipline [23].

After a few lectures (these are usually 4–6 lectures depending on the methodology), it is suggested to actively involve students in the role of future lecturers (teachers). This pedagogical method gives students the opportunity to overcome their own fear of the audience, to try to realize themselves in a quasi-professional educational activity. In this case, we can talk about the pedagogical component. To do this, the "full-time lecturer" on the eve of the lecture sets students tasks for independent work through the head of the study group, determines the topics of the lesson, educational questions that must be worked out by students. Students who are preparing for lectures, study information search, analysis, validation, generally gain practical experience in scientific activities, etc.

The technology of preparing and conducting lectures by students was worked out in the context of a scientific and methodical seminar at the department [24].

Incoming reports, the lecturer observers the practical phase of the development of professional experience, methodological culture of the student.

The lecturer's interaction with students in the process of lecture-scientific-methodical support is based on the model of subject-object interaction. According to the educational component of the scientific and methodological support of students, the theoretical basis is Disterweg's fundamental idea. Disterweg notes that a bad teacher presents the truth, a good one teaches to find it [25, p. 161]. This idea emphasizes the special importance of creating such an educational atmosphere in which the student himself/herself will begin to find answers to questions.

To realize this truth, the principle of independent work of students in the technology of scientific and methodological support should be followed, and the interaction between the lecturer and students according to the algorithm [26], directing its further development and interpreting to solve the task of acquiring quasi-practical experience.

If necessary, the full-time lecturer manages and makes corrections in the pedagogical process. Moreover, this lecturer determines the rotation of the speaker to another student with the continuation of the report on the essence of the educational issue. Thus, the principle of attentiveness of listeners is realized.

### 3.3 Discussion of Research Results

The lecturer's work methods and students' activities are based on the use of pedagogical technologies such as project; programmed training; problem-based learning by transforming ideas [27–30].

The method of teaching educational disciplines of educational components involves assessing the level of formation of the applicant (master's degree) as a future teacher in the field of Information technologies (Computer sciences). Approximate evaluation points are given in Table 3.

**Table 3.** Assessment of the level of teaching by the recipient

| Characteristics | Implementation in the activities of the lecturer |
|---|---|
| Formulation of the topic and definition of the goal | sufficiently clear and understandable for the students |
| | Somewhat blurred |
| | They remained unclear to the students |
| Plan and structure | The structure is clear, meaningful parts are highlighted and interconnected |
| | The general plan is defined, but the transitions from one semantic part to another remain unclear |
| Content | It is difficult to follow the development of the topic, the main ideas were expressed inconsistently |
| The ability to arouse interest in the topic | Theoretical positions were argued, supported by facts and examples |
| | Scientific, evidential, but very difficult to understand |
| | Very popular, empirical material prevailed |
| | The meaning of the topic is revealed convincingly, the material was connected with the personal experience of the student |
| | Only the need to study the topic was indicated, examples were used, there was no reliance on the personal experience of the students |
| | The importance of studying the topic was not motivated in any way, the material was not supported by examples |
| Problem statement | The lecturer drew attention to contradictions, formulated problematic issues, and encouraged the audience to discuss |
| | The lecturer formulated problematic questions and answered them himself/herself |
| | The lecturer expressed the theoretical material as something universally recognized, which does not require proof |

(*continued*)

**Table 3.** (*continued*)

| Characteristics | Implementation in the activities of the lecturer |
|---|---|
| Contact with the audience | The contact was complete, all students worked, the lecturer took into account "feedback" |
| | Sometimes the lecturer lost contact with the audience and the students started to get distracted |
| | The lecturer failed to establish contact with the audience and take into account "feedback" |
| Culture of the lecturer's language | Clear diction, optimal pace of speech, emotional presentation |
| | The diction and pace of speech are normal, but the emotional component was missing |
| | There were problems with diction, language pace, emotionality of presentation |
| The behavior of the lecturer | He/she held himself /herself confidently, reasoned freely on the topic, did not use the lecture notes |
| | He/she held himself/herself confidently, but he/she did not possess the skill of fluent speech, he/she relied on the lecture notes |
| | He/she kept himself/herself rigid, practically did not tear himself/herself away from the text of the lecture |
| Use of visual aids | Graphic methods of presenting the material and other visual aids were actively used |
| | A blackboard was occasionally used |
| | There were no visual aids |
| Good conclusion | The conclusion logically concluded and summarized what was presented |
| | The conclusion was unclear |
| | There was no conclusion |

A large reserve of time budget allocated for independent work will allow students to deeply study the educational material, prepare for the lecture and, as a final result, acquire educational quasi-professional experience for future activities [16–18]. This is confirmed by the results of the study. Student success in studies depending on the time spent on extracurricular activities [31].

We definitely agree with the opinion of the authors that to improve teaching and learning, it is necessary to use assistive technologies [32]. For example, the use of gamification, but in such a way that the method of learning through the game does not turn into a pure game [9, 11, 12, 22].

## 4 Summary and Conclusion

Thus, the method proposed by the authors for teaching educational disciplines of educational components to students of the second (master's) level of higher education of the educational program "Computer Science" of the field of knowledge 12 Information Technologies of the specialty 122 Computer Science provides a practical implementation of the theory of "anticipatory learning". As a result of its application, a large reserve of time budget allocated for independent work will allow students to deeply study the educational material, prepare for a lecture class and, as a final result, acquire educational quasi-professional experience for future activities. In the case of applying the methodology based on gamification, it allows to create such an informational and educational environment that contributes to the independent, active pursuit of higher education students to acquire knowledge, professional skills and abilities, such as critical thinking, the ability to make managerial (managerial) decisions, work in a team, be ready to cooperate; helps reveal abilities and motivates self-education. The methodology reveals excellent properties in the case of a mixed application of distance education and advance training, which will not only ensure the implementation of the curriculum for mastering the educational component, but also provide maximum opportunities for students to preserve life and health in war conditions.

### 4.1 Expanding the Boundaries of the Scientific Field

The scientific result obtained by the authors expands the boundaries of pedagogical sciences in the part related to the scientific specialty "teaching theory and methodology".

### 4.2 Scientific Novelty. Scientific Justification

For the first time, a methodology was developed for teaching educational disciplines of educational components to students of higher (master's) education majoring in "Computer Science" educational program of the field of knowledge 12 Information technologies major 122 Computer science full-time and part-time forms of study based on a game approach, interested in acquiring educational quasi-professional experience pedagogical activity for with the possibility of its application in war conditions due to mixed education.

### 4.3 Practical Use

The proposed technique is fully ready for practical application as advanced anticipatory learning. Since the education system of Ukraine from February 24, 2022, like the entire country, functions in extreme conditions, it is necessary to adapt to the reality of war, during which it is necessary to continue the educational process of training students of higher education for the needs of the national economy.

Thus, the use of distance education and advanced training will not only ensure the implementation of the curriculum for mastering the educational component, but also provide maximum opportunities for students to preserve life and health in war conditions.

For the first time, lecturer I. Kozubtsov at the Department of "Computer Sciences" of the Lutsk National Technical University developed a methodology for teaching educational disciplines of educational components to students of the (master's) level of higher education of the "Computer Science" educational program of the field of knowledge 12 Information technologies of the specialty 122 Computer sciences of full-time and part-time forms of education based on a game approach, interested in acquiring educational quasi-professional experience in teaching activities.

### 4.4 Prospects for Further Research and Study

The research is expected to bring more academic and applicable value.

The methodology needs further improvement if it is necessary to conduct laboratory or practical classes on the use of stationary equipment of higher educational institutions and it is not possible to practice it at home.

**Acknowledgment.** The authors would like to express their respect to the organizers of the 3rd International Conference on Artificial Intelligence and Logistics Engineering (ICAILE2023) (March 11–March 12, 2023, Wuhan, China) and at a tragic time for Ukrainian researchers, Modern Education and Computer Science Press provided a grant to publish their scientific achievements free of charge. We are sincerely grateful.

# References

1. Law of Ukraine. "On higher education" No.1556-VII (2014). http://zakon4.rada.gov.ua/laws/show/1556-18 (in Ukrainian)
2. Schwab, K., Malleret, T.: COVID-19: The Great Reset. Edition 1.0. Switzerland. Cologny/Geneva: Forum publishing World Economic Forum. 2020
3. Churi, P., Rao, N.T.: Teaching cyber security course in the classrooms of NMIMS university. Int. J. Mod. Educ. Comput. Sci. **13**(4), 1–15 (2021)
4. Kozubtsova, L.M., Kozubtsov, I.M.: On the problem of organizing effective distance learning. The First International Scientific and Practical Conference "Social aspects of Military Professional Activity of the Security and Defense Sector: Challenges of our Time": collection of abstracts, (Kharkiv, 20.05). National Academy of the National Guard of Ukraine, pp. 284–286 (2021). (in Ukrainian)
5. Khan, A.A., Madden, J.: Speed learning: maximizing student learning and engagement in a limited amount of time. Int. J. Mod. Educ. Comput. Sci. (IJMECS). **8**(7), 22–30 (2016). https://doi.org/10.5815/ijmecs.2016.07.03
6. Albilali, A.A., Qureshi, R.J.: Proposal to teach software development using gaming technique. Int. J. Mod. Educ. Comput. Sci. (IJMECS). **8**(8), 21–27 (2016). https://doi.org/10.5815/ijmecs.2016.08.03
7. Kozubtsov, I.N.: Teaching electrical engineering disciplines by virtual computer game method. In: Electrical Technologies, Electric Drive and Electrical Equipment of Enterprises: Collection of Scientific Papers of the Second All-Russian Scientific and Technical Conference, vol. **2**, pp. 107–110. Ufa: USNTU Publishing House (2009). (in Russian)
8. Kozubtsov, I.M.: The concept of independent training of cadets of the ground forces on training facilities by playing on a virtual computer. In: Prospects for the Development of Weapons and Military Equipment of the Ground Forces. Second All-Ukrainian Scientific and Technical Conference, p. 77 (2009). (Lviv, April 28-29)

9. Kapp, K.: The Gamification of Learning and Instruction Game-Based Methods and Strategies for Training and Education. Pfeiffer, San Francisco, USA (2012)
10. Bugaeva, V.: Gamification as a way of forming active professional behavior of future IT industry specialists. Pedagogy Psychol. **56**, 129–135 (2018)
11. Tkachenko, O.: Gamification of education: formal and informal space. Top. Issues Humanit. **11**, 303–309 (2015)
12. Petrenko, S.: Gamification as an innovative educational technology. Innov. educ. **2**(7), 177–185 (2018)
13. Nezhivaya, O.: Innovative technologies in the educational process. In: Innovative Trends in Training Specialists in a Multicultural and Multilingual Globalized World: a Collection of Abstracts of Reports of the V All-Ukrainian Scientific Society-Practical Conference (Kiev, April 07). Kiev: KNUTD, pp. 74–77 (2020). (in Ukrainian)
14. Noskov, E.A.: Learning technologies and gamification in educational activities. Yaroslavl Pedagogical Bull. **6**, 138–142 (2018). (in Russian)
15. Volkova, N.P.: Gamification as one of the trends of modern higher education. In: Modern Higher Education: Problems and Prospects: VI All-Ukrainian Scientific and Practical Conference of Students, Postgraduates and Scientists: Abstracts of Reports (Dnipro, 22.03). 2018: 33–35 (in Ukrainian)
16. Nechkina, M.: Increasing the effectiveness of a lesson. Communist. **2**, 51 (1984)
17. Lysenkova, S.M.: The Method of Anticipatory Learning: a Book for Teachers: From Work Experience. Enlightenment, Moscow (1988)
18. Bergmann, J., Sams, A.: Flip Your Classroom: Reach Every Student in Every Class Every Day. International Society for Technology in Education, Washington, DC (2012)
19. Putu Wisna Ariawan, I., Divayana, D.G.H., Wayan Arta Suyasa, P.: Development of blended learning content based on Tri Kaya Parisudha-superitem in Kelase platform. Int. J. Mod. Educ. Comput. Sci. **14**(1), 30–43 (2022)
20. Law of Ukraine "On scientific and technical activities", No. 848-VIII (2015). http://zakon3.rada.gov.ua/laws/show/848-viii (in Ukrainian)
21. Novikov, A.M., Novikov, D.A.: Methodology. SINTEG, Moscow (2007). (in Russian)
22. Lishchina, V., Kozubtsov, I., Kozubtsova, L.: Choice of pedagogical training technologies as a key problem teacher – subject of the educational process. In: International Scientific and Methodological Conference "Innovative Technologies in Military Education", (Odessa, June 25). Odessa: Military Academy, pp. 225–226 (2021). (in Ukrainian)
23. Kozubtsov, I.M.: Method of virtual cognitive presentation of reporting stages of an academic discipline to cadets. In: Fifth Scientific and Technical Conference "Priority Areas for the Development of Telecommunications Systems and Special-purpose Networks", (Kiev, October 20–21). VITI NTUU "KPI", pp. 144–147 (2010). (in Ukrainian)
24. Kozubtsov, I.N., Kozubsova, L.M.: Pedagogicheskaya technology organization nauchno-methodicheskogo workshop chair. Pedagogical skills. Theoretical and scientific-methodical Journal. O'zbekiston Respublikasi Buxoro davlat universiteti. **1**, 24–30 (2016)
25. Disterveg, A.: A guide to the education of German teachers. In: Selected Pedagogical Essays, pp. 136–203 (1956)
26. Mayer, R.V.: Cybernetic pedagogy: simulation modeling of the learning process. Glazov: GGPI, p. 138 (2013). (in Russian)
27. Trubavina, I., Kaplun, S.: Subjectivity of students as a pedagogical condition for the formation of their cognitive independence in learning. In: Fundamental and Applied Research: Modern Scientific and Practical Solutions and Approaches: Proceedings of the Fifth International Scientific and Practical Conference. National Academy of Sciences of Azerbaijan, vol. 5, pp. 288–292 (2019)

28. Trubavina, I., Kaplun, S.: Complex of conditions for applying training as a form of organizing training for students of higher educational institutions of Ukraine. Problèmes et perspectives d'introduction de la recherche cognique innovante: collection de papiers cogniques "ΛÓΓΟΣ" avec des matériaux de la conférence diagnosique et pratique internationale, (Bruxelles, novembre 29). Plateforme scientifique européenne, vol. 5, pp. 21–23 (2019)
29. Trubavina, I.: Formation of skills of independent work among students by methods of problem-based learning in a modern university. Bulletin of Luhansk Taras Shevchenko National University. Pedagogical Sciences **7**(312), 2: 116–125 (2017). (in Ukrainian)
30. Nezhyva, O.: The aspects of smart education in the world. Khazar J. Humanit. Soc. Sci. **24**(3), 62–72 (2021)
31. Sharma, N., Appukutti, S., Garg, U., Mukherjee, J., Mishra, S.: Analysis of student's academic performance based on their time spent on extra-curricular activities using machine learning techniques. Int. J. Mod. Educ. Comput. Sci. (IJMECS). **15**(1), 46–57 (2023). https://doi.org/10.5815/ijmecs.2023.01.04
32. Adebayo, E.O., Ayorinde, I.T.: Efficacy of assistive technology for improved teaching and learning in computer science. Int. J. Educ. Manage. Eng. **12**(5), 9–17 (2022). https://doi.org/10.5815/ijeme.2022.05.02

# Professional Training of Lecturers of Higher Educational Institutions Based on the Cyberontological Approach and Gamification

Oleksii Silko, Lesia Kozubtsova, Ihor Kozubtsov[✉], and Oleksii Beskrovnyi

Kruty Heroes Military Institute of Telecommunications and Information Technology, Kyiv 01011, Ukraine
kozubtsov@gmail.com

**Abstract.** The subject of research in the scientific article was the theoretical and practical foundations of applying the cyberontological approach in the professional training of lecturers. To achieve the research purpose, the following tasks were solved: the current state of research and publications in the area was analyzed; theoretical, practical bases for applying the cyberontological approach in the professional training of lecturers were developed. Cyberontological approach is proposed as an innovative basis of psychological and pedagogical science-cyberpedagogy, which is designed to generalize and systematize scientific knowledge in the field of application of modern information and communication, computer, digital, electronic and Internet technologies in the education system. The paper substantiates the cyberontological model of functional dependence of components and the flow of the information processes in the educational activity of a lecturer-student. Based on the accumulated positive experience of using computer technologies in teaching students, the use of a cyberontological approach in the professional training of lecturers is proposed. Several examples of applying a cyberontological approach with elements of gamification in the training of cybersecurity specialists are considered. The scientific significance of the work is that for the first time a functional cybernetic model of the dependence between the components of the pedagogical system has been developed. The practical significance of its application in practice is that on the basis of the cybernetic model, game mechanics can be developed to increase student motivation in modern conditions. It is possible to use a computer game (gamification) as the basis for building a training complex for training military information and cybersecurity specialists.

**Keywords:** Cyberontological approach · professional training · lecturer · ontology · gamification · higher education institution · cybersecurity · mathematical model

---

© The Author(s), under exclusive license to Springer Nature Switzerland AG 2023
Z. Hu et al. (Eds,): ICAILE 2023, LNDECT 180, pp. 1068–1079, 2023.
https://doi.org/10.1007/978-3-031-36115-9_95

# 1 Introduction

## 1.1 Problem Statement

Both around the world and in our country, a person learns throughout his life. This is training in a general academic school, vocational school, or higher education institution. The state spends a lot of money on financing the education system, so the problem of improving the efficiency of the learning process is urgent. Its optimization requires not only improving the content and methodology of studying individual subjects, but also developing the theoretical foundations of didactics with the involvement of both the humanities (for example, Psychology) and exact sciences (Mathematics, Cybernetics). Now all over the world, when analyzing the educational process, the consideration of the "lecturer – student" system from the point of view of management theory is based. Almost all educational institutions have shifted their academic activities to digital platforms due to the recent COVID-19 epidemic [1].

Given the above, the cybernetic approach is also relevant for the higher education system.

## 1.2 Literature Review. Analysis of Recent Research and Publications

In the scientific research of V. Pleshakov [2, 3] formed and acquired further formation in the works of N. Voloshyn, O. Mukohorenko, L. Zhohin [4], K. Meteshkin, O. Morozov, L. Fedorchenko, N. Khairov [5] the definition of cyberontology as the existence and / or life activity of a person in an innovative alternative reality of cyberspace (cyberreality), determined by the level of development of self-consciousness and motivational needs of the individual, as well as by the complex of objective and subjective micro-, macro-, meso- and megafactors of society. Based on the analysis of human interaction with computer, electronic, digital technology, a cyberontological approach to education is built, according to which training and upbringing of a person is determined by the conditions of its existence, life activity and interaction with computer technology, with other people and the world as a whole, based on the integration of two spaces: real and virtual.

According to V. Bespalko [6], N. Voloshyn [4], A. Ihibaev, A. Toleukhanov [7], I. Kozubtsov [8], R. Maier [9], V. Pleshakov [10], the cyberontological approach is nothing more than a phenomenon of psychological and pedagogical science. It begins the foundations of cyberpedagogy, which is designed to generalize and systematize scientific knowledge in the field of application of modern information and communication, computer, digital, electronic and Internet technologies in the education system.

## 1.3 Highlighting Aspects that are Understudied

The analysis of recent studies has established that the chosen object of study "cyberontological approach" attracted the attention of foreign scientists, but the description of the functional cyberontological model of dependence and the flow of the information processes in human educational activities is not sufficiently reflected.

Based on the above, this current research area was chosen.

## 1.4 Purpose of the Article

To review the theoretical foundations and practical success in applying the cyberontological approach with elements of gamification in the professional training of lecturers of higher educational institutions (HEI).

## 1.5 Research Objectives (Goals)

To achieve this purpose, the following objectives are set:

1. Analyze the current state of research and publications on the keywords of the terms "cyberontological approach", "cyberpedagogy".
2. To review the theoretical foundations and practical success in applying the "cyberontological approach" in the professional training of future lecturers of the HEI.

# 2 Research Methods

## 2.1 Research Tools

The main means of scientific and theoretical research. A set of scientific methods that are comprehensively substantiated and consolidated into a single system. Basic research tools:

methods of theoretical analysis and generalization of scientific literature, on the topic of the research;

generalization to formulate conclusions and recommendations for effectiveness.

## 2.2 Reliability and Accuracy of Results

Reliability of the results of the study is ensured by the correctness of the use of mathematical apparatus and research methods.

## 2.3 Methodological Basis of the Study

The object of scientific and theoretical research is not just a single phenomenon, a specific situation, but a whole class of similar phenomena and situations, their totality. The methodological basis of the research is the ideas of L. Vyhotskyi, P. Halperin, Y. Babanskyi, S. Rubinshtein, (pedagogical psychology), N. Viner, K. Shennon, F. Rozenblatt, A. Kolmohorov, V. Hlushkov, (cybernetics), V. Maier, D. Novikov, (mathematical modeling of learning), V. Bespalko, Y. Mashbits, (cybernetic approach in pedagogy, programmed learning and automated training systems).

Thus, the basic theory for mathematical modeling is cybernetics.

# 3 Research Results

## 3.1 Theoretical Foundations of the "Cyberontological Approach", the "Cyberpedagogy" Component

With the advent of "cyberspace", a virtual dimension, a new environment of life activity and a factor of social change were formed (M. Chitosca [11], S. Petriaiev [12]). As a result, modern person lives and interacts with other people and the world as a whole in parallel in two socializing environments – classical objective (material) reality and alternative innovative reality of cyberspace (cyberreality) – that potentially and objectively affect the formation and transformation of subjective (phenomenon of the psyche) reality. In this regard, it is advisable to talk about the "parallel" existence of a person in cyberspace, about an alternative ontology of modern human civilization – cyberontology.

The types of indirect activities have become the main vectors of socialization (integration) of a person with cyberspace. According to some scientists O. Voinov [13], N. Voloshyn [4], V. Pleshakov [10], S. Petriaiev [12]) in particular:

1. Communication in cyberspace.
2. Leisure in cyberspace.
3. Knowledge in cyberspace.
4. Work in cyberspace.

To effectively solve the actual problems of human education and training, the need to apply a "cyberontological approach" in education and the beginning of an innovative branch of pedagogical science – cyberpedagogy (V. Bespalko [6], N. Voloshyn, L. Zhohin, O. Mukohorenko [4], K. Meteshkin, O. Morozov, L. Fedorchenko, N. Khairov [5]) is justified. At the same time, a new fifth vector of human socialization into education through cyberspace was identified.

So, the fifth vector is **the education of a person in cyberspace** – this is the process of human education, which is determined by the conditions of his life activity and interaction with oneself, with other people and the world as a whole in the context of the integration of classical objective (material) reality and innovative alternative reality of cyberspace (cyberreality), both of which potentially and objectively affect the formation and transformation of subjective (phenomenon of the psyche) reality. The main role of the cyberontological approach is to regulate the development of personality and human life activity in cyberspace, taking into account modern conditions and trends of education, as well as near and far prospects for the evolution of mankind.

Nowadays, the use of cyberspace as an educational environment, a channel of training and educational communications has become commonplace. The human educational process in cyberspace is interdisciplinary, as a result of which "Cyberpedagogy" shows positive and negative trends in education [14]. It was not without reason that the author Y. Mashbits expressed concern at the time: "will the widespread use of the computer lead to the oblivion of the past – our roots, to a less defined future, to the detriment of the cultural and spiritual values of our heritage? The habit of systematically using a computer can make a person neglect his own capabilities, lead him to excessive dependence on the computer, cause atrophy of thinking, isolating a person from the world around them" [15].

A similar opinion was held by (V. Bespalko [6]), noting that a computer cannot effectively teach, relying on "human", expressed in verbal form, pedagogy that is understandable to a living lecturer who has an innate pedagogical intuition and an understanding of polysemous natural language. The computer needs special pedagogy, expressed in the unambiguous language of mathematics and formal logic, and which describes well-defined rules of action (algorithms) in well-defined pedagogical situations (tasks). The author offers the basics of such pedagogy "Cyberpedagogy" (from cybernetics and pedagogy). Cybernetics (V. Hlushkov) refers to the science of managing complex technical, biological, and social systems that can perceive, store, and process information. From the point of view of cyberpedagogy, the processes of teaching and upbringing can be reduced to managing the development of various personal qualities of students through purposeful and coordinated influences on the part of lecturers and parents (R. Mayer [9]). The purpose of training is to transfer a set of knowledge to students, to form skills and abilities, to develop their ability to observe, reason and effectively interact with the world around them.

The main directions of "cyberpedagogy" according to (K. Meteshkin, O. Morozova, L. Fedorchenko, & N. Khairov [5]; D. Novikov [16]):

1. Analysis of the pedagogical system from the point of view of management relations and information flows exchanged between the management and managed subsystems.
2. Optimization of the learning process, finding such forms and methods of organizing the educational process in which the functioning of the education system would be most effective, that is, at the lowest cost, it would bring maximum benefit.
3. Practical use of electronic devices and automated training systems for managing the learning and testing process; programmed learning. Among the modern methods of research of pedagogical systems, a special position is occupied by methods of mathematical and simulation modeling. Their essence lies in the fact that the real pedagogical system is replaced by an abstract model – some idealized object that has the most essential qualities of the system under study. By changing the initial data and parameters of the model, it is possible to investigate the ways of development of the system, determine its state at the end of training [9]. This is the advantage of this approach in comparison with the method of qualitative analysis.

In order to clearly understand the "cyberontological approach", we have developed a functional cyberontological model of dependencies and the flow of the information processes in human educational activities (see Fig. 1). Through "Threat agents", cybernetic destructive information influences on the trained individual are created in order to lead to the loss of "Assets" in the future. Assets include future knowledge, diploma, job, salary, and so on. The challenge is for the learner to understand the need for "Assets" and to make an effort in the game to form quasi-professional solutions that would neutralize the cybernetic destructive information influences created by "Threat agents".

## 3.2 Practice of Applying the Cyberontological Approach in Educational Institutions

The practice of applying the cyberontological approach in education developed on an intuitive level, through gamification (N. Rybka [17]). With the advent of computer technology, the motivation of students and cadets to traditional teaching methods began to decline paradoxically rapidly. Game methods of teaching adults have reached the first milestone.

The phenomenon of the game is that, being entertainment, recreation, it is able to develop into learning, creativity, a model of the type of human relationships and manifestations in work. It is through the gradual introduction of the cyberontological approach into practice that this approach has contributed to the rethinking of the game from the point of view of the learning method. The pedagogical game has a clearly formed goal of teaching students (cadets) and its corresponding pedagogical results, which can be justified, highlighted in an explicit form and are characterized by an educational and cognitive orientation.

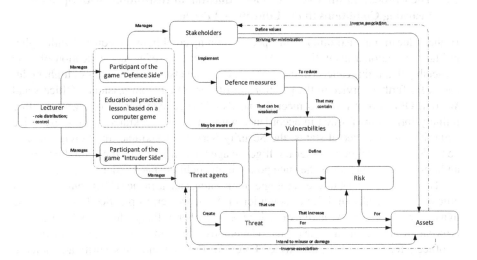

**Fig. 1.** Functional cyber ontological model of dependencies and the flow of the information processes in human educational activity

This is confirmed by the development of an innovative concept of independent training of Ground Forces cadets on training equipment by playing on a virtual computer [18] and methods of teaching electrical engineering disciplines [19].

It is important to correctly identify the motivation of players and forward it in the direction of learning, and not to cybercrime (V. Leshchina, I. Kozubtsov, & L.Kozubtsova [20]), the consequences of which can be catastrophic (I. Kozubtsova, L. Kozubtsov, T. Tereshchenko, & T. Bondarenko [21]).

If the method of pedagogical game is turned into a pedagogical technology, then you can get a wide practical application. Given this circumstance, gamification began to be considered from the angle of game pedagogical technology in education, having a huge potential for a positive impact on the effectiveness of the educational process, which began to attract more and more attention of researchers (K. Tseas, N. Katsioulas, & T.Kalandaridis [22, p. 25]; V. Bugaeva [23, p. 135]; S. Petrenko [24]; N. Rybka[17]).

Publication (D. Kaufmann [25]) confirms that the author has made significant progress in gamifying online higher education.

For the purposes of further practice of applying the cyberontological approach in education, software developers are invited to participate in the creation of a game cybersecurity strategy similar to that used in computer games. Its emulation, as a computer game, is rationally used in the training of military information and cyber security specialists to acquire practical skills in working out the settings of routers, firewalls, etc.on a conditional training information system [26].

As a result of virtual 3D modeling of educational experiments in a in game form, it is impossible for experimental equipment to come out of working order due to erroneous actions.

### 3.3 The Problem in the Use of Higher Educational Institutions with Specific Learning Conditions in the Educational Process

Higher educational institutions with specific training conditions are equated with higher military educational institutions (HMEI) – an institution of higher education of state ownership that trains cadets (trainees, students), adjuncts at certain levels of higher education for further service in the positions of officers (Non-Commissioned Officers and Warrant Officers) in order to meet the needs of the Armed Forces of Ukraine, other military formations formed in accordance with the laws of Ukraine, central executive authorities with a special status, the Security Service of Ukraine, the Foreign Intelligence Service of Ukraine, other intelligence agencies of Ukraine, the central executive authority, what implements the state policy in the field of state border protection.

The result of using game technologies for training cadets in the HMEI confirmed high efficiency, especially in difficult conditions [18]. However, as practice has shown, the dynamic range of creative activity of a lecturer in higher military educational institutions is too small and is limited to the forms and methods of teaching. Organization and training of officers for the needs of the Armed Forces of Ukraine and the security and defense sector is carried out In accordance with the Order of the Ministry of Defense of Ukraine.

### 3.4 Discussion of Research Results

Gamification and a cyberontological approach can not be implemented for all academic disciplines of the educational component. Gamification is clearly prohibited in special professional academic disciplines during which information with restricted access for educational purposes circulates in the classroom.

An interesting combination is the cyberontological approach and gamification when creating a computer game "Become the Head of a field information and communication node" in the course of studying tactical and special disciplines [27].

Assessment and feedback mechanisms are essential components towards effective teaching in higher education and are continuously monitored [28].

Gamification is a good alternative educational practice to promote programming teaching, it allows better engagement of students in their learning. Students acquire a reasonable level of abstraction and logic and develop reflections on various course concepts [29].

Further study and adequate discussion of the results is expected. A more comprehensive study could reveal additional dimensions of the survey and shed light on how students perceive assessment, evaluation and feedback in higher education in general. Demonstrate some important aspects of this and indicate that improved quality of assessment and feedback can have a positive impact on student satisfaction. A more comprehensive study could reveal additional dimensions of the survey and shed light on how students perceive assessment, evaluation and feedback in higher education in general.

As can be clearly seen from Table 1 the following learning methods are the most accepted for the formation of knowledge: didactic games; practical training; teaching others (mutual learning) and independent work [26].

**Table 1.** Comparative characteristics of different teaching methods

| Teaching methods | | Solved tasks | | | | | | |
|---|---|---|---|---|---|---|---|---|
| | | form | | develop | | | | gaining experience |
| | | knowledge | skills | thinking | memory | language | | |
| Verbal | (lecture) | 5% | ++ | – | – | – | ++ | |
| | reading | 10% | | | | | | |
| | listening | 20% | | | | | | |
| Visually | Work with multimedia (audiovisual) | 20% | + | ++ | + | + | – | |
| | Viewing the drawing | 30% | | | | | | |
| | Demonstration | 30% | + | + | + | ++ | – | |
| | Video viewing | 50% | | | | | | |
| **Practical** | Working with a book (reading) | 10% | + | + | + | + | + | |
| | Educational discussions | 50% | ++ | – | ++ | + | ++ | ++ |
| | **Didactic games** | 70% | ++ | – | ++ | + | ++ | ++ |

(continued)

**Table 1.** (*continued*)

| Teaching methods | | Solved tasks | | | | | |
|---|---|---|---|---|---|---|---|
| | | form | | develop | | | gaining experience |
| | | knowledge | skills | thinking | memory | language | |
| | Practical training | 75% | + | ++ | ++ | + | – | ++ |
| | Teaching others is the application of what has been learned | 90% | ++ | – | ++ | + | ++ | ++ |
| Independent work | | 80% | ++ | ++ | ++ | + | + | |
| Oral and written control | | | ++ | – | + | + | ++ | |
| Note: | Note: ++ – solve very well; + – solve partially; – – solve poorly | | | | | | | |

## 4 Summary and Conclusion

Thus, cybersocialization of the individual affects all segments of society, people of different ages, social positions and statuses. The degree of socialization depends on the individual characteristics of people. At the moment, the socialization of people is actively developing in five vectors. Given the accumulated positive experience of using computer technologies in the training of people of different ages, it is considered appropriate to use a cyberontological approach in the professional training of future lecturers.

Cyberontological and game approaches in education aim to create such educational conditions that a person, learning in the course of a computer game, does not even suspect that learning is taking place, obtaining new knowledge. In a computer game, there is no source of knowledge that is easily learned by students. The learning process develops in the language of actions as a result of active contacts with each other, unobtrusively.

### 4.1 Expanding the Boundaries of the Scientific Field

The scientific result obtained in the work expands the scientific boundaries of pedagogical sciences in the field of application of game mechanics in the preparation of higher education students in conditions of low motivation. Thus, the expansion takes place in the systemic unity of the philosophy of education and the theory and methodology of vocational education.

### 4.2 Scientific Novelty. Scientific Justification

For the first time, a functional cyberontological model of dependencies and the flow of the information processes in the educational activities of a lecturer of a higher education institution is proposed, that allows us to find out all the interdisciplinary connections of others in cyberspace and the assets of students.

## 4.3 Practical Use

It is possible to use a computer game (gamification) as the basis for building a training complex for training military information and cybersecurity specialists.

## 4.4 Prospects for Further Research and Study

The research is expected to bring more academic and applicable value. To do this, you need to pay extra: clarification of current problems of applying the gamification of training in higher military educational institutions.

**Acknowledgment.** The authors would like to express their respect to the organizers of the 3rd International Conference on Artificial Intelligence and Logistics Engineering (ICAILE2023) (March 11 – March 12, 2023, Wuhan, China) and at a tragic time for Ukrainian researchers, Modern Education and Computer Science Press provided a grant to publish their scientific achievements free of charge. We are sincerely grateful.

# References

1. Ahmed, N., Nandi, D., Zaman, A.G.M.: Analyzing student evaluations of teaching in a completely online environment. Int. J. Mod. Educ. Comput. Sci. (IJMECS) **14**(6), 13–24. https://doi.org/10.5815/ijmecs
2. Pleshakov, V.A.: Gerontology and psychology of information sphere security: aspect of human cybersocialization in the social networks of the Internet environment. Bull. Orthodox St. Tikhon's Univ. Humanit. Ser. IV Pedagogy Psychol. **4**(19), 131–141 (2010). (in Russian)
3. Pleshakov, V.A.: Prospects of the cyberontological approach in modern education. Bull. Moscow City Pedagogical Univ. Ser. Pedagogy Psychol. **3**(29), 1–18 (2014). (in Russian)
4. Voloshina, N.M., Zhogina, L.M., Mukogorenko, O.S.: On the issue of cybersocialization of Ukrainian society. In: Military Education and Science: Present and Future: XIII International Scientific and Practical Conference, Kiev, VIKNU, pp. 126–127, 24 November 2017. (in Ukrainian)
5. Meteshkin, K.A., Morozov, O.I., Fedorchenko, L.A., Khayrova, N.F.: Cybernetic pedagogy: ontological engineering in teaching and education, p. 207. HNAGH, Kharkiv (2012). (in Russian)
6. Bespalko V.P.: Cyberpedagogics. Pedagogical Foundations of Computer-Controlled Learning (E-Learning), p. 240. T8RUGRAM/Public Education, Moscow (2018). (in Russian)
7. Igibaeva, A.K., Toleukhanova, A.D.: Cyber pedagogy – as an innovative branch of pedagogical science. Bull. KazNPU Named after Abai Ser. Pedagogical Sci. **68**(4), 12–17 (2020). (in Russian)
8. Kozubtsov, I.N., Kozubtsova, L.M.: Cyber pedagogy as a new trend in education of the XXI century: problems and risks. In: International Scientific and Practical Conference: "Pedagogical Education in the XXI Century: Priorities and Searches", Astana, pp. 402–409, 7 October 2022. (in Russian)
9. Mayer, R.V.: Cybernetic Pedagogy: Simulation Modeling of the Learning Process, p. 138. Glazov, GGPI (2013). (in Russian)
10. Pleshakov, V.A.: Cybersocialization as an innovative socio-pedagogical phenomenon. Teacher XXI Century **3**(1), 32–39 (2009). (in Russian)

11. Chitosca, M.I.: The internet as a socializing agent of the "M Generation." J. Soc. Inform. **5**, 3–21 (2006)
12. Yu, P.S.: "Cybersocialization" – A ≠ A. Inf. Law **3**(12), 25–30 (2014). (in Ukrainian)
13. Voinova, O.I., Pleshakov, V.A.: Cyberontological Approach in Education, p. 244. Norilsk, Research Institute (2012). (in Russian)
14. Kozubtsov, I.M., Beskrovny, O.I., Kozubtsova, L.M., Palaguta, A.M., Mironenko, O.V.: Trends in the educational and scientific space: problems and risks. Bull. Mikhail Ostrogradsky Kremenchug National Univ. **1**(132), 40–48 (2022). (in Ukrainian)
15. Mashbits, E.I.: Psychological and Pedagogical Problems of Computerization of Education, p. 192. Pedagogy, Moscow (1988). (in Russian)
16. Novikov, D.A.: Theory of Educational Systems Management, p. 416. Public Education, Moscow (2009). (in Russian)
17. Rybka, N.M.: Graization and experience of using computer games in teaching philosophy in technical institutions of higher education. Inf. Technol. Training Tools **67**(5), 213–225 (2018). (in Ukrainian)
18. Kozubtsov, I.M.: The concept of independent training of cadets of the ground forces on training facilities by playing on a virtual computer. In: Prospects for the Development of Weapons and Military Equipment of the Ground Forces. Collection of Abstracts of the Second All-Ukrainian Scientific and Technical Conference, Lviv, p. 77, 28–29 April 2009. (in Ukrainian)
19. Kozubtsov, I.N.: Teaching electrical engineering disciplines by virtual computer game method. In: Electrical Technologies, Electric Drive and Electrical Equipment of Enterprises: Collection of Scientific Papers of the II All-Russian Scientific and Technical Conference. USNTU Publishing House, Ufa, vol. 2, pp. 107–110 (2009). (in Russian)
20. Leshchina, V., Kozubtsov, I., Kozubtsova, L.: Role of motivative characteristics in cyber security ontology. Sci. Pract. Cyber Secur. J. (SPCSJ) **6**(1), 15–23 (2022)
21. Kozubtsova, L.M., Kozubtsov, I.M., Tereshchenko, T.P., Bondarenko, T.V.: On the cyber security of military personnel playing geolocation games while staying at departmental facilities of critical information infrastructure. Cybersecurity Educ. Sci. Technol. **1**(17), 76–90 (2022). (in Ukrainian)
22. Tseas, K., Katsioulas, N., Kalandaridis, T.: Gamification in higher education. M.S. Thesis, Department, Electrical and Computer Engineering, University of Thessaly, Volos, Greece (2014). https://doi.org/10.4102/hts.v73i3.4527
23. Yu, B.V.: Gamification as a way of forming active professional behavior of future IT industry specialists. Pedagogy Psychol. **56**, 129–135 (2018). https://doi.org/10.5281/zenodo.577567. (inUkrainian)
24. Petrenko, S.V.: Gamification as an innovative educational technology. Innov. Educ. **2**(7), 177–185 (2018). (in Ukrainian)
25. Kaufmann, D.A.: Reflection: benefits of gamification in online higher education. J. Instr. Res. **7**, 125–132 (2018)
26. Kozubtsova, L.M., Kozubtsov, I.M., Lishchina, V.A., Shtanenko, S.S.: Concept of the educational and training complex for training military specialists in information and cybersecurity on the basis of computer games (gamification). Cybersecurity Educ. Sci. Technol. **2**(18), 49–60 (2022). (in Ukrainian)
27. Ponomarev, O.A., Pivovarchuk, S.A., Kozubtsov, I.M.: On the use of the computer game "Become the head of a field information and communication node" in the course of studying tactical and special disciplines. Collection of Reports and Abstracts of Materials of the II International Scientific and Technical Conference on Communication, Informatization and Cybersecurity Systems and Technologies: Current Issues and Development Trends, Kiev, VITI, pp. 174–175, 1 December 2022. (in Ukrainian)

28. Rigopoulos, G.: Assessment and feedback as predictors for student satisfaction in UK higher education. Int. J. Mod. Educ. Comput. Sci. (IJMECS) **14**(5), 1–9 (2022). https://doi.org/10.5815/ijmecs
29. Ouahbi, I., Darhmaoui, H., Kaddari, F.: Gamification approach in teaching web programming courses in PHP: use of KAHOOT application. Int. J. Mod. Educ. Comput. Sci. (IJMECS) **13**(2), 33–39 (2021)

# Exploring the Perceptions of Technical Teachers Towards Introducing Blockchain Technology in Teaching and Learning

P. Raghu Vamsi(✉)

Department of Computer Science and Engineering and IT, Jaypee Institute of Information Technology, A-10, Sector 62, Noida 201307, India
prvonline@yahoo.co.in

**Abstract.** This paper presents a study that assesses technical teachers' perceptions of learning blockchain technology and its applications to plan efficient faculty development programs for industry, academia, and research. With universities and technical councils introducing courses to bridge the gap between academic learning and industry demands, the demand for blockchain developers has been on the rise. The study's findings indicate that technical teachers generally hold positive attitudes towards learning blockchain technology, recognizing its potential applications in various fields such as finance, healthcare, and energy. Quantitative and qualitative study conducted on 79 engineering teachers and professionals who participated in an online faculty development program on Blockchain Technology and its Applications at JIIT University, India by providing survey questionnaire. The study highlights the importance of continuous adaptation and learning of new technologies for teachers to equip students with the skills they need to succeed in the industry. The study also provides insights into technical teachers' perceptions towards integrating blockchain technology in teaching and learning and offers implications for planning efficient faculty development programs. Query ID="Q1" Text="This is to inform you that corresponding author has been identified as per the information available in the Copyright form." '

**Keywords:** Blockchain technology · Faculty development · Motivation · Performance · Professional development · Technical education

## 1 Introduction

The field of teaching requires educators to engage with students to facilitate their learning, which involves various tasks such as planning, delivering, assessing, and reflecting on content. Effective teaching demands a methodical understanding of learning to establish frameworks that encourage students to take responsibility for their own learning [1–3]. As new technological paradigms emerge, teachers must adapt and develop the necessary attitudes, knowledge, skills, and abilities to become competent instructors [4, 5]. Faculty development programs are essential to disseminate expertise from teachers to students. The term "Perceptions" refers to teachers' attitudes, behavior, self-beliefs, and views regarding new technology, which can influence their beliefs about the importance of integrating it into teaching and learning. Continuous adaptation and learning of new technologies is necessary for teachers to equip students with the skills they need to succeed in the industry [6–9]. Prior research has focused on exploring the perceptions of teachers and students towards various technological paradigms, such as ICT [10–12], e-learning systems [13], flipped classrooms [14], and online teaching during the COVID-19 pandemic [15]. However, the rise of blockchain technology as a secure and distributed database technology, which offers solutions for financial transactions, healthcare record protection, and energy and carbon emission trading, among others, demands attention from the academic community. With universities and technical councils introducing courses to bridge the gap between academic learning and industry demands, the demand for blockchain developers has been on the rise. To facilitate efficient faculty development programs for industry, academia, and research, it is crucial to assess technical teachers' perceptions of learning blockchain technology and its applications [16–18]. Therefore, this study aims to assess technical teachers' perceptions of learning blockchain technology and its applications to plan efficient faculty development programs for industry, academia, and research. The remainder of the paper is structured as follows: Sect. 2 presents related work, Sect. 3 presents the methodology of the work, Sect. 4 presents the results with discussion, and Sect. 5 concludes the paper.

## 2 Related Work

Nyme et al. [19] analyzed the student evaluations of teaching in a completely online environment to uncover the key factors that influence the quality of teaching in a virtual classroom environment. The authors utilize Educational Data Mining (EDM) techniques to analyze data gathered from the student evaluations of computer science students in three online semesters at X University. They use Weka, sentimental analysis, and word cloud generator to carry out the research. The decision tree classifies the factors affecting the performance of the teachers, and the authors find that student-faculty relation is the most prominent factor for improving the teaching quality. Tushar et al. [20] showed the status of ICT education and finds the gaps between rural and urban institutions for providing ICT education in secondary and higher secondary institutions in Bangladesh. The authors use primary data collected using a survey questionnaire that is answered by ICT teachers engaged in those institutions. This research found that that there exist low facilities in rural institutions compared to the urban institutions because students-computer ratio (SCR) is 46 in rural areas whereas SCR is 22 in the urban area. Finally, authors presented recommendations to meet the identified gaps. Chen et al. [21] explored the potential applications of blockchain technology in education, highlighting its features, benefits, and current applications. The authors propose innovative applications of blockchain technology and discuss the challenges of implementing it in education. Araujo et al. [22] presented study on the integration of Flipped Classroom and Adaptive Learning techniques for Blockchain courses in higher education. The study evaluates student acceptance and the learning impact of out-of-class material with and without Adaptive Learning techniques, with positive results observed. Chang et al. [23] investigated the determinants affecting the acceptance of blockchain technology in the tourism industry on Jeju Island, Korea. The authors analyze the effect of blockchain trust transparency on performance and effort expectancy, finding that facilitating conditions are the most influential factor in blockchain technology acceptance. The role of blockchain technology in the sustainable development of students' learning process is discussed by Popa et al.[24]. Two specialized studies conducted in Romania are presented, examining teachers' and students' perceptions of technological integration in the learning process. The authors find that implementing a reward system through blockchain technology significantly improves students' motivation and creativity.

## 3 Methodology of Study

The methodology of this study involved the development and distribution of a questionnaire to participants of an online Faculty Development Program (FDP) at Jaypee Institute of Information Technology in July 2022. The questionnaire consisted of 20 questions categorized into four motivational factors: job expectations, social desire, personal desire, and technical capabilities. Each question was surveyed using a five-point scales, with a range of strongly agree to strongly disagree. The questionnaire was distributed for on-spot completion and collection, and the response rate was limited by the FDP participants' attendance on the survey date, which was held on the last day of the FDP and just before the conduction of the final quiz. The collected questionnaires were compiled into a Google Spreadsheet and analyzed using Python programming with the Pandas data science library [25–27]. The analysis involved calculating the mean value for each question, with a mean value of 5 indicating complete agreement with the question merit, and a mean value of 1 indicating complete disagreement with the question merit. Detailed statistical results of the survey were compiled and are presented in Table 1.

## 4 Results and Discussion

Total 79 members participated in the survey, with 51.9% female and 48.1% male participants. Of the total, 48.1% have a Master's degree in Technology, 45.5% have a Doctorate degree in computer science and information technology, and 6.2% have a non-technical degree. Out of the 79 participants, 56.96% are Assistant professors, 18.98% are Associate professors, 18.98% are doctorate research scholars, and 2.6% are software professionals. 31.6% of the participants are in the age range 25–34 years, 59.5% are in the age range of 35–44 years, and 8.9% are in the age range 45–54 years. 36.7% of the participants have more than 15 years of work experience, 26.5% have 10–14 years of work experience, 17.7% have 5–9 years of work experience, and 15.2% have 1–4 years of work experience. When asked about their intention behind learning Blockchain technology, 48.1% answered to improve their knowledge on the topic, 20.5% answered that this technology helps in their ongoing research work, 20.2% answered to become sound in the latest technology, and 6.3% answered to shift their work from academics to industry. Despite women making up the majority of the participants, middle-aged professors are more cognizant of the value of Blockchain technology and understand that it is already crucial for industry 4.0. Many participants believe that learning Blockchain technology would help them progress in their jobs in technology and research because it is a novel subject. 5% of the participants intend to leave academia and work in industry.

From Table 1, the study found that personal desire was the strongest motivator for faculty members to take up blockchain technology courses, while social desire was the lowest. A significant number of participants were Assistant Professors with a Master's degree in Technology and over 15 years of work experience. Job expectations were the second strongest motivator, with respondents showing awareness of the potential employment opportunities in the blockchain sector. The social desire category was significant, with over 80% of respondents believing that blockchain development is a prestigious profession that will maintain their social status. The technical capabilities category was the third motivator, with respondents considering their technical skills as essential for taking a blockchain course. The mean score for the job expectations category was 4.1, while the social desire category had a mean score of 3.5 out of 5. On average, 35.7% of faculty members fell behind in the four categories, while 35.4% strongly agreed with all four. Overall, 71.1% agreed or strongly agreed with the four categories, while 28.9% were neutral, disagreed, or strongly disagreed.

Figure 1 shows the grade distribution of participants who attended the Blockchain FDP, which aims to improve their programming skills in Blockchian and smart contracts. Participants must attend 80% of sessions to qualify for the final quiz, and their grades are used to assess motivation. The course average is 3.5, which is a B + and slightly higher than the normal average of C +. This suggests that participants' technical abilities are above average, which is consistent with their confidence in the technical capabilities category of the survey. This positive observation could be attributed to their prior knowledge in computer networks, cryptography, and programming. Selected participants' evaluations of a Blockchian FDP were analyzed, following All India Council for Technical Education (AICTE) guidelines. 5 items from 16 were chosen, and the average response for each question item was 4.5 out of 5. The participants have a high level of confidence in the knowledge and skills learned in the course. The participants suggested correlating blockchian theoretical aspects with application development and research areas, as well as holding hands-on training workshops on full stack blockchian application development and research directions. The study found that 48.1% of the participants intended to learn Blockchain technology to improve their knowledge, while 20.5% answered that this technology helps in their ongoing research work. 20.2% answered that they intended to become sound in the latest technology, and 6.3% answered that they wanted to shift their work from academics to industry. Interestingly, despite women making up the majority of the participants, middle-aged professors were more aware of the value of Blockchain technology and understood its importance for industry 4.0.

With these results, this study found that faculty members are motivated to learn Blockchain technology for various reasons, including improving their knowledge, ongoing research work, and staying up-to-date with the latest technology. Personal desire was found to be the strongest motivator, while social desire was the weakest. The study also found that academic staff members were well-informed about employment options in the sector, and the potential for new research initiatives and technology to pay off. The

**Table 1.** Technical Teacher's feedback on their motivation towards Blockchain Technology Course at FDP conducted at JIIT Noida. (Scale: 5 = Strongly agree, 4 = Agree, 3 = Neutral, 2 = Disagree, and 1 = Strongly disagree)

| Category | Survey Questions | %of Answers ||||| Mean (out of 5) |
|---|---|---|---|---|---|---|---|
| Job expectation | | 5 | 4 | 3 | 2 | 1 | |
| C1.1 | I want to learn blockchain application development, because this is a well-paid domain compared to most of the other software domains | 40.5 | 40.5 | 13.9 | 3.8 | 1.3 | 4.2 |
| C1.2 | I want to learn blockchain application development, because it will be easy for me to find a good job | 26.6 | 40.5 | 27.8 | 2.5 | 2.5 | 3.9 |
| C1.3 | I want to learn blockchain application development, because it will be easy for complete my research work / publish research papers | 46.8 | 34.2 | 17.7 | 1.3 | 0 | 4.3 |
| C1.4 | I want to learn blockchain application development, because my current job role demanded me learn it | 36.7 | 35.4 | 25.3 | 2.5 | 0 | 4.1 |
| | Category 1 average | | | | | | 4.1 |
| Social desire | | | | | | | |
| C2.1 | I want to learn blockchain application development, because this is going to be one of the prestigious professions that will maintain my class level in the society | 39.2 | 32.9 | 22.8 | 2.5 | 2.5 | 4 |
| C2.2 | I want to learn blockchain application development, because my family (parents) told me to do so since cryptocurrencies like Bitcoin is gaining importance | 31.6 | 22.8 | 25.3 | 5.1 | 15.2 | 3.5 |

(*continued*)

**Table 1.** (*continued*)

| Category | Survey Questions | %of Answers | | | | | Mean (out of 5) |
|---|---|---|---|---|---|---|---|
| C2.3 | I want to learn blockchain application development, because my friends/peers told me to do so | 32.9 | 21.5 | 25.3 | 7.6 | 12.7 | 3.5 |
| | Category 2 average | | | | | | 3.6 |
| Personal desire | | | | | | | |
| C3.1 | I want to learn blockchain application development, because this will contribute to society by solving community/world problems | 40.5 | 41.8 | 17.7 | 0 | 0 | 4.2 |
| C3.2 | Blockchain application development is interesting and fun and I feel good when I deal with trust and privacy issues | 38 | 45.6 | 15.2 | 1.3 | 0 | 4.2 |
| C3.3 | I chose blockchain application development because I like to build and fix trust and privacy issues in business chain | 40.5 | 39.2 | 20.3 | 0 | 0 | 4.2 |
| C3.4 | I chose blockchain application development because I like to develop and design new systems | 34.2 | 48.1 | 17.7 | 0 | 0 | 4.2 |
| C3.5 | I chose blockchain application development because I like to work on projects in teams to solve problems | 44.3 | 36.7 | 16.5 | 2.5 | 0 | 4.2 |
| C3.6 | I chose blockchain application development because I feel I have the ability to present new systems/products to the public/students than software professionals | 40.5 | 36.7 | 20.3 | 2.5 | 0 | 4.2 |
| | Category 3 average | | | | | | 4.2 |

(*continued*)

**Table 1.** (*continued*)

| Category | Survey Questions | %of Answers | | | | | Mean (out of 5) |
|---|---|---|---|---|---|---|---|
| Technical capabilities | | | | | | | |
| C4.1 | I chose blockchain application development because I like networking and security subjects | 38 | 35.4 | 22.8 | 3.8 | 0 | 4.1 |
| C4.2 | I chose blockchain application development because I have been always doing great in distributed networking and have great grip on subject | 26.6 | 41.8 | 22.8 | 8.9 | 0 | 3.9 |
| C4.3 | I chose blockchain application development because I have been always doing great in network security and have great grip on subject | 30.4 | 35.4 | 24.1 | 10.1 | 0 | 3.9 |
| C4.4 | I chose blockchain application development because I have been always doing great in full stack web development and have great grip on subject | 31.6 | 35.4 | 22.8 | 10.1 | 0 | 3.9 |
| C4.5 | I chose blockchain application development because I have been always doing great in networking, security and full stack development and have great grip on subjects | 25.3 | 41.8 | 25.3 | 7.6 | 0 | 3.8 |
| C4.6 | I chose blockchain application development because I like to develop security applications, analyze and address the trust and privacy issues | 34.2 | 40.5 | 19 | 6.3 | 0 | 4 |
| C4.7 | I chose blockchain application development because I like to develop cryptocurrencies | 25.3 | 38 | 25.3 | 11.4 | 0 | 3.8 |

(*continued*)

**Table 1.** (*continued*)

| Category | Survey Questions | %of Answers | Mean (out of 5) |
|---|---|---|---|
| | Category 4 average | | 3.9 |

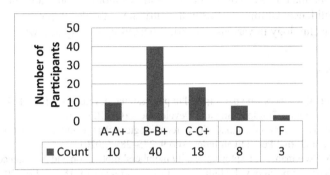

**Fig. 1.** Final grades of the participants in FDP

findings of this study could be used to design effective training programs for faculty members in Blockchain technology. Important observations of this study are.

- This study examined the perceptions and motivations of faculty members towards learning Blockchain technology. Results showed that most participants were assistant professors with a master's degree in technology and over 15 years of work experience.
- Personal desire was found to be the primary motivator, followed by job expectations and technical capabilities. Social desire was the lowest motivator. Middle-aged faculty members had a higher level of awareness about the importance of Blockchain technology in industry 4.0.
- Many participants considered leaving academia to work in the industry due to the high demand for blockchain developers and better pay. Policymakers and educators can use these findings to design curriculum and training programs that focus on personal motivation and job expectations.
- Finally, this study suggests promoting the social value of learning Blockchain technology to maintain social class status. The study identified an opportunity for future research to explore additional motivators for taking Blockchain technology courses.

## 5 Conclusion

The purpose of this study was to examine the motivational behaviors and performance of participants in online faculty development programs focused on blockchain technology. Based on the survey results, it is clear that individuals pursuing careers in teaching and research recognize the importance of academic professionals in solving technical challenges facing society. Furthermore, respondents understand that programming, networking, and security concepts are essential for successful blockchain technology research. However, to maintain motivation and confidence in pursuing blockchain technology as a primary focus, participants require adequate support in terms of teaching blockchain application development and research tools. The study recommends that participants be informed about the correlation between basic computer science courses and full-stack application development with blockchain technology at the beginner and intermediate levels of academic teaching. This knowledge will enable them to effectively address research problems related to or with blockchain technology and maintain their motivation and confidence. Ultimately, this will contribute to the development and adoption of blockchain technology in various industries, benefiting society as a whole.

## References

1. Allocation of teachers who participated in professional development in ICT skills, Organisation for Economic Co-Operation and Development (OECD) (2022)
2. Mardiana, H.: Lecturers' adaptability to technological change and its impact on the teaching process. JPI (Jurnal Pendidikan Indonesia) **9**, 275–289 (2020)
3. Darby, A., Newman, G.: Exploring faculty members' motivation and persistence in academic service-learning pedagogy. J. High. Educ. Outreach Engagem. **18**, 91–119 (2014)
4. Gomez, D.R., Swann, W., Willms Wohlwend, M., Spong, S.: Adapting under pressure: a case study in scaling faculty development for emergency remote teaching. J. Comput. High. Educ. **35**, 1–20 (2022)
5. Phuong, T.T., Foster, M.J., Reio, T.G., Jr.: Faculty development: a systematic review of review studies. New Horiz. Adult Educ. Hum. Resour. Dev. **32**, 17–36 (2020)
6. Daniels, E.: Logistical factors in teachers' motivation. Clearing House J. Educ. Strat. Issues Ideas **89**, 61–66 (2016)
7. Welch, M., Plaxton-Moore, S.: Faculty development for advancing community engagement in higher education: current trends and future directions. J. High. Educ. Outreach Engagem. **21**, 131–166 (2017)
8. Sadiku, S.: Factors affecting teacher motivation. Int. Sci. J. Monte (ISJM) **4**, 98–113 (2021)
9. Penprase, B.E.: The fourth industrial revolution and higher education. High. Educ. Era Fourth Ind. Revolution **10**, 978–981 (2018)
10. Saito, H., Umeda, K.: The development of teaching skills using ICT in teacher training: practices in first-year introduction for ICT. In: Proceedings of the 2019 3rd International Conference on Education and Multimedia Technology, New York, NY, USA (2019)
11. Liesa-Orús, M., Latorre-Cosculluela, C., Vázquez-Toledo, S., Sierra-Sánchez, V.: The technological challenge facing higher education professors: perceptions of ICT tools for developing 21st century skills. Sustainability **12**, 5339 (2020)
12. Razak, S., Khan, S., Hussein, N., Alshikhabobakr, H., Gedawy, H., Yousaf, A.W.: Integrating computer science and ICT concepts in a cohesive curriculum for middle school - an experience report. In: Proceedings of the 52nd ACM Technical Symposium on Computer Science Education, New York, NY, USA (2021)
13. Hashim, N.A., Mukhtar, M., Safie, N.: Factors affecting teachers' motivation to adopt cloud-based E-learning system in Iraqi deaf institutions: a pilot study. In: Proceedings of the 2019 International Conference on Electrical Engineering and Informatics (ICEEI) (2019)
14. Mishra, C.: Faculty perceptions of digital information literacy (DIL) at an Indian University: an exploratory study. New Rev. Acad. Librarianship **25**, 76–94 (2019)
15. Redmond, P., Heffernan, A., Abawi, L., Brown, A., Henderson, R.: An online engagement framework for higher education. Online Learn. **22**, 183–204 (2018)
16. Ravindran, U., Bhardwaj, P., Vamsi, P.R.: Blockchain design for securing supply chain management in coffee retailer network. Int. J. Sci. Res. Sci. Technol. **7**, 492–502 (2021)

17. IEEE Standard for Application Technical Specification of Blockchain-based E-Commerce Transaction Evidence Collecting. IEEE
18. Ravindran, U., Vamsi, P.R.: A secure blockchain based finance application. In: Proceedings of the 2021 Thirteenth International Conference on Contemporary Computing (IC3–2021) (2021)
19. Ahmed, N., Nandi, D., Zaman, A.G.M.: Analyzing student evaluations of teaching in a completely online environment. Int. J. Mod. Educ. Comput. Sci. (IJMECS) **14**(6), 13–24 (2022). https://doi.org/10.5815/ijmecs.2022.06.02
20. Saha, T.K., Shahrin, R., Prodhan, U.K.: A survey on ICT education at the secondary and higher secondary levels in Bangladesh. Int. J. Mod. Educ. Comput. Sci. (IJMECS) **14**(1), 17–29 (2022). https://doi.org/10.5815/ijmecs.2022.01.02
21. Chen, G., Xu, B., Lu, M., Chen, N.-S.: Exploring blockchain technology and its potential applications for education. Smart Learn. Environ. **5**(1), 1 (2018). https://doi.org/10.1186/s40561-017-0050-x
22. Araujo, P., Viana, W., Veras, N., Farias, E.J., de Castro Filho, J.A.: Exploring students perceptions and performance in flipped classroom designed with adaptive learning techniques: a study in distributed systems courses. In: Brazilian Symposium on Computers in Education (Simpósio Brasileiro de Informática na Educação-SBIE), vol. 30, no. 1, p. 219 (2019)
23. Chang, M., Arachchilage, C.S.M.W., Kim, M., Lim, H.: Acceptance of tourism blockchain based on UTAUT and connectivism theory. Technol. Soc. **71**, 102027 (2022)
24. Popa, I.-C., Orzan, M.-C., Marinescu, C., Florescu, M.S., Orzan, A.-O.: The role of blockchain technologies in the sustainable development of students' learning process. Sustainability **14**(3), 1406 (2022)
25. Punch, K.F., Oancea, A.: Introduction to Research Methods in Education. Sage (2014)
26. Cohen, L., Manion, L., Morrison, K.: Research Methods in Education, 7th ed., Routledge (2011)
27. Ouda, O.K.M., AL-Asad, J.F., Asiz, A.: An assessment of student's motivations to join College of Engineering: Case study at Prince Mohammad Bin Fahd University - Saudi Arabia (2014)

# Author Index

**A**

Acheme, Ijegwa David  152, 165
Aderounmu, Ganiyu A.  239
Adigwe, Wilfred  165
Ahmed, Aiman Magde Abdalla  944
Akhatov, Akmal  437
Akhter, Mshura  129
Akhund, Tajim Md. Niamat Ullah  178
Akinyemi, Bodunde O.  239
Akter, Lubana  129
Alifov, Alishir A.  101
Anabtawi, Wasim  271
Anton, Yakovlev  416

**B**

Bai, Chuntao  514, 876, 911
Bazilo, Constantine  120
Beskrovnyi, Oleksii  1068
Bilanovych, Alisa  310
Borhan, Rownak  299

**C**

Cai, Qi  573, 734
Cao, Jingjing  681
Chang, Ming  1006
Chen, Chen  983
Chen, Hongbao  963
Chen, Jinming  963
Chen, Liping  963
Chen, Ning  640, 754
Chen, Qingfeng  719
Chen, Xi  628
Chen, Yan  65, 963
Chen, Yi  598
Chen, Yuan  25
Chimir, Igor  35
Cui, Bin  524
Cui, Zhiyu  1028

**D**

Dai, Jinshan  65, 899
Das, Prantho kumar  1039

Degila, Jules  239
Delyavskyy, Mykhaylo  382
Deresh, Olha  217
Dey, Samrat Kumar  129, 1039
Ding, Yao  833, 954
Dong, Pan  854
Dong, Sijie  899
Doronina, Iryna  404
Du, Lijing  3, 573, 734
Du, Weihui  651, 696
Du, Yong  514
Durnyak, Bohdan  187, 197

**E**

Efiong, John E.  239

**F**

Fan, Simeng  983
Fang, Can  651
Fang, Liang  777
Feng, Bin  886
Feng, Gong  681
Feng, Wei  1006
Filimonov, Sergey  120
Filimonova, Nadiia  120
Frontoni, Emanuele  288, 310

**G**

Gao, Ying  598
Ge, Jingjing  822
Gumen, Olena  448

**H**

Havryliuk, Myroslav  372
He, Anqi  524
Hong, Huawei  109
Honsor, Oksana  393
Huang, Bihui  14
Huang, Saixiao  25
Huang, Yixuan  944
Huang, Yujie  844
Huang, Zhicheng  65

## I
Ibrahim, Isa A. 239

## J
Jarin, Mahnaz 334
Jiang, Chun 974
Jiang, Hong 764, 899
Joseph, Isabona 488
Ju, Chengwei 79
Jumanov, Isroil I. 478

## K
Kaminskyi, Andrii 426
Kaminskyy, Roman 372
Kandiy, Sergey 288, 310
Katerynchuk, I. 206
Khan, Saikat Islam 299
Kobylianska, Olena 288
Kopych, A. 206
Korniyenko, Bogdan 55
Kozubtsov, Ihor 1054, 1068
Kozubtsova, Lesia 1054, 1068
Kubytskyi, Volodymyr 321
Kuno, I. 206
Kuznetsov, Oleksandr 288, 310, 360
Kuznetsova, Kateryna 360

## L
Lavrenchuk, Svitlana 382
Li, Bin 524
Li, Liwei 931
Li, Shujun 524
Li, Wenhui 719
Li, Xin 764, 899
Li, Yan 764
Li, Yani 899
Li, Zhengxie 558, 664
Lian, Jinxiang 777
Liang, Chengquan 865
Liang, Tingting 90
Liang, Zhaomin 90
Liaskovska, Solomiya 448
Lin, Qian 719
Lishchyna, Nataliia 251, 382
Lishchyna, Valerii 251, 382
Lisovych, Taras 372
Liu, Bing 14
Liu, Ke 801

Liu, Li 833, 954
Liu, Mingfei 944
Liu, Minwei 811
Liu, Qian 681
Liu, Shan 708
Liu, Yanhui 777
Lu, Huiting 787
Lu, Liuqing 1016
Lukianchuk, Iurii 251
Luo, Wei 25
Luo, Ximei 719
Lutskiv, Mikola 187
Lv, Peng 109

## M
Ma, Jie 854
Mai, Liqiang 833, 954
Makinde, Ayodeji S. 152
Maqboul, Ahmad 271
Martyn, Yevgen 448
Melnyk, Roman 458
Mori, Margherita 3
Mutovkina, Nataliya 921, 995
Myronenko, Olha 1054

## N
Nazarov, Fayzullo 437
Nehrey, Maryna 404, 426
Neroda, Tetyana 197
Nesteruk, Andrii 55
Nezhyva, Olha 1054
Noman, S. M. 178
Nwankwo, Chukwuemeka Pascal 152, 165
Nwankwo, Wilson 152, 165

## O
Oberyshyn, Roksolana 393
Ogbonda, Clement 488
Oghorodi, Duke 165
Ojei, Emmanuel 165
Olajubu, Emmanuel A. 239
Olayinka, Akinola S. 152
Oleh, Lisovychenko 416
Othman Othman, M. M. 271

## P
Pan, Zhikang 754
Panchenko, Taras 321

# Author Index

Pang, Rongrong 764
Pang, Weinan 886
Pasieka, Mykola 187
Pasieka, Nadiia 187, 197
Perelygin, Sergey 229
Prokopov, Serhii 310
Putrenko, Viktor 404

## Q

Qi, Xiaoyang 514
Qi, Xin 583
Qian, Xiaorui 109
Qian, Yuntong 618
Qiao, Lichen 608
Qin, Fengcai 974
Qin, Shanyong 811
Qiu, Suzhen 90

## R

Rahaman, A. S. M. Mostafizur 334
Rahman, Elias Ur 299, 1039
Rahman, Md. Hafizur 178
Rahman, Md. Mahbubur 468
Rahman, Md. Mokhlesur 468
Rahman, Sabbir 468
Rashid, Md. Mamun Ar 1039
Rashidov, Akbar 437
Risi, Ikechi 488
Romanyshyn, Yulia 197
Rong, Donglin 696

## S

Safarov, Rustam A. 478
Saha, Dola 129
Salehin, Imrus 178
Semotiuk, O. 206
Sharmin, Fateha 299
Sharmin, Nusrat 468
Sheketa, Vasyl 187
Shepita, Petro 187, 197
Shmyhelskyy, Ya. 206
Silko, Oleksii 1068
Smirnov, Oleksii 288
Smirnova, Olga 995
Smirnova, Tetiana 310
Stetsenko, Inna V. 344
Su, Lifang 801
Su, Rengshang 79

Sulim, Viktoriia 382
Sun, Jiawei 696, 719
Sveleba, N. 206
Sveleba, S. 206

## T

Tang, Jiwei 25
Tavrov, Danylo 141
Temnikov, Andrii 141
Temnikov, Volodymyr 141
Temnikova, Olena 141
The, Quan Trong 229
Tian, Jia 628
Tian, Jintian 745
Tikhonov, Andrii 360
Tolokonnikov, G. K. 261
Tong, Gang 109
Tong, Xinshun 536
Tulashvili, Yurii 251
Tupychak, Lyubov 197
Tushnytskyy, Ruslan 458

## U

Uddin, Khandaker Mohammad Mohi 129, 299, 1039
Ulianovska, Yuliia 288
Umezuruike, Chinecherem 165

## V

Vamsi, P. Raghu 1080
Vdovyn, Mariana 217
Velgosh, S. 206

## W

Wang, Enshi 14
Wang, Haiyan 618
Wang, Jing 573, 734
Wang, Lulu 503
Wang, Meng 764, 899
Wang, Mengqiu 899
Wang, Xiujuan 514, 876, 911
Wang, Ying 573, 734
Wang, Zaitao 876, 911
Wei, Puzhe 583
Wen, Jin 833, 954
Wu, Hongyu 628
Wu, Xia 45
Wu, Yunyue 547

## X

Xiao, Kai 109
Xie, Mei E. 608
Xie, Sida 899
Xie, Xiaoling 65
Xing, Saipeng 651, 696
Xiong, Kevin 573, 734
Xiong, Shuiping 45
Xiong, Yanbing 573, 734
Xu, Yajie 536
Xv, Li 628

## Y

Yan, Yang 719
Yang, Haifeng 854, 963
Yang, Lingxue 719
Yang, Xiaozhe 787
Yang, Xue 640
Yao, Zhenhua 1006
Yashchenko, Sergei 120
Yassine, Radouani 681
Ye, Fang 524
Ye, Hui 608
Ye, Yanxin 844
Yemets, Kyrylo 372
Yezhov, Anton 360
Yin, Guanchao 833, 954

You, Jiawei 944
Yu, Xunran 628
Yuan, Jun 503

## Z

Zarichkovyi, Alexander 344
Zdolbitska, Nina 382
Zhan, Xiangpeng 109
Zhang, Cheng 583
Zhang, Chenyu 65
Zhang, Geng E. 79, 1016
Zhang, Lei 745
Zhang, Long 558, 664
Zhang, Peilin 503
Zhang, Xiaoyu 651
Zhang, Yao 503, 608
Zhao, Ting 876, 911
Zhelykh, Vasyl 448
Zheng, Zhong 708, 931
Zhou, Qilai 833, 954, 1006
Zhou, Xiaofen 764, 899
Zhou, Xiaoguang 777
Zhou, Xiaoying 708
Zhu, Lingling 109
Zhu, Shixiong 931
Zomchak, Larysa 217
Zuo, Jing 844